VOLUME SEVENTY NINE

ADVANCES IN INORGANIC CHEMISTRY

Recent Highlights II

ADVISORY BOARD

VOLUME SEVENTY NINE

ADVANCES IN INORGANIC CHEMISTRY

Recent Highlights II

Edited by

RUDI van ELDIK
University of Erlangen–Nuremberg, Germany
Nicolaus Copernicus University in Torun, Poland

COLIN D. HUBBARD
Department of Chemistry,
University of New Hampshire,
Durham, NH, United States

Academic Press is an imprint of Elsevier
50 Hampshire Street, 5th Floor, Cambridge, MA 02139, United States
525 B Street, Suite 1650, San Diego, CA 92101, United States
The Boulevard, Langford Lane, Kidlington, Oxford, OX5 1GB, United Kingdom
125 London Wall, London, EC2Y 5AS, United Kingdom

First edition 2022

Notices
Knowledge and best practice in this field are constantly changing. As new research and experience broaden our understanding, changes in research methods, professional practices, or medical treatment may become necessary.

Practitioners and researchers must always rely on their own experience and knowledge in evaluating and using any information, methods, compounds, or experiments described herein. In using such information or methods they should be mindful of their own safety and the safety of others, including parties for whom they have a professional responsibility.

To the fullest extent of the law, neither the Publisher nor the authors, contributors, or editors, assume any liability for any injury and/or damage to persons or property as a matter of products liability, negligence or otherwise, or from any use or operation of any methods, products, instructions, or ideas contained in the material herein.

ISBN: 978-0-323-99972-4
ISSN: 0898-8838

For information on all Academic Press publications
visit our website at https://www.elsevier.com/books-and-journals

Publisher: Zoe Kruze
Acquisitions Editor: Sam Mahfoudh
Developmental Editor: Federico Paulo Mendoza
Production Project Manager: Vijayaraj Purushothaman
Cover Designer: Vicky Pearson Esser

Typeset by STRAIVE, India

Working together
to grow libraries in
developing countries

www.elsevier.com • www.bookaid.org

Contents

Contributors

Lucjan Chmielarz
Jagiellonian University in Kraków, Faculty of Chemistry, Kraków, Poland

Edwin C. Constable
University of Basel, Basel, Switzerland

Janusz M. Dąbrowski
Faculty of Chemistry; Małopolska Centre of Biotechnology, Jagiellonian University, Kraków, Poland

Geyla C. Dubed Bandomo
Institute of Chemical Research of Catalonia (ICIQ), The Barcelona Institute of Science and Technology; Departament de Química Física i Inorgànica, Universitat Rovira i Virgili, Tarragona, Spain

István Fábián
Department of Inorganic and Analytical Chemistry, University of Debrecen, Egyetem tér 1; MTA-DE Redox and Homogeneous Catalytic Reaction Mechanisms Research Group, Egyetem tér 1, Debrecen, Hungary

Sergio Fernández
Institute of Chemical Research of Catalonia (ICIQ), The Barcelona Institute of Science and Technology; Departament de Química Física i Inorgànica, Universitat Rovira i Virgili, Tarragona, Spain

Krzysztof Fic
Institute of Chemistry and Technical Electrochemistry, Poznan University of Technology, Poznan, Poland

Elżbieta Frąckowiak
Institute of Chemistry and Technical Electrochemistry, Poznan University of Technology, Poznan, Poland

Gernot Frenking
Fachbereich Chemie, Philipps-Universität Marburg, Marburg, Germany; Institute of Advanced Synthesis, School of Chemistry and Molecular Engineering, Nanjing Tech University, Nanjing, China

Yurii V. Geletii
Department of Chemistry, Emory University, Atlanta, GA, United States

Craig L. Hill
Department of Chemistry, Emory University, Atlanta, GA, United States

Aleksandra Jankowska
Jagiellonian University in Kraków, Faculty of Chemistry, Kraków, Poland

Anna Kusior
AGH University of Science and Technology, Faculty of Materials Science and Ceramics, Krakow, Poland

Tianquan Lian
Department of Chemistry, Emory University, Atlanta, GA, United States

Norbert Lihi
Department of Inorganic and Analytical Chemistry, University of Debrecen; MTA-DE Redox and Homogeneous Catalytic Reaction Mechanisms Research Group, Egyetem tér 1, Debrecen, Hungary

Julio Lloret-Fillol
Institute of Chemical Research of Catalonia (ICIQ), The Barcelona Institute of Science and Technology, Tarragona; Catalan Institution for Research and Advanced Studies (ICREA), Barcelona, Spain

Julia Mazurków
AGH University of Science and Technology, Faculty of Materials Science and Ceramics, Krakow, Poland

Kinga Michalec
AGH University of Science and Technology, Faculty of Materials Science and Ceramics, Krakow, Poland

Djamaladdin G. Musaev
Department of Chemistry; Cherry L. Emerson Center for Scientific Computation, Emory University, Atlanta, GA, United States

Wolfgang Petz
Fachbereich Chemie, Philipps-Universität Marburg, Marburg, Germany

Justyna Piwek
Institute of Chemistry and Technical Electrochemistry, Poznan University of Technology, Poznan, Poland

Anetta Płatek-Mielczarek
Laboratory for Multiphase Thermofluidics and Surface Nanoengineering, Department of Mechanical and Process Engineering, ETH Zurich, Sonneggstrasse 3, Zurich, Switzerland

Barbara Pucelik
Faculty of Chemistry; Małopolska Centre of Biotechnology, Jagiellonian University, Kraków, Poland

Marta Radecka
AGH University of Science and Technology, Faculty of Materials Science and Ceramics, Krakow, Poland

Alexander B. Sorokin
Institut de Recherches sur la Catalyse et l'Environnement de Lyon IRCELYON, UMR 5256, CNRS—Université Lyon 1, Einstein, France

Milena Synowiec
AGH University of Science and Technology, Faculty of Materials Science and Ceramics, Krakow, Poland

Anita Trenczek-Zajac
AGH University of Science and Technology, Faculty of Materials Science and Ceramics, Krakow, Poland

Q. Yin
Department of Chemistry, Emory University, Atlanta, GA, United States

Preface

The inception of this series (founding editors H.J. Emeléus and A.G. Sharpe) dates back to six decades, and, undoubtedly, the content and illustrations within the review contributions were very different from those currently published. Early volumes, until about 1980, included reviews of radiochemistry research. Dramatic advances in a range of spectroscopic techniques, particularly nuclear magnetic resonance, improved methodology for X-ray structure determinations, maturity of fast reaction methods, vast improvements in data acquisition and analysis, and computational methods, among other developments led to a broad diversification of inorganic chemistry-based subjects, since that time. The ingenuity of the inorganic chemistry community has generated some remarkable applications of inorganic compounds and materials, as witnessed in some of the thematic volumes of the past few years, exemplified in "Insights from Imaging in Bioinorganic Chemistry," Volume 68, 2016; "Water Oxidation Catalysts," Volume 74, 2019; "Medicinal Chemistry," Volume 75, 2020; and "Catalysis in Biomass Conversion," Volume 77, 2021. Volume 78, the nonthematic "Recent Highlights I," 2021, was no exception, and there was an increasing awareness of environmental and sustainability issues in, for example, a chapter authored by Michele Aresta and colleagues titled "Stepping Toward the Carbon Circular Economy."

The current volume (Volume 79), "Recent Highlights II," contains contributions in fundamental inorganic chemistry research and contributions in inorganic chemistry with potential practical applications, within the context of the environment and sustainability. In addition, there are contributions with relevance to energy storage and public health.

An aptly titled chapter, considering the first word of the title and the contents, "Insight into the Thermodynamic and Catalytic Features of NiSOD Related Metallopeptides," is Chapter 1 of Volume 79, authored by Norbert Lihi and István Fábián. Chapter 2, by Alexander B. Sorokin, "Cleavage of C—F Bonds in Oxidative Conditions Mediated by Transition Metal Complexes," describes defluorination of fluorinated aromatic compounds by severing inert C—F bonds, among other species, using a large range of iron-active sites from enzymatic species, yielding absorbing mechanistic chemistry. The topic "Photodynamic Inactivation (PDI) as a Promising Alternative to Current Pharmaceuticals for the Treatment of Resistant

Microorganisms" is discussed in Chapter 3, authored by Barbara Pucelik and Janusz M. Dąbrowski. The authors describe efforts to provide an alternative to the use of antibiotics, regarded as overprescribed, for bacterial resistance. The method involves metal-based macrocycles as reagents in a light-based process. Edwin C. Constable, in Chapter 4, stimulates readers' interest with "The Secret Life of Oligopyridines: Complexes of Group 1 Elements" and does not disappoint in this elegant exposure of the subject. As the origin of energy requirements changes over to increasing use of renewable sources from dependence on fossil fuel sources, there will be a need for further energy storage and conversion technology. This topic, "Advanced Characterization Techniques for Electrochemical Capacitors," is addressed in Chapter 5 by Elżbieta Frąckowiak and colleagues, with an emphasis on the inorganic and other materials involved. Lucjan Chmielarz and Aleksandra Jankowska, in Chapter 6, present "Mesoporous Silica-Based Catalysts for Selective Catalytic Reduction of NO_x with Ammonia—Recent Advances." This is a comprehensive account of efforts to optimize conditions for a successful process of selective catalytic reduction of nitrogen oxide species with ammonia, by deposition of active metal species containing, for example, copper, iron, or manganese, in combination with other components, onto the surface of mesoporous silicates. Chapter 7, by Wolfgang Petz and Gernot Frenking, "Neutral and Charged Group 13–16 Homologs of Carbones EL_2 ($E=B^-$–In^-; Si–Pb; N^+–Bi^+; O^{2+}–Te^{2+})," is a veritable feast of chemical bonding description through experimental evidence supported by computation of these low-valent main group homologs of classical carbones, written by participant experts of this developing, exciting field of inorganic chemistry. Chapter 8, "Recent Advances in Electrocatalytic CO_2 Reduction with Molecular Complexes," is authored by Julio Lloret-Fillol and colleagues. Comprehensive efforts to reduce CO_2 electrocatalytically to useful chemicals, using a variety of metal-based potential catalysts, and using renewable energy, are described. Besides the environmental benefits that would accrue from this contribution is a banquet of catalytic and mechanistic chemistry. In Chapter 9, "Polyoxometalate Systems to Probe Catalyst Environment and Structure in Water Oxidation Catalysis," Craig L. Hill and colleagues assess polyoxometalates as participants in photoelectrochemical cells in water oxidation catalysis, in both homogeneous and heterogeneous systems. This is a lucid account of the key properties and parameters that could lead to an eventual commercial viability in this environmentally vital context. Anita Trenczek-Zajac and Anna Kusior are the corresponding authors for their team for a two-chapter contribution, Chapters 10 and 11, "Interface Design, Surface-Related Properties, and Their Role in Interfacial Electron Transfer" with the subtitles "Part I: Materials-Related Topics" and "Part II: Photochemistry-Related Topics." The overall purpose is development of systems for heterogeneous photoelectrochemical decomposition of water or photoelectrocatalytic decomposition of pollutants. A wide range of inorganic materials, particularly oxides of titanium and tin, titanium and iron, or titanium and copper paired in various nanostructures, were examined for their suitability to the objective of this contribution, within a green chemistry context.

The endeavors of the corresponding authors to produce their contributions, sometimes in less-than-ideal circumstances, are most gratefully acknowledged by the editors. The editors are appreciative of the efforts of their coauthors for their support in the form of their contributions by participating in the experimental research reported, as this contributes to the success of the volume. The high quality of the narrative, illustrations, and schemes is manifestly obvious within the chapters herein. Readers, whether experienced investigators or relative newcomers, can enjoy a selection from among this eclectic collection of inorganic chemistry research topics, those that are fundamental in nature or those steering toward applications in perhaps an industry context, within the current landscape of societal interests. Overall, this volume is an excellent demonstration of the breadth and scientific health of inorganic chemistry research.

RUDI VAN ELDIK
Series Editor of *Advances in Inorganic Chemistry*
Emeritus Professor of Inorganic Chemistry
University of Erlangen-Nuremberg, Germany
Research Professor of Inorganic Chemistry
Nicolaus Copernicus University in Torun, Poland

COLIN D. HUBBARD
Coeditor of Volume 79
Emeritus Professor of Chemistry
University of New Hampshire
Durham, NH, United States of America

CHAPTER ONE

Insight into the thermodynamic and catalytic features of NiSOD related metallopeptides

Norbert Lihi[a,b] and István Fábián[a,b,*]

[a]Department of Inorganic and Analytical Chemistry, University of Debrecen, Egyetem tér 1, Debrecen, Hungary

[b]MTA-DE Redox and Homogeneous Catalytic Reaction Mechanisms Research Group, Egyetem tér 1, Debrecen, Hungary

*Corresponding author: e-mail address: ifabian@science.unideb.hu

Contents

Abstract

The most recently discovered superoxide dismutase enzymes, NiSODs, contain nickel in their active center. This paper surveys the most relevant recent results in this field. Literature data confirm that designated Ni-containing metallopeptides are appropriate mimics of NiSOD. By varying the amino acid sequence of the binding hook, it was unequivocally confirmed that the dismutase activity is related to the coordination of cysteine moieties to Ni via the thiolate groups. While these metallopeptides are essential in modeling the structure, and electronic properties of NiSOD enzymes, their redox stability is not satisfactory to deliver sustained dismutase activity. The redox degradation of the model compounds is kinetically coupled with the dismutation cycle and the conventional first-order approach cannot be used for the interpretation of the decay

Advances in Inorganic Chemistry, Volume 79
ISSN 0898-8838
https://doi.org/10.1016/bs.adioch.2021.12.001

of $O_2^{-\bullet}$. For the very same reason, the use of the McCord-Fridovich assay is of limited value in these systems. Arguments are presented to demonstrate that direct kinetic studies, utilizing the sequential stopped-flow method, are required for studying the dismutation process under real catalytic conditions.

1. Introduction

Reactive oxygen species (ROS) form during the incomplete reduction of molecular dioxygen in biological systems. The collective term, ROS, includes the superoxide anion radical ($O_2^{-\bullet}$), hydrogen peroxide (H_2O_2), singlet oxygen ($^1O_2$), and hydroxyl radical ($OH^\bullet$).[1] Although dioxygen acts as a four-electron oxidizing agent, its activation is not a trivial task. Basically, the net reduction process (1), which yields water as a product, consists of four sequential one-electron steps involving the three most frequently ROS, $O_2^{-\bullet}$, H_2O_2 and $OH^\bullet$ (2–5)[2,3]

$$O_2 + 4\,H^+ + 4\,e^- = 2\,H_2O \quad (1)$$

$$O_2 + e^- = O_2^{-\bullet} \quad (2)$$

$$O_2^{-\bullet} + 2\,H^+ + e^- = H_2O_2 \quad (3)$$

$$H_2O_2 + H^+ + e^- = H_2O + OH^\bullet \quad (4)$$

$$OH^\bullet + H^+ + e^- = H_2O \quad (5)$$

These species have generated great interest as potential green oxidants in advanced oxidation processes (AOPs).[4–6] and vast efforts have been devoted to explore the potential benefits of these species in ROS-related therapeutics.[7] Indeed, cellular respiration requires molecular oxygen in all living systems, however, a considerable amount of dioxygen is metabolized through ROS which can act as signal agents,[8] toxic species, or harmless intermediates.[9] A human body generates approximately 5 g of ROS/day with two primary products, $O_2^{-\bullet}$ and H_2O_2.[10] As mentioned, $O_2^{-\bullet}$ forms as a byproduct of mitochondrial respiration but it is also generated from phagocytic nicotinamide adenine dinucleotide phosphate oxidase (NADPH oxidase) as a key component of the immune defense system.[11] Under normal circumstances, biological systems release two different types of enzymes to control ROS level, superoxide reductase (SOR) and superoxide dismutase (SOD).[12] The former enzymes catalyze the one-electron reduction of superoxide anion radical to hydrogen peroxide, while SODs are capable to assist the decomposition of superoxide anion radical to molecular oxygen and hydrogen peroxide.[13,14] In the absence of these enzymes, the elevated level of ROS can cause oxidative stress, cellular damages, and several inflammatory diseases.[15–17] ROS have also crucial role in the development of human cancer due to the alteration of DNA structure or the inactivation of tumor suppressor genes, moreover, several neurodegenerative disorders are also associated with elevated ROS level.[18,19]

All of the SORs and a huge number of SOD enzymes are metalloenzymes containing a redox active metal ion which facilitates the electron involved in the reduction/oxidation processes. While SORs contain exclusively iron in the active center, metalloenzymes of copper, iron and manganese form the biggest family of SODs. So far, the only nickel containing SODs expressed by the *sodN* gene were found in marine cyanobacteria and Streptomyces species.[20–23] Although numerous papers and reviews report the features of these enzymes, the results obtained in model systems and considerations regarding their potential application as drugs highlight several fundamental issues to be addressed in further studies.[7,12,13,15,24] In this paper, we discuss recent results on the equilibrium and kinetic features of the reactions of NiSOD related metallopeptides. In the first part, the general characteristics of the NiSOD enzyme are detailed, subsequently the methods for generating superoxide anion and measuring the SOD activity are reviewed and completed with recent results from our group.

2. The nickel SOD (NiSOD) enzymes

The NiSOD enzymes facilitate the disproportionation of superoxide ion by cycling between Ni^{II} and Ni^{III} oxidation states. To some extent, the presence of the Ni^{III} oxidation state in aqueous media is unexpected, because the estimated redox potential of the $Ni_{aq}^{III/II}$ couple is around +2.26 V.[25] However, EPR studies on the first native isolated enzyme unambiguously proved the presence of low-spin Ni^{III} in the active center.[23]

The protein structure and the active site of NiSOD are shown in Fig. 1. The NiSOD built in an 80-kDa homohexamer consisting of antiparallel four-helix bundle subunits and each monomer contains a nickel ion in the active site. The hexamer structure is stabilized by hydrophobic, hydrogen bonding and salt bridges interactions. Nickel is located at the N-terminal part of the enzyme and the first six amino acid residues are involved in binding Ni. This N-terminal part is buried in the space between two four-helix bundle subunits, but it is effectively protected from the bulk solvents by the so-called NiSOD binding hook. This motif is characteristic

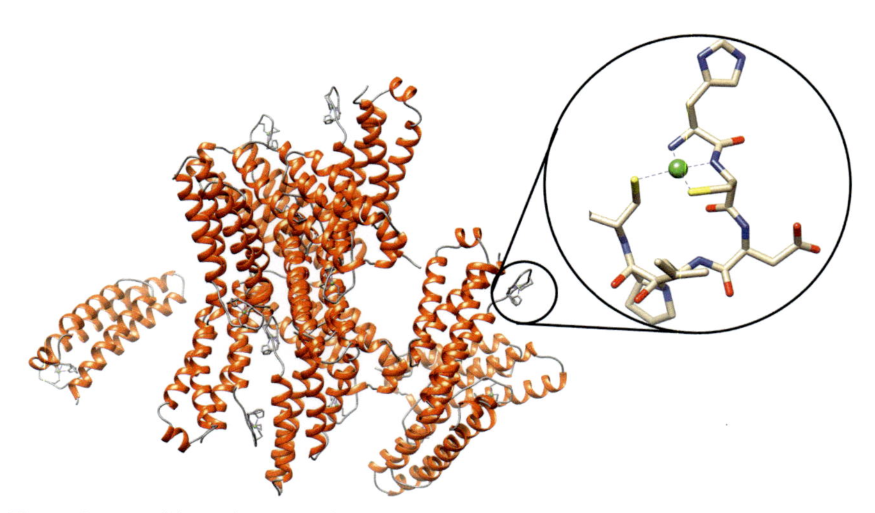

Fig. 1 Stereo ribbon diagram of NiSOD (PDB code: 1T6U) expressed from *Escherichia coli*.[22] The helix conformations are colored by red and the random coil conformations involving the NiSOD binding hook are gray. The diagram is generated by Chimera.[26] Inset: schematic view of the reduced form of NiSOD binding hook.

for all the NiSOD enzymes and contains a consensus sequence (HisCysXaaYaaProCys) where the role of proline is crucial in order to form *cis*-peptide conformation.[22,27] (Xaa and Yaa are amino acids within the sequence, which are varied to test certain concepts). NiSODs exhibit a unique coordination environment which incorporates cysteine-S donors. In the reduced form of the enzyme, nickel(II) is coordinated via the N-terminal amino group, the first peptide nitrogen and the two thiolates of cysteine residues in a square-planar coordination environment and exhibits diamagnetic character. The oxidation of Ni^{II} yields a Ni^{III} paramagnetic complex and the apical position of the metal ion is occupied by the imidazole-*N* of His1 resulting in a five-coordinated pyramidal Ni^{III} center.[20] The binding of the imidazole-*N* in the axial position exhibits a typical nitrogen hyperfine splitting in the EPR spectrum (Fig. 2). This rhombic signal is characteristic for a low-spin Ni^{III} complex with $S = 1/2$ and the unpaired electron is being localized in the d_{z^2} orbital (SOMO). Computational studies confirmed that this molecular orbital is involved in the oxidation of $O_2^{-\bullet}$, while the Ni $d_{x_2-y_2}$ orbital promotes the reduction of $O_2^{-\bullet}$.[20]

NiSODs (and in general SOD enzymes) degrade the superoxide anion via the ping-pong type dismutation reaction in which the metal centers oscillate between two oxidation states that differ by one electron (6 and 7).[13]

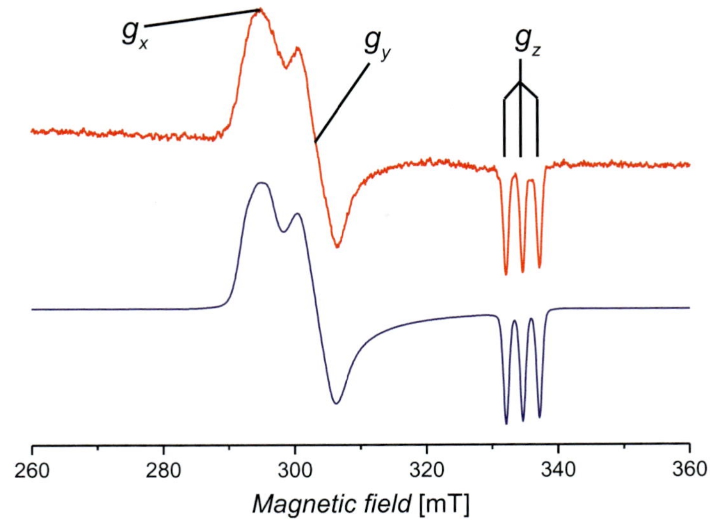

Fig. 2 Top: X-Band EPR spectrum of the $Ni^{III}L$ complex formed with the NiSOD binding hook (L: HisCysAspLeuProCysGlyValTyr-NH_2). Bottom: Simulated EPR spectrum from the spin-Hamiltonian parameters.[28]

$$Ni^{2+} + O_2^{-\bullet} + 2\,H^+ = Ni^{3+} + H_2O_2 \qquad (6)$$

$$Ni^{3+} + O_2^{-\bullet} = Ni^{2+} + O_2 \qquad (7)$$

NiSODs possess a redox potential which falls into the optimum range of superoxide dismutation (0.29 V).[29] The reduction potential is controlled by the thiolate donors of cysteine residues. These electron-rich donors play a significant role of pushing electron density to the Ni center in order to promote electron-transfer reactions.[30] Indeed, the biochemistry of nickel is closely associated with the thiolate groups as shown in the case of [NiFe] hydrogenase,[31–33] methyl-CoM reductase [34,35] and CO dehydrogenase.[36,37] Theoretical calculations also indicate that these anionic ligands in the equatorial plane are capable to promote filled/filled antibonding π-interactions with nickel resulting in ideal redox potential for the dismutation of superoxide.[38]

The rate constant of the dismutation reaction approaches the diffusion controlled limit, $k_{cat} > 2 \times 10^9\ M^{-1}\,s^{-1}$ and occurs through a proton-coupled electron-transfer mechanism. Since the rate constant is independent of pH at around physiological conditions, the protons which are required to the H_2O_2 formation originates from the protein backbone.[39] However, the exact source of the proton cannot be identified and there is a considerable disagreement between the mechanisms proposed for the dismutation process (outer- vs inner-sphere).[40–43] Theoretical calculations suggested the coordination of the superoxide anion (inner-sphere mechanism) to the open coordination site of nickel. However, the experimental observation that the azide ion cannot bind to the nickel center challenges the inner-sphere mechanism and favors the outer-sphere interaction.[22,43] The NiSOD mimics derived from the primary sequence may be suitable for exploring the source of the proton and the nature of electron transfer. So far, there is no solid evidence for supporting any of the disproportionation mechanisms assisted by the NiSOD enzyme.

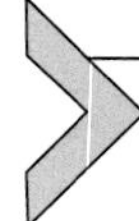

3. Sources of superoxide anion and measuring SOD activity

One electron reduction of molecular dioxygen yields the superoxide anion radical. This species is highly soluble in water and solvated by four hydrogen bonded water molecules.[44,45] Protonation of $O_2^{-\bullet}$ yields the hydroperoxyl radical, HO_2, which is a weak acid with a pK_a of 4.8.[46] Consequently, $O_2^{-\bullet}$ is the major species under biological conditions which disproportionates in water as follows:

$$O_2^{-\bullet} + 2\,H^+ = H_2O_2 + O_2 \tag{8}$$

The need of the proton to stabilize the peroxide anion as hydrogen-peroxide or hydroperoxide-anion, implies that the disproportionation strongly depends on the pH.[47] Indeed, at acidic pH-range, the hydroperoxyl radical acts both as a reductant and an oxidant (9).

$$HO_2 + HO_2 = H_2O_2 + O_2 \tag{9}$$

The basic form, $O_2^{-\bullet}$, is relatively stable at high pH, the corresponding second-order rate constant is 3.9×10^4 M^{-1} s^{-1}. The fastest self-disproportionation was observed at pH = pK_a = 4.8. At this pH, the concentrations of hydroperoxyl and superoxide are the same and HO_2 acts as an oxidizing species while $O_2^{-\bullet}$ acts as a reducing agent (10).

$$HO_2 + O_2^{-\bullet} + H^+ = H_2O_2 + O_2 \tag{10}$$

Due to the disproportionation processes, the preparation of superoxide stock solution is not a trivial task. Therefore, the McCord-Fridovich assay is one of the most frequently used methods to study the SOD activity of enzymes and model complexes.[48,49] In this assay, the xanthine-xanthine oxidase (X/XO) system constantly produces superoxide anion at a low steady-state concentration level. The superoxide anion reduces the reporter or detector molecule (usually ferricytochrome *c* or nitroblue tetrazolium (NBT)) and the formation of the reduced form of the reporter can be followed by spectrophotometric or fluorometric methods. In the presence of an SOD mimic, the mimic quenches the superoxide anion and, as a consequence, partly eliminates the response of the reporter molecule. A typical kinetic trace of this assay is shown in Fig. 3 together with a scheme of the chemical reactions involved in the generation and consumption of superoxide in the X/XO/NBT system.

This test is based on the kinetic competition for superoxide ion between the SOD mimic and the detector molecule. It is also assumed that the mimic consumes the superoxide anion faster than the detector molecule. Since the superoxide anion is present at a low steady-state concentration level, true catalytic conditions cannot be applied, and the SOD mimic is always present in excess. Moreover, there are several preliminary trials that must be performed to ensure the applicability of the method in the system tested.[50,51]

(i) The SOD mimics should not inhibit the production of superoxide by xanthine-oxidase. The rate of conversion of xanthine to uric acid must be tested in the presence and absence of SOD mimics.

(ii) The SOD mimics should not reduce the detector molecule. Independent measurements need to be performed to explore the possibility of reactions between the SOD mimics and the detector molecule.

(iii) The SOD mimics should not react with H_2O_2 formed during the dismutation of superoxide ion. Independent measurements are required to confirm that the SOD mimics do not react with H_2O_2, or catalase needs to be added for eliminating H_2O_2 (another potential source of cross reactions).

(iv) The SOD mimics should not dissociate under the experimental conditions applied. However, they are typically used at very low concentration levels in the McCord-Fridovich assay, and such conditions favor the dissociation of these compounds into the metal ion and the corresponding ligand. It is well-known that $Cu^{2+}{}_{aq}$ and $Mn^{2+}{}_{aq}$ exhibit SOD activity. Consequently, the dissociated metal ion may contribute to the observed catalytic effect in related systems leading to the misinterpretation of SOD activity. In this case, preliminary equilibrium studies are required in order to estimate the stability of

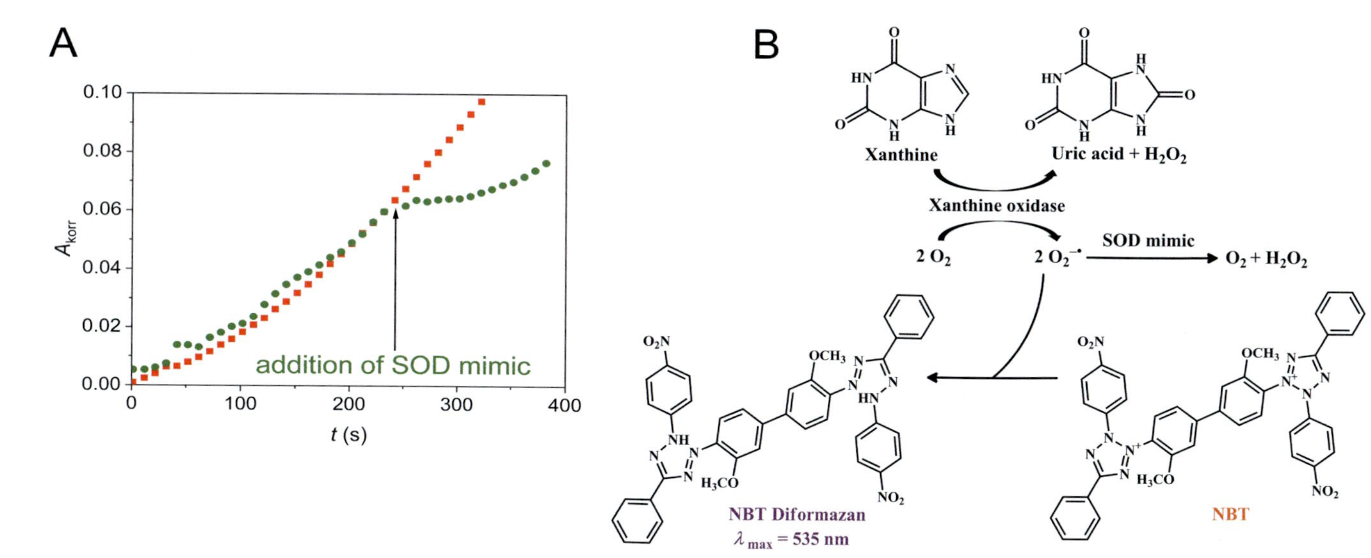

Fig. 3 Panel A: Kinetic curves recorded in the X/XO/NBT system in the absence (red) and in the presence of SOD mimic (green) at 535 nm. Panel B: Production of superoxide by X/XO and its reaction with the NBT.[48]

the complexes and to establish the concentration distributions of the complexes under the actual experimental conditions.

Once the above issues are satisfactorily addressed, the inhibition percentage can be calculated for the SOD mimics and expressed as an IC_{50} value. However, the estimated IC_{50} values strongly depend on the concentration and nature of detector molecule, and direct comparison of IC_{50} values for different assays is not feasible.[50] This problem can be circumvented by calculating the rate constant (k_{McCF}) for the dismutation reaction which is independent from the reporter molecule and its concentration, as shown in Eq. (11).

$$k_{McCF} = k_{detector} \bullet [detector] / IC_{50} \tag{11}$$

It needs to be emphasized that the McCord-Fridovich assay is not perfect, but it provides essential information on the SOD activity under certain conditions. It also serves as a useful preliminary test for studying the details of the dismutation kinetics in the presence of SOD mimics.

The problems mentioned in the previous paragraphs can be avoided by studying the reactions of superoxide ion with SOD mimics by direct kinetic methods. Aqueous solution with a relatively high superoxide ion concentration can be generated by pulse radiolysis which is a frequently used direct method to study SOD and SOR enzymes and their mimics.[52,53] The method allows precise measurements of the rate of the dismutation. The absorption maximum of superoxide anion is at 245 nm ($\varepsilon = 2350\ M^{-1}\ cm^{-1}$) while HO_2 exhibits its maximum at 225 nm ($\varepsilon = 1400\ M^{-1}\ cm^{-1}$) and the reaction can readily be followed by spectrophotometry.[47] However, highly specialized equipment is required to perform pulse radiolysis experiments which limits its applicability.[53]

Another possible direct method utilizes the stopped-flow technique which rapidly mixes the stock solution of superoxide anion with the putative SOD mimics and the absorption of the superoxide can be followed by spectrophotometry.[54] In these experiments, potassium superoxide (KO_2) is dissolved in dry dimethyl sulfoxide (DMSO). This solution may also contain 18-crown-6 which acts as a carrier molecule and increases the solubility of KO_2. Such an $O_2^{-\bullet}$ solution is stable for an extended period of time and makes possible to design and perform useful experiments. The concentration of KO_2 stock solutions can easily be determined by spectrophotometry.

At this point, it should be noted, that the mixing of aqueous and DMSO solutions generates some inhomogeneity due to the different viscosities of these solvents. This leads to spectral interference, and pre-equilibration of the system is necessary for ~40 s after mixing in a 1:1 DMSO/water solution before performing spectrophotometric measurements.[55] Because the half-life of $O_2^{-\bullet}$ sharply decreases by increasing the water content, the use of the sequential stopped-flow technique is required for studying the dismutation reaction in DMSO/water mixture. In such an experiment, first the DMSO solution of $O_2^{-\bullet}$ is mixed with water and the mixture is incubated for the required period of time. In the second step, this reactant solution is mixed with the DMSO/water solution of the SOD mimic. The absorbance of DMSO is relatively high at 245 nm. This prevents precise spectrophotometric measurements at the absorbance maximum of the superoxide ion, therefore, the dismutation reaction needs to be followed at 260 nm where the interference of DMSO is negligible.[55]

Unlike pulse radiolysis, the initial concentration of superoxide can be high compared to the SOD mimics in the stopped-flow experiments and the dismutation reactions can directly be monitored under real catalytic conditions.[24] In the simplest case, first-order decay of superoxide is obtained after mixing the solutions of KO_2 and the buffer containing the SOD mimics, and the fundamental features of the catalysis (effects of catalyst concentration, pH, temperature, ionic strength) can easily be investigated.[56] Consequently, this method allows the direct determination of the catalytic rate constant (k_{cat}) under a given set of conditions. However, such a simple kinetic feature diminishes in the case of mimics with relatively low catalytic activity—$k_{cat} < 10^{5.5}$ M^{-1} s^{-1} (at pH 7.4)—due to the competing second-order self-dismutation of the superoxide ion. Degradation of the catalyst may also complicate the kinetic features.

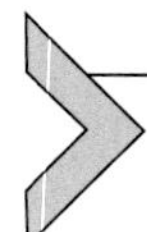

4. Selected NiSOD related metallopeptides: Equilibrium, spectroscopic and kinetic features

In recent years, considerable efforts have been devoted to explore the intimate details of the dismutation mechanism catalyzed by NiSOD enzymes (i.e., inner/outer sphere mechanism, the source of the proton, etc.). Still, only a few metallopeptides containing short chain ligands have been investigated to mimic the NiSOD binding hook.[28,40,42,55,57–60] Several low molecular weight functional mimics of NiSOD have also been developed, which are capable to model the dismutation cycle and the coordination environment of nickel.[61–66]

First, the Ni^{II} complex of a relatively short enzyme fragment (HisCysAspLeuProCys) was studied. Unlike NiSOD, that was isolated as a mixture of Ni^{II}/Ni^{III} oxidation states, the first-generation of NiSOD models was obtained exclusively in a Ni^{II} oxidation state. So far, metallopeptides of the NiSOD binding hook containing Ni^{III} oxidation state were not isolated. This allows the proper spectroscopic and electrochemical characterizations of the coordination mode of the NiSOD binding hook. It is important to note that several Ni^{II} complexes decompose in a few hours when they are exposed to air. Such decomposition has not been observed in the case of the wild type NiSOD enzyme.

The redox potential of the metallopeptide involving the 12 residues form the N-terminal sequence falls into the appropriate range for superoxide dismutation and the spectroscopic parameters are similar to those obtained for the native enzyme. The SOD activity of the complex was tested in the reaction with KO_2/NBT. The results confirmed that the complex hinders the reaction between the NBT and KO_2 up to 2800 eq. of KO_2. Upon increasing the amount of KO_2, the complex loses its SOD activity presumably due to irreversible redox transformation. Specific details of the dismutation mechanism have not been addressed in this study.[67]

NiSOD enzymes possess a unique coordination environment via the incorporation of redox-active cysteine residues as illustrated in Fig. 4.[57]

In order to gain insight into the superoxide disproportionation mechanism, the role of the cysteine residues as well as the role of further amino acids involved in the NiSOD binding motif have been studied

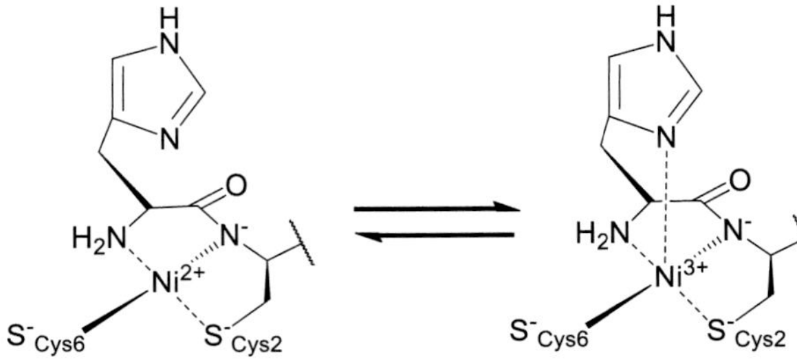

Fig. 4 Structure of the active sites of the reduced and oxidized forms of NiSOD related metallopeptides mimicking the NiSOD binding hook.

systematically. It was concluded that the metal center and not the ligand is the active partner in the dismutation reaction.[67,68]

4.1 The role of axial coordination in the Ni^{III} oxidation state

EPR spectra of the oxidized form of native enzyme and its model metallopeptides unambiguously confirm the axial coordination of Ni(III) via the imidazole-*N* of His(1) residue. It was demonstrated that the position of histidine in the peptide sequence is essential in optimizing the rate of the superoxide disproportionation. Site-directed mutagenesis yielded a new protein containing glutamine in the first position (H1Q mutant). Although the mutant is capable of accelerating the disproportionation of superoxide anion, its catalytic activity is two orders of magnitude lower than that of the native enzyme. The EPR spectrum of the mutant also exhibits hyperfine structure, therefore it was expected that the H1 residue is not responsible for the hyperfine interaction; however, the binding of the side chain amide of glutamine ($CONH_2$) was not considered. This demonstrates the importance of H1 in the fine-tuning of the electronic and catalytic features in the NiSOD enzyme.[69,70]

The axially coordinating moiety was altered by incorporating aspartic acid and alanine residues into the peptide chain (XaaCysAspLeuProCysGly, Xaa = His, Asp or Ala). The results confirms that the axial imidazole enhances the activity of NiSOD models in a number of ways:

(i) Since the coordination of imidazole leads to overall structural rearrangement, it increases the rate of the metal-centered redox process instead of ligand-based oxidations.

(ii) The binding of imidazole-*N* decreases the S-character of the redox active molecular orbital (RAMO), thus ensures the Ni-centered oxidation process.

(iii) It alters the Ni^{II}/Ni^{III} redox potential of the five-coordinated nickel center which becomes more negative relative to a four-coordinate Ni-center and to the midpoint of the oxidation/reduction potential of superoxide anion.[67]

4.2 The carboxamide nitrogen in the coordination sphere

Nickel induced deprotonation and coordination of peptide-*N* (or carboxamide group) yields the formation of (5,5)-membered chelate systems both in the oxidized and reduced forms of NiSOD. The binding of the amide nitrogen plays a significant role in the electronic structure of Ni^{III} via the stabilization of the oxidation product. Native chemical coordination of a pentapeptide (HisCysAspLeuPro) produces a variant wherein the amide-*N* of Cys2 is changed to a secondary amine group. The results confirm drastic changes both in the electronic and kinetic features. The EPR spectrum of the Ni^{III} metallopeptide exhibits more axial character than that obtained for the native enzyme. Analysis of the electronic structure leads to the conclusion that the binding of the amide-*N* donor raises the energy of the π-symmetry of the Ni 3d orbital via the π–π interactions. This interaction is not present for amine ligands which affects different d_{xz} and d_{yz} orbital contributions resulting in a rhombic EPR signal.[69]

The incorporation of a secondary amino group also increases the redox potential, i.e., the formation of the Ni^{III} state and, as a consequence, the reduction of superoxide to hydrogen peroxide are less favorable. Thus, the catalytic activity significantly decreases. This leads to the conclusion that the presence of the amide nitrogen has an important role in controlling the catalytic effect.

4.3 The role of the cysteine residues

The catalytic center of NiSOD possesses two thiolate groups of cysteine residues. One of them (Cys2) is the member of a (5,5)-membered chelate system which is supported by macrochelation via the thiolate group of Cys6 (Fig. 4). The formation of the macrochelate system is further supported by the proline residue which exhibits *cis*-peptide conformation. The cysteine residues in the catalytic center are likely to influence the Ni^{II}/Ni^{III} redox potential. Earlier studies demonstrated that the redox potential of Ni^{II}/Ni^{III} with N2O2 donor set is higher than with N2S2 donors.[71,72] This was further supported by theoretical calculations, and the noted trend in the redox potentials was explained by the π-interaction between the Ni 3d orbital and π-based S orbital. Moreover, it was emphasized that simultaneous existence of both cysteinyl residues is essential in order to stabilize the low spin state of Ni^{II}.[73]

On the basis of thorough analysis of spectroscopic data, very recent studies confirmed, that the presence of only one cysteine residue (essentially Cys2) is sufficient to induce spin pairing and establish the low spin state with square-planar geometry. In addition, stopped-flow experiments, and SOD activity studies on the basis of the McCord-Fridovich assay confirmed that both cysteine donors are essential in the coordination sphere of nickel to produce effective dismutation of the superoxide anion. In the absence of one of the cysteine moieties, the metallopeptide is capable to assist the dismutation, but the catalytic activity is considerably smaller than in the case of the metallopeptide with two cysteine residues.[57] It was also reported that the protonation of cysteine and the water hydrogen bond network surrounding the two cysteine residues are crucial for the reduction of superoxide anion. This reduction process takes place through an outer-sphere hydrogen atom transfer, however, the source of the proton could not be identified.[40] It is notable that a low-molecular weight pseudopeptide containing a 3S donor ligand was also synthesized and its nickel(II) complexes were characterized. At physiological pH, nickel(II) is coordinated by the $3S + O_{water}$ coordination environment and the complex shows notable SOD activity.[74]

Recently, we have thoroughly investigated the complex formation reactions between nickel(II) and the NiSOD related apopeptides as a function of pH.[28,55,57,59] The overall stability constant of the complexes was determined providing essential information on the metal binding ability of the peptides. The coordination modes of the nickel(II) complexes in aqueous solution were characterized by several spectroscopic methods and compared with those obtained for the native enzyme fragment. The feasibility of the oxidation of the nickel(II) complexes by KO_2 were also explored. The formation of the nickel(III) transient species was followed by EPR spectroscopy which was also used to characterize the donor groups involved in the binding of Ni^{III}. Finally, the SOD activity of the complexes was studied by dedicated stopped-flow experiments at physiological pH. The complementary use of these methods provided further insights into the thermodynamic, electronic and catalytic features of NiSOD models which are detailed in the subsequent part of the paper.

4.4 The role of the terminal amino group

Since nickel is accommodated in the N-terminal part of the enzyme and the terminal amino group is also involved in the metal binding, the essential role of this group is obvious. Acetylation of the N-terminal amino group of the native enzyme fragment significantly reduces the nickel binding ability of the peptide.[28] In comparison, the zinc(II) binding affinity of the peptide is enhanced in this case and the stability of the zinc(II) complex of the acetylated peptide was higher than that formed with the native enzyme fragment.[59] This is illustrated in Fig. 5, where the pM ($=-\log[M^{2+}]$) values are calculated at pH 7.6 and compared for different donor sets.

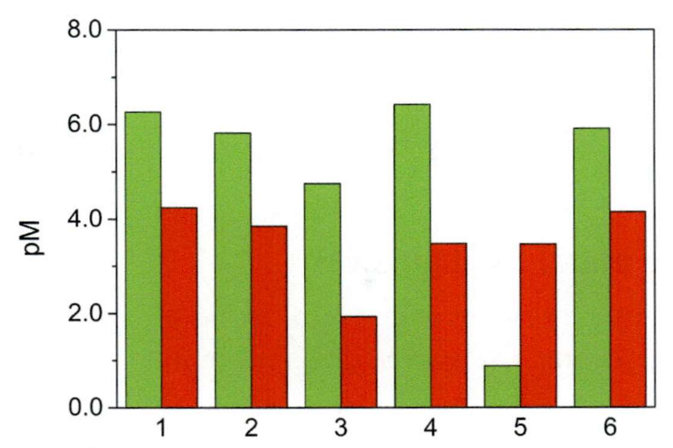

Fig. 5 pM ($=-\log[M^{2+}]$) values calculated for the NiSOD related peptides at pH 7.6 at 10 μM metal and 50 μM peptide concentrations; M^{2+} stands for Ni^{2+} (green) and Zn^{2+} (red). The results are shown for the complexes formed with the following peptides (the corresponding donor sets are in parenthesis): 1: HisCysAspLeuProCysGlyValTyr-NH_2; (NH_2,N^-,S^-,S^-); 2: HisCysAspLeuProHisGlyValTyr-NH_2, (NH_2,N^-,S^-,N_{Im}); 3: HisHisAspLeuProCysGlyValTyr-NH_2 (NH_2,N_{Im},S^-,S^-); 4: HisCysAspLeuProCysGly, (NH_2,N^-,S^-,S^-); 5: Ac-HisCysAspLeuProCysGlyValTyr-NH_2 (N_{Im}, S^-,S^-); 6: HisCysAspLeuAlaCysGlyValTyr-NH_2 (NH_2,N^-,S^-,S^-).[28,57,59]

At physiological pH, nickel(II) is bound via the (N_{Im},S^-,S^-) donor set and features the low spin state. KO_2 induced oxidation yields a paramagnetic Ni^{III} species, however the apical position of the metal ion is not occupied by the imidazole-*N*. Indeed, EPR spectra clearly proved the lack of the hyperfine interactions, and the axial position(s) is/are coordinated by the thiolate group(s) in a square-pyramidal or an elongated octahedral coordination environment. Thus, the EPR parameters as well as the coordination sphere of the oxidized form considerably differ from the native enzyme, and the acetylated complex shows negligible SOD activity.

4.5 The effect of the proline/alanine point mutation

The consensus sequence of the NiSOD binding hook contains proline. It is expected that this provides the binding of the distant cysteine residue via the formation of *cis*-peptide conformation. This phenomenon was further confirmed by thorough equilibrium studies. Point mutation of proline to alanine (HisCysAspLeuAlaCysGlyValTyr-NH_2) yields a new metallopeptide and the results confirmed that the equilibrium constant for the formation of the active site of NiSOD is higher in the native fragment than that obtained in the alanine counterpart. Moreover, the formation of the active structure is shifted into the less alkaline pH range (pK_a values are estimated to be 8.18 for HisCysAspLeuAlaCysGlyValTyr-NH_2 and 6.51 for the native enzyme fragment). Interestingly, the oxidation of the corresponding nickel(II) complex yields the same Ni^{III} species that is formed with the native enzyme fragments.[28]

4.6 Multiple metal binding sites

The NiSOD enzyme consists of several nickel ions which can act as an independent active center for superoxide dismutation. In an attempt, two NiSOD binding motifs (HisCysAspLeuProCysGlyValTyr) were linked via the lysine residue.[55] This pseudo-metallopeptide exhibits remarkable nickel binding affinity. Equilibrium studies confirmed an unusual thermodynamic feature; i.e., the formation of a dinuclear nickel(II) complex is preferred even in equimolar solution. This conclusion was also corroborated by capillary electrophoresis-mass spectrometry (CE-MS) results. Spectroscopic studies confirmed that the coordination modes of the Ni^{II} centers are exactly the same and are consistent with the structure of the reduced form of the NiSOD enzyme. The structures of the Ni^{III} complexes strongly depend on the pH and the metal to peptide ratio applied. In general, when the two nickel binding sites are occupied by the metal ion, the oxidation yields an exclusive Ni^{III} complex, where both Ni^{III} possess the same coordination environment which is consistent with the structure of the oxidized form of the native fragment. Interestingly, the interaction between the two Ni^{III} centers leads to significant line broadening in the EPR spectra, however, the distance between the paramagnetic centers is not close enough to cause antiferromagnetic coupling. Kinetic studies indicated that the dinuclear nickel complex exhibits superior SOD activity, unambiguously, the two nickel centers are kinetically indistinguishable.

4.7 Redox degradation process

SOD activity of SOD mimics can readily be studied by monitoring the catalytic decay of the superoxide anion in sequential stopped flow experiments. As shown in Fig. 6, the kinetic traces correspond to a very fast dismutation process in the presence of Ni_2L which models multiple metal binding sites of NiSOD. The ligand, $(HisCysAspLeuProCysGlyValTyr)_2Lys$, includes two arms of the binding hook of the native enzyme. Although about 10% of the superoxide radical decomposes within the dead-time of the instrument ($\sim$1.5 ms), these experiments are suitable for the estimation of the rate constant of the dismutation process.

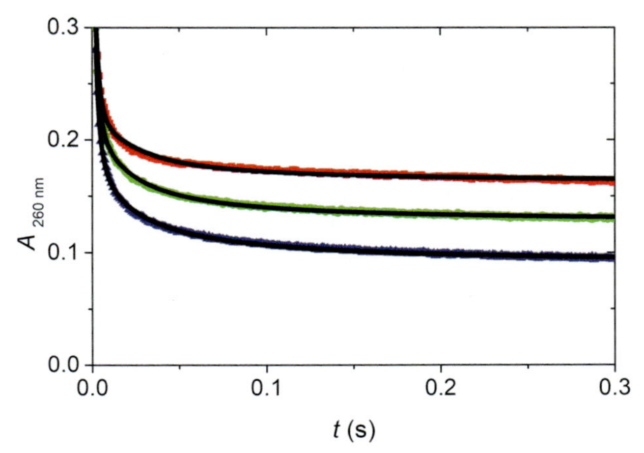

Fig. 6 The decomposition of superoxide anion in the presence of the Ni_2L complex where L = $(HisCysAspLeuProCysGlyValTyr)_2Lys$. The kinetic traces were recorded by using the sequential stopped-flow method. Solid black lines represent the kinetic traces fitted according to the proposed kinetic model. $c(Ni_2L)_0 = 10\ \mu M$ (red), 7.6 μM (green), 5.0 μM (blue); $c(O_2^{-\bullet})_0 = 877\ \mu M$; solvent: 1:1 aqueous buffer of HEPES (20 mm, pH 7.8): DMSO mixture; 2 mm optical path.

As a first approach, it is assumed that steady-state conditions apply for the Ni(III) form and a simple first-order expression can be derived for the interpretation of the data. In reality, kinetic traces for several NiSOD models cannot be fitted with such a simple expression.[55,75] After the fast initial phase, the rate of the absorbance change slowly decreases, i.e., the rate of $O_2^{-\bullet}$ decomposition declines and reaches the non-catalytic limit at longer reaction time. This feature is due to the kinetic coupling of the dismutation steps with the redox degradation of the catalyst. Since the corresponding nickel(II) complexes are stable for a long period of time, losing the catalytic activity is associated with the spontaneous degradation of the Ni^{III} complex as shown in Fig. 7. Presumably, it is an intramolecular redox process yielding an unidentified product (shown as Ni* in Fig. 7).[55] (For sake of simplicity, only the redox state but not the composition of the complexes is indicated in the kinetic model and in Eqs. (12–20). In fact, each complex is a dinuclear Ni species.) The exact composition of the product(s) was not determined; however, nickel complexes containing sulfoxide, sulfone and sulfonic acid derivatives were identified on the basis of their mass spectra.

The experimental data were evaluated by fitting each kinetic trace according to the postulated kinetic models shown in Eqs. (12–15).

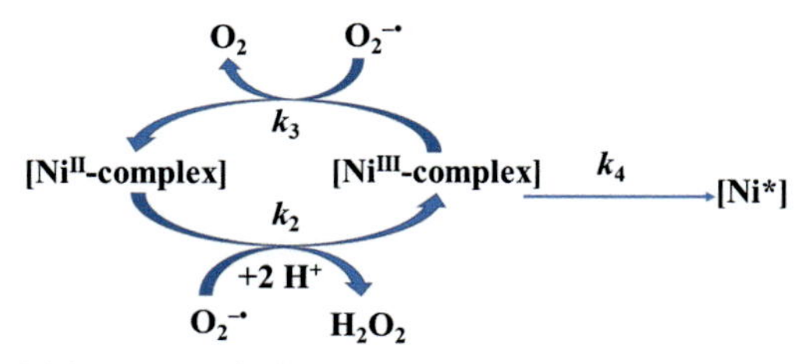

Fig. 7 Kinetic model for SOD including the kinetically coupled degradation of the Ni(III) form.[55]

The corresponding differential equation system (16–20) was solved using a nonlinear least square routine.

$$2\,H^+ + 2\,O_2^{-\bullet} \rightarrow H_2O_2 + O_2 \qquad k_1 \qquad (12)$$

$$Ni^{II}L + 2\,H^+ + O_2^{-\bullet} \rightarrow Ni^{III}L + H_2O_2 \qquad k_2 \qquad (13)$$

$$Ni^{III}L + O_2^{-\bullet} \rightarrow Ni^{II}L + O_2 \qquad k_3 \qquad (14)$$

$$Ni^{III}L \rightarrow Ni^{*}L \qquad k_4 \qquad (15)$$

$$\frac{d\left[O_2^-\right]}{dt} = -k_1 \times \left[O_2^{-\bullet}\right]^2 - k_2 \times \left[Ni^{II}L\right]\left[O_2^{-\bullet}\right] - k_3 \times \left[Ni^{III}L\right]\left[O_2^{-\bullet}\right] \qquad (16)$$

$$\frac{d[H_2O_2]}{dt} = k_2 \times \left[Ni^{II}L\right]\left[O_2^{-\bullet}\right] \qquad (17)$$

$$\frac{d[Ni^{II}L]}{dt} = -k_2 \times \left[Ni^{II}L\right]\left[O_2^{-\bullet}\right] + k_3 \times \left[Ni^{III}L\right]\left[O_2^{-\bullet}\right] \qquad (18)$$

$$\frac{d[Ni^{III}L]}{dt} = k_2 \times \left[Ni^{II}L\right]\left[O_2^{-\bullet}\right] - k_3 \times \left[Ni^{III}L\right]\left[O_2^{-\bullet}\right] - k_4 \times \left[Ni^{*}L\right] \qquad (19)$$

$$\frac{d\left[Ni^{*}L\right]}{dt} = k_4 \times \left[Ni^{III}L\right] \qquad (20)$$

In the fitting procedure, the individual kinetic parameters were estimated as follows $k_2 > 1.0\ 10^8\ M^{-1}\ s^{-1}$, $k_3 = 1.9\ 10^7\ M^{-1}\ s^{-1}$ and $k_4 = 215\ s^{-1}$ and the molar absorption coefficients of the absorbing species were also calculated.[55] The concentration profiles as a function of time confirm that a substantial portion of the superoxide ion is dismutated in the first 10 ms of the reaction. In the initial phase, the Ni^{III} form of the catalyst rapidly accumulates, and the relatively fast degradation of the Ni^{III} form leads to the elimination of the SOD activity (Fig. 8). It needs to be emphasized that the steady-state approach assumed in several SOD models for Ni(III) is not applicable here. The modification of the active site by introducing a penicillamine moiety confirmed these conclusions. The electron donating

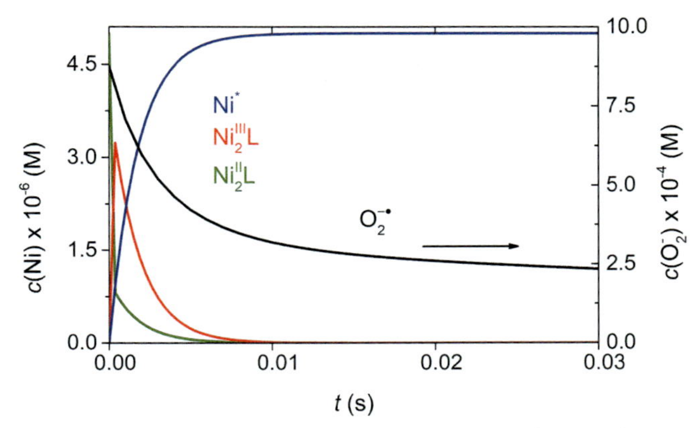

Fig. 8 Calculated concentration profiles as a function of time during the metal assisted decomposition of superoxide anion at pH 7.6 at 1:1 DMSO aqueous HEPES buffer solution. The catalyst is the dinuclear complex of nickel, Ni_2L, formed with the model peptide of the multiple metal binding sites of the native enzyme. L = (HisCysAspLeuProCysGlyValTyr)$_2$Lys.

substituents of the methyl groups of the penicillamine moiety increases the lifetime of the transient Ni^{III} complex and leads to a more pronounced self-degradation process than that observed for the native enzyme fragment.

5. Concluding remarks

Studying the superoxide dismutase activity is not a trivial task since complicated experimental problems need to be addressed. This is the consequence of the high reactivity of the superoxide anion, especially under physiological conditions. The McCord-Fridovich assay provides a possibility to compare the SOD activity of various enzymes and their mimics on the basis of competitive kinetics. However, this method is not suitable for exploring the mechanisms of dismutation processes. In order to gain information on the chemical background of SOD activity, the use of advanced and direct kinetic techniques is required, such as pulse radiolysis or stopped-flow methods. The stopped-flow method is particularly useful because it makes possible to monitor the dismutation process under real catalytic conditions.

NiSOD, the relatively novel type of superoxide dismutase enzyme family, possesses unique coordination environment because of the thiolate groups of cysteine moieties in the coordination sphere. Several Ni containing metallopeptides have been synthesized and studied earlier to gain information on the structure of the active site, the role of each donor set involved in the metal ion coordination, and the intimate details of the dismutation mechanism. This paper surveys the most relevant results in this field.

The SOD activity studies on NiSOD related metallopeptides clearly demonstrate that the classical first-order approach cannot always be used for the interpretation of the data and the kinetically coupled redox degradation of the catalyst needs to be included in the kinetic models. The mechanism of this redox degradation is not completely understood. Noticeably, such a degradation process was not observed for the native NiSOD enzymes indicating that a protective mechanism is operative in this case. Most likely, the hydrogen bonds of cysteine sulfur atoms or the surrounding water hydrogen bond network enhance the robustness of NiSOD against oxidative damage.

In summary, NiSOD metallopeptides are useful mimics for understanding the structure, and electronic properties of NiSOD enzymes. However, it remains a challenging issue why, in contrast to the native enzyme, the mimics undergo redox degradation. Further studies should focus on model systems with improved redox stability in order to explore the intimate features of the redox chemistry of NiSODs.

Acknowledgment

N.L. and I.F. are grateful for the financial support of the Hungarian National Research, Development and Innovation Office (NKFIH PD-128326 and K-139140).

References

1. D'Autréaux, B.; Toledano, M. B. *Nat. Rev. Mol. Cell Biol.* **2007**, *8*, 813–824.
2. Burrell, C. J.; Blake, D. R. *Br. Heart J.* **1989**, *61*, 4–8.
3. Nordberg, J.; Arnér, E. S. J. *Free Radical Biol. Med.* **2001**, *31*, 1287–1312.
4. Fernández-Castro, P.; Vallejo, M.; San Román, M. F.; Ortiz, I. *J. Chem. Technol. Biotechnol.* **2015**, *90*, 796–820.
5. He, Y.; Grieser, F.; Ashokkumar, M. *J. Phys. Chem. A* **2011**, *115*, 6582–6588.
6. Yang, X.-j.; Xu, X.-m.; Xu, J.; Han, Y.-f. *J. Am. Chem. Soc.* **2013**, *135*, 16058–16061.
7. Yang, B.; Chen, Y.; Shi, J. *Chem. Rev.* **2019**, *119*, 4881–4985.
8. Reczek, C. R.; Chandel, N. S. *Curr. Opin. Cell Biol.* **2015**, *33*, 8–13.
9. Nathan, C.; Cunningham-Bussel, A. *Nat. Rev. Immunol.* **2013**, *13*, 349–361.
10. Halliwell, B. *Nutr. Rev.* **1997**, *55*, S44–S49 (discussion S49-52).
11. Su, J.; Groves, J. T. *Inorg. Chem.* **2010**, *49*, 6317–6329.
12. Sheng, Y.; Abreu, I. A.; Cabelli, D. E.; Maroney, M. J.; Miller, A.-F.; Teixeira, M.; Valentine, J. S. *Chem. Rev.* **2014**, *114*, 3854–3918.
13. Abreu, I. A.; Cabelli, D. E. *Biochim. Biophys. Acta Protein. Proteomics* **1804**, *2010*, 263–274.
14. Zámocký, M.; Koller, F. *Prog. Biophys. Mol. Biol.* **1999**, *72*, 19–66.
15. Miller, A.-F. *Curr. Opin. Chem. Biol.* **2004**, *8*, 162–168.
16. El-Kenawi, A.; Ruffell, B. *Cancer Cell* **2017**, *32*, 727–729.

17. Umeno, A.; Biju, V.; Yoshida, Y. *Free Radical Res.* **2017**, *51*, 413–427.
18. WISEMAN, H.; HALLIWELL, B. *Biochem. J.* **1996**, *313*, 17–29.
19. Newsholme, P.; Cruzat, V. F.; Keane, K. N.; Carlessi, R.; de Bittencourt, P. I., Jr. *Biochem. J.* **2016**, *473*, 4527–4550.
20. Wuerges, J.; Lee, J.-W.; Yim, Y.-I.; Yim, H.-S.; Kang, S.-O.; Carugo, K. D. *Proc. Natl. Acad. Sci. U. S. A.* **2004**, *101*, 8569–8574.
21. Palenik, B.; Brahamsha, B.; Larimer, F. W.; Land, M.; Hauser, L.; Chain, P.; Lamerdin, J.; Regala, W.; Allen, E. E.; McCarren, J.; Paulsen, I.; Dufresne, A.; Partensky, F.; Webb, E. A.; Waterbury, J. *Nature* **2003**, *424*, 1037–1042.
22. Barondeau, D. P.; Kassmann, C. J.; Bruns, C. K.; Tainer, J. A.; Getzoff, E. D. *Biochemistry* **2004**, *43*, 8038–8047.
23. Youn, H. D.; Kim, E. J.; Roe, J. H.; Hah, Y. C.; Kang, S. O. *Biochem. J.* **1996**, *318*, 889–896.
24. Riley, D. P. *Chem. Rev.* **1999**, *99*, 2573–2588.
25. Zilbermann, I.; Maimon, E.; Cohen, H.; Meyerstein, D. *Chem. Rev.* **2005**, *105*, 2609–2626.
26. Pettersen, E. F.; Goddard, T. D.; Huang, C. C.; Couch, G. S.; Greenblatt, D. M.; Meng, E. C.; Ferrin, T. E. *J. Comput. Chem.* **2004**, *25*, 1605–1612.
27. Dupont, C. L.; Neupane, K.; Shearer, J.; Palenik, B. *Environ. Microbiol.* **2008**, *10*, 1831–1843.
28. Lihi, N.; Csire, G.; Szakács, B.; May, N. V.; Várnagy, K.; Sóvágó, I.; Fábián, I. *Inorg. Chem.* **2019**, *58*, 1414–1424.
29. Herbst, R. W.; Guce, A.; Bryngelson, P. A.; Higgins, K. A.; Ryan, K. C.; Cabelli, D. E.; Garman, S. C.; Maroney, M. J. *Biochemistry* **2009**, *48*, 3354–3369.
30. Pelmenschikov, V.; Siegbahn, P. E. M. *J. Am. Chem. Soc.* **2006**, *128*, 7466–7475.
31. Volbeda, A.; Garcin, E.; Piras, C.; de Lacey, A. L.; Fernandez, V. M.; Hatchikian, E. C.; Frey, M.; Fontecilla-Camps, J. C. *J. Am. Chem. Soc.* **1996**, *118*, 12989–12996.
32. Peters, J. W.; Schut, G. J.; Boyd, E. S.; Mulder, D. W.; Shepard, E. M.; Broderick, J. B.; King, P. W.; Adams, M. W. W. *Biochim. Biophys. Acta Mol. Cell Res.* **1853**, *2015*, 1350–1369.
33. Marr, A. C.; Spencer, D. J. E.; Schröder, M. *Coord. Chem. Rev.* **2001**, *219–221*, 1055–1074.
34. Ermler, U.; Grabarse, W.; Shima, S.; Goubeaud, M.; Thauer, R. K. *Science* **1997**, *278*, 1457–1462.
35. Shima, S.; Krueger, M.; Weinert, T.; Demmer, U.; Kahnt, J.; Thauer, R. K.; Ermler, U. *Nature* **2012**, *481*, 98–101.
36. Can, M.; Armstrong, F. A.; Ragsdale, S. W. *Chem. Rev.* **2014**, *114*, 4149–4174.
37. Fesseler, J.; Jeoung, J.-H.; Dobbek, H. *Angew. Chem., Int. Ed.* **2015**, *54*, 8560–8564.
38. Fiedler, A. T.; Bryngelson, P. A.; Maroney, M. J.; Brunold, T. C. *J. Am. Chem. Soc.* **2005**, *127*, 5449–5462.
39. Bryngelson, P. A.; Arobo, S. E.; Pinkham, J. L.; Cabelli, D. E.; Maroney, M. J. *J. Am. Chem. Soc.* **2004**, *126*, 460–461.
40. Shearer, J.; Peck, K. L.; Schmitt, J. C.; Neupane, K. P. *J. Am. Chem. Soc.* **2014**, *136*, 16009–16022.
41. Prabhakar, R.; Morokuma, K.; Musaev, D. G. *J. Comput. Chem.* **2006**, *27*, 1438–1445.
42. Shearer, J. *Angew. Chem., Int. Ed.* **2013**, *52*, 2569–2572.
43. Tietze, D.; Breitzke, H.; Imhof, D.; Kothe, E.; Weston, J.; Buntkowsky, G. *Chemistry* **2009**, *15*, 517–523.
44. Narayana, P. A.; Suryanarayana, D.; Kevan, L. *J. Am. Chem. Soc.* **1982**, *104*, 3552–3555.
45. Martins-Costa, M. T. C.; Anglada, J. M.; Francisco, J. S.; Ruiz-Lopez, M. F. *Angew. Chem., Int. Ed.* **2012**, *51*, 5413–5417.
46. Behar, D.; Czapski, G.; Rabani, J.; Dorfman, L. M.; Schwarz, H. A. *J. Phys. Chem.* **1970**, *74*, 3209–3213.

47. Bielski, B. H. J. *Photochem. Photobiol.* **1978**, *28*, 645–649.
48. Beyer, W. F.; Fridovich, I. *Anal. Biochem.* **1987**, *161*, 559–566.
49. Faulkner, K. M.; Stevens, R. D.; Fridovich, I. *Arch. Biochem. Biophys.* **1994**, *310*, 341–346.
50. Durot, S.; Policar, C.; Cisnetti, F.; Lambert, F.; Renault, J.-P.; Pelosi, G.; Blain, G.; Korri-Youssoufi, H.; Mahy, J.-P. *Eur. J. Inorg. Chem.* **2005**, *2005*, 3513–3523.
51. Garda, Z.; Molnár, E.; Hamon, N.; Barriada, J. L.; Esteban-Gómez, D.; Váradi, B.; Nagy, V.; Pota, K.; Kálmán, F. K.; Tóth, I.; Lihi, N.; Platas-Iglesias, C.; Tóth, É.; Tripier, R.; Tircsó, G. *Inorg. Chem.* **2021**, *60*, 1133–1148.
52. Goto, J. J.; Gralla, E. B.; Valentine, J. S.; Cabelli, D. E. *J. Biol. Chem.* **1998**, *273*, 30104–30109.
53. Kobayashi, K. *Chem. Rev.* **2019**, *119*, 4413–4462.
54. Riley, D. P.; Rivers, W. J.; Weiss, R. H. *Anal. Biochem.* **1991**, *196*, 344–349.
55. Kelemen, D.; May, N. V.; Andrási, M.; Gáspár, A.; Fábián, I.; Lihi, N. *Chem. Eur. J.* **2020**, *26*, 16767–16773.
56. Domergue, J.; Guinard, P.; Douillard, M.; Pécaut, J.; Proux, O.; Lebrun, C.; Le Goff, A.; Maldivi, P.; Delangle, P.; Duboc, C. *Inorg. Chem.* **2021**, *60*, 12772–12780. https://doi.org/10.1021/acs.inorgchem.1c00899.
57. Lihi, N.; Kelemen, D.; May, N. V.; Fábián, I. *Inorg. Chem.* **2020**, *59*, 4772–4780.
58. Johnson, O. E.; Ryan, K. C.; Maroney, M. J.; Brunold, T. C. *J. Biol. Inorg. Chem.* **2010**, *15*, 777–793.
59. Csire, G.; Kolozsi, A.; Gajda, T.; Pappalardo, G.; Várnagy, K.; Sóvágó, I.; Fábián, I.; Lihi, N. *Dalton Trans.* **2019**, *48*, 6217–6227.
60. Shearer, J. *Acc. Chem. Res.* **2014**, *47*, 2332–2341.
61. Broering, E. P.; Dillon, S.; Gale, E. M.; Steiner, R. A.; Telser, J.; Brunold, T. C.; Harrop, T. C. *Inorg. Chem.* **2015**, *54*, 3815–3828.
62. Gale, E. M.; Patra, A. K.; Harrop, T. C. *Inorg. Chem.* **2009**, *48*, 5620–5622.
63. Gale, E. M.; Simmonett, A. C.; Telser, J.; Schaefer, H. F.; Harrop, T. C. *Inorg. Chem.* **2011**, *50*, 9216–9218.
64. Jenkins, R. M.; Singleton, M. L.; Almaraz, E.; Reibenspies, J. H.; Darensbourg, M. Y. *Inorg. Chem.* **2009**, *48*, 7280–7293.
65. Nakane, D.; Wasada-Tsutsui, Y.; Funahashi, Y.; Hatanaka, T.; Ozawa, T.; Masuda, H. *Inorg. Chem.* **2014**, *53*, 6512–6523.
66. Steiner, R. A.; Dzul, S. P.; Stemmler, T. L.; Harrop, T. C. *Inorg. Chem.* **2017**, *56*, 2849–2862.
67. Neupane, K. P.; Gearty, K.; Francis, A.; Shearer, J. *J. Am. Chem. Soc.* **2007**, *129*, 14605–14618.
68. Shearer, J.; Long, L. M. *Inorg. Chem.* **2006**, *45*, 2358–2360.
69. Campeciño, J. O.; Dudycz, L. W.; Tumelty, D.; Berg, V.; Cabelli, D. E.; Maroney, M. J. *J. Am. Chem. Soc.* **2015**, *137*, 9044–9052.
70. Shearer, J.; Neupane, K. P.; Callan, P. E. *Inorg. Chem.* **2009**, *48*, 10560–10571.
71. Krueger, H. J.; Holm, R. H. *Inorg. Chem.* **1987**, *26*, 3645–3647.
72. Kruger, H. J.; Peng, G.; Holm, R. H. *Inorg. Chem.* **1991**, *30*, 734–742.
73. Ryan, K. C.; Johnson, O. E.; Cabelli, D. E.; Brunold, T. C.; Maroney, M. J. *J. Biol. Inorg. Chem.* **2010**, *15*, 795–807.
74. Domergue, J.; Pécaut, J.; Proux, O.; Lebrun, C.; Gateau, C.; Le Goff, A.; Maldivi, P.; Duboc, C.; Delangle, P. *Inorg. Chem.* **2019**, *58*, 12775–12785.
75. Tietze, D.; Sartorius, J.; Koley Seth, B.; Herr, K.; Heimer, P.; Imhof, D.; Mollenhauer, D.; Buntkowsky, G. *Sci. Rep.* **2017**, 7, 17194.

CHAPTER TWO

Cleavage of C—F bonds in oxidative conditions mediated by transition metal complexes

Alexander B. Sorokin*
Institut de Recherches sur la Catalyse et l'Environnement de Lyon IRCELYON, UMR 5256, CNRS—Université Lyon 1, Einstein, France
*Corresponding author: e-mail address: alexander.sorokin@ircelyon.univ-lyon1.fr

Contents

Abstract

The chemistry of C—F bonds continues to attract significant attention since fluorinated organic compounds are increasingly used as pharmaceuticals, agrochemicals, special products and in materials chemistry. Fluorinated organic compounds are particularly inert and their transformation is challenging. Traditionally, electron-rich species, such as complexes of low and zero-valent transition metals, strong nucleophiles and reductants, are employed for the activation of C—F bonds. However, some enzymes and

Advances in Inorganic Chemistry, Volume 79
ISSN 0898-8838
https://doi.org/10.1016/bs.adioch.2021.12.002

transition metal complexes are capable of transforming C—F bonds via oxidative pathways. This reactivity is in somewhat counter-intuitive since electron-deficient oxidizing species are involved in reaction with a C—F bond formed by the most electronegative element. In this chapter we will review oxidative defluorination of fluorinated aromatic compounds in biological and chemical systems containing principally iron active sites. The μ-nitrido diiron phthalocyanine complexes that show very high catalytic efficiency in oxidative transformation of poly- and even perfluorinated aromatics under mild conditions will be described in more details.

1. Introduction

Fluorinated organic compounds find wide and rapidly expanding utilization in a great variety of applications due to the unique properties of C—F bonds, e.g., C—F bond strength, high thermal and chemical stability, low polarity, weak intermolecular interactions, lipophilicity. Because of the small size of the fluorine atom and its low atomic polarizability volume, fluorine can replace hydrogen in molecular design to provide a huge variety of fluorinated organic molecules with specific biological, physical and electronic properties. The global production of fluorinated chemicals attained 4.2 million tons in 2018.[1] Fluorine substitution enhances the bioactivity and metabolic stability and plays a key role in the development of pharmaceuticals and agrochemicals. Various environmental and metabolic aspects connected to the advances in fluorinated pharmaceuticals and agrochemicals have recently been reviewed.[2]

Consequently, among a variety of chemical reactions catalyzed by transition metal complexes, the activation and transformation of C—F bonds are of particular interest.[3,4] For a long time the catalytic chemistry of C—F bonds has been principally focused on the development of synthetic approaches to the preparation of fluorinated compounds.[5–7] The selective functionalization of readily accessible polyfluorinated arenes via transition metal-mediated defluorination, can be used to access functionalized fluorinated aromatic compounds.[8,9] Traditional approaches to these reactions involve electron-rich reagents, e.g., low- and zero-valent transition metal complexes, strong reductants and nucleophiles.

Owing to the highest electronegativity of fluorine, the transformation of C—F bonds under oxidative conditions using electron-deficient oxidizing species appears to be highly challenging. Indeed, the interaction of an electrophilic oxidant with strong and electron-poor carbon atom bearing a fluorine substituent is not immediately evident. Nevertheless, despite the extraordinary stability of fluoro-organic compounds, several enzymes are capable of cleaving C—F bonds via oxidative and non-oxidative routes.[1,10,11] In general, the active sites of these enzymes feature a mononuclear iron center in heme or non-heme coordination environment. Several synthetic transition metal complexes were reported to cleave C—F bonds in intramolecular and stoichiometric reactions. The recent discovery of the remarkably strong oxidizing properties of μ-nitrido diiron phthalocyanine and porphyrin complexes operating via high-valent oxo species, opens new possibilities in the challenging oxidations including oxidative transformation of C—F bonds.[12–17] In this chapter, the defluorination reactivity of μ-nitrido diiron complexes will be described in details with emphasis on the reaction scope, mechanistic aspects and possible practical application in disposal of persistent heavily fluorinated pollutants. An overview of enzymatic oxidative defluorination as well as intramolecular and stoichiometric cleavages of a C—F bond carried out by transition metal complexes, will be also given to follow the progress in this research area.

2. Enzymatic defluorination

Several enzymes are capable of performing the cleavage of aromatic C—F bonds.[1] Specifically, the following heme and non-heme enzymes have been shown to be competent in the oxidative defluorination of fluorinated phenols:

- **(i)** cytochrome P450 family with a thiolate-ligated iron porphyrin active site
- **(ii)** horseradish peroxidase
- **(iii)** histidine-ligated heme-dependent dihaloperoxidase
- **(iv)** histidine-ligated heme-dependent tyrosine hydroxylase
- **(v)** tetrahydrobiopterin-dependent aromatic amino acid hydroxylase
- **(vi)** Rieske dioxygenase
- **(vii)** thiol dioxygenase

Below we describe the essential catalytic properties of these enzymes in defluorination of aromatic compounds. For more details on structural and spectroscopic features of these enzymes as well as their reactivity with different halogenated substrates, we recommend reading the excellent review by Wang and Liu and original papers cited therein.[1]

2.1 Cytochrome P450

The ability of cytochrome P450[18] and methane monooxygenase (MMO)[19,20] enzymes to perform challenging oxidation of strong C—H bonds is well documented in the literature. Cytochrome P450 enzymes were also investigated in the reaction with fluorinated aromatic compounds.[21,22,26–31] Transformation of substituted fluorobenzenes mediated by cytochrome P450 enzymes occurs mainly by hydroxylation of C—H bonds rather than via C—F cleavage. However, the hydroxylation efficiency of C_6H_5F by engineered P450 BM3 M2 was much lower compared to C_6H_5Cl, C_6H_5Br, C_6H_5I and $C_6H_5CH_3$, whereas wild-type P450 BM3 did not convert C_6H_5F at a detectable level.[21] In vivo cytochrome P450-mediated biotransformation of fluorobenzene, 1,2-difluorobenzene, 1,3-difluorobenzene, 1,2,3-trifluorobenzene and 1,2,4-trifluorobenzene led to different fluorophenols hydroxylated at all nonhalogenated positions.[22] Phenols resulting from hydroxylation accompanied by either defluorination or an NIH shift (migration of the substituent originally present at the site of hydroxylation to an adjacent aromatic carbon atom)[23–25] were not observed.

Cytochrome P450 carried out in vivo defluorination of C_6F_6 to C_6F_5OH but further oxidation to tetrafluorohydroquinone derivatives was not observed.[26] Cytochrome P450 dependent monooxygenation of C_6F_5Cl resulted in preferential elimination of the fluoride atom in the para position with respect to the chlorine substituent. The preferential elimination of the fluorine atom was also observed in 1,3,5-trifluoro-2,4,6-trichlorobenzene.[26]

The regioselectivity of the aromatic hydroxylation of monofluorinated anilines performed by cytochrome P450 depends on the substitution pattern.[27] *o*-Fluoroaniline and *m*-fluoroaniline were predominantly hydroxylated at the *para*-C4, and *ortho*-C6 and *para*-C4 positions, respectively. Only *p*-fluoroaniline underwent the significant defluorination to provide fluoride anion and the reactive benzoquinoneimine as primary reaction products. It is noteworthy that, the fluorine substituent at the C4-position was more easily eliminated from the aromatic cycle than other halogen substituents (Cl, Br, I) in a system with purified reconstituted cytochrome P450 IIB1, in a ^{t}BuOOH supported microsomal cytochrome P450 system, and in a system with microperoxidase 8.[28] The elimination of the *para*-fluorine substituent became more difficult upon the introduction of additional electron-withdrawing fluorine substituents resulting in the switch to *ortho*-C-H hydroxylation. However, pentafluorophenol could still be *para*-defluorinated by cytochrome P450.[29]

Importantly, cytochrome P450 mediated oxidation of 2,3,5, 6-tetrafluorophenol and pentafluorophenol resulted in different reaction products.[30] While 2,3,5,6-tetrafluorophenol afforded expected tetrafluorohydroquinone, oxidative defluorination of pentafluorophenol led to the formation of tetrafluorobenzoquinone as the primary reaction product using O_2/NADPH, PhIO or cumene hydroperoxide (Fig. 1).

Microsomal incubation in the presence of NADPH and O_2 resulted in turnover numbers (TON) of 36 and 17 for the oxidation of 2,3,5,6-tetrafluorophenol and pentafluorophenol, respectively.[30]

The aromatic hydroxylation of fluorobenzenes with a different substitution pattern mediated by cytochrome P450 was studied in vivo in order to evaluate the importance of the fluorine NIH shift, i.e., the migration of the substituent from the hydroxylation site of arene to the adjacent aromatic position (Fig. 2).[31] The so-called NIH shift arises from the National Institutes of Health where this phenomenon that occurred in aromatic hydroxylation was discovered.

Careful product analysis by ^{19}F NMR showed a very strong preference for C—H hydroxylation products versus products obtained from C—F bond cleavage. The ratios $yield_{C\text{-}H}/yield_{C\text{-}F}$ were determined to be 76.8%:0.6% for 1,4-difluorobenzene, 37.8%:0.4% for 1,2,3-trifluorobenzene, 38.9%: 1.8% for 1,2,4-trifluorobenzene, 49.0%:1.6% for 1,2,3,4-tetrafluoro-benzene and 30.3%:5.9% for 1,2,3,5-tetrafluorobenzene. Such a product composition is typical for the oxidative transformation of fluorinated aromatic compounds mediated by cytochrome P450 enzymes. Minor amounts of NIH-shifted phenols were observed in the oxidation of 1,4-difluorobenzene and 1,2,3, 5-tetrafluorobenzene, whereas the possible yields of oxidation products which could be obtained via an NIH shift from other substrates were less than the detection limit of 0.1%. This result is in agreement with a previous observation that a fluorine substituent is preferentially lost from the molecule rather than giving rise to a fluorine NIH shift.[31]

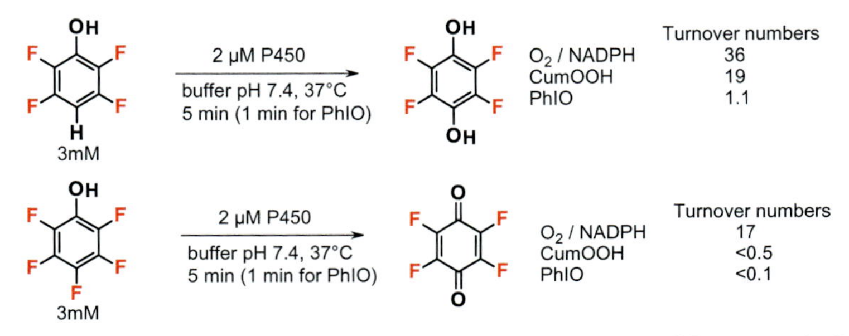

Fig. 1 Influence of the oxidant on the efficiency of oxidation of fluorinated phenols mediated by cytochrome P450.[30]

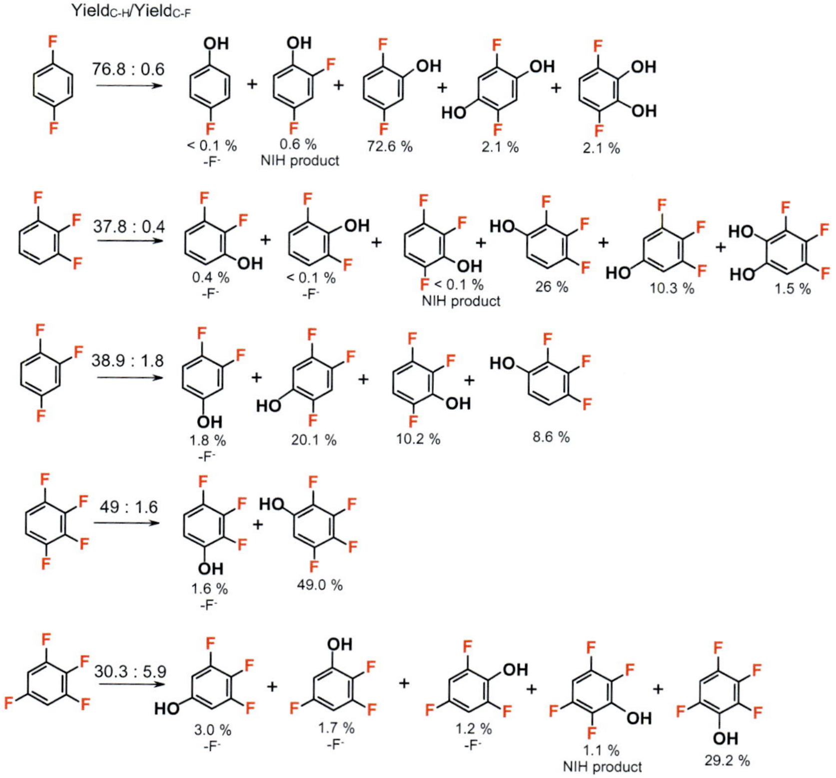

Fig. 2 Product composition of oxidation of fluorinated aromatic compounds by cytochrome P450 in vivo.[31]

Oxidative dehalogenation of perfluorinated and perchlorinated benzenes by Compound I of cytochrome P450 was investigated by DFT calculations.[32] Unexpectedly, the barriers for the addition of Compound I to fluorine-substituted carbon atoms of C_6F_6 and C_6F_5OH are comparable to those calculated for the oxidation of benzene through an analogous reaction route. However, the following steps of the reaction, namely, fluorine migration to generate cyclohexadienone intermediates and elimination of the fluoride anion from hexafluorocyclohexadiene radical anion and deprotonated hydroxypentafluorocyclohexadie
none occur with significant barriers.[32]

To conclude, the monooxygenase cytochrome P450 enzymes exhibit a sluggish activity in transformation of fluorinated aromatics. Their limited efficiency in this reaction explains the slow biodegradation of fluorinated aromatic compounds.

2.2 Horseradish peroxidase

Horseradish peroxidase in combination with H_2O_2 carried out the oxidation of 4-fluorophenol accompanied by the production of 4-fluorophenoxy radicals.[33] The coupling of these radicals resulted in the formation of products with high molecular weight which were identified by mass- spectrometry (Fig. 3).

Although several cleavages of C—F bonds occurred during the formation of the coupling products, they still contain several fluorophenyl moieties.

2.3 Histidine-ligated heme-dependent dehaloperoxidase

Isoenzymes A and B of histidine-ligated heme-dependent dehaloperoxidase (DHP) mediate *para*-defluorination of phenols including the oxidation of trifluorophenol to the corresponding1,4-benzoquinones using H_2O_2 as the oxidant.[1,34] The reaction mechanism of DHP is believed to involve high-valent iron oxo species performing the two-step oxidation.[35–38] The crystal structures of the DHP–substrate complexes show that the phenol hydroxyl group is pointed toward the heme iron, whereas the *para*-halogen to be eliminated points away.[1,37] Thus, the direct attack of the phenolic *para*-position by the Compound I-like species formed via coordination of H_2O_2 and heterolytic cleavage of the O—O bond of the Fe(III) hydroperoxo complex should be excluded. Instead, the high-valent oxo species with a histidine axial ligand abstracts the hydrogen atom from the hydroxyl group to give the phenoxy radical (Fig. 4).

A second electron transfer results in the formation of a phenoxy cation, followed by the nucleophilic attack by a water molecule at the *para*-position, and elimination of HF. The $^{18}O_2$-labeling study using 4-fluoroguaiacol confirmed that the oxygen atom incorporation to yield quinone originated

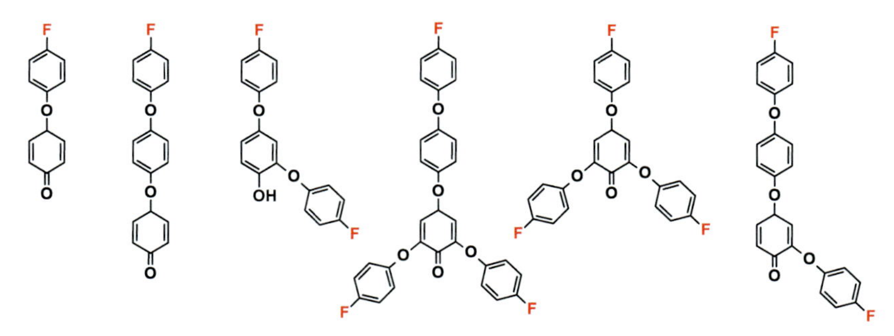

Fig. 3 Products of the oxidation of 4-fluorophenol by H_2O_2 catalyzed by horseradish peroxidase.

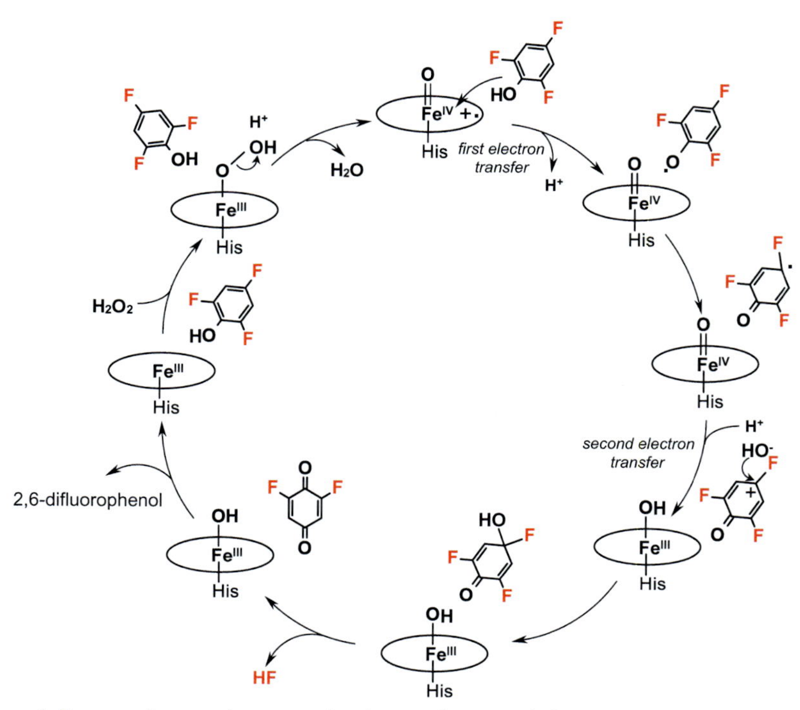

Fig. 4 Proposed reaction mechanism of *para*-defluorination mediated by histidine-ligated heme-dependent dihaloperoxidase (DHP).

exclusively from water.[37] This finding is consistent with a peroxidase mechanism and binding orientation of the substrate in the enzyme cavity. It should be noted that fluorinated phenols exhibit weak binding affinities and limited turnovers.[1,39]

2.4 Histidine-ligated heme-dependent tyrosine hydroxylase

Histidine-ligated heme-dependent tyrosine hydroxylase (TyrHs) catalyzes the oxidation of L-tyrosine compounds to corresponding *L*-dihydroxyphenylalanine derivatives including 3-fluorotyrosine (Fig. 5).[39,40]

Comparable amounts of products resulting from the cleavage of C—F and C—H bonds, were obtained via two mechanistically different reactions. The C—F bond cleavage requires two additional electrons compared to C—H cleavage because of the loss of fluoride anion instead of elimination of the proton. These electrons are provided during oxidation of H_2O_2 to dioxygen in a catalase-like reaction. The ^{18}O-labeling studies using $H_2^{18}O$ or $H_2^{18}O_2$ revealed the incorporation of an oxygen atom from hydrogen

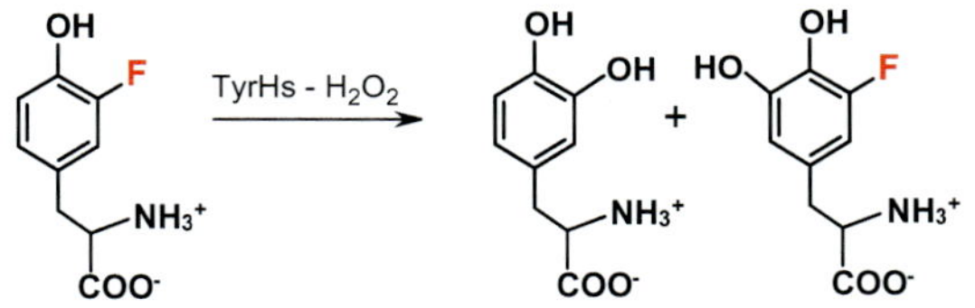

Fig. 5 Formation of two products originated from the cleavages of C—F or C—H bonds mediated by histidine-ligated heme-dependent tyrosine hydroxylase (TyrHs).

peroxide to the hydroxylated products.[39] Hydroxylation of an aromatic C—H bond occurs via either a P450-like rebound mechanism or via initial electron transfer with formation of a phenoxy radical-cation, followed by the formation of σ-complex and rearomatization to a dihydroxyphenyl derivative. In contrast, an iron hydroperoxo complex of TyrHs was proposed to perform a nucleophilic attack at the partially positive C3 atom bearing a fluorine substituent accompanied by the formation of Compound I-like species (Fig. 6).

Elimination of a fluoride anion and re-aromatization provide *L*-dihydroxyphenylalanine. In turn, high-valent iron oxo species can oxidize a H_2O_2 to O_2 with regeneration of the Fe(III) resting state of TyrHs. Recent crystal structure resolution of the TyrHs complex with 3-fluorotyrosine showed one substrate binding conformation but two orientations of the fluorine atom with a ratio of 7:3.[40] Thus, the product distribution depends on the substrate orientation: the conformation with a fluorine atom pointed toward the heme furnishes the defluorinated product. Comparison with other halogen-substituted analogs showed that 3-fluoro-L-tyrosine afforded the highest yield of L-3,4-dihydroxyphenylalanine (DOPA) due to the cleavage of a C—F bond.

2.5 Tetrahydrobiopterin-dependent aromatic amino acid hydroxylase

This non-heme enzyme carries out the aerobic oxidation of L-tyrosine to DOPA via high-valent iron oxo species generated during hydroxylation of tetrahydrobiopterin co-substrate.[1] The active Fe(II) site contains one glutamate, two histidine residues and two water ligands. In contrast to heme-dependent tyrosine hydroxylase, non-heme TyrH selectively oxidizes 3-fluorotyrosine at the C3 position providing only a defluorinated DOPA product.[41] Dehalogenation of chlorinated and brominated analogs was less efficient. Non-heme TyrH is also capable of defluorinating

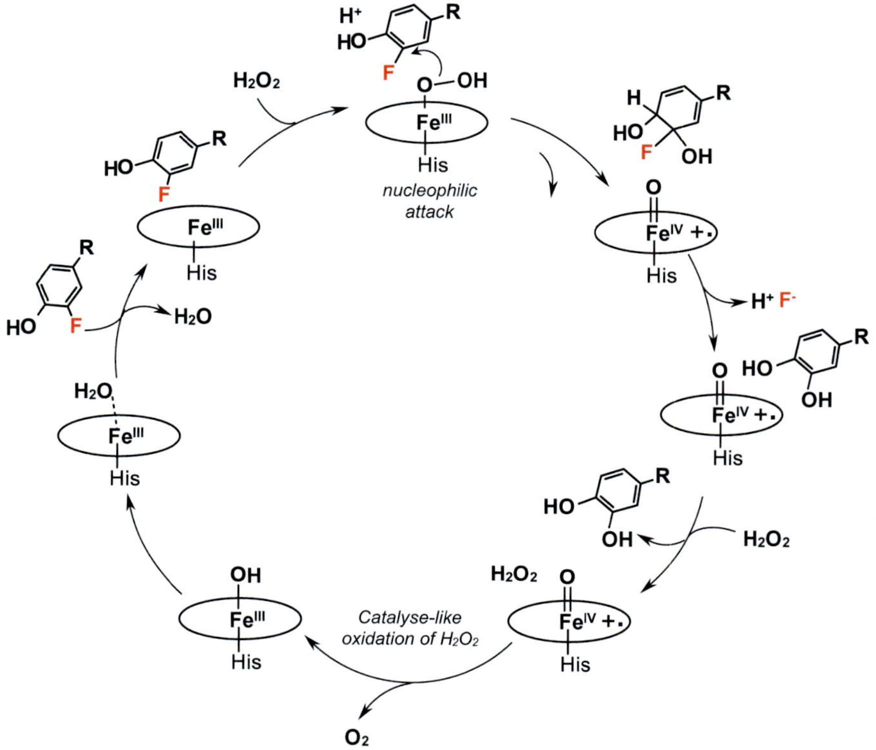

Fig. 6 Proposed nucleophilic mechanism of aromatic C—F cleavage by histidine-ligated heme-dependent tyrosine hydroxylase.

4-fluorophenylalanine with selective formation of tyrosine. In turn, the reaction with 4-chlorophenylalanine and 4-bromophenylalanine provided mixtures of hydroxylation products at the C3 and C4 positions involving halogen NIH shift.[1,41] It was noteworthy that in the course of oxidation of 4-fluorophenylalanine no fluorine NIH shift was detected. An iron(IV) oxo species was proposed to be responsible for this defluorination activity, but further studies should be performed to obtain insight into the mechanism.[1]

2.6 Rieske dioxygenases

Rieske dioxygenases catalyze aerobic bio-oxidation of various substrates including *cis*-hydroxylation of aromatic compounds. In addition to the Fe(II) active site with two histidine and one carboxylate fragments (similar to that of tetrahydrobiopterin-dependent aromatic amino acid hydroxylase), these enzymes contain an iron-sulfur cluster [2Fe—2S] responsible for the electron transfer from NADPH. Along with aromatic dihydroxylation, some Rieske dioxygenases mediate a defluorination reaction. For instance, *Pseudomonas* sp. Strain T-12 cells containing toluene-2,3-dioxygenase catalyzed oxidation of several 3-fluorosubstituted benzenes to the corresponding 2,3-catechols (Fig. 7).[42]

While 3-fluorotoluene and 3-fluoroanisole afforded only defluorinated catechols, other compounds produced a mixture of defluorinated and fluorinated catechols in a ratio depending on the size of substituent R.

2-Halobenzoate-1,2-dioxygenase exhibited a very broad substrate scope converting substituted benzoates to corresponding catechols including fluorinated benzoates (Fig. 8).[43]

The mechanism of Rieske non-heme dioxygenase was proposed to include the initial formation of superoxo species followed by its reduction to form an Fe(III) side-on hydroperoxo complex (Fig. 9).[1]

Homolytic cleavage of the O—O bond accompanied by $HO^{\bullet}$ radical addition to the aromatic double bond, resulted in formation of the Fe(IV)=O species which, in turn, carried out one-electron oxidation of the substrate radical intermediate to dihydroxylated 2-fluorobenzoate. The resting Fe(II) state was regenerated after second electron transfer and dihydroxylated 2-fluorobenzoate underwent re-aromatization with release of CO_2 and fluoride anion.

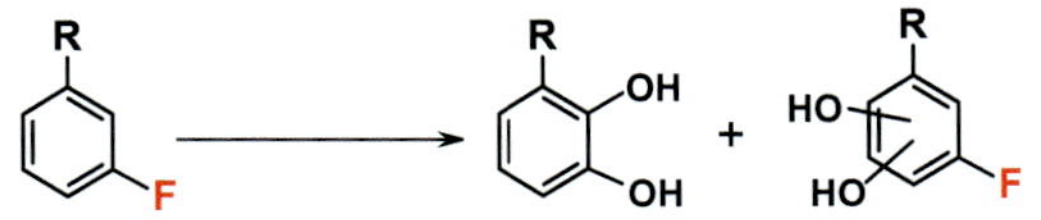

R	Product ratio	
CH_3	100 %	-
CF_3	67 %	33 %
OCH_3	100 %	-
F	90 %	10 %
Cl	74 %	26 %
Br	81 %	19 %
I	55 %	45 %
CN	82 %	18 %

Fig. 7 Transformation of fluorinated aromatics by *Pseudomonas* sp. Strain T-12 cells containing toluene-2,3-dioxygenase.

Relative reactivity

130	105	100	30	25	10

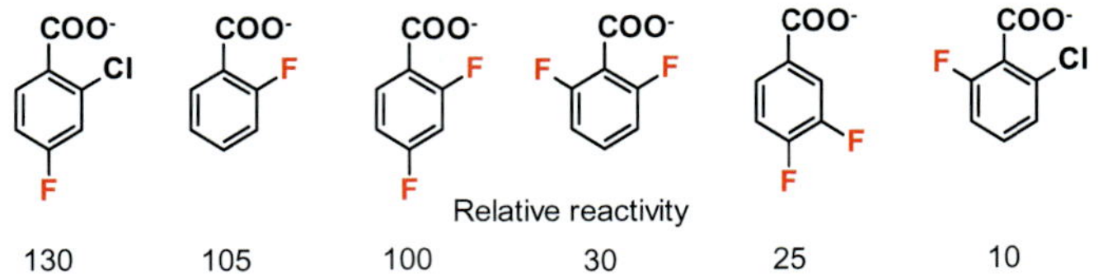

Fig. 8 Order of reactivity in oxidation of fluorinated benzoates mediated by 2-halobenzoate-1,2-dioxygenase.

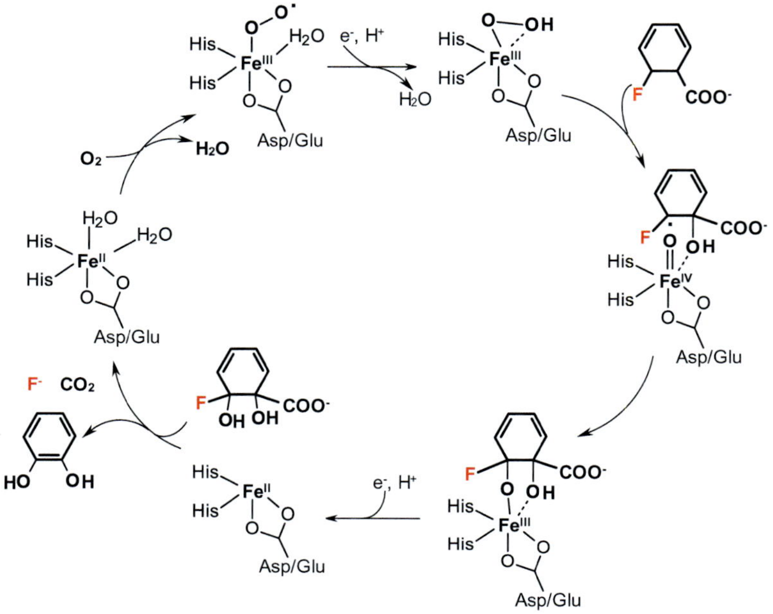

Fig. 9 Proposed mechanism of the defluorination of 2-fluorobenzoate mediated by 2-halobenzoate-1,2-dioxygenase.

2.7 Thiol dioxygenases

Cysteamine dioxygenase and cysteine dioxygenase are non-heme enzymes with an Fe(II) center bearing three histidine ligands. The first coordination sphere is completed with three water molecules and a cysteine-tyrosine cofactor is located in the second coordination sphere. Liu and co-workers have prepared an engineered cysteine dioxygenase by incorporating a 3,5-difluorotyrosine residue in place of tyrosine.[44] This enzyme catalyzed a stoichiometric defluorination with the formation of Cys-Tyr cofactor and release of fluoride anion (Fig. 10).

In contrast to other iron-dependent enzymes operating via high-valent iron oxo species, cysteine dioxygenase proceeds via the formation of a thiyl radical by hydrogen atom abstraction from cysteine by superoxo complex Fe(III)-O-O$^\bullet$. The thiyl radical forms a thioether bond with a carbon atom bearing a fluorine substituent, thus generating a difluorotyrosyl radical. The subsequent deprotonation and fluoride anion elimination leads to a semiquinone species which abstracts a hydrogen atom from Fe(III)-O-OH providing the monofluorinated Cys-His cofactor and regenerated superoxo complex Fe(III)-O-O$^\bullet$ which is not consumed during the defluorination reaction.[1]

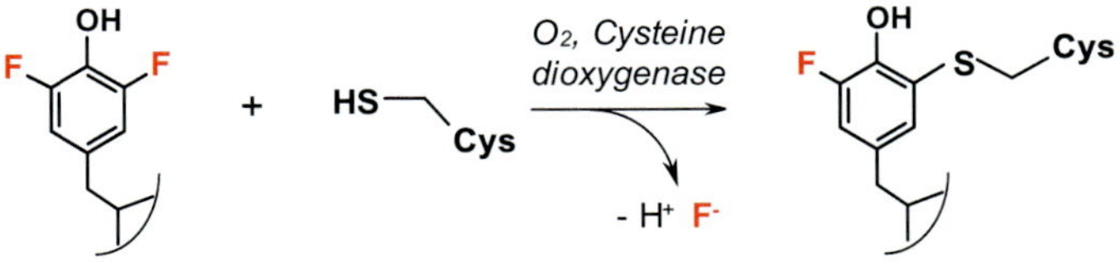

Fig. 10 Cross-linked formation of Cys-Tyr cofactor mediated by engineered cysteine dioxygenase involving the cleavage of a C—F bond.

To summarize, within this short overview of the enzymes that perform transformation of fluorinated aromatic compounds via oxidation pathways, it should be mentioned that there is a relative paucity of data on catalytic efficiency in terms of turnover numbers, conversions and product yields of defluorination reactions. Some studies report on the stoichiometric chemistry with a narrow substrate scope. Many studies have been devoted to mechanistic aspects of these reactions and involve the special substrates. In general, defluorination of monofluorinated compounds has been investigated, while studies involving heavily fluorinated aromatic compounds are scarce.

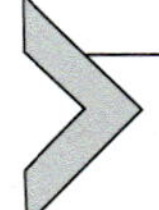

3. Transformation of aromatic C—F bonds under oxidative conditions

3.1 Stoichiometric defluorination mediated by μ-nitrido diiron phthalocyanines

In recent years, we have disclosed remarkable catalytic properties of μ-nitrido diiron phthalocyanine and porphyrin complexes in the oxidation of difficult-to-oxidize compounds.[12–17,45,46] In particular, $(FePc^tBu_4)_2N$ catalyzed the oxidation of methane by H_2O_2 to methanol, formaldehyde and formic acid with high TON in water under mild conditions.[13,15] Moreover, $(FePc^tBu_4)_2N$ in combination with ^tBuOOH was capable of cleaving the C—Cl bond in CH_2Cl_2 routinely used as a solvent for oxidation reactions due to its stability to oxidation.[16] An inspection of reactivity of different halogenated compounds in the presence of $(FePc^tBu_4)_2N$ and oxidant showed that the dehalogenated reactivity can be extended even to the fluorinated aromatic compounds.[17]

During treatment of $(FePc^tBu_4)_2N$ with 30 equiv. of ^tBuOOH in an equimolar C_6F_6/C_6H_6 mixture at 60 °C, the UV–vis spectrum indicated the formation of a new complex with bands at 549, 615 and 666 nm attesting to the phthalocyanine cation-radical. The disappearance of an EPR signal at $g = 2.091$ and the appearance of the strong symmetric and narrow signal at $g = 2.0038$, supports the presence of the cation-radical at one phthalocyanine ligand and the $Fe^{IV}Fe^{IV}$ configuration. The ^{19}F NMR spectrum of the isolated complex contained one signal at −132 ppm indicating the presence of fluoride anion. Energy-dispersive X-ray fluorescence analysis showed a 1:1 Fe:F atomic ratio compatible with $[(Pc^tBu_4)F\text{-}Fe^{IV}(\mu\text{-}N)Fe^{IV}\text{-}F(Pc^tBu_4{}^{+\cdot})]$ formulation (Fig. 11).

The K-edge XANES spectra showed a significant increase of the pre-edge peak energy from 7114.7 eV in the initial complex to 7116.2 eV in the isolated complex as well as a characteristic splitting of the pre-edge peak, indicating the oxidation state change from $Fe^{III}Fe^{IV}$ to $Fe^{IV}Fe^{IV}$. The EXAFS analysis of $[(Pc)F\text{-}Fe^{IV}(\mu\text{-}N)Fe^{IV}\text{-}F(Pc^{+\cdot})]$ showed a symmetrical structure with Fe-μN and Fe—F bond lengths of 1.64 and 1.99 Å, respectively.

3.2 Catalytic defluorination mediated by μ-nitrido diiron phthalocyanines

The $[(Pc)F\text{-}Fe^{IV}(\mu\text{-}N)Fe^{IV}\text{-}F(Pc^{+\cdot})]$ complex formed in a stoichiometric reaction is stable under reaction conditions. If $[(Pc)F\text{-}Fe^{IV}(\mu\text{-}N)Fe^{IV}\text{-}F(Pc^{+\cdot})]$ could be reduced back to $(FePc^tBu_4)_2N$, an interesting possibility would arise for the catalytic oxidative defluorination. Gratifyingly, the addition of H_2O_2 to $[(Pc)F\text{-}Fe^{IV}(\mu\text{-}N)Fe^{IV}\text{-}F(Pc^{+\cdot})]$ led to the evolution of O_2 and regeneration of the starting $(FePc^tBu_4)_2N$ complex. Therefore, H_2O_2 serves as the oxidant to generate the active high-valent oxo species $[(Pc)Fe^{IV}(\mu\text{-}N)Fe^{IV}{=}O(Pc^{+\cdot})]$ which reacts with the fluorinated compound and as reductant of $[(Pc)F\text{-}Fe^{IV}(\mu\text{-}N)Fe^{IV}\text{-}F(Pc^{+\cdot})]$ to the initial $(FePc^tBu_4)_2N$, thus completing the catalytic cycle (Fig. 12).

The scope of the $(FePc^tBu_4)_2N$–H_2O_2 system was evaluated in reaction with a series of poly- and perfluorinated aromatic compounds having diverse fluorination patterns as well as bearing electron-donating and electron-withdrawing functional groups (Table 1).[17]

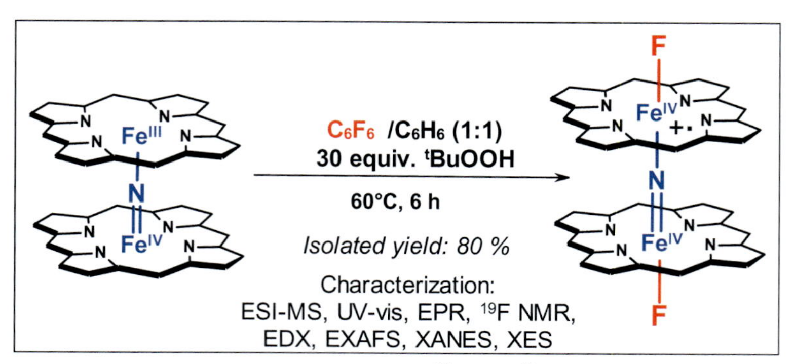

Fig. 11 Stoichiometric formation of $[(Pc)(F)Fe^{IV}(\mu\text{-}N)Fe^{IV}(F)(Pc^{+\cdot})]$ upon incubation of $(FePc^tBu_4)_2N$ in a C_6F_6/C_6H_6 mixture in the presence of ^tBuOOH.

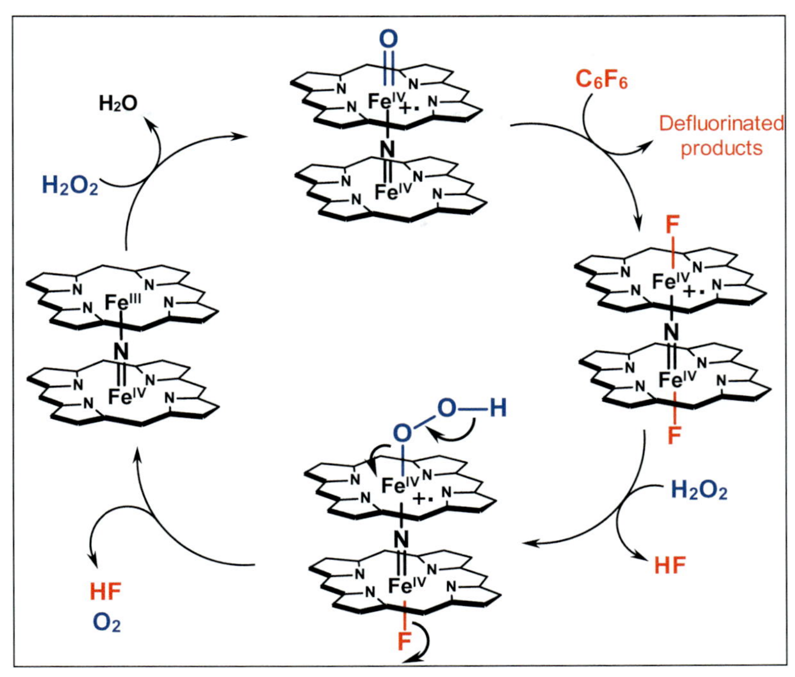

Fig. 12 Proposed mechanism of the catalytic defluorination reaction.

Table 1 Catalytic defluorination of fluorinated aromatic compounds by $(FePc^tBu_4)_2N$-H_2O_2 system.[a]

Entry	Fluorinated substrate	Conversion, %	TON[b]	Degree of defluorination,[c] %	Conversion in Fenton system,[d] %	TON_{Fenton}[e]
1		29	226	52	0	0
2	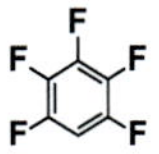	33	152	37	n.d.[f]	n.d.
3	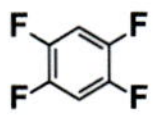	59	341	56	0	0

Continued

Table 1 Catalytic defluorination of fluorinated aromatic compounds by $(FePc^tBu_4)_2N$-H_2O_2 system.[a]—cont'd

Entry	Fluorinated substrate	Conversion, %	TON[b]	Degree of defluorination,[c] %	Conversion in Fenton system,[d] %	TON_{Fenton}[e]
4	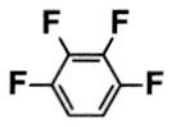	41	336	83	0	0
5	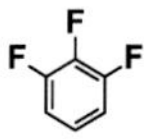	41	169	57	n.d.	n.d.
6	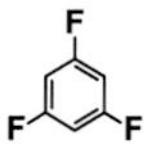	66	351	70	n.d.	n.d.
7	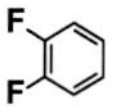	47	120	50	0	0
8	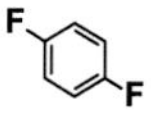	69	240	70	0	0
9	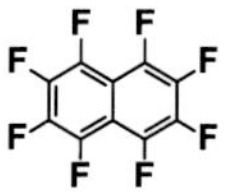	29	150	26	11	78
10	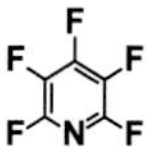	15	69	38	2	17
11	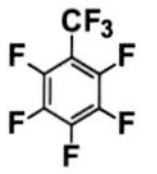	30	242	64	2	15
12	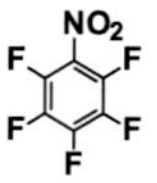	45	376	66	2	15
13	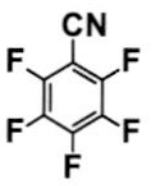	20	141	56	4	9

Table 1 Catalytic defluorination of fluorinated aromatic compounds by $(FePc^{t}Bu_4)_2N$-H_2O_2 system.[a]—cont'd

Entry	Fluorinated substrate	Conversion, %	TON[b]	Degree of defluorination,[c] %	Conversion in Fenton system,[d] %	TON_{Fenton}[e]
14	3,4,5,6-tetrafluorophthalonitrile (F, F, CN, CN, F, F)	13	136	99	n.d.	n.d.
15	CH_3, F, F, F, F, F	45	282	50	7	43
16	OCH_3, F, F, F, F, F	40	188	40	3	17
17	COOH, F, F, F, F, F	31	178	52	9	70
18	F, F, COOH, COOH, F, F	20	88	45	n.d.	n.d.
19	OH, F, F, F, F, F	82	818	80	7	45
20	O, F, F, F, F, O	99	986	99	98	750

[a]Conditions:catalyst:substrate:oxidant = 1:250:4000, CD_3CN, 60 °C, 15 h.
[b]Turnover numbers (TON) were calculated as moles of fluoride anion per mole of catalyst.
[c]Degree of defluorination is the ratio of the formed fluoride amount to the total amount of fluorine in the converted substrate.
[d]Conversion obtained with $FeSO_4$ catalyst under the same conditions.
[e]TON_{Fenton}: obtained with $FeSO_4$ catalyst under the same conditions.
[f]n.d.—not determined.

The results indicate that perfluorinated aromatics such as hexafluorobenzene, pentafluoropyridine and octafluoronaphthalene, and functionalized polyfluorinated compounds, reacted with high turnover numbers resulting in elevated degrees of defluorination, i.e., in mineralization of organic fluorine. Since perfluorinated species cannot be transformed via oxidative enzymatic routes including by cytochrome P450,[32] this reactivity is unprecedented.

Strong electron-withdrawing groups (CF_3, CN, NO_2) were well tolerated and C_6F_5CN, $C_6F_5NO_2$ and $C_6F_5CF_3$ underwent defluorination reaction of the aromatic ring. No products of CF_3 conversion were observed during the course of reaction with perfluorotoluene. The study of regioselectivity of the first step in the reaction with perfluorinated compounds containing strong electron-withdrawing substituents, showed a clear preference for the transformation of the *meta*-C-F bond (Fig. 13).[47]

This trend is typical for electrophilic aromatic substitution reactions where electron-withdrawing substituents operate as *meta*-directing groups. The DFT calculations of epoxidation of $C_6F_5CF_3$ showed that attack on the *ortho*-, *meta*- and *para*-positions are close in energy, but the *meta*-attack configuration is slightly lower by 0.7 kcal mol^{-1}. Therefore, the mixture of products should be obtained with dominant *meta*-phenol as indeed observed experimentally.[47]

The reactivity of the $(FePc^tBu_4)_2N$-H_2O_2 system, in particularly, in the oxidation of methane, increases in the presence of low acid concentration due to the more efficient formation of oxo species $(Pc)Fe^{IV}(\mu\text{-}N)Fe^{IV}{=}O(Pc^{+})$ from the hydroperoxo complex $(Pc)Fe^{IV}(\mu\text{-}N)Fe^{III}\text{-}O\text{-}OH(Pc)$.[12,13] The protonation of peroxide oxygen facilitates the heterolytic cleavage of the O—O bond with release of a water molecule. Indeed, the conversion of perfluorobenzene increased from 29% to 49% when the concentration of trifluoroacetic acid was increased from 0 to 100 mM with significant gain in TON from 226 to 514.[47]

The homogeneous defluorination reactions were conducted using acetonitrile solvent which can be oxidized by the $(FePc^tBu_4)_2N$-H_2O_2 system under reaction conditions as was previously demonstrated.[13] Thus, the

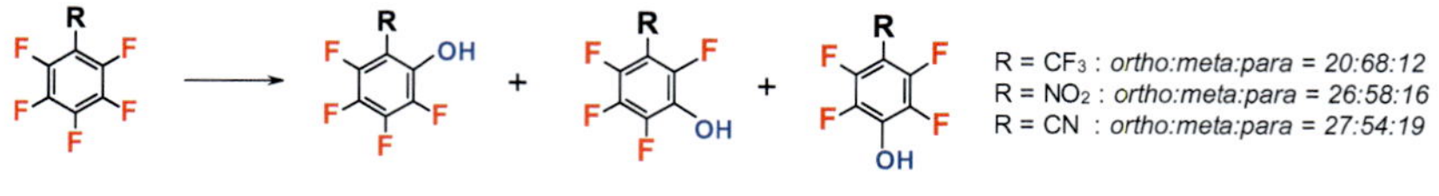

R = CF_3 : *ortho:meta:para* = 20:68:12
R = NO_2 : *ortho:meta:para* = 26:58:16
R = CN : *ortho:meta:para* = 27:54:19

Fig. 13 Regioselectivity of oxidation of fluorinated compounds with strong electron-withdrawing substituents.

oxidation of solvent might partially hide the true defluorination performance of $(FePc^tBu_4)_2N$. To avoid possible oxidation of organic solvent and to further explore the practical potential of the catalytic system, we have evaluated the possibility to use water as a green and stable solvent in combination with a supported catalyst. Recognizing that HF readily reacts with siliceous materials with formation of hexafluorosilicate, identified by ^{19}F NMR and IR spectroscopies, $(FePc^tBu_4)_2N$ was supported onto carbon (surface area = 300 m^2/g, 10 μmol/g).

In sharp contrast to conventional reductive and organometallic systems operating in strict inert and anhydrous conditions, the $(FePc^tBu_4)_2N$-H_2O_2 system was not deactivated in water and a very high efficiency in the transformation of water-soluble polyfluorinated compounds was observed using a low catalyst loading of 0.1 mol% (Table 2).

Moreover, heterogeneous transformation of C_6F_5OH in water with 0.1 mol% catalyst loading was much more efficient than the homogeneous reaction with 0.4 mol% catalyst loading: 98% vs. 82% conversions and 4825 TON vs. 818 TON, respectively (Table 2, entry 1). Turnover frequency of heterogeneous oxidation achieved 322 h^{-1} (5.4 min^{-1}) compared to 54.5 h^{-1} in the case of homogeneous oxidation. The improvement of catalytic performance was much more significant for C_6F_5COOH: 76% and 31% conversions and TON of 2950 and 178 in water and acetonitrile, respectively (Table 2, entry 2), and for tetrafluorophthalic acid: 56% and 20% conversions and TON of 1730 and 88 in water and acetonitrile, respectively (Table 2, entry 4). Hexafluorobenzene and pentafluoropyridine, which are only sparingly soluble in water, were converted with 75% and 85% conversions, respectively, showing a high efficiency of the catalytic system even in dilute aqueous solutions.

The oxidative defluorination of C_6F_5OH by H_2O_2 in water carried out by supported $(FePc^tBu_4)_2N$ catalyst was studied in detail with identification and quantification of the reaction products by GC–MS, ^{19}F NMR and total organic carbon determination (Fig. 14).[17]

The principal reaction products were hydrogen fluoride, oxalic, difluoromaleic and fluorooxaloacetic acids. The total organic carbon determinations indicated a 54% loss of the organic carbon due to the formation of CO and CO_2 in a 1:15 ratio. In addition, 89% of the organic fluorine was converted to inorganic fluoride anions. Thus, the diiron supported catalyst in combination with the green oxidant H_2O_2, efficiently mineralizes in water such a recalcitrant compound as pentafluorophenol. The proposed reaction sequence is shown in Fig. 15 where identified intermediate and final products are in red.

Table 2 Catalytic heterogeneous defluorination by carbon-supported $(FePc^{t}Bu_4)_2N$-H_2O_2 system in water.[a]

Entry	Water-soluble fluorinated compound	Conversion, %	Turnover number[b]	Degree of defluorination,[c] %
1	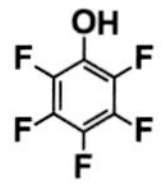	98 (82)[d]	4825 (818)[d]	98 (80)[d]
2	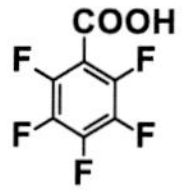	76 (31)[d]	2950 (178)[d]	78 (52)[d]
3	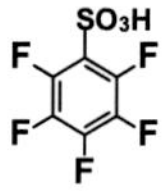	53	2140	80
4	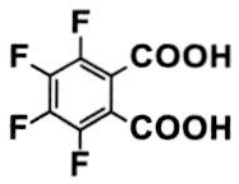	56 (20)[d]	1730 (88)[d]	78 (45)[d]
5[e]	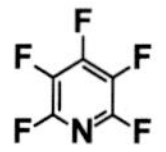	85	152	72
6[e]	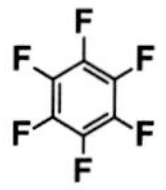	75	98	44

[a]Conditions:catalyst:substrate:oxidant = 1:1000:26,000, 0.1 mol% catalyst loading, D_2O, 60 °C, 15 h.
[b]Turnover numbers (TON) were calculated as moles of fluoride anion per mole of catalyst.
[c]Degree of defluorination is the ratio of the formed fluoride amount to the total amount of fluorine in the converted substrate.
[d]In parentheses: catalytic efficiency of homogeneous oxidation with 0.4 mol% catalyst loading.
[e]Reactions were performed by stirring the supported catalyst (20 mg, 0.2 μmol of complex), substrate (0.005 M for C_5F_5N and 0.003 M for C_6F_6) and H_2O_2 (0.1 M) in D_2O at 60 °C for 15 h.

The oxidation of pentafluorophenol via tetrafluorohydroquinone leads to tetrafluorobenzoquinone, which undergoes aromatic cycle cleavage via an epoxidation-epoxide hydrolysis-HF elimination sequence previously proposed for the oxidation of chlorinated phenols by the iron phthalocyanine–H_2O_2 system[48,49] to form difluoromaleic, fluorooxaloacetic, oxalic acids as well as HF and CO_2.

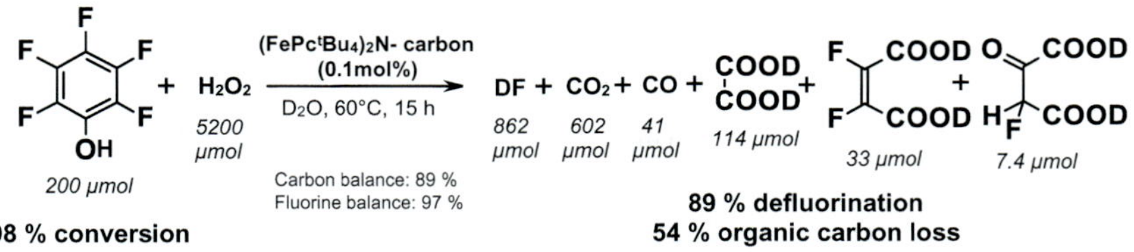

Fig. 14 Catalytic mineralization of C_6F_5OH by H_2O_2 in water mediated by the carbon-supported $(FePc^tBu_4)_2N$ complex.

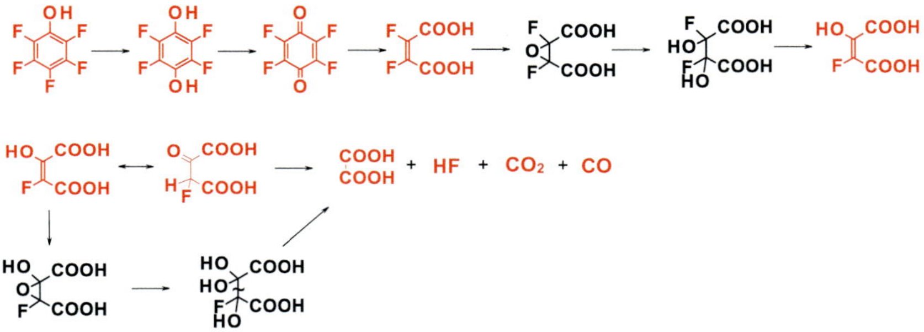

Fig. 15 Proposed reaction sequence for the oxidative defluorination of fluorinated aromatics. Identified intermediates and products are shown in *red*.

The product composition can be changed by adjusting the amount of hydrogen peroxide. While large oxidant excess leads to the considerable mineralization of fluorinated aromatic compounds, the limited amount of oxidant (five H_2O_2 equivalents) results in the reaction that can be used for the preparation of perfluorinated acids. The oxidation of octafluoronaphthalene and pentafluorophenol furnished tetrafluorophthalic and difluoromaleic acids with 66% and 62% yields, respectively.[17]

This oxidative strategy for the degradation of recalcitrant pollutants appears to be particularly convenient from a practical viewpoint. Indeed, (i) the catalyst based on earth abundant and non-toxic iron is industrially relevant and accessible on a large scale; (ii) the μ-nitrido diiron complex can be easily heterogenized using different supports; (iii) environmentally compatible H_2O_2 is an industrially available oxidant; (iv) the catalytic system exhibits a large substrate scope and tolerates well air and water; (v) the process is flexible and can be performed in concentrated organic solutions (for instance, the treatment of industrial wastes) or in water including diluted solutions (treatment of contaminated water). This new oxidative method holds a great promise for the development of novel remediation processes.

3.3 Reactivity of heteroleptic μ-nitrido diiron complex

The oxidative defluorination reactivity was further explored using the heteroleptic μ-nitrido diiron complex PcFe(μN)FePz with phthalocyanine and octapropylporphyrazine ligands.[50] The formation of the high-valent diiron oxo species was investigated by cryospray mass-spectrometry which showed comparable affinity of the oxo group to iron porphyrazine and phthalocyanine sites with slight preference for the iron phthalocyanine center. The PcFe(μN)FePz–H_2O_2 system was less efficient than $(FePc^tBu_4)_2N$–H_2O_2 providing a 11% conversion of C_6F_6 and $TON_F = 34.5$ compared with 29% conversion and $TON_F = 226$ observed in case of $(FePc^tBu_4)_2N$.[17] No C_6F_6 conversion was observed when ^tBuOOH was used as the oxidant. In contrast, both PcFe(μN)FePz–H_2O_2 and PcFe(μN)FePz–^tBuOOH systems were very efficient in the reaction with perfluoro(allylbenzene) (Fig. 16).

Moreover, the PcFeNFePz–^tBuOOH system was even more active affording a 97% $C_6F_5CF_2CF{=}CF_2$ conversion compared to a 58% conversion observed with H_2O_2 as oxidant. $C_6F_5CF_2COOH$ and C_6F_5COOH were the principal products, indicating a much higher defluorination activity of the perfluoroolefinic group compared with the perfluoroaromatic ring. It was noteworthy that while the PcFeNFePz–H_2O_2 system furnished $C_6F_5CF_2COOH$ with a 37% yield, the PcFeNFePz–^tBuOOH system was more selective toward C_6F_5COOH providing a 51% yield. Very high defluorination turnover numbers were achieved with both oxidants: 538 and 1465 using H_2O_2 and ^tBuOOH, respectively.[50]

3.4 Mechanism of oxidative defluorination

The counter-intuitive oxidative defluorination of the C—F bonds raises a question regarding the mechanistic background of this striking reactivity,

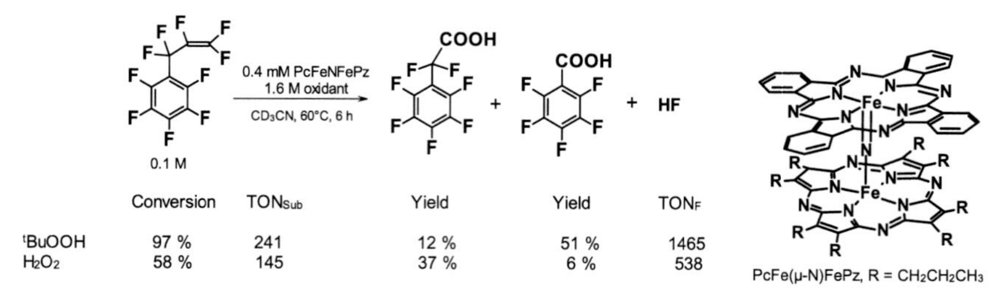

Fig. 16 Oxidative defluorination of perfluoro(allylbenzene) by PcFeNFePz–H_2O_2 and PcFeNFePz–^tBuOOH systems. TON_{Sub} was defined as molar amount of converted substrate per mole of the catalyst. TON_{F-} was calculated as molar amount of F^- formed per mole of the catalyst.

which should be associated with highly electrophilic (Pc)Fe^{IV}-N-Fe^{IV}($Pc^{+\cdot\cdot}$) = O species. Such ultra-high-valent diiron oxo species (L) Fe^{IV}-N-Fe^{IV}($L^{+\cdot}$) =O (L = phthalocyanine, porphyrin) have been detected at low temperatures and characterized by cryospray MS, UV–vis, EPR, X-ray absorption spectroscopies and Mössbauer techniques[12–15,45] as well as by DFT calculations.[15,51–55] Yet, how the extremely electron deficient (Pc)Fe^{IV}(μ-N)Fe^{IV}(Pc^{+}) = O species attacks the C—F bond formed by the most electronegative element, is indeed puzzling? Four mechanistic hypotheses have been examined: (i) free-radical Fenton oxidation; (ii) nucleophilic substitution; (iii) electrophilic attack and (iv) pathway through the epoxidation of the aromatic cycle.

3.4.1 Free-radical Fenton oxidation

Interaction of iron complexes with H_2O_2 can result either in the formation of high-valent iron oxo species or hydroxyl radicals $OH^{\bullet}$ which can also oxidize recalcitrant compounds such as alkanes having strong C—H bonds. Despite the ubiquitous occurrence of $OH^{\bullet}$ radicals in biology and their usage in remediation processes and other related extensive studies, mechanistic aspects of Fenton chemistry still remain a matter of intensive debate in the literature.[56,57] Depending on the reaction conditions, hydroxyl radicals and/or high-valent metal oxo species can be formed in the presence of the same metal complex and H_2O_2.[58]

Hydroxyl radicals can attack C—F bonds of aliphatic and aromatic molecules under harsh conditions. For instance, $OH^{\bullet}$ radicals generated by thermal decomposition of H_2O_2 reacted with octafluoronaphthalene and heptafluoronaphthols to produce dimeric coupling products and hexafluoronaphthoquinone.[59] A report on the defluorination of C_6F_5OH and C_6F_5COOH under photo-Fenton conditions in the presence of ferrous sulfate or ferric oxalate and H_2O_2, subjected to UV-C light (200–300 nm), has been published.[60,61] However, $OH^{\bullet}$ radicals generated by thermolysis of 90% H_2O_2 in CH_3CN did not react with C_6F_6 even at 100 °C.[62]

To probe the possible involvement of $OH^{\bullet}$ radicals, a series of substituted fluorinated benzenes and functionalized fluorinated aromatics were incubated in the presence of $FeSO_4$ and H_2O_2 under conditions used for the $(FePc^{t}Bu_4)_2N$–H_2O_2 system. All fluorinated benzenes were stable under Fenton conditions (Table 1). When the $FeSO_4$ loading was increased to 10 mol% (25 times more than the $(FePc^{t}Bu_4)_2N$ amount), no conversion of C_6F_6 was observed. The fluoroaromatic compounds bearing functional groups showed conversions at the detection limit. Fluoranil was the only compound which showed comparable reactivity with $(FePc^tBu_4)_2N$ and $FeSO_4$. However, fluoranil reacted with H_2O_2 even in the absence of $(FePc^tBu_4)_2N$ or $FeSO_4$ catalyst via a nucleophilic attack of H_2O_2 at the *ortho*-position of fluoranil.[63] Thus, comparison of conversions and TONs obtained in the two catalytic systems, unambiguously indicated that $OH^•$ radicals are not involved in defluorination of fluorinated aromatics performed by the $(FePc^tBu_4)_2N$–H_2O_2 system.

3.4.2 Nucleophilic substitution

In principle, the observed higher reactivity of the C—F bond in comparison with the C—Cl bond is consistent with a nucleophilic substitution because the reactivity of the halogenated arenes increases in the order of Ar-I < Ar-Br < Ar-Cl < Ar-F. However, nucleophilic substitution of C—F occurs under basic conditions and/or involves electron-rich transition metal complexes (Rh, Ni, Pd).[64–67] The order of reactivity of fluorinated benzenes is not compatible with the nucleophilic mechanism: conversions are higher for less fluorinated compounds (Fig. 17).

The observed higher reactivity of C_6F_5H compared with C_6F_6 is consistent with the relative energy of HOMOs.[17]

A competition reaction between $C_6F_5CH_3$ and $C_6F_5CF_3$ led to 49% and 31% conversions, respectively, whereas these compounds exhibit similar reactivity in S_NAr reactions.[68] The conversion of $C_6F_5(CN)_2$ (20%) is higher than that of $C_6F_5(CN)_2$ (13%), results which are also not compatible with the nucleophilic mechanism.

3.4.3 Electrophilic attack

The high-valent diiron oxo species $(Pc)Fe^{IV}$-(μN)-$Fe^{IV}(Pc^{+•})$=O are particularly powerful oxidants with strong electrophilic properties. To probe the possible electrophilic attack, the relative reactivity of C—F bonds with

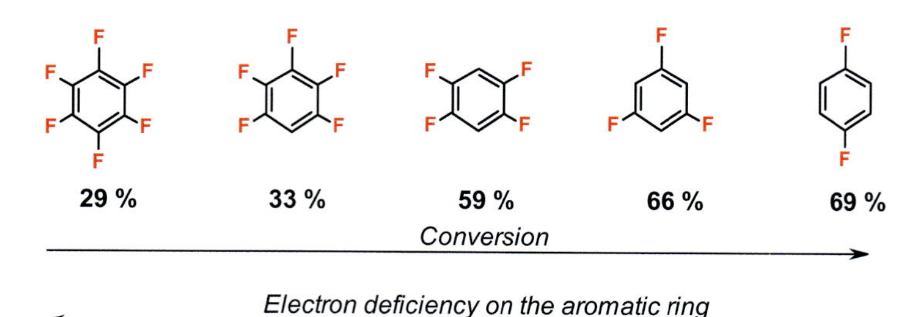

Fig. 17 Order of reactivity of fluorinated benzenes in reaction with $(FePc^tBu_4)_2N$ and H_2O_2.

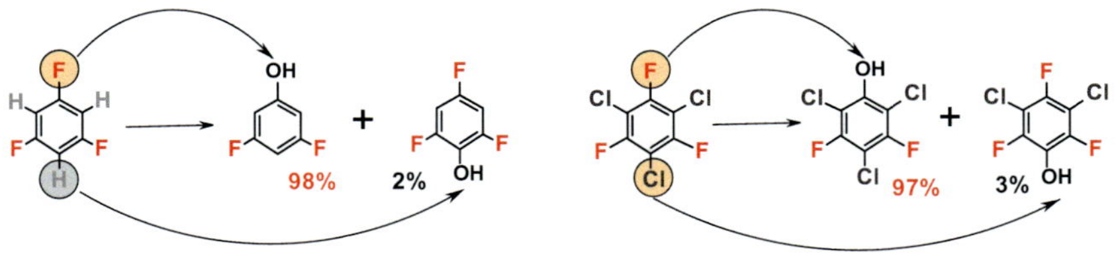

Fig. 18 Intramolecular determination of relative reactivity of C-F, C-H and C-Cl bonds.

respect to C—H and C—Cl bonds was studied in intramolecular competitive experiments using 1,3,5-trifluorobenzene and 1,3,5-trichloro-2,4,6-trifluorobenzene in large excess (Fig. 18).

The ratio of initial phenol products of 1,3,5-trifluorobenzene oxidation determined at short reaction time was 49:1 in favor of 3,5-difluorophenol obtained from the attack on a C—F bond. Similarly, transformation of 1,3,5-trichloro-2,4,6-trifluorobenzene afforded phenols obtained from F and Cl elimination in a 32:1 ratio. Such a strong preference of the $(FePc^{t}Bu_4)_2N\text{-}H_2O_2$ system for a C—F bond compared to C—H and C—Cl bonds is not consistent with electrophilic attack of oxo species on the C-X bond, because in this case the more electron-rich C—Cl bond and especially the C—H bond should be preferred.

3.4.4 Pathway through the epoxidation of aromatic cycle

Cytochrome P450 catalyzes the oxidation of aromatic compounds[23–25] including monohalogenated arenes[68] through an epoxidation route accompanied by migration of the substituent to the adjacent position (NIH shift). Since the $(Pc)Fe^{IV}\text{-}N\text{-}Fe^{IV}(Pc^{+\bullet}){=}O$ species also oxidizes benzene with intermediate formation of benzene epoxide accompanied by a NIH shift,[69] the initial formation of fluoroarene epoxide appears to be a plausible pathway. Analysis of the energies and shapes of the molecular orbitals of C_6F_5H indicates the HOMO has the maximum density at the CF=CF bonds rather than at the CF=CH bond, and therefore epoxidation should be primarily directed on these bonds. Indeed, only tetrafluorophenols resulting from fluorine elimination were obtained in the oxidation of C_6F_5H.

Experimental evidence for epoxidation of fluoroarenes and for occurrence of a fluorinated NIH shift is more challenging to obtain than it is for regular aromatic compounds. Although epoxide of hexafluorobenzene could not be detected owing to its instability under GC–MS conditions and low ESI-MS response, a minor amount of epoxide has been detected in the oxidation of octafluoronaphthalene.[17] In contrast to brominated and chlorinated aromatic compounds, featuring a NIH shift,[70] the fluorine NIH migration is uncommon[71] as a consequence of dominant fluoride release.[68] A rare experimental evidence for a fluorine NIH shift was reported for cytochrome P450-mediated oxidation of 1,4-difluorobenzene.[31]

A careful analysis of the products of oxidation of 1,4-difluorobenzene by the $(FePc^tBu_4)_2N$–H_2O_2 system allowed detection of the fluorinated NIH shift. Along with the regular *p*-fluorophenol product, the NIH-shifted 2,4-difluorophenol was detected (Fig. 19).

Mechanistic features of the oxidation of fluorinated aromatic compounds by the $(FePc^tBu_4)_2N$–H_2O_2 system, namely fluoroarene epoxide formation accompanied by a fluorinated NIH shift are typical for the involvement of high-valent oxo species and are reminiscent of biological oxidation.

The details of the reaction mechanism have been investigated by DFT methods using N-bridged diiron phthalocyanine (**B**) and shortened N-bridged diiron porphyrazine (**A**) models.[47] Their oxo species were calculated to be in a doublet spin ground state that is in agreement with experimental EPR and Mössbauer studies.[12,45] The quartet and sextet spin states are much higher in energy. The principal mechanisms of reaction with C_6F_6 take place on the doublet spin state surface. The activation of C_6F_6 by oxo species of both models leading to epoxide product, occurs stepwise through an electrophilic transition state **TS1,** leading to intermediate **IM1** followed by a ring-closure transition state **TS2** to generate epoxide, as shown in Fig. 20 for the porphyrazine model A.

The transition state structures and energies of the larger phthalocyanine model **B** are similar to those for the shortened porphyrazine model **A** with **TS1** barriers of 14.7 and 18.6 kcal mol^{-1}, respectively. Thus, a shift from porphyrazine to the phthalocyanine macrocycle results in a small stabilizing effect on the structures and energetics, whereby the **TS1** barrier is lowered

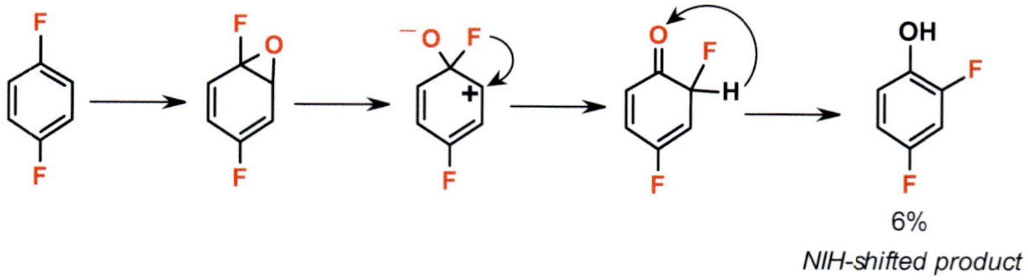

Fig. 19 Fluorine NIH shift in the oxidation of 1,4-difluorobenzene.

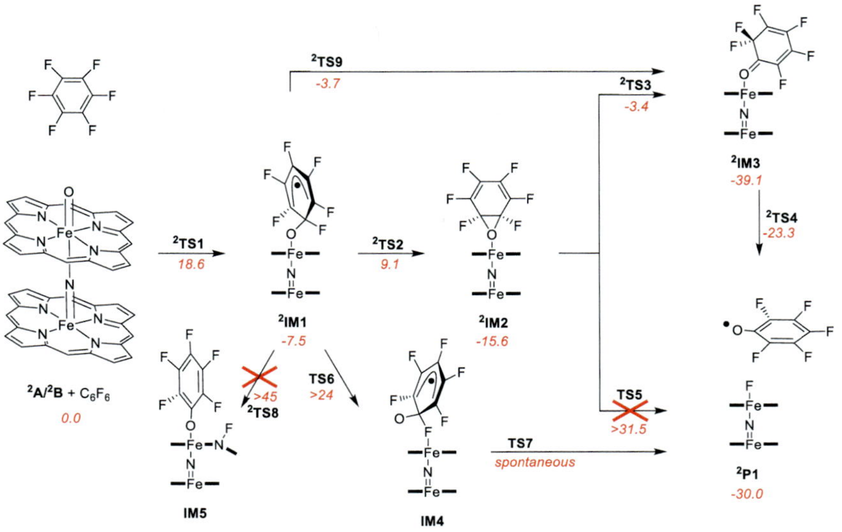

Fig. 20 Reaction mechanisms for the reaction of hexafluorobenzene with high-valent oxo species of N-bridged diiron porphyrazine (model A) as calculated in Gaussian 09. Energies refer to UB3LYP/BS2//UB3LYP/BS1 + ZPE values in kcal mol^{-1}.[47] *Reprinted with permission from Colomban, C.; Tobing, A. H.; Mukherjee, G.; Sastri, C. V.; Sorokin, A. B.; de Visser, S. P.* Chem. Eur. J. ***2019,*** 25, *14320–14331. Copyright 2019 John Wiley and Sons.*

by 3.9 kcal mol^{-1}. However, this does not affect the overall results. The intermediates **IM1**$_{\mathbf{A}}$ and **IM1**$_{\mathbf{B}}$ can generate the epoxide intermediates **IM2**$_{\mathbf{A}}$ and **IM2**$_{\mathbf{B}}$ via a ring-closure step with transition state **TS2**. The epoxide intermediates have the lowest energy and −15.5 and −21.8 kcal mol^{-1} below **IM2**$_{\mathbf{B}}$ are very similar showing Fe1—N and Fe2—N distances of 1.655 and 1.78 Å, respectively, for both structures. Therefore, they possess very similar electronic configurations and properties suggesting the negligible effect of the macrocyclic ligand on the structure and chemical properties of the epoxide intermediates.

The intermediate **IM1**$_{\mathbf{A}}$ can lead to different possible products via several pathways (Fig. 20). The formation of epoxide intermediate **IM2**$_{\mathbf{A}}$ occurs with a ring-closure barrier **TS2** of 16.6 kcal mol^{-1}. The intermediate **IM1**$_{\mathbf{A}}$ can also generate the ketone intermediate **IM3** via a 1,2-fluoride shift with a barrier **TS9** of 3.8 kcal mol^{-1}. Several possible mechanisms can lead to phenol. The fluoride NIH shift from C1 to C2 of the epoxide with cyclic opening generates the ketone intermediate **IM3** via transition state **TS3** with a barrier of 12.2 kcal mol^{-1}. This step is highly exothermic, −39.1 kcal mol^{-1} with respect to isolated reactants. The aliphatic fluoride atom of this ketone intermediate can be abstracted by the iron site to form **P1** product with the **TS4** barrier of 15.8 kcal mol^{-1}. This step is endothermic by 9.1 kcal mol^{-1}. However, the rate-determining step in the activation of C_6F_6 via epoxide and ketone, is the C—O bond formation via **TS1** of 18.6 kcal mol^{-1} and all following steps are much lower in energy. Several other routes considered for the oxygen atom transfer to C_6F_6 to generate C_6F_5OH via **TS5**, **TS6** and **TS8** are much higher in energy and can be ruled out as viable reaction pathways (Fig. 20).

The mechanism of oxidation of fluorinated arenes differs from that for benzene and its derivatives, though the initial step of both mechanisms is the electrophilic addition of oxo to the arene carbon atom. The calculations show that the μ-nitrido diiron phthalocyanine intermediate can isomerize into ketone directly or via epoxide. This ketone can readily release a fluoride anion from the aliphatic C—F bond. Thus, defluorination products are formed in a sequential process, whereas in aromatic arene hydroxylation all products arise from a common intermediate.[24,72] This mechanism can be compared to that for oxidation of fluorinated aromatic compounds by cytochrome P450 proposed by Hadad and co-workers.[32]

The oxidative defluorination of perfluorinated aromatic compounds performed by the $(FePc^tBu_4)_2N$-H_2O_2 system proceeds through a different mechanism compared with that of aromatic C—H hydroxylation by high-valent iron oxo species (Fig. 21).

Escape of a ketone intermediate formed from aromatic hydrocarbons into solution, followed by the proton loss, results in the formation of phenol. In the case of a fluorinated ketone intermediate, the loss of a fluoride anion generates a cationic species, followed by the addition of an OH^- group and elimination of HF to produce a quinone product. Therefore, there is a considerable difference in the behavior of the ketone intermediates formed from aromatic hydrocarbons and perfluorinated aromatic compounds. In the former case, the phenol product is formed in a two-electron reaction whereas the fluorinated ketone intermediate directly generates fluoranil in a six-electron process without intermediate formation of fluorinated phenol (Fig. 21). However, fluorinated phenols have been identified as reaction products by GC–MS and ^{19}F NMR methods.[17] The formation of phenols from fluoroarenes can be explained by retention of the ketone intermediate in the coordination sphere of the μ-nitrido diiron phthalocyanine complex. The loss of fluoride anion from *ortho*-position furnishes a coordinated cationic intermediate, followed by a two-electron transfer from (Pc) Fe^{IV}(μ-N)Fe^{III}(Pc) to the coordinated cationic species yielding the phenol,

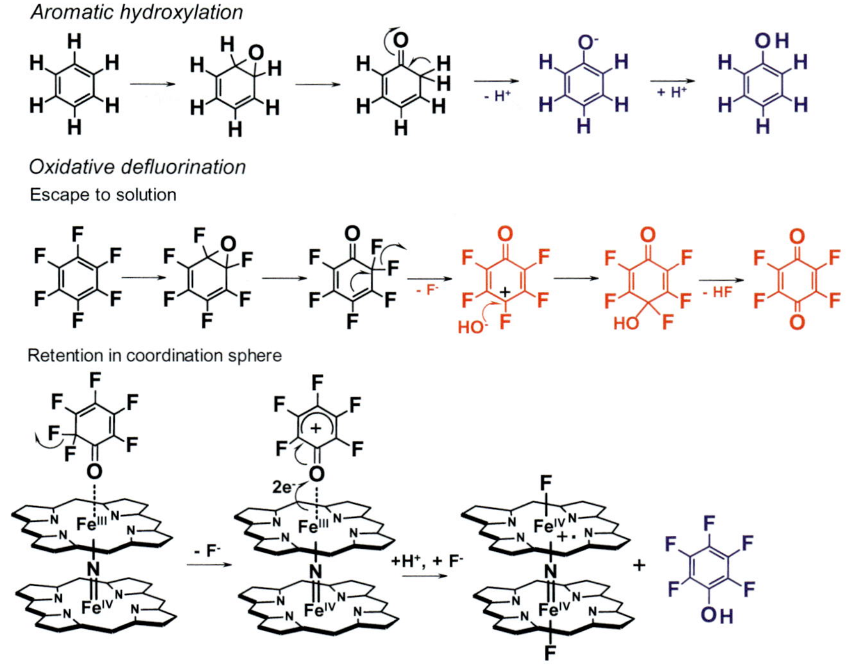

Fig. 21 Different mechanistic issues in evolution of ketone intermediates generated from aromatic hydrocarbons and perfluorinated aromatics.

C_6F_5OH, and a high-valent (Pc)F-Fe^{IV}(μ-N)Fe^{IV}-F($Pc^{+\cdot}$) complex which was isolated and completely spectroscopically characterized.[17] An intramolecular two electron transfer and therefore, the formation of phenol can be promoted by the protonation of the carbonyl group, thus explaining the improvement of the progress of the reaction in the presence of acid.

4. Other chemical systems for C—F bond activation

4.1 Catalytic defluorination

One of the first examples was published by Hirobe and co-workers.[73] The iron *meso*-tetrakis(2,6-difluorophenyl)porphyrin–*meta*-chloroperbenzoic acid (*m*-CPBA) system performed substituent elimination from various para-substituted phenols, including *p*-fluorophenol, to furnish *p*-benzoquinone with a TON ~3.3. When ^{18}O-labeled *m*-CPBA was employed, a 92% ^{18}O incorporation to the quinone was observed suggesting a rebound mechanism (Fig. 22).[73]

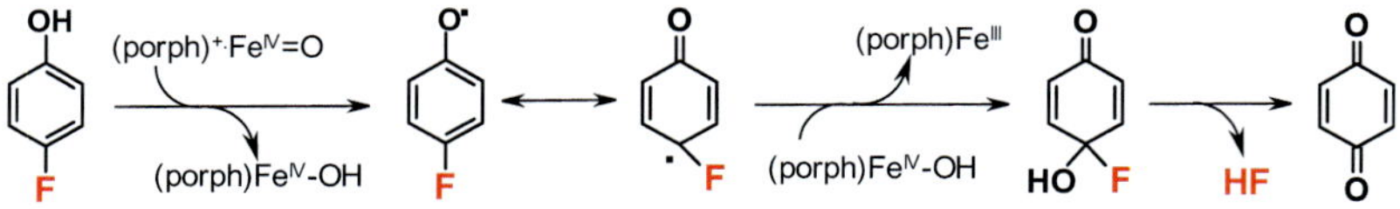

Fig. 22 Proposed rebound mechanism for the oxidation of *p*-fluorophenol by the iron *meso*-tetrakis(2,6-difluorophenyl)porphyrin–*m*-CPBA system.

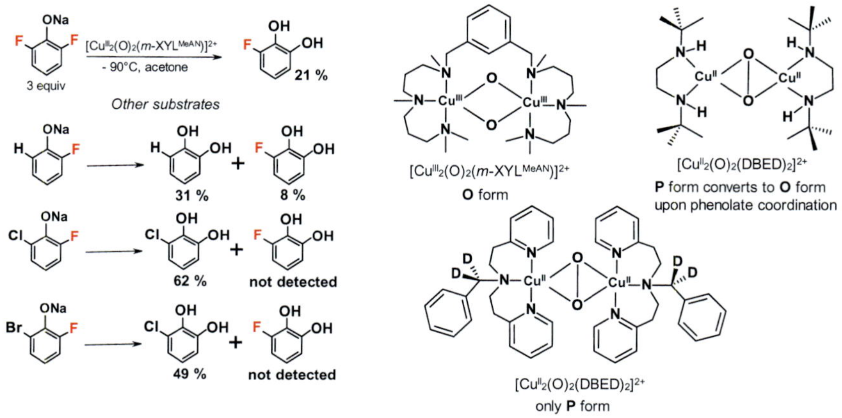

Fig. 23 Stoichiometric ortho-defluorination-hydroxylation reactions of sodium fluorophenolates promoted by a dicopper complex in acetone at −90 °C.

Interestingly, that in contrast to the two-electron C—H hydroxylation, the oxidative defluorination of *p*-fluorophenol proceeds via a four-electron reaction.

4.2 Stoichiometric defluorination

Costas, Company, and co-workers have prepared three dicopper complexes by the reaction of dicopper(I) precursors with O_2 at −90 °C.[74] These thermally unstable compounds exist in (η^2:η^2-peroxo)dicopper(II) or in bis(μ-oxo)dicopper(III) forms depicted as **P** and **O** species, respectively (Fig. 23).

Only bis(μ-oxo)dicopper(III) species (**O** form) reacts with sodium 2-fluorophenolates in stoichiometric fashion leading to the selective *ortho*-defluorination. Depending on the substrate structure, corresponding catechols were obtained with 21–62% yields based on the amount of bis(μ-oxo)dicopper(III) species. It was noteworthy that the cleavage of C—F bond in the *ortho*-position was the major reaction pathway for sodium 2-phenolate bearing H, Cl or Br substituents in another ortho-position (Fig. 23). Moreover, no reaction was observed with the phenolic compounds, 2,6-dichlorophenolate and 2,6-dimethylphenolate. Low temperature UV–vis, Raman and cryospray mass spectrometry studies showed the initial binding of phenolate to the copper site of the bis(μ-oxo)dicopper core. A Hammett correlation gave a negative parameter $\rho = -2.4$ indicating that the decay of this intermediate to form catechol occurred via an electrophilic attack. The preference for the C—F bond even in the presence of competing substituents with weaker C—X bonds (X = Br, Cl) at the other *ortho*-position of the phenolate was explained by the interaction between an unpaired pair of electrons of the *ortho*-F atom and the adjacent Cu(III) site, so that the carbon atom of the C—F bond became properly oriented for the electrophilic attack by the bis(μ-oxo) core.[74] The further combined experimental and theoretical study of the *ortho*-defluorination-hydroxylation of 2-fluorophenolates by the bis(μ-oxo)dicopper(III) complex $[Cu_2O_2(DBED)_2]^{2+}$ formed upon phenolate coordination, provided more details on this interesting reaction and confirmed the proposed mechanism.[75] Mechanistically, this reaction resembles defluorination of 2-fluorophenols mediated by flavin adenine dinucleotide (FAD)-containing phenol hydroxylases.

The conversion between (η^2:η^2-peroxo)dicopper(II) (**P**) and bis(μ-oxo) dicopper(III) (**O**) species can be controlled through ligand design, substrate coordination and reaction conditions (temperature, solvent).[75] As well, the addition of 1 or 2 equiv. of a Lewis acid such as $Sc(CF_3SO_3)_3$, $B(C_6F_5)_3$ or Brønsted acid, $DMF{\cdot}CF_3SO_3H$, resulted in the formation of **O**-type dicopper species with increased oxidizing activity. In the presence of 2,6-difluorophenolate, the dicopper complex **P** was transformed to the **O** species which oxidized the substrate to 3-fluorocatechol in a 30% yield (Fig. 24).

However, the defluorination step ***A*** demands a 2-electron reduction and the origin of these electrons is not clear.

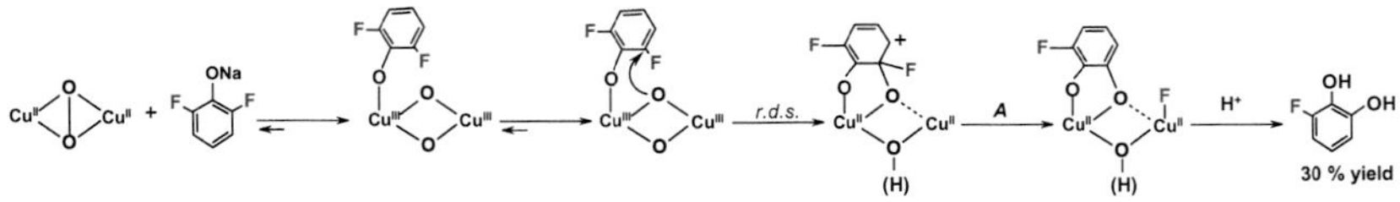

Fig. 24 Reaction mechanism for the *ortho*-defluorination-hydroxylation of 2,6-difluorophenalate proposed in Ref. [75].

4.3 Intramolecular defluorination

Tetranuclear iron complexes $[LFe_3(PhPz)_3OFe](OTf)_2$ (PhPz = 3-phenylpyrazolate) and its fluorinated analog $[LFe_3(F_2ArPz)_3Ofe](Otf)_2$ (F_2ArPz = 3-[2,6-difluorophenyl]pyrazolate) in combination with PhIO or $^nBu_4NIO_4$, underwent intramolecular hydroxylation of C—H and C—F bonds of the supporting ligand (Fig. 25).[76,77]

Detailed XRD, Mössbauer and ESI MS studies showed the regioselective hydroxylation of a bridging pyrazolate ligand. The similar reactivity observed with both PhIO and $^nBu_4NIO_4$ oxidants suggests the formation of a terminal iron-oxo species. The C—F versus C—H selectivity was determined using the related iron complex bearing 3-(2-fluorophenyl) pyrazolate ligand containing *ortho*-C-F and *ortho*-C-H bonds.[78] In the presence of 2-(*tert*-butylsulfonyl)-iodosobenzene the Fe_3Fe^{II} complex underwent the intramolecular *ortho*-hydroxylation with a 8:1C-F/C-H selectivity. Interestingly, the related Fe_3Mn^{II} complex provided comparable amounts of C—F and C—H hydroxylated products whereas Mn_3Mn^{II} complex cleaved the C—H bond preferentially. However, when both *ortho*-positions in 3-phenylpyrazolate ligands were replaced with fluorine atoms, $[LMn_3(F_2ArPz)_3OMn](OTf)_x$ (x = 1 or 2) in combination with PhIO was capable of hydroxylating the C—F bond in intramolecular fashion.[79] The species after fluorine atom loss and oxygen atom addition was detected by ESI MS.

Single crystal XRD showed the interaction between the manganese apical center and one of the *ortho* fluorine substituents with the Mn4-F4 distance of 2.469 Å which is significantly shorter than the sum of corresponding van der Waals radii (3.55 Å).

Non-heme iron(IV) oxo complexes often suffer from instability resulting in intramolecular hydroxylation of arene ligand instead of the desired intermolecular oxidation of aromatic substrates.[80] To mitigate this intramolecular ligand oxidation, Goldberg and co-workers have replaced four C—H bonds in the aromatic ligand fragments with C—F bonds. Indeed, this modification has allowed for trapping of the yellow $Fe^{IV}{=}O$ species at −20 °C (Fig. 26).

This yellow intermediate was transformed with a first-order rate constant of $k = 0.9 \times 10^{-3}\ s^{-1}$ into a dark green complex at 23 °C within 50 min, which was isolated and crystallized. Single crystal XRD revealed the structure with an Fe—O distance of 1.787 Å. DFT calculations showed an electrophilic attack upon the aromatic ring by $Fe^{IV}{=}O$ in the triplet spin state with electron transfer from the arene to the iron atom to form an arene

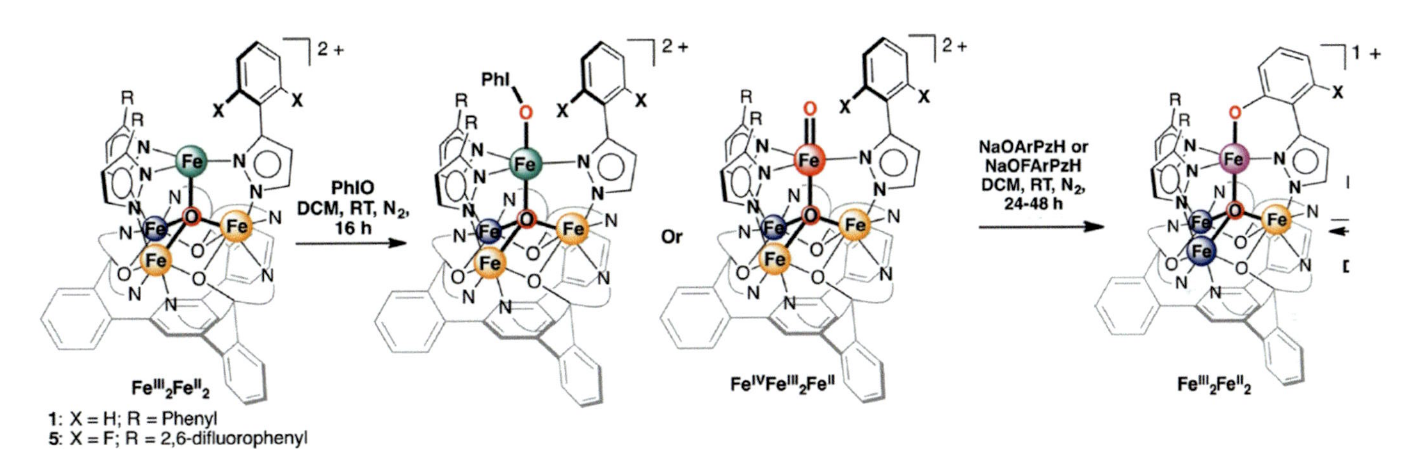

Fig. 25 Intramolecular defluorination of 3-(2,6-difluorophenyl)pyrazolate ligand mediated by tetranuclear iron complex. *Adapted with permission from de Ruiter, G.; Thompson, N. B.; Takase, M. K.; Agapie, T.* J. Am. Chem. Soc. ***2016,*** 138, *1486–1489.*

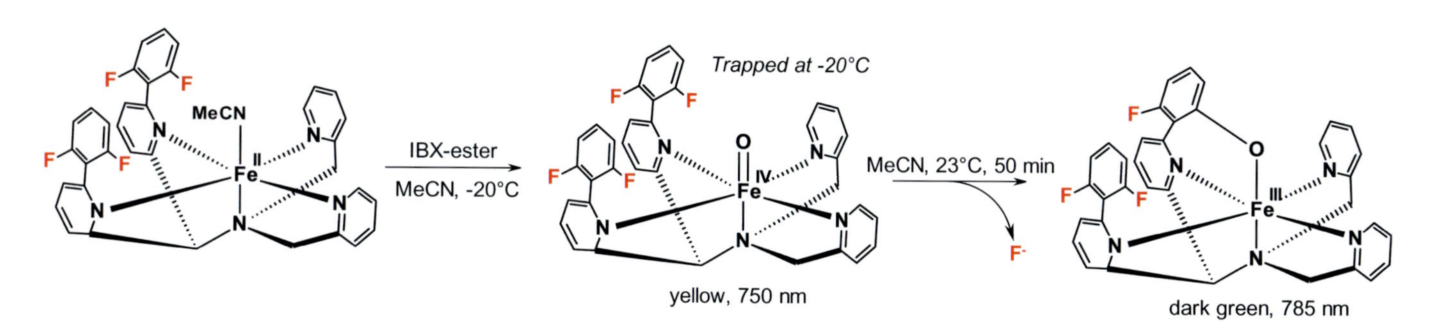

Fig. 26 Intramolecular hydroxylation of aromatic C—F bonds of $[Fe^{II}(N4Py^{2PhF}{}_2)(CH_3CN)](BF_4)_2$ by isopropyl-2-iodoxybenzoate (IBX-ester) via $Fe^{IV}{=}O$ species.

radical with C—O formation. The triplet spin pathway is the lowest in energy (18.1 kcal mol^{-1}), which is unusual in non-heme iron reactivity since the high-spin quintet barrier is typically lower in energy.[80] In fact, the optimal orientation of the arene unit in the coordination sphere of the Fe^{IV}=O complex with an angle of Fe-O-C of 116.8° resulted in the dramatic lowering of the triplet pathway. Since the release of the fluorine radical is unfavorable, it was suggested that the radical intermediate should be reduced by an unidentified one-electron reductant, allowing for the fluoride anion elimination and formation of the dark green complex.[80]

To investigate the intramolecular C—F hydroxylation in more detail, two complexes $[Fe^{II}(N4Py_2^{2Ar})(CH_3CN)](ClO_4)_2$ (Ar_2 = 2,6-difluoro-4-methoxyphenyl) and $[Fe^{II}(N4Py_3^{2Ar})(CH_3CN)](ClO_4)_2$ (Ar_3 = 2,6-difluoro-3-methoxyphenyl) were synthesized.[81] Both complexes reacted with the IBX-ester to give corresponding oxo species which provided the C—F hydroxylated Fe^{III}-OAr complexes. While the former complex showed a C—F hydroxylation rate similar to that of the previously studied $[Fe^{IV}(O)(N4Py_2^{2PhF})(CH_3CN)](ClO_4)_2$ species, the complex bearing a methoxy substituent in *ortho*−/*para*-positions relative to the fluorine substituents displayed a significant rate enhancement for C—F hydroxylation (Fig. 27).

Two possible pathways for the initial attack of Fe^{IV}=O species on the difluorophenyl fragment have been considered (Fig. 28).

Addition of an electron-donating methoxy group to the *ortho*-/*para*-position relative to the fluorine substituents should increase the rate of the electrophilic pathway **A** due to stabilization of the electron-deficient radical, whereas the rate of the nucleophilic pathway **B** should be decreased. The observed 120–160 fold increase of the reaction constant for the complex having the methoxy group in the *ortho*−/*para*-position with respect to the

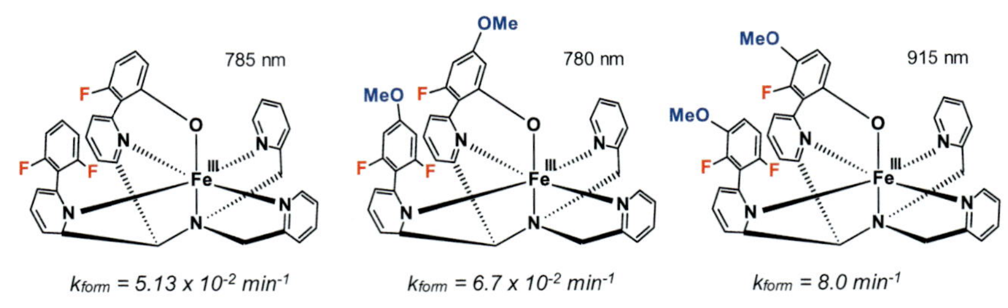

Fig. 27 Complexes obtained by intramolecular defluorination-hydroxylation of the corresponding oxo species. The first-order rate constants of formation were determined from stopped-flow UV–vis kinetics.

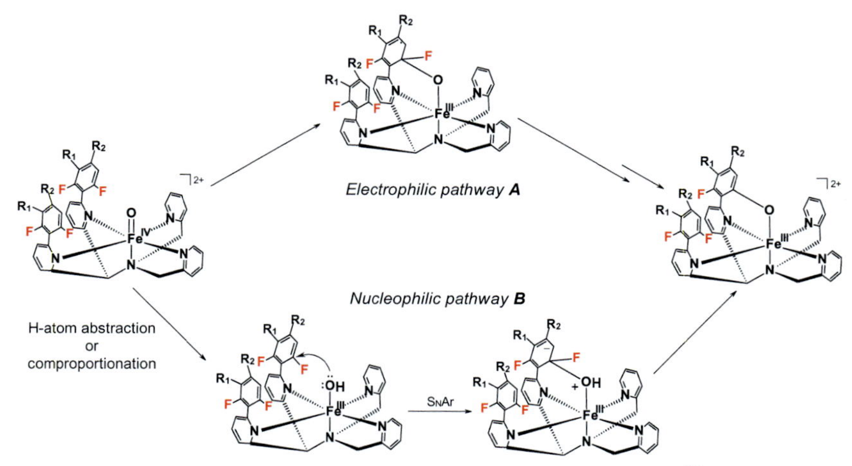

Fig. 28 Possible mechanisms for aromatic C—F hydroxylation by Fe^{IV}=O species.

fluorine substituents, clearly indicates the electrophilic mechanism **A**. The lack of the effect for the *meta*-OMe-substitution is also consistent with the electrophilic mechanism.[81]

The studies on the mechanistic background of the oxidative defluorination of aromatic C—F bonds and origin of the rare preference for C(sp^2)-F cleavage with respect to C(sp^2)-H hydroxylation are still rather limited, most likely owing to the lack of experimental findings until recently.[17] Hadad and co-workers proposed a rate-limiting electrophilic attack of the hexafluorobenzene π-system by a high-valent iron oxo species followed by a fluorine NIH shift.[32]

5. Conclusion and outlook

Compared to the activation of C—F bonds based on electron-rich organometallic complexes, strong reductants and nucleophiles, the oxidative pathways involving fluorinated compounds are much less investigated in chemistry and biology. Several enzymes are capable of transforming molecules containing limited amounts of fluorine substituents. However, the oxidative metabolism of poly- and perfluorinated compounds involving oxygenases is still unknown. In recent years, a variety of engineered metalloenzymes showing remarkable catalytic efficiency in carbene transfer and other reactions have been developed.[82,83] However, the application of these artificial enzymes for the cleavage of C—F bonds has not yet been reported.

In recent years a limited number of transition metal complexes able to oxidatively cleave C—F bond of supporting ligand or to carry out stoichiometric chemistry, have been published. However, mononuclear metal complexes performing catalytic oxidative defluorination reactions, are not yet known. In this context, the discovery and development of a novel catalytic oxidative route for the activation of the strong aromatic C—F bonds by N-bridged diiron phthalocyanine complexes are important from both fundamental and industrial perspectives. μ-Nitrido diiron phthalocyanine complexes in combination with H_2O_2 can form ultra-high-valent diiron oxo species $(Pc)Fe^{IV}(\mu\text{-}N)Fe^{IV}{=}O(Pc^{+\cdot})$ which are extremely potent oxidants and their reactivity exceeds that of the usual mononuclear biomimetic oxo complexes. In addition, μ-nitrido diiron phthalocyanine complexes mediate the oxidative defluorination of poly- and perfluoroaromatic compounds: this is not observed utilizing enzymes. Importantly, these reactions can be carried out in clean and mild conditions with high conversions and turnover numbers. μ-Nitrido diiron species demonstrate similar mechanistic features (^{18}O labeling, formation of benzene epoxide and NIH shift in the aromatic oxidation) as monooxygenase enzymes operating via high-valent iron oxo species. Comprehensive spectroscopic and reactivity studies confirm the participation of high-valent diiron oxo species in these efficient catalytic reactions.

For decades, the development of fluorine chemistry has been focused on the preparation of fluorinated compounds for different applications. The worldwide production of the most important fluorinated aromatic compounds was 35,000 tons per year in 2000.[84] There has been an increase in output since that time. About 40% of agrochemicals and 25% of pharmaceuticals currently used contain fluorine atoms. Owing to the expansion of utilization of fluorinated compounds in many domains and their extreme persistence to biodegradation, especially of perfluorinated entities, they are rapidly accumulated in the environment.[85] The disposal of emerging fluorinated pollutants is very important and is an urgent problem to solve, in the context of sustainable development. Fenton chemistry is widely used in remediation processes for elimination of toxic compounds.[86] However, heavily fluorinated aromatic compounds are very stable under Fenton conditions[59,62] and this approach cannot be applied for their efficient elimination. In turn, reductive approaches involving hydrodefluorination to dispose of fluorinated aromatics requires the use of sophisticated reagents in strict inert, anhydrous conditions which compromises the practical application of reductive processes.[87] The oxidative approach involving μ-nitrido diiron phthalocyanines and H_2O_2 represents an attractive strategy, for several reasons, for the treatment of fluorinated contaminants with very low biodegradability. This catalytic system consists of an accessible, earth abundant and non-toxic iron-based catalyst available in bulk quantities, and industrial environmentally compatible H_2O_2 oxidant. The diiron complex can be readily immobilized onto different supports, thus realizing a catalytic system tolerant to water and air, that can operate upon a large scope of substrates. μ-Nitrido diiron complexes are also efficient in the degradation of toxic chlorinated phenols[88] and the Orange II reagent[89] in water. These catalysts can be applied in both homogeneous and heterogeneous catalytic processes either in organic or aqueous solutions to eliminate a wide range of recalcitrant pollutants, including heavily fluorinated aromatic compounds. For all these reasons we believe that this novel oxidative approach might be useful for the development of novel remediation processes.

Acknowledgment

The author is grateful to all co-workers whose names are given in the references for their valuable contributions to the development of this topic. Dr. P. Afanasiev, Dr. C. Colomban and Dr. E.V. Kudrik are particularly acknowledged for the fruitful collaboration and stimulating discussions. Research support was provided by the Agence Nationale de Recherche (ANR, France, grants ANR-08-BLANC-0183-01 and ANR-16-29CE-0018-01), by CNRS and RFBR (PICS project 6295, RFBR project 14-03-91054) and by Région Rhône-Alpes.

References

1. Wang, Y.; Liu, A. *Chem. Soc. Rev.* **2020**, *49*, 4906–4925.
2. Han, J.; Kiss, L.; Mei, H.; Remete, A. M.; Ponikvar-Svet, M.; Sedgwick, D. M.; Roman, R.; Fustero, S.; Moriwaki, H.; Soloshonok, V. A. *Chem. Rev.* **2021**, *121*, 4678–4742.
3. Clot, E.; Eisenstein, O.; Jasim, N.; Macgregor, S. A.; Mcgrady, J. E.; Perutz, R. N. *Acc. Chem. Res.* **2011**, *44*, 333–348.
4. Fujita, T.; Fuchibe, K.; Ichikawa, J. *Angew. Chem. Int. Ed.* **2019**, *58*, 390–402.
5. Ahrens, T.; Kohlmann, J.; Braun, T. *Chem. Rev.* **2015**, *115*, 931–972.
6. Grushin, V. V. *Acc. Chem. Res.* **2010**, *43*, 160–171.
7. Amii, H.; Uneyama, K. *Chem. Rev.* **2009**, *109*, 2119–2183.
8. Eisenstein, O.; Milani, J.; Perutz, R. N. *Chem. Rev.* **2017**, *117*, 8710–8753.
9. Li, X.; Fu, B.; Zhang, Q.; Yuan, X.; Zhang, Q.; Xiong, T.; Zhang, Q. *Angew. Chem. Int. Ed.* **2020**, *59*, 23056–23060.
10. Natarajan, R.; Azerad, R.; Badet, B.; Copin, E. *J. Fluor. Chem.* **2005**, *126*, 425–436.
11. Murphy, C. D. *Biotechnol. Lett.* **2010**, *32*, 351–359.
12. Sorokin, A. B. In *Adv. in Inorg. Chem*; van Eldik, R., Ed.; vol. 70; Elsevier, 2017; pp. 107–165. Chapter 3.
13. Sorokin, A. B.; Kudrik, E. V.; Bouchu, D. *Chem. Commun.* **2008**, 2562–2564.

14. Kudrik, E. V.; Afanasiev, P.; Alvarez, L. X.; Blondin, G.; Clémancey, M.; Latour, J.-M.; Bouchu, D.; Albrieux, F.; Nefedov, S. E.; Sorokin, A. B. *Nat. Chem.* **2012**, *4*, 1024–1029.
15. İşci, Ü.; Faponle, A. S.; Afanasiev, P.; Albrieux, F.; Briois, V.; Ahsen, V.; Dumoulin, F.; Sorokin, A. B.; de Visser, S. P. *Chem. Sci.* **2015**, *6*, 5063–5075.
16. Afanasiev, P.; Bouchu, D.; Kudrik, E. V.; Millet, J.-M. M.; Sorokin, A. B. *Dalton Trans.* **2009**, 9828–9836.
17. Colomban, C.; Kudrik, E. V.; Afanasiev, P.; Sorokin, A. B. *J. Am. Chem. Soc.* **2014**, *136*, 11321–11330.
18. de Montellano, P. R. O. *Chem. Rev.* **2010**, *110*, 932–948.
19. Tinberg, C. E.; Lippard, S. J. *Acc. Chem. Res.* **2011**, *44*, 280–288.
20. Banerjee, R.; Proshlyakov, Y.; Lipscomb, J. D.; Proshlyakov, D. A. *Nature* **2015**, *518*, 431–434.
21. Dennig, A.; Lülsdorf, N.; Liu, H.; Schwaneberg, U. *Angew. Chem. Int. Ed.* **2013**, *52*, 8459–8462.
22. Rietjens, I. M. C. M.; Soffers, A. E. M. F.; Veeger, C.; Vervoort, J. *Biochemistry* **1993**, *32*, 4801–4812.
23. Jerina, D. M.; Daly, J. W. *Science* **1974**, *185*, 573–582.
24. de Visser, S. P.; Shaik, S. *J. Am. Chem. Soc.* **2003**, *125*, 7413–7424 (and references therein).
25. Mitchell, K. H.; Rogge, C. E.; Gierahn, T.; Fox, B. G. *Proc. Natl. Acad. Sci. U. S. A.* **2003**, *100*, 3784–3789.
26. Rietjens, I. M. C. M.; Vervoort, J. *Chem. Res. Toxicol.* **1992**, *5*, 10–19.
27. Cnubben, N. H. P.; Vervoort, J.; Veeger, C.; Rietjens, I. M. C. M. *Chem. Biol. Interact.* **1992**, *85*, 151–172.
28. Cnubben, N. H. P.; Vervoort, J.; Boersma, M. G.; Rietjens, I. M. C. M. *Biochem. Pharmacol.* **1995**, *49*, 1235–1248.
29. Rietjens, I. M. C. M.; Vervoort, J. *Chem. Biol. Interact.* **1991**, 77, 263–281.
30. den Besten, C.; van Bladeren, P. J.; Duizer, E.; Vervoort, J.; Rietjens, I. M. C. M. *Chem. Res. Toxicol.* **1993**, *6*, 674–680.
31. Koerts, J.; Soffers, A. E. M. F.; Vervoort, J.; De Jager, A.; Rietjens, I. M. C. M. *Chem. Res. Toxicol.* **1998**, *11*, 503–512.
32. Hackett, J. C.; Sanan, T. T.; Hadad, C. M. *Biochemistry* **2007**, *46*, 5924–5940.
33. Pirzad, R.; Newman, J. D.; Dowman, A. A.; Cowell, D. C. *Analyst* **1994**, *119*, 213–218.
34. McGuire, A. H.; Carey, L. M.; de Serrano, V.; Dali, S.; Ghiladi, R. A. *Biochemistry* **2018**, *57*, 4455–4468.
35. D'Antonio, J.; Ghiladi, R. A. *Biochemistry* **2011**, *50*, 5999–6011.
36. Davydov, R.; Osborne, R. L.; Shanmugam, M.; Du, J.; Dawson, J. H.; Hoffman, B. M. *J. Am. Chem. Soc.* **2010**, *132*, 14995–15004.
37. Zhao, J.; de Serrano, V.; Zhao, J.; Le, P.; Franzen, S. *Biochemistry* **2013**, *52*, 2427–2439.
38. D'Antonio, J.; D'Antonio, E. L.; Thompson, M. K.; Bowden, E. F.; Franzen, S.; Smirnova, T.; Ghiladi, R. A. *Biochemistry* **2010**, *49*, 6600–6616.
39. Wang, Y.; Davis, I.; Shin, I.; Wherritt, D. J.; Griffith, W. P.; Dornevil, K.; Colabroy, K. L.; Liu, A. *ACS Catal.* **2019**, *9*, 4764–4776.
40. Wang, Y.; Davis, I.; Shin, I.; Xu, H.; Liu, A. *J. Am. Chem. Soc.* **2021**, *143*, 4680–4693.
41. Hillas, P. J.; Fitzpatrick, P. F. *Biochemistry* **1996**, *35*, 6969–6975.
42. Renganathan, V. *Appl. Environ. Microbiol.* **1989**, *55*, 330–334.
43. Fetzner, S.; Müller, R.; Lingens, F. *J. Bacteriol.* **1992**, *174*, 279–290.
44. Li, J.; Griffith, W. P.; Davis, I.; Shin, I.; Wang, J.; Li, F.; Wang, Y.; Wherritt, D. J.; Liu, A. *Nat. Chem. Biol.* **2018**, *14*, 853–860.
45. Afanasiev, P.; Sorokin, A. B. *Acc. Chem. Rev.* **2016**, *49*, 583–593.
46. Sorokin, A. B. *Catal. Today* **2021**, *373*, 38–58.

47. Colomban, C.; Tobing, A. H.; Mukherjee, G.; Sastri, C. V.; Sorokin, A. B.; de Visser, S. P. *Chem. Eur. J.* **2019**, *25*, 14320–14331.
48. Sorokin, A.; Séris, J.-L.; Meunier, B. *Science* **1995**, *268*, 1163–1166.
49. Sorokin, A.; Meunier, B. *Chem. Eur. J.* **1996**, *2*, 1308–1317.
50. Colomban, C.; Kudrik, E. V.; Sorokin, A. B. *J. Porphyrins Phthalocyanines* **2017**, *21*, 345–353.
51. Colomban, C.; Kudrik, E. V.; Briois, V.; Shwarbrick, J. C.; Sorokin, A. B.; Afanasiev, P. *Inorg. Chem.* **2014**, *53*, 11517–11530.
52. Quesne, M. G.; Senthilnathan, D.; Singh, D.; Kumar, D.; Maldivi, P.; Sorokin, A. B.; de Visser, S. P. *ACS Catal.* **2016**, *6*, 2230–2243.
53. Silaghi-Dumitresku, R.; Makarov, S. V.; Uta, M.-M.; Dereven'kov, I. A.; Stuzhin, P. A. *New J. Chem.* **2011**, *35*, 1140–1145.
54. Ansari, M.; Vyas, N.; Ansari, A.; Rajaraman, G. *Dalton Trans.* **2015**, *44*, 15232–15243.
55. Phung, Q. M.; Pierloot, K. *Chem. Eur. J.* **2019**, *25*, 12491–12496.
56. Rachmilovich-Calis, S.; Masarwa, A.; Meyerstein, N.; Meyerstein, D.; van Eldik, R. *Chem. Eur. J.* **2009**, *15*, 8303–8309.
57. Gozzo, F. *J. Mol. Catal. A Chem.* **2001**, *171*, 1–22.
58. Shi, F.; Tse, M. K.; Li, Z.; Beller, M. *Chem. Eur. J.* **2008**, *14*, 8793–8797.
59. Bogachev, A. A.; Kobrina, L. S.; Yakobson, G. G. *Russ. J. Org. Chem.* **1986**, *22*, 2307–2313.
60. Ravichandran, L.; Selvam, K.; Swaminathan, M. *J. Photochem. Photobiol. A: Chem.* **2007**, *188*, 392–398.
61. Ravichandran, L.; Selvam, K.; Swaminathan, M. *Desalination* **2010**, *260*, 18–22.
62. Bogachev, A. A.; Kobrina, L. S.; Yakobson, G. G. *Russ. J. Org. Chem.* **1986**, *22*, 2313–2317.
63. Osman, A. M.; Posthumus, M. A.; Veeger, C.; van Bladeren, P. J.; Laane, C.; Rietjens, I. M. C. M. *Chem. Res. Toxicol.* **1998**, *11*, 1319–1325.
64. Kuehnel, M. F.; Lentz, D.; Braun, T. *Angew. Chem. Int. Ed.* **2013**, *52*, 3328–3348.
65. Nova, A.; Mas-Ballesté, R.; Lledós, A. *Organometallics* **2012**, *31*, 1245–1256.
66. Arisawa, M.; Suzuki, T.; Ishikawa, T.; Yamaguchi, M. *J. Am. Chem. Soc.* **2008**, *130*, 12214–12215.
67. Liu, C.; Cao, L.; Yin, X.; Xu, H.; Zhang, B. *J. Fluor. Chem.* **2013**, *156*, 51–60.
68. Rietjens, I. M.; den Besten, C.; Hanzlik, R. P.; van Bladeren, P. J. *Chem. Res. Toxicol.* **1997**, *10*, 629–635.
69. Kudrik, E. V.; Sorokin, A. B. *Chem. Eur. J.* **2008**, *14*, 7123–7126.
70. Bogaards, J. P.; Van Ommen, B.; Wolf, C. R.; Van Bladeren, P. *J. Toxicol. Appl. Pharmacol.* **1995**, *132*, 44–52.
71. Rietjens, I. M. C. M.; Vervoort, J. *Xenobiotica* **1989**, *19*, 1297–1305.
72. Faponle, A. S.; Quesne, M. G.; Sastri, C. V.; Banse, F.; de Visser, S. P. *Chem. Eur. J.* **2015**, *21*, 1221–1236.
73. Ohe, T.; Mashino, T.; Hirobe, M. *Tetrahedron Lett.* **1995**, *36*, 7681–7684.
74. Serrano-Plana, J.; Garcia-Bosch, I.; Miyake, R.; Costas, M.; Company, A. *Angew. Chem. Int. Ed.* **2014**, *53*, 9608–9612.
75. Besalú-Sala, P.; Magallón, C.; Costas, M.; Company, A.; Luis, J. M. *Inorg. Chem.* **2020**, *59*, 17018–17027.
76. Garcia-Bosch, I.; Cowley, R. E.; Diaz, D. E.; Peterson, R. L.; Solomon, E. I.; Karlin, K. D. *J. Am. Chem. Soc.* **2017**, *139*, 3186–3195.
77. de Ruiter, G.; Thompson, N. B.; Takase, M. K.; Agapie, T. *J. Am. Chem. Soc.* **2016**, *138*, 1486–1489.
78. de Ruiter, G.; Carsch, K. M.; Takase, M. K.; Agapie, T. *Chem. Eur. J.* **2017**, *23*, 10744–10748.
79. Carsch, K. M.; de Ruiter, G.; Agapie, T. *Inorg. Chem.* **2017**, *56*, 9044–9054.

80. Sahu, S.; Quesne, M. G.; Davies, C. G.; Dürr, M.; Ivanović-Burmazović, I.; Siegler, M. A.; Jameson, G. N. L.; de Visser, S. P.; Goldberg, D. P. *J. Am. Chem. Soc.* **2014**, *136*, 13542–13545.
81. Sahu, S.; Zhang, B.; Pollock, C. J.; Dürr, M.; Davies, C. G.; Confer, A. M.; Ivanović-Burmazović, I.; Siegler, M. A.; Jameson, G. N. L.; Krebs, C.; Goldberg, D. P. *J. Am. Chem. Soc.* **2016**, *138*, 12791–12802.
82. Brandenberg, O. F.; Fasan, R.; Arnold, F. H. *Curr. Opin. Biotechnol.* **2017**, *47*, 102–111.
83. Arnold, F. H. *Angew. Chem. Int. Ed.* **2019**, *58*, 14420–14426.
84. Baumgartner, R.; McNeill, K. *Environ. Sci. Technol.* **2012**, *46*, 10199–10205.
85. Lim, X. *Nature* **2019**, *566*, 27–29.
86. Pignatello, J. J.; Oliveros, E.; MacKay, A. *Crit. Rev. Environ. Sci.* **2006**, *36*, 1–84.
87. Douvris, C.; Nagaraja, C. M.; Chen, C.-H.; Foxman, B. M.; Ozerov, O. V. *J. Am. Chem. Soc.* **2010**, *132*, 4946–4953.
88. Colomban, C.; Kudrik, E. V.; Afanasiev, P.; Sorokin, A. B. *Catal. Today* **2014**, *235*, 14–19.
89. Makarova, A. S.; Kudrik, E. V.; Makarov, S. V.; Koifman, O. I. *J. Porphyrins Phthalocyanines* **2014**, *18*, 604–613.

CHAPTER THREE

Photodynamic inactivation (PDI) as a promising alternative to current pharmaceuticals for the treatment of resistant microorganisms

Barbara Pucelik and Janusz M. Dąbrowski*
Faculty of Chemistry, Jagiellonian University, Kraków, Poland
Małopolska Centre of Biotechnology, Jagiellonian University, Kraków, Poland
*Corresponding author: e-mail address: jdabrows@chemia.uj.edu.pl

Contents

Abstract

Although the whole world is currently observing the global battle against COVID-19, it should not be underestimated that in the next 30 years, approximately 10 million people per year could be exposed to infections caused by multi-drug resistant bacteria. As new antibiotics come under pressure from unpredictable resistance patterns and relegation to last-line therapy, immediate action is needed to establish a radically different approach to countering resistant microorganisms. Among the most widely explored alternative methods for combating bacterial infections are metal complexes and nanoparticles, often in combination with light, but strategies using monoclonal antibodies and bacteriophages are increasingly gaining acceptance. Photodynamic inactivation (PDI) uses light

Advances in Inorganic Chemistry, Volume 79
ISSN 0898-8838
https://doi.org/10.1016/bs.adioch.2021.12.003

and a dye termed a photosensitizer (PS) in the presence of oxygen to generate reactive oxygen species (ROS) in the field of illumination that eventually kill microorganisms. Over the past few years, hundreds of photomaterials have been investigated, seeking ideal strategies based either on single molecules (e.g., tetrapyrroles, metal complexes) or in combination with various delivery systems. The present work describes some of the most recent advances of PDI, focusing on the design of suitable photosensitizers, their formulations, and their potential to inactivate bacteria, viruses, and fungi. Particular attention is focused on the compounds and materials developed in our laboratories that are capable of killing in the exponential growth phase (up to seven logarithmic units) of bacteria without loss of efficacy or resistance, while being completely safe for human cells. Prospectively, PDI using these photomaterials could potentially cure infected wounds and oral infections caused by various multidrug-resistant bacteria. It is also possible to treat the surfaces of medical equipment with the materials described, in order to disinfect them with light, and reduce the risk of nosocomial infections.

1. Introduction

The global pandemic of COVID-19 has been raging during the last 2 years, but medical experts warn another health crisis is looming—antibiotic resistance of bacteria and development of "superbugs." Since the penicillin discovery, antibiotics have been named the "silver bullets" of medicine. However, less than a century following, the future impact of antibiotics is decreasing at a pace that no one expected, with more microbes out-smarting and out-evolving these "miracle" medicines. Unfortunately, bacteria resistance is still in progress and, consequently, if this problem remains unresolved, it could kill an estimated 10 million people each year by 2050.[1–3] Thus, the current situation requires rapid intervention and the main challenge of modern medicine is to develop a novel, innovative antimicrobial treatment that will overcome bacterial resistance mechanisms.

Antimicrobial resistance (AMR) of bacteria, driven by antibiotic consumption, is a global problem and a major threat to public health. Globalization and the spread of long-distance travel have contributed to the spread of pathogens on an unprecedented scale.[4] In parallel, the genes responsible for resistance have also become widespread. This has led to the emergence of superbugs - microorganisms that are very difficult, if not impossible, to eradicate with existing drugs. These include methicillin-resistant *Staphylococcus aureus* (MRSA) and extremely resistant *Mycobacterium tuberculosis* (TB). Another urgent problem is the growing hospital infections (nosocomial) associated with medical devices such as ventilator-associated pneumonia (VAP), central-catheter bloodstream infection, and catheter-associated urinary tract infection, accounting for approx. 26% of nosocomial infections, followed by surgical-site infections of approximately 22%.[5] It is also worth noting that projections of global antibiotic consumption in the near future, assuming no policy changes, look unlikely to be optimistic. It has been estimated that global antibiotic consumption will increase from 42 billion defined daily doses (DDD) in 2015 to as much as 128 billion DDD in 2030.[6] This is consistent with the increase in the number of infections resistant to antibiotics. The demand for antibiotics effective against multidrug-resistant microorganisms is not reflected in the pipelines of pharmaceutical industries. Marketed antibiotics are very inexpensive, so it is difficult to predict the emergence of antibiotic resistance for a newly approved antibiotic. The global market increased human mobility and facilitated access to medicines to accelerate the onset of resistance. At least 700,000 people currently die each year worldwide from untreatable infections. Moreover, it has been estimated that by 2050 drug-resistant strains of TB, malaria, HIV and several bacterial infections could claim 10 million lives annually. This will come at an economic cost of $100 trillion from global gross domestic product (GDP) over the next 35 years.[7] There is no doubt, therefore, that all these problems discussed above force both scientists and politicians to urgently investigate and promote alternative methods of combating bacterial infections. This chapter describes some state-of-the-art approaches to this problem, particularly the employment of (light-activated) metal complexes and nanoparticles or monoclonal antibodies and bacteriophages. Our strategy to control infections is to combine a non-toxic photosensitizer with visible light, which in the presence of oxygen leads to the formation of reactive oxygen species (ROS) that are cytotoxic but have a very small (nanometer size) diffusion radius and can overcome multi-drug resistance. This approach is known as photodynamic inactivation of microorganisms (PDI) and, as the main topic of this chapter, has been described most extensively. We have mainly focused on exploring the physicochemical and pharmacological properties of new photosensitizing drugs/materials and elucidating the unique mechanisms of PDI, which make this method an alternative to the current treatments of multidrug-resistant pathogens. Furthermore, strategies that combine multiple approaches to increase antimicrobial efficacy will be presented.

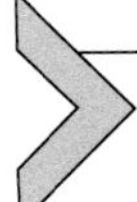

2. Resistance of microorganisms to antibiotics

2.1 Antibiotics

Antibiotics are naturally occurring compounds produced by microorganisms (and their semi-synthetic and synthetic derivatives) to destroy (bactericidal effect) or inhibit the growth (bacteriostatic effect) of other microorganisms. There are several classes of antibiotics according to their targets. The most significant among them are: (i) inhibitors of the cell wall biosynthesis, (ii) proteins, and (iii) nucleic acids.[8] The cell wall provides the shape and appropriate rigidity to bacteria and protects them from adverse effects of the external environment (Fig. 1).

Its main component is peptidoglycan (PGN), a heteropolymer consisting of a sugar molecules, composed of *N*-acetylglucosamine and *N*-acetylmuramic acid and a protein molecule that forms cross-links between the sugar chains. This unique structure of PGN imparts rigidity and mechanical strength to the cell wall. Biosynthesis of PGN is a multi-step process involving about 30 different enzymes. Many of these have no counterparts in the human body, making them attractive targets for antibiotics.[8–10] Antibiotics bind covalently to the enzyme active site, blocking its activity and ultimately reducing the availability of the PGN precursor.[11] A PGN monomer is transported across the cytoplasmic membrane to the outside of the cell, becoming a target for glycopeptide antibiotics.[12] β-lactam derived antibiotics, including penicillins, cephalosporins and carbapenems, inhibit transpeptidases activity. Thus, they contribute to the weakening of the cell wall and, consequently, the inhibition of bacterial growth and often to bacterial death.[9,13] Antibiotics that inhibit protein biosynthesis act by blocking ribosomes. This class includes aminoglycosides, tetracyclines, macrolides, and lincosamides.[14–16] Antibiotics that affect nucleic acid biosynthesis are inhibitors of enzymes involved in these processes. They inhibit the action of DNA gyrase and topoisomerase IV—enzymes crucial in DNA replication,[17] whereas rifamycin binds to bacterial RNA polymerase, disrupting the transcription process.[18] The last class includes polymyxins-antibiotics that are not directly involved in DNA replication, transcription, translation, and cell wall synthesis. They bind to lipopolysaccharide (LPS), a component of the outer membrane of Gram-negative bacteria,

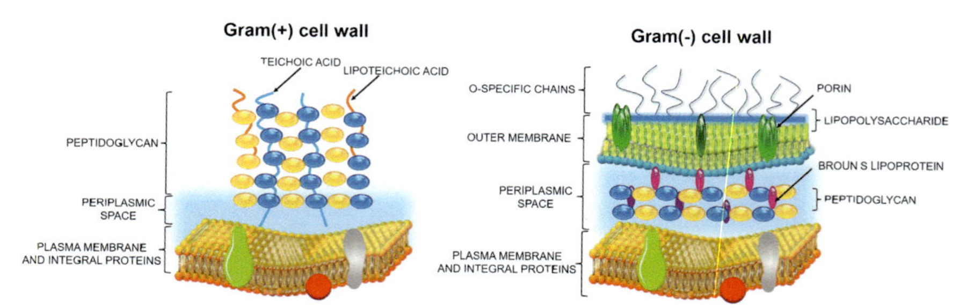

Fig. 1 Scheme showing the composition of bacterial cell wall structure: *left*—Gram-positive bacteria possessing a thick peptidoglycan layer combined with teichoic acid molecules and no external lipid membrane; *right*—Gram-negative bacteria with a thin peptidoglycan layer and an external lipid membrane that contains porin proteins.

destabilizing the membrane and increasing its permeability, ultimately leading to cell death.[19] This class also includes gramicidin, which impairs cell membrane function by generating defects within it.[20]

Antibiotics, as compounds produced by living organisms, have been present in the environment long before the appearance of humans, and some of the defense mechanisms of bacteria against them have evolved as far back as ancient times.[21] These mechanisms allow bacteria to share ecological niches with organisms that secrete bactericidal substances. The progression of bacterial resistance to new antibiotics continues due to increasing environmental pressures resulting from the careless, unnecessary, and excessive use of antibiotics.[21] There are two main routes by which bacteria acquire resistance: mutations and horizontal gene transfer. In the first case, there are accidental changes in genes related to the antibiotic uptake from the environment or the protein synthesis that is its biological target. In the second situation, the bacteria acquire resistance through the incorporation of host DNA by one of the following mechanisms: (i) transformation (uptake of free DNA present in the environment), (ii) transduction (acquisition of host DNA via bacteriophage), and (iii) conjugation (exchange of mobile genetic elements, e.g. plasmids between two cells in direct contact).[21] Acquired genes can induce resistance through four main mechanisms: (i) modification or synthesis of a novel biological target, (ii) enzymatic inactivation of the antibiotic, (iii) decreasing bacterial envelope permeability, and (iv) increasing active efflux of the antibiotic from the cell by proteins known as efflux pumps.[8]

2.2 Mechanisms of resistance

2.2.1 Structural modification of a biological target: synthesis of novel molecular entities

Modification of the biological target represents a mechanism of resistance to β-lactam and glycopeptide-type antibiotics. Resistance of some bacterial strains, including methicillin-resistant *S. aureus*, to β-lactams, results from an acquired gene encoding a modified version of the transpeptidase designated PBP2a. Kinetic studies indicate that the binding rate constant of PBP2a to β-lactams is reduced by 3–4 orders of magnitude compared to other transpeptidases.[22] This significant decrease is attributed to the structure of PBP2a, which exhibits significant conformational changes before covalent binding to a β-lactam can occur.[23] Glycopeptides inhibit bacterial cell wall biosynthesis by binding to the D-alanine-D-alanine terminal fragment of the peptidoglycan precursor. The resistance of most bacteria to antibiotics in this group may be due to their acquisition of genes encoding enzymes whose concerted activity leads to the formation of a peptidoglycan precursor with a terminal fragment in the form of D-alanine-D-lactate.[24] Glycopeptides show a much lower affinity for such a modified peptidoglycan precursor than for its standard form.[12]

2.2.2 Enzymatic inactivation of antibiotics

Inactivation of antibiotics usually occurs by enzymatically catalyzed hydrolysis, group transfer, or redox reactions. These result either in modification of the drug molecule or in its complete decomposition.[25] Antibiotics containing ester, amide, or epoxy groups in their structure are susceptible to inactivation by hydrolysis which can easily occur both inside and outside the bacterial cell if the bacterium releases enzymes into the external environment, as the only necessary co-substrate for this reaction is a water molecule.[26]

2.2.3 Reduction in permeability of bacterial membranes

This mechanism is particularly important for Gram-negative bacteria and is one of the reasons for the increased resistance to many antibiotics of Gram-negative species compared to Gram-positive ones. This is due to the differences in the structure of their outer shell: gram-negative bacteria are additionally surrounded by an outer membrane. It is composed of phospholipids, ensuring its hydrophobic character, hindering the diffusion of hydrophilic moieties, LPS, which determines the membrane impermeability for many hydrophobic compounds, and proteins involved in the transport of molecules.[27,28] The general structures of the Gram-positive and Gram-negative bacteria cell walls are presented in Fig. 1. Although the outer membrane of Gram-negative bacteria is itself a good barrier to many xenobiotics, in some resistant strains, an additional reduction of its permeability is observed. It is achieved by reducing the number of porins, changing their structure, or modifying LPS. Porins are protein channels in the cell membrane loaded with water molecules allowing diffusion of hydrophilic substances. Antibiotics such as β-lactams, chloramphenicol or fluoroquinolones only penetrate the outer membrane of Gram-negative bacteria. Therefore, a decrease in the number of porins, a reduction in the diameter of their channels, or other modifications of their structure can limit the penetration of specific antibiotics into the cell, leading to a decrease in bacterial susceptibility. LPS is also important for the proper functioning of the outer membrane of Gram-negative bacteria. Its presence in the outer membrane of Gram-negative bacteria is mainly responsible for its impermeability to hydrophobic antibiotics. Several characteristic features of the LPS structure contribute to this phenomenon. First, there are saturated fatty acid residues in its structure, which are responsible for its gel-like structure and low fluidity, making the diffusion of small molecules much more difficult. Second, O-antigen is endowed with a strong negative charge. Third, there are numerous cross-links within the central region of LPS. Phosphate groups and divalent cations are involved, further contributing to reducing bacterial membrane permeability to hydrophobic substances.[28] In addition, LPS can undergo modifications that contribute to bacterial resistance to specific groups of antibiotics, such as polymyxins, which are the last line of defense for infections with strains exhibiting multidrug resistance. LPS is the target of polymyxins. Its most common modifications are changes in lipid A, which lead to a reduction of its negative charge. As a result, the positively charged polymyxins bind more weakly with it decreasing their effectiveness. Other common changes in LPS structure include deacylation, hydroxylation, and palmitoylation.[29,30]

2.2.4 The activity of efflux pumps

Efflux pumps are membrane proteins responsible for transporting substances across the cell membrane. They are found in almost all prokaryotic and eukaryotic organisms. They are involved in various processes, such as maintaining an appropriate potential and pH gradient across the cell membrane, intercellular signaling, processes associated with microbial virulence, and removal of unwanted metabolites and toxic substances from the cell. Thus, they contribute to the maintenance of cell homeostasis.[31–33] The activity of efflux pumps is one of the reasons for bacterial resistance to certain antibiotics and bactericides. This occurs when the substance structurally resembles the pump's natural substrate or when the selectivity of the pump is modest (multidrug resistance, MDR pumps). Chromosomal DNA usually encodes pumps with a broader spectrum of substrates. In contrast, mobile genetic elements, such as plasmids, typically contain genes encoding pumps with greater substrate selectivity.[32,33] Efflux pumps contribute to antibiotic resistance according to three fundamental mechanisms: natural, acquired, and phenotypic. Natural resistance results from the constitutive expression of pumps. Inhibition of their expression in bacteria considered sensitive to a given antibiotic leads to the development of hypersensitivity. Higher levels of bacteria resistance can be acquired through horizontal gene transfer and mutations, leading to overexpression of chromosomally encoded pumps. Phenotypic resistance, on the other hand, is based on transient overexpression of pump-encoding genes triggered by specific external conditions or the presence of an appropriate inducer.[33] Various pumps differ in their selectivity, structure, source of energy used, and occurrence. There are five basic families of efflux pumps: SMR (Small Multidrug Resistance), MATE (Multi Antimicrobial Extrusion), MFS (Major Facilitator Superfamily), RND (Resistance Nodulation and cell Division), and ABC (ATP Binding Cassette). The ABC family transporters utilize energy derived from ATP hydrolysis. On the other hand, pumps belonging to the other four families are second-order transporters—they utilize a proton gradient.[31,33] The structure of efflux pumps is diverse. They can be constructed from one or several subunits. In Gram-positive bacteria, pumps built from a single polypeptide chain are predominant. In contrast, in Gram-negative bacteria, the pumps consist of three subunits: an inner membrane protein, a periplasmic linker protein, and an outer membrane protein (Fig. 1). Most transporters of the RND family and representatives of the ABC, MFS, and MATE families contain such a structure. This type of pump contributes to the increased resistance of Gram-negative bacteria to antibiotics more than single-subunit transporters (belonging mainly to the SMR and MFS families).[31]

2.2.5 Biofilm formation

In addition to the antibiotic resistance mechanisms described above, biofilm formation is highly important for the survival of bacteria. It is an aggregation of bacterial cells attached to the substrate and immersed in the matrix they form. It comprises polysaccharides, proteins, lipids, and DNA collectively referred to as extracellular polymeric substances (EPS).[34] Biofilm is the most common form of bacteria, found on various biotic and abiotic surfaces, including pipelines, drinking water distribution systems, and medical devices. It poses a severe threat to human health and life. It is assumed that about 65–80% of infections are related to biofilm.[35] These include hospital-acquired pneumonia, infections associated with the insertion of non-sterile implants, surgical wound and burn infections, chronic urinary tract infections, sinus infections, middle ear infections, and periodontitis.[11,13] Biofilm formation is a survival strategy for bacteria in adverse conditions. Immersed cells in the matrix are protected from factors such as: temperature and pH extremes, high salinity, dehydration, and UV radiation. The matrix also provides them with a reservoir of oxygen and nutrients.[34,36,37] In addition, these unique physical properties protect it from mechanical removals (Fig. 2).

Bacteria in a biofilm can be up to 1000 times less sensitive to a given antibiotic than their counterparts of the same species found in the planktonic form.[38] It is recognized that there are some specific properties of the biofilm

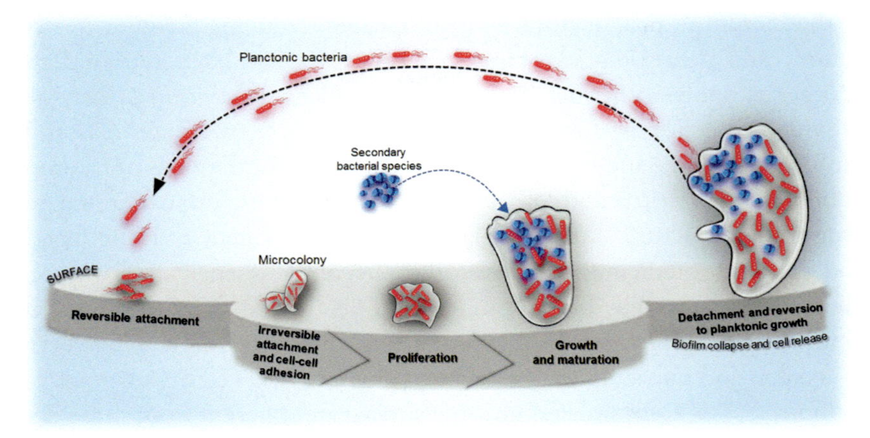

Fig. 2 Schematic representation of steps leading to bacterial biofilm growth. *Adapted and modified from Pinto, A. M.; Cerqueira, M. A.; Bañobre-Lópes, M.; Pastrana, L. M.; Sillankorva, S.* Viruses ***2020,*** 12 *(2), 235 and Maunders, E.; Welch, M.* FEMS Microbiol. Lett. ***2017,*** 364 *(13).*

responsible for antibiotic treatment failures and chronic and recurrent inflammation. The complete removal of biofilm from its occupied surface represents a major challenge for modern medicine. Biofilm resistance to antibiotics is an extremely complex phenomenon. Many factors contribute to its overall condition: the previously described mechanisms common to both the biofilm and the planktonic form, specific properties of the matrix surrounding the cells, the high heterogeneity of the biofilm, and changes in gene expression profile and metabolism compared to the planktonic form.[39,40] The matrix provides some structure to the biofilm and to a certain extent, protects the cells within it from antibiotic access.[41] It provides a physical and chemical barrier that impedes the diffusion of drugs deep into the biofilm, and components of the matrix may even bind some antibiotics. However, it applies only to a few types of antibiotics (e.g., aminoglycosides), and its effectiveness depends on several factors such as species and strain of bacteria, the composition of the matrix, age, and biofilm growth conditions.[39,40,42] However, limited matrix permeability is not considered the leading cause of biofilm resistance.[40] Instead, it is worth emphasizing the role of the interaction of the matrix components with the host immune system. Staphylococcus bacteria are able to produce some extracellular polymeric substances (EPS), most notably polysaccharide intercellular adhesin (PIA). The presence of PIA may protect the biofilm from the host immune response. Numerous studies suggest that PIA can reduce the susceptibility of the biofilm to phagocytosis by macrophages, reducing granulocyte activation, and decreasing cytokine secretion.[43] It has been reported that another exopolysaccharide, alginate, produced by *P. aeruginosa*, may also provide protection against phagocytosis.[39] However, the main reason for biofilm persistence and resistance lies elsewhere. Its formation is not a simple combination of matrix and cells. It represents an extremely dynamic and heterogeneous structure.[41] The key factor of this heterogeneity is the oxygen and nutrient gradient. As the biofilm becomes deeper, their content decreases because they are consumed by cells that lie close to the surface before they can reach the deepest layers of the biofilm. Differences in access to oxygen and nutrients are the cause of physiological cell heterogeneity in biofilms.[39,40,44] Cells situated near the surface are different from those deep within the biofilm. This is due to changes in gene expression in response to stress, such as hypoxia. These changes contribute to a decrease in the metabolic activity of the cells and their transition to a stationary-like state. The so-called persisters, representing up to 1% of all biofilm cells, are most responsible for recurrent, difficult-to-control infections.[45]

These are dormant, non-dividing cells in which metabolic processes have been significantly or completely inhibited. Due to the arrest of certain metabolic pathways, antibiotics lose their target. For this reason, surviving cells can persist through antibiotic therapy. However, once the therapy is completed, they regain metabolic activity and proliferate, leading to population recovery and recurrence of the infection.[46,47]

3. Alternative methods for controlling bacterial infections

3.1 Small-molecule metal complexes

One common approach to combat bacterial resistance to antibiotics is the use of small-molecule metal complexes. The antibacterial activity of metals has been known for a long time and was used long before the discovery of microbes. Specific metals can bind to biomolecules present in the cell. This leads to changes in the structure of these molecules and, consequently, to their dysfunction. Some metals also bind to cell membranes, compromising their integrity and affecting their membrane potential. In addition, redox-active metals catalyze the Fenton reaction, which results in the formation of reactive oxygen species (ROS) capable of oxidizing proteins, lipids, and nucleic acids.[25,48,49] Metal complexes can simultaneously exhibit several of the mechanisms of action mentioned above, and additionally combine them with the ligands' activity.[50–52] Due to the synergism of action of the antibiotic and desired metal ions, complexes obtained in this way often show enhanced antimicrobial activity compared to the antibiotic alone. This effect is mainly observed with resistant bacterial strains.[25] The second approach is to synthesize completely new metal complexes and to test their antimicrobial activity. Silver, copper, ruthenium, and iron complexes are particularly interesting in this aspect.[25]

3.2 Antibacterial oligonucleotides

The mode of antibacterial oligonucleotides action is based on inhibiting the expression of specific bacterial genes. These may be genes essential for bacterial survival, as well as those related to virulence or antibiotic resistance. The latter approach may be applicable in combined therapy with an appropriate antibiotic.[53] Oligonucleotides with antibacterial activity include antisense nucleotides (ASOs) and short interfering RNA (siRNA). The sequence of both ASOs and siRNA is complementary to the bacterial mRNA. Hybridization of the oligonucleotide to the mRNA inhibits the translation process. This occurs by degrading the mRNA or blocking its binding to the ribosome.[54,55] The most significant challenge to the therapeutic success of antibacterial oligonucleotides is their delivery to the site of action, which is the interior of the bacterial cell.[56] To avoid premature excretion by the kidney and degradation by nucleases present in the blood, nucleotides must be chemically modified. Additionally, due to their large size, oligonucleotides are unable to penetrate the bacterial cell wall. This problem can be solved by combining them with peptides capable of penetrating the wall.[53] Moreover, the concern may be whether the APOs or siRNAs used will also interact with the host mRNA. Nevertheless, since they target specific bacterial genes, the effect on human gene expression should be minimal.[55] Furthermore, the use of bioanalytical screening allows the identification and elimination of molecules that have a high risk of unwanted cross-reactivity.[57] The development of antimicrobial oligonucleotides is still in its early stages as none of these compounds have yet been approved for therapeutic use by the Food and Drug Administration (FDA). Yet there is no doubt that therapies based on silencing bacterial genes have considerable potential. Blocking gene expression may produce faster and longer-lasting effects than conventional therapies. However, the development of such an approach requires a large initial investment to understand the pharmacological profile of oligonucleotides and methods to modify them. Anyhow, costs are likely to decrease significantly as the pharmacological and toxicological profile of oligonucleotides within a particular class is very similar.

3.3 Monoclonal antibodies

Monoclonal antibody therapy has its origins in the use of serum as the treatment of choice for numerous infections, including tetanus, scarlet fever, and pneumonia, in the early 20th century. It involved the passive introduction of animal serum into the patient's body. Humanized or fully human antibodies characterized by increased selectivity and reduced toxicity are used.[58] The mechanism of antibacterial activity of antibodies varies depending on the antibody class, the type of targeted molecule, and the role it plays in the pathogenesis process. The direct action of antibodies is based on their binding to proteins and polysaccharides present on the surface of the bacterial cell and inducing a host immune response. On the other hand, the indirect action is based on binding to virulence factors secreted by bacteria (exotoxins, proteases, and signaling molecules, among others). After binding to antibodies, these factors are neutralized—they lose their ability to bind to elements of the host organism. Thus, the pathogenesis process is inhibited, but the organism itself must fight the infection in parallel.[59–61] Currently, all FDA-approved monoclonal antibodies for antimicrobial therapy have been shown to neutralize virulence factors. The advantage of this strategy is that virulence factors are highly conserved and essential to pathogenesis.[53] There are also antibodies that are completely independent of the host immune system and exhibit direct bactericidal activity.[62]

3.4 Nanoparticles

The use of metal nanoparticles and metal oxides is still considered one of the most promising strategies to combat resistant microbes. The probability of microbes developing resistance to these types of pharmaceuticals is low since they exhibit various mechanisms of action simultaneously, such as cell wall damage, generation of ROS, indirect effects on respiratory chain inhibition, transcription and translation, and disruption of nutrient uptake from the environment. The antibacterial and antifungal properties of nanoparticles are mainly influenced by their size and distribution, shape, and morphology.[25] Nanoparticles generally exhibit more favorable morphological, catalytic, optical, and magnetic properties than their micro-sized counterparts. Moreover, their high surface area to volume ratio allows for increased contact with the pathogen surface. Additionally, the surface of nanoparticles can be easily functionalized with polymers, peptides, antibodies, etc. Thus, their biocompatibility, selectivity, and colloidal stability can be enhanced, and they can acquire an additional mechanism of antibacterial action.[63,64] Nanoparticles can also serve as carriers for other antibacterial drugs and photosensitizers.[25]

3.5 Bacteriophages

Bacteriophages are viruses that selectively attack bacteria. They bind to the surface receptors of bacterial cells and then insert their own genetic material into the cells. New viruses-copies of the virus that attacked the cell are assembled from the elements synthesized in this way. The newly formed virions leave the cell and can attack other cells, starting another infection cycle. Thus, bacteriophages are self-replicating pharmaceuticals.[65] Most bacteriophages are characterized by their selectivity toward the bacterial strains attacked. This can be considered as both an advantage and a disadvantage. It may allow selective attack of pathogenic bacteria while not harming the natural microflora.[66] However, there are currently no effective immediate methods to assess the affinity of a bacteriophage for a bacterial strain isolated from the patient's body. Thus, it is often necessary to deliver a mixture of different bacteriophages.[65,67]

3.6 Photodynamic inactivation of microorganisms (PDI)

Photodynamic inactivation of microorganisms (PDI), also referred to as antimicrobial photodynamic therapy (aPDT), photodynamic antimicrobial chemotherapy (PACT) or photodynamic disinfection (PDDI), is a method of destroying microorganisms by inducing oxidative stress in their cells.[68–72] It is used to treat localized infections, as well as disinfection and sterilization.[73] The mechanism of action of PDI is based on the generation of reactive oxygen species (ROS), including singlet oxygen, which trigger a cascade of oxidation reactions within the microbial cell. The main molecules targeted by ROS are cell membrane proteins and lipids, as well as other cell wall components.[74,75] Such oxidative stress-induced damage is often irreversible and then ultimately leads to cell death.[73,76,77] The cascade of oxidative reactions can also lead to the inactivation of bacterial enzymes such as NADH dehydrogenase, lactate dehydrogenase, ATPase, and succinate dehydrogenase. This results in cell cycle inhibition and the death of the microorganism (Fig. 3).[75]

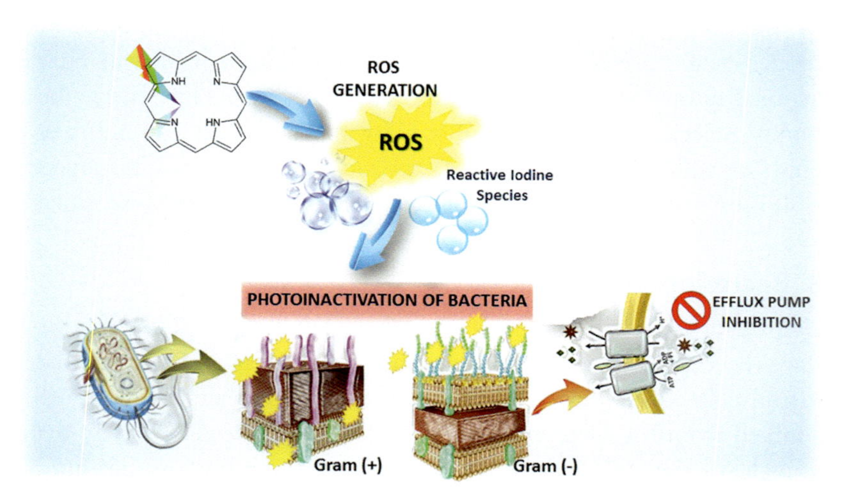

Fig. 3 Mechanisms of action of photodynamic inactivation (PDI) of resistant microorganism including efflux pump inhibition.

The diversity of photogenerated ROS and their high reactivity toward multiple biomolecules therefore makes PDI a multi-target approach with a non-specific mechanism of action, so that the development of bacterial resistance to this treatment regimen is highly unlikely. However, it should be recognized that bacterial cells are equipped with scavenger enzymes and various antioxidantsthat combat oxidative stress. These include, in particular, superoxide dismutase, catalase, cysteine and glutathione.[78] Nevertheless, bacteria have not yet developed an adequate defense system against singlet oxygen. An additional advantage is a possibility of modifying the molecular pathways of bacteria, e.g., by blocking proteins belonging to the efflux pumps system (Fig. 3). Thus, the effectiveness of PDI is influenced by the selection of the appropriate structure of the photosensitizer, its dose, as well as the light dose, so that the required ROS are generated at the right time in the right place with a sufficiently high efficiency.

3.6.1 Mechanisms of PDI

There are three individually non-toxic elements involved in PDI: a chemical compound called a photosensitizer, light from the visible or near-infrared (NIR) range of electromagnetic radiation, and molecular oxygen present within or around the bacterial cells. When a photon with energy tailored to this absorption band is absorbed, the PS is excited to one of the excited electronic states presented in Fig. 4. Among the processes illustrated in a

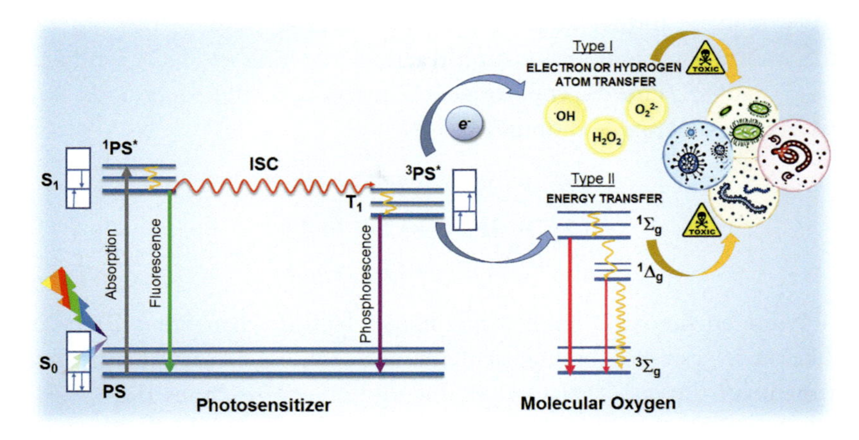

Fig. 4 Jablonski Diagram and mechanisms of ROS generation as crucial processes and reactions in photoinactivation of microorganisms.

Jablonski Diagram, intersystem crossing (ISC) is the most important transition in PDI, even if it is forbidden according to selection rules.[79] The occurrence of such a phenomenon is possible due to spin-orbital coupling. It occurs predominantly in the presence of a heavy atom (the so-called heavy atom effect).[80] The photosensitizer in the triplet excited state can transfer an electron/hydrogen atom (type I mechanism) or energy (type II mechanism) to the molecular oxygen in its ground state.[73,79] The photosensitizer then returns to the singlet ground state, while the oxygen molecule undergoes excitations and transformations to various ROS.[80] The transitions and possible reactions described above are illustrated in Fig. 4.

3.6.1.1 Type I mechanism

This mechanism involves the transfer of an electron (or hydrogen atom) to the π^*2p orbital of an oxygen molecule, resulting in the formation of an oxygen-centered radical and a cascade of subsequent redox reactions. The electron transfer to the oxygen molecule can occur directly (Eq. 1) from the excited photosensitizer or indirectly with the participation of a reducing agent (e.g., NADH, Eqs. 2 and 3). In both cases, a superoxide ion is formed.

$$PS^* + O_2 \rightarrow PS^{\bullet+} + O_2^{\bullet-} \quad (1)$$

$$2PS^* + NAD(P)H \rightarrow 2PS^{\bullet-} + NAD(P)^+ + H^+ \quad (2)$$

$$PS^{\bullet-} + O_2 \rightarrow PS + O_2^{\bullet-} \quad (3)$$

$O_2^{\bullet-}$ can be further converted to a perhydroxyl radical (Eq. 4). Both ROS undergo a disproportionation reaction, resulting in H_2O_2 production (Eqs. 5 and 6). The disproportionation reaction of the superoxide ion is catalyzed by superoxide dismutase.[79]

$$PS^{\bullet+} + O_2^{\bullet-} \rightarrow PS^{\bullet} + HO_2^{\bullet} \quad (4)$$

$$2O_2^{\bullet-} + 2H^+ \rightarrow O_2 + H_2O_2 \quad (5)$$

$$2HO_2^{\bullet} \rightarrow O_2 + H_2O_2 \quad (6)$$

Hydrogen peroxide has a much longer half-life than other ROS and, unlike others, can cross biological membranes, causing damage to other compartments of the cell.[81] H_2O_2 can undergo several reactions (Eqs. 7–9) to produce highly reactive hydroxyl radicals. The particularly important one is the so-called Fenton reaction which occurs in the presence of Fe^{2+} ions (Eq. 9). In the cellular environment, the Fe^{3+} ion can be reduced back to Fe^{2+} by the hydroxyl anion radical (Eq. 10).[79] The combination of these two reactions is referred to as the Haber-Weiss reaction.[82]

$$H_2O_2 + O_2^{\bullet-} \rightarrow HO^{\bullet} + O_2 + OH^- \quad (7)$$

$$H_2O_2 + HO_2^{\bullet} \rightarrow HO^{\bullet} + O_2 + H_2O \quad (8)$$

$$H_2O_2 + Fe^{2+} \rightarrow HO^{\bullet} + OH^- + Fe^{3+} \quad (9)$$

$$Fe^{3+} + O_2^{\bullet-} \rightarrow Fe^{2+} + O_2 \quad (10)$$

The hydroxyl radical formed by the aforementioned reactions readily oxidizes biologically important molecules, especially proteins, lipids, and carbohydrates. In addition, it can inactivate naturally occurring antioxidants, such as tocopherol.[79]

3.6.1.2 Type II mechanism

According to the following equations, this mechanism is driven by the direct transfer of energy from the excited photosensitizer in the triplet excited state to the oxygen molecule.

$$PS + h\upsilon \rightarrow {}^1PS^* \quad (11)$$

$${}^1PS^* \rightarrow {}^3PS^* \quad (12)$$

$${}^3PS^* + {}^3O_2 \rightarrow PS + {}^1O_2 \quad (13)$$

The presence of an unfilled π^*2p orbital in the singlet oxygen molecule is responsible for its reactivity toward electron-rich compounds. When exposed to ${}^1O_2$, lipids and the amino acids (tryptophan, tyrosine, histidine, methionine, cysteine, and cystine) are oxidized. These reactions are responsible for the cytotoxic effects of singlet oxygen.[79,83] The mechanisms described above can occur simultaneously, and the contribution of each depends on several factors such as the type of photosensitizer, its electronic structure, and photophysical properties, as well as the concentration of oxygen in the reaction medium.[84] Nevertheless, increasing evidence, including our work, suggests that mechanism II is more relevant for effective PDI because bacteria have not yet developed resistance to singlet oxygen.

3.6.2 Light excitation

Since ancient times, sun and light have always been associated with health, wellness, and vitality. The use of sunlight to treat, among others, tuberculosis was popular already in the 19th century. However, the 20th century brought the development of various therapeutic lamps, as well as other light sources. Before the invention of antibiotics, light, especially from ultraviolet radiation, was widely used to eradicate several types of microorganisms. In clinical practice, phototherapy refers not only to therapeutic, but also prophylactic and diagnostic applications of various ranges of electromagnetic radiation—from ultraviolet (UV) through visible, to near-infrared (NIR) light.[85]

Light is also a fundamental part of photodynamic therapy whereby it is most frequently associated with the concept of the "phototherapeutic window." It refers to photons of relatively low energy in the visible/near-infrared range corresponding to wavelengths in the 630–850 nm range.[73,86–88] This energy is high enough for efficient ROS generation but not as high to be harmful to the body. More importantly, photons in this range can penetrate much deeper into tissue than those in the blue range. Moreover, the light of wavelengths below 630 nm is absorbed by several endogenous chromophores, including hemoglobin and melanins.[79] This is actually more important in the anticancer approach, and the treatment of localized superficial infections usually does not require high penetration depth. Therefore, higher energy light such as blue light can be used in some of the applications discussed.[89]

Nevertheless, the use of longer wavelengths may be crucial for chronic infections that, in many cases, involve built-up microbial biofilms.[90] Bacterial biofilms are not only more difficult to destroy than planktonic forms, but also bacteria in a biofilm form are much more prone to become resistant to antimicrobial agents.[91] This reinforced strength of biofilm-growing bacteria against antimicrobials is related to differences in molecular mechanisms compared with the planktonic counterparts, e.g., horizontal gene transfer, genetic diversity, and alterations in gene expression while occurring only in a biofilm state. Moreover, drug diffusion is slowed down by the higher viscosity of the biofilm matrix due to the EPS network that is a 3D structure surrounding the bacteria within the biofilm and acts as a physically rugged barrier to protect the biofilm. The presence of the negatively charged EPS may protect the bacteria from the photodynamic action of positively charged photosensitizers (no sufficient attachment, no uptake to induce a photodynamic reaction after light activation). In this regard, the photodynamic inactivation procedure against biofilms generally requires longer preincubation times (up to 24 h), higher concentrations (up to 25 times), and light exposure times (up to 30 min) to reach relevant phototoxicity.

As already mentioned, PDT is used to treat localized infections, as well as disinfection and sterilization.[73] The PDI procedure leads to the generation of high amounts of ROS on the site of infection due to local irradiation. PDI can be repeated at least 25 times with the same microorganism (under subtherapeutic conditions to allow the survivors to grow back again) without significant loss of phototoxicity.[92] There are several potential milieus, in which PDI can replace or complement conventional antibacterial therapies for bacterial and fungal, viral, and parasitic infections. The easiest targets are: skin, oral and periodontal infections. Further development will allow for the treatment of infections where endoluminal illumination is possible, including nasal, ear, throat, lung and urinary tract infections (Fig. 5). It has been employed to treat acne vulgaris, biofilms associated with chronic periodontitis, burn infections, nasal decolonization of MRSA, surgical wound infections or infected wounds (e.g., venous, pressure, or diabetic ulcers).[72] PDI has proven to be an effective therapeutic strategy in the treatment of fungal (*Candida albicans*), viral (*Condylomata acuminata*, *Molluscum contagiosum*) and protozoan (*Leishmania*) diseases. [93]

The light sources employed in PDI should be suitably adapted to the areas to be illuminated. Nowadays there are more and more solutions in the form of flexible fiber optics that can adapt to different anatomical structures of the body, allowing uniform illumination of large areas. The cost of light sources, recently considered a limitation, is becoming lower due to the availability of LEDs characterized by relatively narrow emission bands and fluence rates of tens of mW/cm^2. In most preclinical protocols, this is sufficient to deliver light doses up to ca. $10\,J/cm^2$ within a few minutes of illumination.

3.6.3 Photosensitizers for PDI

The effectiveness of PDI depends, among others, on the selection of a suitable photosensitizer. A promising photosensitizer (PS) is characterized by

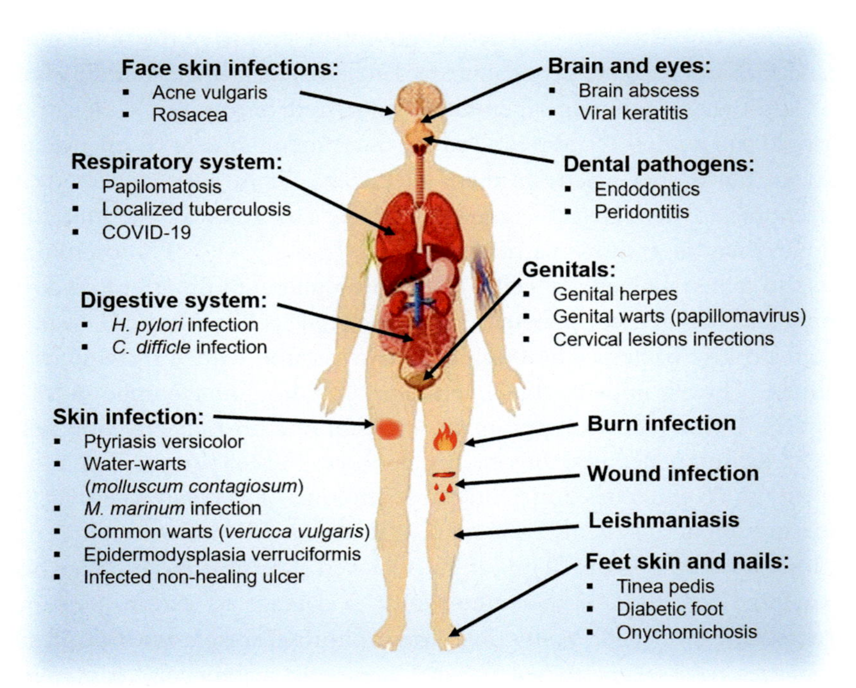

Fig. 5 Schematic illustration of sites of infection that are or could potentially be treated with PDI along with the characteristic types of microorganisms.

high purity, stability, low toxicity in the dark, and high molar absorption coefficient in the visible or near-infrared light range.[79] Another important aspect to be considered when selecting a PS for PDI is to ensure the appropriate interaction of the PS with the bacterial cell. The PS can: (i) bind/interact with the bacterial cell membrane, (ii) penetrate the bacterial cell, and (iii) affect the bacterial cell without direct contact if $^1O_2$ is generated with high efficiency.[94]

The mechanism of PS binding to the bacterial cell membrane differs for Gram-positive and Gram-negative bacteria. This is due to the differences in cell wall structure described before. The cell wall of Gram-positive bacteria, made up of a thick layer of porous peptidoglycan, is relatively easy for PSs to cross, making it easy for them to reach the cell membrane, which is their site of action. The situation is quite different in the case of Gram-negative bacteria (Fig. 1). Gram-negative bacteria are equipped with an additional membrane, a physical and functional barrier for PSs, making it difficult to reach the cell membrane. Thus, many PSs show greater efficacy in inactivating Gram-positive than Gram-negative bacteria. Positively charged PSs generally lead to the more effective inactivation of Gram-negative bacteria.[89,94]

They bind to the negatively-charged phosphate groups of the outer membrane and, when excited, contribute to damage to membrane-building lipids and proteins, including membrane-associated enzymes.[95] Negatively charged and neutral PSs also bind to the outer membrane of Gram-negative bacteria but do not inactivate them effectively. Gram-positive bacteria are susceptible to both negatively and positively charged PSs. Neutral compounds may also show particularly good results.[89,94,96] Photosensitizers can also penetrate bacterial cells. The hydrophilicity of the PS often determines the ability to penetrate the bacterial cells. The cell wall of Gram-positive bacteria is hydrophilic due to its carbohydrate and amino acid content. Thus, it impedes the penetration of hydrophobic compounds into the cell.[94] There are also reports of successful microbial inactivation when the PS has not penetrated the cell or is bound to the cell surface. This is possible if the PS generates sufficiently large amounts of singlet oxygen. Its lifetime in biological systems averages 0.04 μs[97], roughly corresponding to an action radius of 0.02 μm. Thus, if PS is located at or near this distance from a bacterial cell, there is the opportunity to damage its external, essential structures.[94] In recent years, many new photosensitizers with promising properties have been studied.[88,98–104] Synthetic macrocyclic compounds such as porphyrins, chlorins, bacteriochlorins, and phthalocyanines are of greatest interest.[73,86] Fig. 6.

Porphyrins are a class of heterocyclic aromatic compounds constituted by four subunits of pyrrole type linked by methynic bridges. Synthetic meso-aryl-substituted porphyrins are particularly versatile starting materials to design new PSs as either ionic or nonionic moieties can be equally positioned on the periphery of the tetrapyrrole ring, thus modulating the photosensitizer polar character.[73,101,103,105] In order to obtain PSs with significantly longer absorption wavelength for optimal penetration in tissue, reduction of a single pyrrole double bond on the porphyrin periphery affords the chlorin core, and further reduction of a second pyrrole double bond on the chlorin periphery gives the bacteriochlorin derivative (Fig. 6). Therefore, both classes of molecules possess electronic absorption at longer wavelengths (λ_{max} = 650–670 nm for chlorins and λ_{max} = 730–800 nm for bacteriochlorins) than porphyrins and yet remain efficient ROS generators.[79,86,106–108] Phthalocyanines also exhibit several features making them potentially good photosensitizers for PDI.[109,110] They are characterized by high molar absorption coefficients within the phototherapeutic window, low toxicity in the dark, high photo- and thermal stability, and the ability to generate singlet oxygen efficiently due to the high quantum yield of the triplet state and its relatively long lifetime. The properties of phthalocyanine

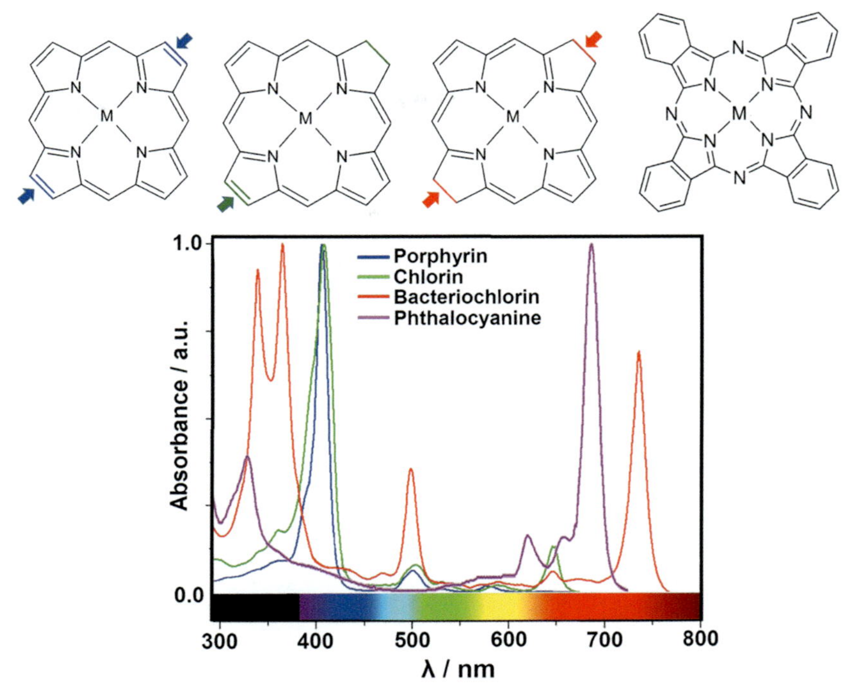

Fig. 6 The general structures of porphyrins, chlorins, bacteriochlorins, and phthalocyanines—the macrocyclic photosensitizers that are promising candidates for PDI, and their electronic absorption spectra.

derivatives depend largely on the nature of the coordinated metal ion.[111] Moreover, the introduction of specific functional groups into their structure at the axial and peripheral positions allows control of their solubility in biological media.[112] Unsubstituted phthalocyanines are characterized by very high hydrophobicity, resulting in poor solubility in the physiological environment and the tendency to form aggregates. The introduction of hydrophilic groups into their structure, e.g., carboxylic or tertbutoxysulfonyl groups, reduces this phenomenon, at least partially.[113–115] However, this approach is sometimes insufficient. Then it becomes necessary to use drug delivery systems, such as liposomes, micelles, microemulsions or nanocrystalline TiO_2, because aggregation of photosensitizer adversely affects the efficiency of ROS generation (Fig. 7).[114,116,117]

3.6.4 Tetrapyrrolic derivatives as antibacterial photoagents—a proof of concept examples

The scientific interests of our research group are broadly focused on the use of modified halogenated (metallo)tetrapyrroles for both photodynamic

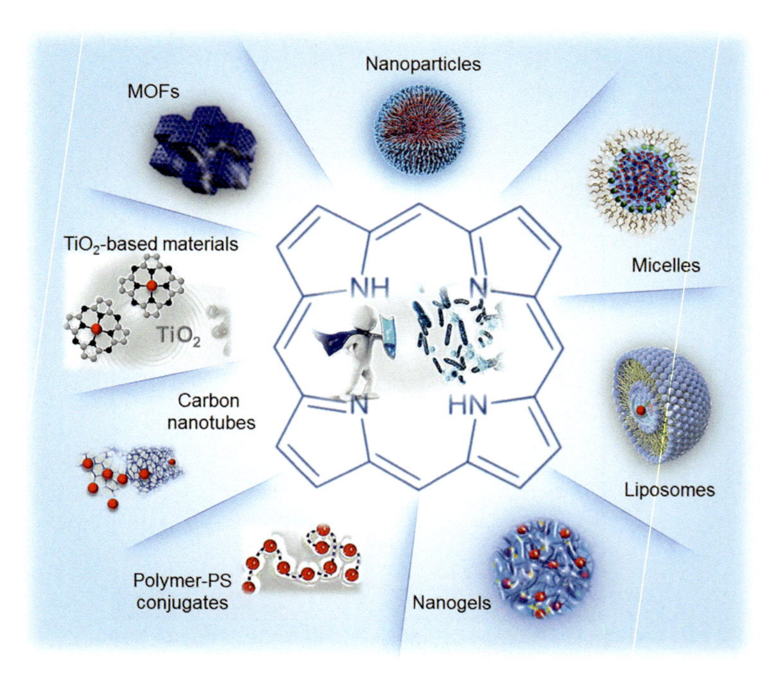

Fig. 7 Types of drug delivery systems and materials that can be used to improve photosensitizer' pharmacological optical and photosensitizing properties.

therapy of cancer (PDT) and PDI. Recently, we have developed the library of synthetic halogenated porphyrins and their derivatives.[118,119] The preparation of these compounds consists of several steps: (i) the synthesis of appropriate porphyrins by a modified nitrobenzene method that involves the condensation of pyrrole with the desired aromatic aldehyde in the presence of acetic acid and nitrobenzene; (ii) chlorosulfonation to the corresponding chlorosulfophenylporphyrins; (iii) hydrolysis or nucleophilic attack with amines to yield either hydrophilic sulfonated or amphiphilic sulfonamide halogenated porphyrins.[99,119–121] The key advantages are: efficiency, simplicity, low environmental impact and the ability to provide a library of multigram pure, stable and versatile UV–Vis/NIR absorbing dyes with favorable physicochemical properties.[118,120–125] In addition to the effect on the photophysical properties of these porphyrin derivatives, the introduction of halogen atoms into the structure also plays an important role in pharmacological properties. The following possible modification of the new halogenated tetrapyrroles involves functionalizing with various peripheral groups and coordinating with certain metal ions (i.e., Zn^{2+}, Pd^{2+}).[73,103] Substitution of polarity-tunable groups with higher electron-receptor properties allows control of the hydrophobicity of molecules and increases their stability by the effect of steric protection. It affects its interaction with biological membranes and biologically important molecules, leading to a remarkable increase in the photodynamic efficacy of such photosensitizers.[98,100,106,126–128]

The unprecedented success of our collaboration with the group of Professor Arnaut and Professsor Mariette M. Pereira from the University of Coimbra (Portugal) resulted in the development of Redaporfin (NCT02070432).[87,129–133] During the last decade, we have also concentrated our scientific efforts on the design of effective tetrapyrrolic-based photosensitizers for antimicrobial PDT. We mainly focus on antibiotic resistance, which contributes to one of the leading healthcare problems in a clinically vicious cycle related to the overuse of antibiotics and the development of multidrug resistant microbes. The use of higher doses of antibiotics to fight infections also leads to their increased toxicity to the human body. A consequence of this is that emerging superbugs could develop the ability to resist commonly prescribed medications. As a further consequence, new antibiotic development is not as successful. Therefore, developing novel alternative antibacterial strategies such as photodynamic inactivation still remains a top challenge for scientists worldwide.

The photosensitizers commonly used in antimicrobial therapy are phenothiazine dyes (toluidine blue O, methylene blue) that absorb radiation in the 600–700 nm range. In an aqueous solution, phenothiazines have a positive charge that provides the desired affinity for bacterial cell walls. The FDA has approved them for the therapy of periodontal diseases (at a concentration of 0.01% (0.1 mg/mL).[95,134,135] A group of our tetrapyrrolic photosensitizers for PDI, appropriately modified, could also be targeted against G(+), G(−) bacteria (groups with a positive or negative charge), and fungi (sulfonamide substituents) and to show selectivity toward microorganisms cells in comparison to normal skin cells.[73,86] For instance, by investigating a series of modified tetraphenylporphyrin derivatives in the context of PDI, we have shown that the substitution of halogen atoms (—F or —Cl) into the peripheral phenyl rings increases the ISC leading to enhanced photophysical properties. Moreover, the further introduction of sulfonic or sulfonamide groups provides water solubility that moderates their possible interactions with biological membranes. All of these substituents also prevent PS from aggregation and increase their photostability.

It was reported that the sulfonated porphyrin derivatives (e.g., 5,10,15,20-tetrakis(2,6-difluorosulfonylophenyl)porphyrin [F_2POH] and 5,10,15, 20-tetrakis(2,6-dichlorosulfonylophenyl)porphyrin [Cl_2POH]) as well as sulfonamide derivatives (e.g., 5,10,15,20-tetrakis(2,6-dichloro-3-*N*-ethyl-sulfamoylphenyl)porphyrin [Cl_2PEt]) combine properties mentioned above with low dark cytotoxicity and efficient accumulation in both cancer and microbial cells. However, the photobiological properties in vitro (i.e., cellular uptake, ROS generation, and photodynamic efficacy) of hydrophilic F_2POH were significantly increased after its encapsulation in Pluronic L121 micelles.[89] PDI experiments showed that halogenated porphyrins investigated by us indicate various antimicrobial activity after a short incubation time and 10 J/cm^2 blue light irradiation. As expected, due to the differences in the microbial cell wall structure, the complete eradication of Gram-negative *E. coli* was more difficult to achieve than that for the Gram-positive species. The PDI susceptibility of *C. albicans* was intermediate between both types of microorganisms. Sulfonic acid derivatives were shown to inactivate effectively planktonic *S. aureus* related to the negative molecular charge of these PS derived from the peripheral sulfonic groups (–SO_3H). On the other hand, only the sulfonamide conjugate (Cl_2PEt) indicated the synergistic effect of chemo- and photodynamic inactivation and was toxic to fungal yeast in the dark (ca. one order of magnitude extra killing). Thus, these results clearly show that the molecular design of porphyrin-based PSs, including their substitution patterns and formulation, may improve PDI efficacy toward each species of microorganisms.[89]

In recent years, PDI with porphyrin derivatives has been proposed as an alternative treatment for localized infections in response to the ever-growing problem of antibiotic resistance.[25] Among the various molecular and biochemical mechanisms of antibiotic resistance, active efflux of antibiotics in bacteria plays an important role in both intrinsic and acquired multidrug resistance of clinical relevance. It also interplays with other resistance mechanisms, such as the membrane permeability barrier, enzymatic inactivation/modification of drugs, and/or antibiotic target changes/protection, in significantly increasing the levels and profiles of resistance. Thus, in our work, we also reported that PDI could be more effective by combining the PS with an efflux pump inhibition (EPI).[28,31,136]

The current efforts to make the "ideal photosensitizer" focus on binding and permeability through the bacterial membrane, its accumulation at the cytoplasmic membrane, and possessing high quantum efficiency for generating singlet oxygen or other ROS. Therefore, PSs used in PDI may include cationic, neutral, and anionic *meso*-tetraphenylporphyrins with targeted antimicrobial properties. The presence of a negatively or positively charged group alters the physicochemical properties of PSs, making them more amphiphilic and interfering with the balance between hydrophilic and hydrophobic moieties with modulation of cell membranes permeability and, overall, the PDI effect. Our recent work demonstrated that negatively charged sulfonyl porphyrins (TPPS and Cl_2TPPS) had increased cellular uptake in Gram-positive bacteria. In contrast, positively charged TMPyP is required for proper electrostatic adhesion to, or penetration of Gram-negative bacterial cell walls.[137] (Fig. 8A and B). Moreover, the introduction of halogen atoms stabilizes the porphyrin macrocycle and increases activity due to the high singlet oxygen quantum yield ($\Phi\Delta = 0.95$). This study shows that the most appropriate PDI photosensitizers should be charged, water-soluble, and photostable. Nevertheless, an affinity for the microbial cell wall is an essential factor affecting PDI performance. Low diffusion of the PS limits the efficacy of PDI against the fungal yeast into the cytoplasm due to their cell walls, which also contain β-glucan. Thus, the positively charged PS or the addition of substances that can significantly increase the permeability of the outer membrane and may enhance the inactivation efficiency.

The other challenge for effective PDI is the ability to overcome antibiotic resistance. Many studies have revealed that a PDI + EPI combination may result in (i) increase in the PS cellular uptake that often is excreted by the efflux system; (ii) reduced antibiotic/PS resistance; (iii) diminish the resistance mechanism derived from efflux pumps overexpression and (iv) decrease the development of more resistant superbugs. Therefore, we combined porphyrin-based PDI with verapamil (a well-known efflux pump inhibitor) (Fig. 8).

We indicated that the combination of PDI with verapamil reduced the Gram-negative bacteria survival—for *E. coli*, by even three orders of magnitude of extra killing. For *S. aureus*, the verapamil addition did not influence the PDI efficacy, and the bactericidal effect (six orders of magnitude reduction, >99.9999% killed bacteria) was achieved after each PDI mode. Nevertheless, the investigated PSs did not yield significant activity against *E. coli* after PDI with low light doses (up to 20 J/cm^2). Thus, KI was used, leading to strong potentiation of TPPS-, Cl_2TPPS-, and TMPyP-mediated PDI (three to four orders of magnitude of extra killing), suggesting the contribution of hypoiodite and other reactive iodine species to the PDI mechanism. Due to the high stability of halogenated PS, the PDI might also be

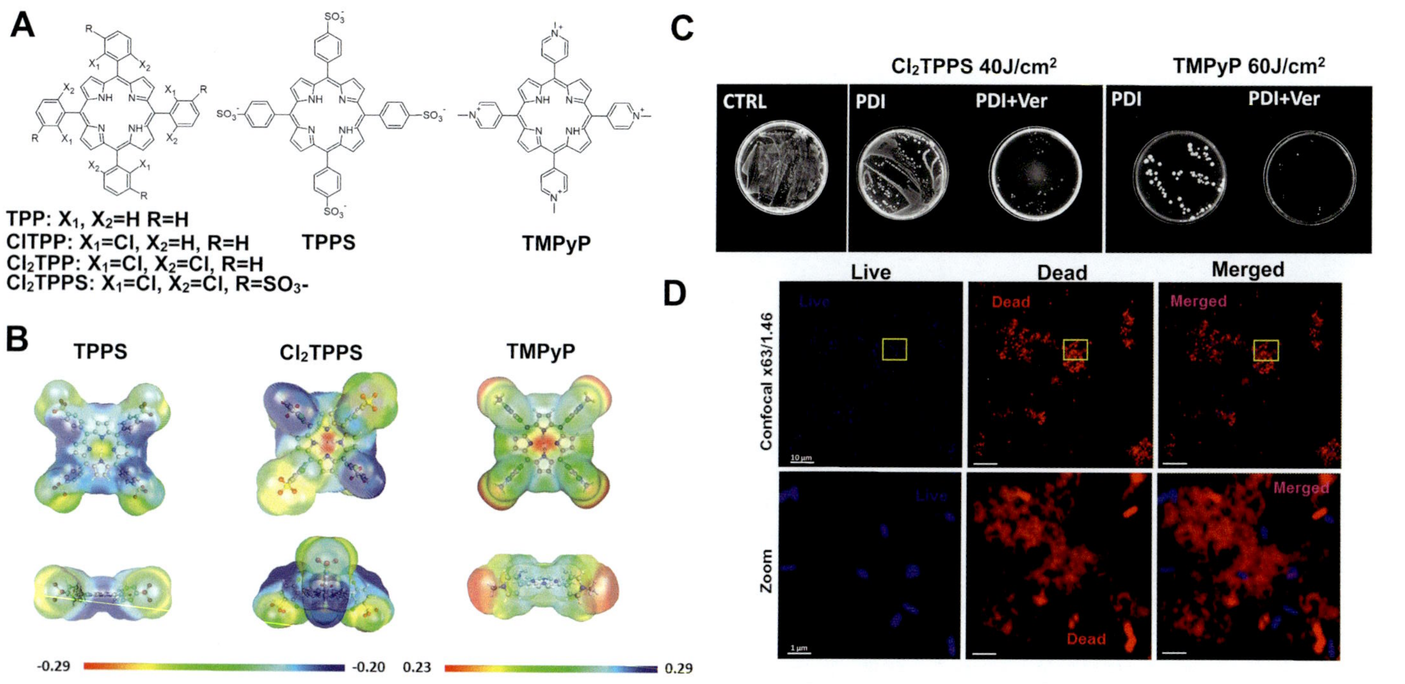

Fig. 8 Porphyrin-mediated PDI—the effect of charge and lipophilicity on their biological activity in vitro: (A) the chemical structures of tetraphenylporphyrin derivatives with various charge; (B) electronic density maps of the photosensitizers from the total self-consistent field density mapped with the electrostatic potential; (C) the effect of the efflux pump inhibition on high-dose porphyrin-mediated PDI with and without verapamil (Ver) addition; (D) the antibacterial effect of PDI confirmed by confocal microscopy images of the non-treated PDI-treated *E. coli*.[130]

performed in higher light doses (up to 120 J/cm^2), but satisfactory results were obtained after using 60 J/cm^2. Thus, these data suggest that for high light-dose-PDI, its combination with EPI may enhance the inactivation. However, for low light-dose-PDI, the KI addition with the generation of reactive iodide species gives more promising results.[137]

Keeping in mind the importance of the molecular charge of a PS, we also described the synthesis of a family of cationic tetra-imidazolyl phthalocyanine-based photosensitizers with varying structural features including the size of the cationized alkylic chain, degree of cationization, and the type of central coordinating metal (Fig. 9).[109]

The influence of these structural modifications on their biological performance was assessed based upon their spectroscopic and photophysical properties, as well as ROS generation. The antimicrobial activity tests (PDI with white light) reveal the remarkable differences in efficacy between tested metallophthalocyanines. Some examples were highly active, especially in killing Gram-negative bacteria (*E. coli*, *P. aeruginosa*) and fungi (*C. albicans*). Among others, the Zn(II) tetra-ethyl cationic phthalocyanine derivative showed PDI activity against tested Gram-negative bacteria and fungi, with six orders of magnitude reduction in viability of those species at PS concentrations as low as 100 nM and 1 μM, respectively.[109] The cationic PS evaluated in this study showed low toxicity toward human keratinocytes cells (HaCaT) at concentrations effective in killing Gram-negative bacteria, which raises the potential for in vivo utility for developing treatments of localized infections.

3.6.5 Porphyrin-based hybrid materials

The next-generation of nanomaterials that can be activated with visible light represent an exciting new step in progress toward PDI. Furthermore, the surface chemistry of the nanomaterials also plays a crucial role in PS uptake, which can be modulated through the addition of different targeted substituents. While there is significant research on PSs accumulation in mammalian cells, the cellular uptake pathways in bacteria and fungi are less well studied. Cellular uptake of metal nanomaterials can occur when the materials are sufficiently small to cross the cellular membrane. In the case of mammalian cells, it has been suggested that particles below 100 nm are most efficient for cellular uptake. Moreover, some metal oxides and semiconductors (e.g., ZnO, TiO_2) may be functionalized with tetrapyrroles toward better antimicrobial activity. This group of hybrid materials was found to be very stable in the biological environment and active in the PDI against a broad

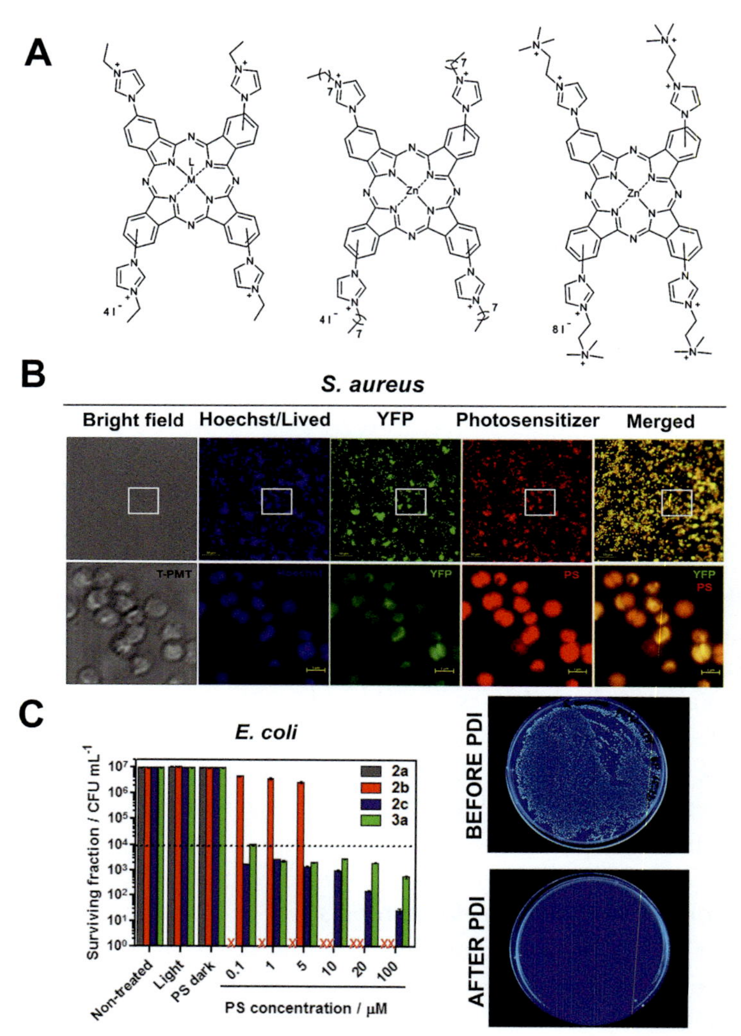

Fig. 9 Metallophthalocyanines as promising antimicrobial agents for PDI: (A) the structures of modified imidazolium metallophthalocyanines and their biological activity: (B) cellular accumulation in bacteria investigated by confocal microscopy imaging and (C) bacteria survival curves and Petri dishes images before and after PDI.[102]

spectrum of microbes. Especially, the physicochemical properties of TiO_2 ensure the effective interaction between its surface and porphyrin derivatives. It results in a suitable absorption profile and the possibility of using visible light to photoinactivate the bacteria more efficiently than employing the corresponding porphyrin alone.[25,73,138,139] TiO_2 has been successfully used in heterogeneous photocatalysis. It owes its popularity mainly to high photocatalytic activity, high chemical and photochemical stability, low toxicity, and low cost.[140] The possibility of using TiO_2 for photocatalytic water and air purification is being intensively studied. Photocatalysis, based on the oxidation of pollutants by ROS generated by excited TiO_2, is classified as one of the so-called advanced oxidation technologies (AOTs), which are an alternative to the commonly used water treatment methods.[115] The properties of TiO_2 mentioned above, its biocompatibility and unique surface properties as well as effective ROS generation, make it attractive as a nanomaterial also for biomedical applications, including PDI.[116] However, a factor that significantly limits the clinical application of TiO_2 is its large energy gap (3.2 eV). Its absorption, therefore, does not fall within the phototherapeutic window. One strategy to cope with this problem is to impregnate TiO_2 with organic dyes, including porphyrins and phthalocyanines.[141] Due to their redox properties, phthalocyanines are very well suited for sensitizing broadband semiconductors.[115] The process of excitation of TiO_2 modified with organic dyes proceeds as follows: upon irradiation, the dye (e.g., porphyrin) is first excited and $e-$/dye$+$ pairs are generated, followed by electron transfer from the dye to the conduction band of TiO_2.[117] The excited semiconductor reacts with molecular oxygen, water, and hydroxyl ions in the immediate environment, resulting in ROS generation and eventually leading to oxidative stress in cells.[115] The fabrication of a dye-TiO_2 hybrid material also carries other benefits. It may help overcome the limitations that diminish the widespread use of phthalocyanines in clinical practice: their high hydrophobicity, low solubility in biological media, and low penetration into cells.[141] The synergistic action of dyes and TiO_2 may also help to enhance the photodynamic effect and produce more effective photomaterials than the compound alone.[142] In addition, the fluorescent properties of porphyrins and phthalocyanines may allow the use of such hybrid materials in bioimaging.[126,140,143,144]

Unfortunately, the bandgap of bulky TiO_2 lies in the UV range (3.2 eV for anatase and 3.0 eV for rutile, respectively), which may be intrinsically harmful and carcinogenic. Among the various dyes, particular attention should be paid to transition metal complexes with macrocyclic ligands (mainly metallophthalocyanines and metalloporphyrins) that are capable of injecting electrons from their elctronic excited states into the conduction band of titanium dioxide.[142] Such hybrid TiO_2-based materials may be obtained via impregnation of the semiconductor surface with the sulfonic porphyrins due to the presence of -SO_3H groups, which interact directly with the surface of TiO_2 (P25), Fig. 10.[138] The photophysical properties

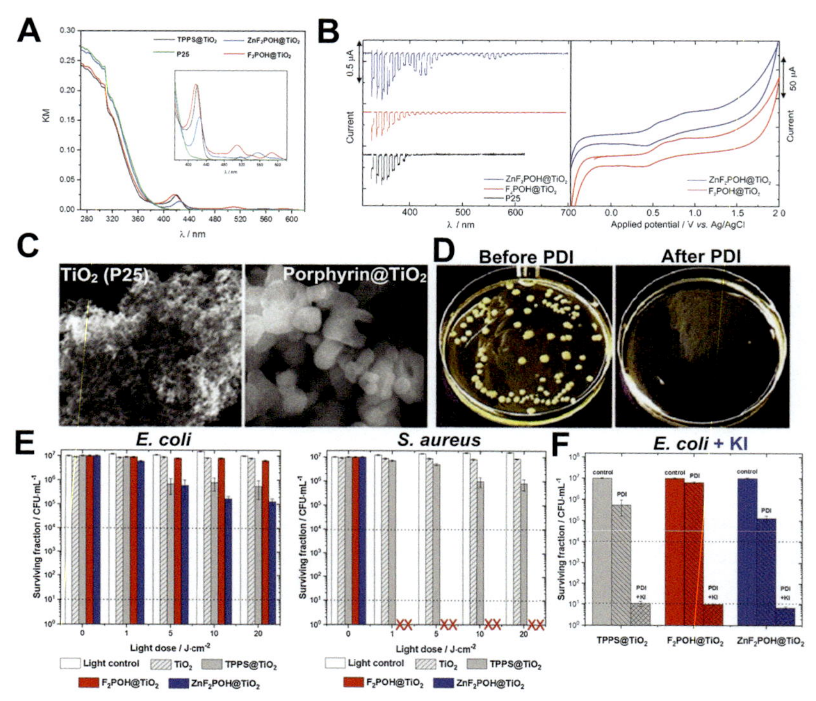

Fig. 10 Antimicrobial activity of (metallo)porphyrin@TiO_2 hybrid materials: (A) diffuse reflectance spectra in the representation of the Kubelka-Munk function of bare P25 and surface modified TiO_2 materials; (B) photocurrent generation of TiO_2 and TiO_2 with adsorbed (metallo)porphyrins and their cyclic voltammograms; (C) SEM images of porphyrin modified TiO_2 surface; (D) petri dishes before and after PDI; (E) photodynamic inactivation of *E.coli* and *S. aureus* mediated by TiO_2-based hybrid materials under visible light irradiation and (F) the potentiation effect of KI on the phototoxicity of TiO_2-based hybrid materials against *E. coli.*[131]

(derived from, e.g., spectroscopic data and photocurrents measurements) confirmed the electron transfer from the excited PS molecule to the valence band (VB) of TiO_2. It has been shown that the modification of TiO_2 with halogenated (metallo)porphyrins—difluorinated sulfonic derivative F_2POH and its complex ZnF_2POH moved the absorption to the visible part of the electromagnetic radiation (>400 nm) and consequently improved the photoproperties and photoactivity mediated by these hybrid materials. They indicated high antimicrobial activity, which can be further potentiated by KI addition. The most important feature of their increased PDI efficacy toward bacteria derived from their ability to generate various ROS (singlet oxygen and oxygen-centered radicals) and generation of reactive iodide species upon visible light irradiation, because only materials able to act via multiple mechanisms prove the best efficiency against drug-resistant microorganisms.[138,139]

Porphyrins and metalloporphyrins, mainly with sulfonyl groups that can act as anchors, thus improving the interaction between the semiconducting support and the sensitizer, were also successfully used to sensitize nanostructured $qTiO_2$. Well designed nanomaterials can provide better affinity toward microbial cells over host cells, enhance ROS formation, and overcome antibiotic-resistance mechanisms. Highly active surface-modified nanoparticles were synthesized and fully characterized in our laboratories. The morphology and crystal structure of the (metallo)porphyrin@$qTiO_2$ materials obtained were examined by several techniques, including absorption and fluorescence spectroscopies as well as scanning electron microscopy (SEM) imaging (Fig. 11).[139]

These 20–70 nm-sized, modified TiO_2 nanoparticles were found to be effective ROS generators under blue light irradiation (420 ± 20 nm). Their PDI potency against Gram-negative and Gram-positive bacteria was also investigated and revealed that the PDI with porphyrin@$qTiO_2$ nanoparticles (1 g/L) and irradiation with a light dose of $10\,J/cm^2$ led to reduced bacteria survival up to three logarithmic units for *E. coli* and up to four logarithmic units for *S. aureus*, respectively. A more pronounced antibacterial effect was observed when PDI was potentiated with H_2O_2 or KI, leading to complete eradication of tested species, even with lower concentrations of nanomaterials (starting from 0.1 g/L). Imaging of bacteria morphology after PDI suggested distinct mechanisms of cell destruction, which depends on the type of generated ROS and/or reactive iodine species. These data indicate that TiO_2-based materials modified with sulfonated porphyrins are promising photoagents that may be useful in several biomedical strategies, especially in PDI. Taken together, all of the studies we performed clearly show that hybrid porphyrin-based nanomaterials represent bioinorganic photoactive agents with proven biological activity. Notably, the in vitro antibacterial efficacy makes them worth further investigation in vivo for the treatment of localized infections and wound-healing therapy.

4. Summary and future perspectives

The era of antibiotics is inevitably coming to an end and even mild infections can be potentially deadly in the future. According to the WHO report, results obtained for *E. coli*, *Klebsiella pneumoniae* and *S. aureus* showed

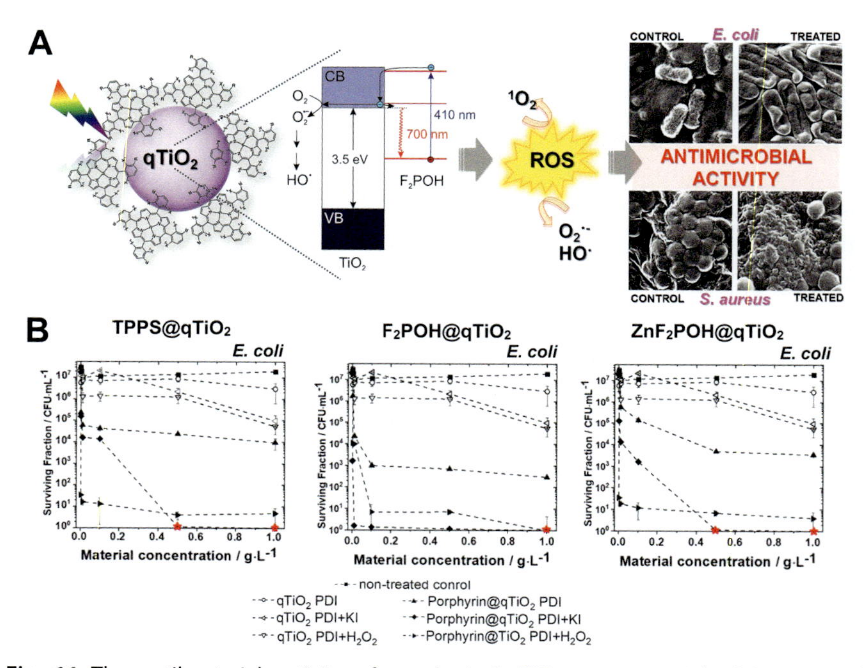

Fig. 11 The antibacterial activity of porphyrin@qTiO_2 nanomaterials: (A) Proposed mechanism of the TiO_2 photosensitization, and possible pathway of ROS generation after excitation of porphyrin@qTiO_2 nanomaterials with their antimicrobial activity. (B) Photodynamic inactivation of *E. coli* and *S. aureus* mediated by qTiO_2-based hybrid materials (porphyrin@qTiO_2) and KI and H_2O_2 PDI potentiation after 2 h incubation and irradiated by 420 ± 20 nm LED light.[132]

that the proportion resistance to commonly used antibacterial drugs exceeded 50% in many settings.[65,145,146] The most common cause of surgical wound infections is *S. aureus* a Gram-positive bacterium that can be a part of the normal flora on the skin. Strains of *S. aureus* resistant to penicillins, including ampicillin and amoxicillin, have acquired a gene (mecA) that codes for a novel penicillin-binding protein with low affinity for β-lactams and thus, leads to the resistance. These strains are termed methicillin-resistant *S. aureus* (MRSA) and are resistant to all penicillins. The future perspectives are rather daunting, due to the low chances of developing new antibacterial drugs as a result of the difficulty of finding alternative mechanisms of action for the antibiotics. The number of infections by multidrug-resistant bacteria in Europe was 400,000 in 2007, and there were 25,000 attributable deaths.[147] Unless proper action is taken, this number can reach 10 million by 2050. Antibiotic resistance is exacerbated in bacteria biofilms where gene mutation frequency is significantly increased compared to planktonically growing isogenic bacteria. However, antibiotic resistance mechanisms are not sufficient to explain most cases of antibiotic-resistant biofilm infection. The same bacteria can become up to 1000 times more resistant to antibiotics when grown in biofilm.[39] The biofilm matrix reduces the bioavailability of antibiotics by physical and chemical processes. The penetration of the antibiotics is impaired, leading to diffusion coefficients that are a factor of 2–3 lower than in pure water, and the chemistry of the biofilm microenvironment alters the growth of the bacteria and the antibiotic potency. The reduced bioavailability of the antibiotics and the diversity of the microenvironment enable a timely adaptive response of some biofilm cells to an antimicrobial challenge. Finally, biofilms favor the occurrence of subpopulations of cells that entered a highly protected, spore-like state—the persisters.[148] Thus, there is an urgent need to find alternatives to deal with microbial infections. Ideally, such alternatives should minimize the risk of developing new resistance mechanisms, spare healthy tissue, and stimulate the host immune system to deal with the infection. The conventional treatment consists in the administration of topical and systemic antibiotics for long periods of time and may be responsible for the increased microbial strains resistant to available drugs. Therefore, it is important to develop a treatment regimen specifically targeting pathogens without causing any side effects on the host tissue.

The compounds that are used in PDT are characterized by strong absorption in the phototherapeutic window, long triplet state lifetime, high ROS generation quantum yields, no toxicity in the dark, and favorable pharmacokinetics.[70,149–152] The most frequently studied photosensitizers for PDT are porphyrins, chlorins, bacteriochlorins or phthalocyanines—organic compounds generally characterized by a high molecular weight. The ideal photosensitizer/material for PDI of microorganisms should rather have a low molecular weight/small size, a strong light absorption in the visible and appropriate lipophilicity to facilitate the diffusion of non-aggregated molecules in the biofilms. Additionally, the ideal PS for PDI may require the presence of positive charges to inactivate Gram-negative bacteria. Biofilms may reduce the amount of oxygen available for PDI, but this can be compensated with PSs with long triplet lifetimes and high rates of triplet-state quenching by molecular oxygen. Finally, the photostability of the PS is critical to the success of efficient photodynamic treatment. It is therefore important to find the appropriate balance between the potential toxicity of the material itself, photostability and efficiency of ROS generation when developing photosensitizers/materials for PDI application.

This brings up another difference between anticancer and antitumor photodynamic treatments. Our ongoing studies have revealed that in order to achieve adequate therapeutic success manifested not only by the destruction of the primary tumor but also by controlling distant metastases, Mechanism I seems to be more efficient, because such photogenerated oxygen-centered radicals are well-known inflammatory mediators and guarantee an adequate immune response. Whereas the opposite is true for PDI and Mechanism II is more preferred, for the simple reason that bacteria have not yet developed resistance to the main product of photoinduced energy transfer—singlet oxygen. Looking ahead, it may be possible in the near future to apply PDI using the compounds and materials described in this paper to treat infections of Gram-positive and Gram-negative multidrug-resistant bacteria in relevant animal models of infected burns and ulcers. Compared to other strategies mentioned in this paper, PDI is safe, cheap, and can be repeated hundreds of times without significant loss of efficacy. We believe that with the current development of nanotechnology and the availability of light sources, PDI can become the first line of treatment for localized infections. Moreover, PDI can reduce antibiotic use and make hospital infections more manageable through the disinfection of surface-modified medical devices.

Acknowledgments

This research was funded by the National Science Center (NCN), Poland, Sonata Bis grant number 2016/22/E/NZ7/00420 given to J.M.D.

References

1. Lesho, E. P.; Laguio-Vila, M. *Mayo Clinic Proceedings*, Vol. 94; Elsevier, 2019; pp. 1040–1047.
2. Tagliabue, A.; Rappuoli, R. *Front. Immunol.* **2018**, *9*, 1068.
3. Abadi, A. T. B.; Rizvanov, A. A.; Haertlé, T.; Blatt, N. L. *BioNanoScience* **2019**, *9* (4), 778–788.
4. Bryan-Wilson, J. *Artforum Int.* **2016**, *54* (10), 113–114.
5. Kollef, M. H.; Hamilton, C. W.; Ernst, F. R. *Infect. Control Hosp. Epidemiol.* **2012**, *33*, 250–256.
6. Klein, E. Y.; van Boeckel, T. P.; Martinez, E. M.; Pant, S.; Gandra, S.; Levin, S. A.; Laxminarayan, R. *Proc. Natl. Acad. Sci. U. S. A.* **2018**, *115*, E3463–E3470.
7. O'Niell, J. M, Infection Prevention, Control and Surveillance: Limiting the Development and Spread of Drug Resistance, Government and the Wellcome Trust, London, 2016.
8. Perichon, B.; Courvalin, P.; Stratton, C. *The Desk Encyclopedia of Microbiology*; Elsevier, 2009.
9. Lima, L. M.; da Silva, B. N. M.; Barbosa, G.; Barreiro, E. J. *Eur. J. Med. Chem.* **2020**, 112829.
10. Sobhanifar, S.; King, D. T.; Strynadka, N. C. *Curr. Opin. Struct. Biol.* **2013**, *23* (5), 695–703.

11. Falagas, M. E.; Athanasaki, F.; Voulgaris, G. L.; Triarides, N. A.; Vardakas, K. Z. *Int. J. Antimicrob. Agents* **2019**, *53* (1), 22–28.
12. Allen, N. E.; Nicas, T. I. *FEMS Microbiol. Rev.* **2003**, *26* (5), 511–532.
13. Rice, L. B. *Mayo Clinic Proceedings*, Vol. 87; Elsevier, 2012; pp. 198–208.
14. Tenson, T.; Lovmar, M.; Ehrenberg, M. *J. Mol. Biol.* **2003**, *330* (5), 1005–1014.
15. Samanta, I.; Bandyopadhyay, S. Antimicrobial Resistance in Agriculture, Academic Press, 2019.
16. Krause, K. M.; Serio, A. W.; Kane, T. R.; Connolly, L. E. *Cold Spring Harb. Perspect. Med.* **2016**, *6* (6), a027029.
17. Koteva, K.; Cox, G.; Kelso, J. K.; Surette, M. D.; Zubyk, H. L.; Ejim, L.; Stogios, P.; Savchenko, A.; Sørensen, D.; Wright, G. D. Cell Chem. Biol. 2018, 25 (4), 403–412. e405.
18. Rodríguez-Martínez, J. M.; Cano, M. E.; Velasco, C.; Martínez-Martínez, L.; Pascual, A. *J. Infect. Chemother.* **2011**, *17* (2), 149–182.
19. Bakthavatchalam, Y. D.; Pragasam, A. K.; Biswas, I.; Veeraraghavan, B. *J. Glob. Antimicrob. Resist.* **2018**, *12*, 124–136.
20. Ashrafuzzaman, M.; Andersen, O.; McElhaney, R. *Biochim. Biophys. Acta (BBA)-Biomembr.* **2008**, *1778* (12), 2814–2822.
21. Wright, G. D. *Nat. Rev. Microbiol.* **2007**, *5* (3), 175–186.
22. Fuda, C.; Suvorov, M.; Vakulenko, S. B.; Mobashery, S. *J. Biol. Chem.* **2004**, *279* (39), 40802–40806.
23. Gordon, E.; Mouz, N.; Duee, E.; Dideberg, O. *J. Mol. Biol.* **2000**, *299* (2), 477–485.
24. Lambert, P. A. *Adv. Drug Deliv. Rev.* **2005**, *57* (10), 1471–1485.
25. Regiel-Futyra, A.; Dąbrowski, J. M.; Mazuryk, O.; Śpiewak, K.; Kyzioł, A.; Pucelik, B.; Brindell, M.; Stochel, G. *Coord. Chem. Rev.* **2017**, *351*, 76–117.
26. Wright, G. D. *Adv. Drug Deliv. Rev.* **2005**, *57* (10), 1451–1470.
27. Poole, K. *J. Appl. Microbiol.* **2002**, *92*, 55S–64S.
28. Kumar, A.; Schweizer, H. P. *Adv. Drug Deliv. Rev.* **2005**, *57* (10), 1486–1513.
29. Olaitan, A. O.; Morand, S.; Rolain, J.-M. *Front. Microbiol.* **2014**, *5*, 643.
30. Velkov, T.; Thompson, P. E.; Nation, R. L.; Li, J. *J. Med. Chem.* **2010**, *53* (5), 1898–1916.
31. Auda, I. G.; Salman, I. M. A.; Odah, J. G. *Gene Rep.* **2020**, *20*, 100666.
32. Schindler, B. D.; Kaatz, G. W. *Drug Resist. Updat.* **2016**, *27*, 1–13.
33. Hernando-Amado, S.; Blanco, P.; Alcalde-Rico, M.; Corona, F.; Reales-Calderón, J. A.; Sánchez, M. B.; Martínez, J. L. *Drug Resist. Updat.* **2016**, *28*, 13–27.
34. Gędas, A.; Olszewska, M. A. *Recent Trends in Biofilm Science and Technology*; Elsevier, 2020; pp. 1–21.
35. Van Acker, H.; Van Dijck, P.; Coenye, T. *Trends Microbiol.* **2014**, *22* (6), 326–333.
36. Pinto, A. M.; Cerqueira, M. A.; Bañobre-Lópes, M.; Pastrana, L. M.; Sillankorva, S. *Viruses* **2020**, *12* (2), 235.
37. Maunders, E.; Welch, M. *FEMS Microbiol. Lett.* **2017**, *364* (13), 1–10.
38. Gebreyohannes, G.; Nyerere, A.; Bii, C.; Sbhatu, D. B. *Heliyon* **2019**, *5* (8), e02192.
39. Mah, T.-F. *Future Microbiol.* **2012**, 7 (9), 1061–1072.
40. Hall, C. W.; Mah, T.-F. *FEMS Microbiol. Rev.* **2017**, *41* (3), 276–301.
41. Bi, Y.; Xia, G.; Shi, C.; Wan, J.; Liu, L.; Chen, Y.; Wu, Y.; Zhang, W.; Zhou, M.; He, H. *Fundamen. Res.* **2021**,.
42. Proctor, R. A.; Von Eiff, C.; Kahl, B. C.; Becker, K.; McNamara, P.; Herrmann, M.; Peters, G. *Nat. Rev. Microbiol.* **2006**, *4* (4), 295–305.
43. Nguyen, H. T.; Nguyen, T. H.; Otto, M. *Comput. Struct. Biotechnol. J.* **2020**,.
44. Crabbé, A.; Jensen, P.Ø.; Bjarnsholt, T.; Coenye, T. *Trends Microbiol.* **2019**, *27* (10), 850–863.
45. Yan, J.; Bassler, B. L. *Cell Host Microbe* **2019**, *26* (1), 15–21.

46. Ranieri, M. R.; Whitchurch, C. B.; Burrows, L. L. *Curr. Opin. Microbiol.* **2018**, *45*, 164–169.
47. Parastan, R.; Kargar, M.; Solhjoo, K.; Kafilzadeh, F. *J. Glob. Antimicrob. Resist.* **2020**, *22*, 379–385.
48. Lemire, J. A.; Harrison, J. J.; Turner, R. J. *Nat. Rev. Microbiol.* **2013**, *11* (6), 371–384.
49. Kuncewicz, J.; Dąbrowski, J. M.; Kyzioł, A.; Brindell, M.; Łabuz, P.; Mazuryk, O.; Macyk, W.; Stochel, G. *Coord. Chem. Rev.* **2019**, *398*, 113012.
50. Kong, B.; Joshi, T.; Belousoff, M. J.; Tor, Y.; Graham, B.; Spiccia, L. *J. Inorg. Biochem.* **2016**, *162*, 334–342.
51. Guerra, W.; de Andrade Azevedo, E.; de Souza Monteiro, A. R.; Bucciarelli-Rodriguez, M.; Chartone-Souza, E.; Nascimento, A. M. A.; Fontes, A. P. S.; Le Moyec, L.; Pereira-Maia, E. C. *J. Inorg. Biochem.* **2005**, *99* (12), 2348–2354.
52. Drevenšek, P.; Košmrlj, J.; Giester, G.; Skauge, T.; Sletten, E.; Sepčić, K.; Turel, I. *J. Inorg. Biochem.* **2006**, *100* (11), 1755–1763.
53. Streicher, L. M. *J. Glob. Antimicrob. Resist.* **2021**, *14*, 285–295.
54. Watts, J. K.; Corey, D. R. *J. Pathol.* **2012**, *226* (2), 365–379.
55. Chi, X.; Gatti, P.; Papoian, T. *Drug Discov. Today* **2017**, *22* (5), 823–833.
56. Xue, X.-Y.; Mao, X.-G.; Zhou, Y.; Chen, Z.; Hu, Y.; Hou, Z.; Li, M.-K.; Meng, J.-R.; Luo, X.-X. *Nanomed. Nanotechnol. Biol. Med.* **2018**, *14* (3), 745–758.
57. Frazier, K. S. *Toxicol. Pathol.* **2015**, *43* (1), 78–89.
58. Casadevall, A.; Scharff, M. D. *Antimicrob. Agents Chemother.* **1994**, *38* (8), 1695–1702.
59. Pelfrene, E.; Mura, M.; Sanches, A. C.; Cavaleri, M. *Clin. Microbiol. Infect.* **2019**, *25* (1), 60–64.
60. Bebbington, C.; Yarranton, G. *Curr. Opin. Biotechnol.* **2008**, *19* (6), 613–619.
61. Hansel, T. T.; Kropshofer, H.; Singer, T.; Mitchell, J. A.; George, A. J. *Nat. Rev. Drug Discov.* **2010**, *9* (4), 325–338.
62. Wang-Lin, S. X.; Balthasar, J. P. *Antibodies* **2018**, *7* (1), 5.
63. Kango, S.; Kalia, S.; Celli, A.; Njuguna, J.; Habibi, Y.; Kumar, R. *Prog. Polym. Sci.* **2013**, *38* (8), 1232–1261.
64. Sperling, R. A.; Parak, W. J. *Philos. Trans. Roy. Soc. A Math. Phys. Eng. Sci.* **1915**, *2010* (368), 1333–1383.
65. Ghosh, C.; Sarkar, P.; Issa, R.; Haldar, J. *Trends Microbiol.* **2019**, *27* (4), 323–338.
66. Clark, J. R.; March, J. B. *Trends Biotechnol.* **2006**, *24* (5), 212–218.
67. Abedon, S. T.; Kuhl, S. J.; Blasdel, B. G.; Kutter, E. M. *Bacteriophage* **2011**, *1* (2), 66–85.
68. Tunçel, A.; Öztürk, İ.; Ince, M.; Ocakoglu, K.; Hoşgör-Limoncu, M.; Yurt, F. *J. Porphyrins Phthalocyanines* **2019**, *23* (01n02), 206–212.
69. Pérez-Laguna, V.; Gilaberte, Y.; Millán-Lou, M. I.; Agut, M.; Nonell, S.; Rezusta, A.; Hamblin, M. R. *Photochem. Photobiol. Sci.* **2019**, *18* (5), 1020–1029.
70. Hamblin, M. R.; Hasan, T. *Photochem. Photobiol. Sci.* **2004**, *3* (5), 436–450.
71. Sabino, C.; Garcez, A.; Núñez, S.; Ribeiro, M.; Hamblin, M. *Lasers Med. Sci.* **2015**, *30* (6), 1657–1665.
72. Kawczyk-Krupka, A.; Pucelik, B.; Międzybrodzka, A.; Sieroń, A. R.; Dąbrowski, J. M. *Photodiagn. Photodyn. Ther.* **2018**, *23*, 132–143.
73. Dąbrowski, J. M.; Pucelik, B.; Regiel-Futyra, A.; Brindell, M.; Mazuryk, O.; Kyzioł, A.; Stochel, G.; Macyk, W.; Arnaut, L. G. *Coord. Chem. Rev.* **2016**, *325*, 67–101.
74. Alves, E.; Esteves, A. C.; Correia, A.; Cunha, A.; Faustino, M. A.; Neves, M. G.; Almeida, A. *Photochem. Photobiol. Sci.* **2015**, *14* (6), 1169–1178.
75. Koch, G. *CHIMIA Int. J. Chem.* **2017**, *71* (10), 643(1).
76. Tavares, A.; Dias, S. R.; Carvalho, C. M.; Faustino, M. A.; Tomé, J. P.; Neves, M. G.; Tome, A. C.; Cavaleiro, J. A.; Cunha, Â.; Gomes, N. C. *Photochem. Photobiol. Sci.* **2011**, *10* (10), 1659–1669.

77. Lopes, D.; Melo, T.; Santos, N.; Rosa, L.; Alves, E.; Gomes, M. C.; Cunha, Â.; Neves, M. G.; Faustino, M. A.; Domingues, M. R. M. *J. Photochem. Photobiol. B Biol.* **2014**, *141*, 145–153.
78. Mitoraj, D.; Jańczyk, A.; Strus, M.; Kisch, H.; Stochel, G.; Heczko, P. B.; Macyk, W. *Photochem. Photobiol. Sci.* **2007**, *6* (6), 642–648.
79. Dąbrowski, J. M. *Adv. Inorg. Chem.* **2017**, *70*, 343–394.
80. Whittaker, A.; Mount, A.; Heal, M.; Galus, M. Wydawnictwo Naukowe PWN, 2012.
81. Oszajca, M.; Brindell, M.; Orzeł, Ł.; Dąbrowski, J. M.; Śpiewak, K.; Łabuz, P.; Pacia, M.; Stochel-Gaudyn, A.; Macyk, W.; van Eldik, R. *Coord. Chem. Rev.* **2016**, *327*, 143–165.
82. Kehrer, J. P. *Toxicology* **2000**, *149* (1), 43–50.
83. DeRosa, M. C.; Crutchley, R. J. *Coord. Chem. Rev.* **2002**, *233*, 351–371.
84. Agostinis, P.; Berg, K.; Cengel, K. A.; Foster, T. H.; Girotti, A. W.; Gollnick, S. O.; Hahn, S. M.; Hamblin, M. R.; Juzeniene, A.; Kessel, D. *CA Cancer J. Clin.* **2011**, *61* (4), 250–281.
85. Mead, M. N., Benefits of Sunlight: A Bright Spot for Human Health, Secondary National Institute of Environmental Health Sciences: 2008.
86. Pucelik, B.; Sułek, A.; Dąbrowski, J. M. *Coord. Chem. Rev.* **2020**, *416*, 213340.
87. Pucelik, B.; Arnaut, L. G.; Stochel, G. Y.; Dabrowski, J. M. *ACS Appl. Mater. Interfaces* **2016**, *8* (34), 22039–22055.
88. Rocha, L. B.; Gomes-da-Silva, L. C.; Dąbrowski, J. M.; Arnaut, L. G. *Eur. J. Cancer* **2015**, *51* (13), 1822–1830.
89. Pucelik, B.; Paczyński, R.; Dubin, G.; Pereira, M. M.; Arnaut, L. G.; Dąbrowski, J. M. *PLoS One* **2017**, *12* (10), e0185984.
90. Vinagreiro, C. S.; Zangirolami, A.; Schaberle, F. A.; Nunes, S. C.; Blanco, K. C.; Inada, N. M.; da Silva, G. J.; Pais, A. A.; Bagnato, V. S.; Arnaut, L. G. *ACS Infect. Dis.* **2020**, *6* (6), 1517–1526.
91. Davies, D. *Nat. Rev. Drug Discov.* **2003**, *2* (2), 114–122.
92. Pedigo, L. A.; Gibbs, A. J.; Scott, R. J.; Street, C. N. *Proc. SPIE* **2009**, *7380*, 73803H.
93. Biel, M. A. *Adv. Exp. Med. Biol.* **2015**, *831*, 119–136.
94. Nagata, J. Y.; Hioka, N.; Kimura, E.; Batistela, V. R.; Terada, R. S. S.; Graciano, A. X.; Baesso, M. L.; Hayacibara, M. F. *Photodiagn. Photodyn. Ther.* **2012**, *9* (2), 122–131.
95. Usacheva, M. N.; Teichert, M. C.; Biel, M. A. *Lasers Surg. Med. Official J. Am. Soc. Laser Med. Surg.* **2001**, *29* (2), 165–173.
96. Wood, S.; Metcalf, D.; Devine, D.; Robinson, C. *J. Antimicrob. Chemother.* **2006**, *57* (4), 680–684.
97. Ochsner, M. *J. Photochem. Photobiol. B Biol.* **1997**, *39* (1), 1–18.
98. Pucelik, B.; Arnaut, L. G.; Dąbrowski, J. M. *J. Clin. Med.* **2020**, *9* (1), 8.
99. Pereira, M. M.; Monteiro, C. J.; Simões, A. V.; Pinto, S. M.; Arnaut, L. G.; Sá, G. F.; Silva, E. F.; Rocha, L. B.; Simões, S.; Formosinho, S. J. *J. Porphyrins Phthalocyanines* **2009**, *13* (04n05), 567–573.
100. Pucelik, B.; Sułek, A.; Drozd, A.; Stochel, G.; Pereira, M. M.; Pinto, S.; Arnaut, L. G.; Dąbrowski, J. M. *Int. J. Mol. Sci.* **2020**, *21* (8), 2786.
101. Arnaut, L. G.; Pereira, M. M.; Dąbrowski, J. M.; Silva, E. F.; Schaberle, F. A.; Abreu, A. R.; Rocha, L. B.; Barsan, M. M.; Urbańska, K.; Stochel, G. *Chemistry – A Eur. J. Dermatol.* **2014**, *20* (18), 5346–5357.
102. Dąbrowski, J. M.; Pereira, M. M.; Arnaut, L. G.; Monteiro, C. J.; Peixoto, A. F.; Karocki, A.; Urbańska, K.; Stochel, G. *Photochem. Photobiol.* **2007**, *83* (4), 897–903.
103. Dąbrowski, J. M.; Pucelik, B.; Pereira, M. M.; Arnaut, L. G.; Stochel, G. *J. Coord. Chem.* **2015**, *68* (17–18), 3116–3134.
104. Luz, A. F.; Pucelik, B.; Pereira, M. M.; Dąbrowski, J. M.; Arnaut, L. G. *Lasers Surg. Med.* **2018**, *50* (5), 451–459.

105. Arnaut, L. G. *Adv. Inorg. Chem.* **2011**, *63*, 187–233.
106. Silva, E. F.; Schaberle, F. A.; Monteiro, C. J.; Dąbrowski, J. M.; Arnaut, L. G. *Photochem. Photobiol. Sci.* **2013**, *12* (7), 1187–1192.
107. Dabrowski, J. M.; Arnaut, L. G.; Pereira, M. M.; Monteiro, C.; Urbanska, K.; Simões, S.; Stochel, G. *ChemMedChem* **2010**, *5* (10), 1770–1780.
108. Dąbrowski, J. M.; Krzykawska, M.; Arnaut, L. G.; Pereira, M. M.; Monteiro, C. J.; Simões, S.; Urbańska, K.; Stochel, G. *ChemMedChem* **2011**, *6* (9), 1715–1726.
109. Aroso, R. T.; Calvete, M. J.; Pucelik, B.; Dubin, G.; Arnaut, L. G.; Pereira, M. M.; Dąbrowski, J. M. *Eur. J. Med. Chem.* **2019**, *184*, 111740.
110. Mete, E.; Kabay, N.; Dumoulin, F.; Ahsen, V.; Tuncel Kostakoğlu, S.; Ergin, Ç. *Biotech. Histochem.* **2021**, *96* (4), 311–314.
111. Lo, P.-C.; Rodríguez-Morgade, M. S.; Pandey, R. K.; Ng, D. K.; Torres, T.; Dumoulin, F. *Chem. Soc. Rev.* **2020**, *49* (4), 1041–1056.
112. Topal, S. Z.; İşci, Ü.; Kumru, U.; Atilla, D.; Gürek, A. G.; Hirel, C.; Durmuş, M.; Tommasino, J.-B.; Luneau, D.; Berber, S. *Dalton Trans.* **2014**, *43* (18), 6897–6908.
113. Tunç, G.; Albakour, M.; Ahsen, V.; Gürek, A. E. G. *Porphyrin Science by Women*, Vol. 3; World Scientific, 2019; pp. 698–707.
114. Santos, K. L. M.; Barros, R. M.; da Silva Lima, D. P.; Nunes, A. M. A.; Sato, M. R.; Faccio, R.; de Lima Damasceno, B. P. G.; Junior, J. A. O. *Photodiagn. Photodyn. Ther.* **2020**, 102032.
115. Güzel, E.; Şişman, İ.; Gül, A.; Kocak, M. B. *J. Porphyrins Phthalocyanines* **2019**, *23* (03), 279–286.
116. Mantareva, V.; Eneva, I.; Kussovski, V.; Borisova, E.; Angelov, I. In 18th International School on Quantum Electronics: Laser Physics and Applications, 2015; International Society for Optics and Photonics: Vol. 9447, p 94470W.
117. Yurt, F.; Ince, M.; Colak, S. G.; Ocakoglu, K.; Er, O.; Soylu, H. M.; Gunduz, C.; Avci, C. B.; Kurt, C. C. *Int. J. Pharm.* **2017**, *524* (1–2), 467–474.
118. Simoes, A. V.; Adamowicz, A.; Dąbrowski, J. M.; Calvete, M. J.; Abreu, A. R.; Stochel, G.; Arnaut, L. G.; Pereira, M. M. *Tetrahedron* **2012**, *68* (42), 8767–8772.
119. Pinto, S. M.; Henriques, C. A.; Tomé, V. A.; Vinagreiro, C. S.; Calvete, M. J.; Dąbrowski, J. M.; Piñeiro, M.; Arnaut, L. G.; Pereira, M. M. *J. Porphyrins Phthalocyanines* **2016**, *20* (01n04), 45–60.
120. Pereira, M. M.; Monteiro, C. J.; Simoes, A. V.; Pinto, S. M.; Abreu, A. R.; Sá, G. F.; Silva, E. F.; Rocha, L. B.; Dąbrowski, J. M.; Formosinho, S. J. *Tetrahedron* **2010**, *66* (49), 9545–9551.
121. Pereira, M. M.; Abreu, A. R.; Goncalves, N. P.; Calvete, M. J.; Simões, A. V.; Monteiro, C. J.; Arnaut, L. G.; Eusebio, M. E.; Canotilho, J. *Green Chem.* **2012**, *14* (6), 1666–1672.
122. Dąbrowski, J. M.; Urbanska, K.; Arnaut, L. G.; Pereira, M. M.; Abreu, A. R.; Simões, S.; Stochel, G. *ChemMedChem* **2011**, *6* (3), 465–475.
123. Saavedra, R.; Rocha, L. B.; Dąbrowski, J. M.; Arnaut, L. G. *ChemMedChem* **2014**, *9* (2), 390–398.
124. Gonçalves, N. P.; Simões, A. V.; Abreu, A. R.; Abrunhosa, A. J.; Dąbrowski, J. M.; Pereira, M. M. *J. Porphyrins Phthalocyanines* **2015**, *19* (08), 946–955.
125. Dąbrowski, J. M.; Arnaut, L. G.; Pereira, M. M.; Urbańska, K.; Simões, S.; Stochel, G.; Cortes, L. *Free Radic. Biol. Med.* **2012**, *52* (7), 1188–1200.
126. Pucelik, B.; Gürol, I.; Ahsen, V.; Dumoulin, F.; Dąbrowski, J. M. *Eur. J. Med. Chem.* **2016**, *124*, 284–298.
127. Silva, E. F.; Serpa, C.; Dąbrowski, J. M.; Monteiro, C. J.; Formosinho, S. J.; Stochel, G.; Urbanska, K.; Simões, S.; Pereira, M. M.; Arnaut, L. G. *Chemistry–A* Eur. J. Dermatol. 2010, 16 (30), 9273–9286.

128. Soares, H. T.; Campos, J.; Gomes-da-Silva, L. C.; Schaberle, F. A.; Dabrowski, J. M.; Arnaut, L. G. *Chembiochem* **2016**, *17* (9), 836–842.
129. Santos, L. L.; Oliveira, J.; Monteiro, E.; Santos, J.; Sarmento, C. *Case Rep. Oncol.* **2018**, *11* (3), 769–776.
130. Pucelik, B.; Sułek, A.; Barzowska, A.; Dąbrowski, J. M. *Cancer Lett.* **2020**, *492*, 116–135.
131. Krzykawska-Serda, M.; Dąbrowski, J. M.; Arnaut, L. G.; Szczygieł, M.; Urbańska, K.; Stochel, G.; Elas, M. *Free Radic. Biol. Med.* **2014**, *73*, 239–251.
132. Karwicka, M.; Pucelik, B.; Gonet, M.; Elas, M.; Dąbrowski, J. M. *Sci. Rep.* **2019**, *9* (1), 1–15.
133. Rocha, L. B.; Schaberle, F.; Dąbrowski, J. M.; Simões, S.; Arnaut, L. G. *Int. J. Mol. Sci.* **2015**, *16* (12), 29236–29249.
134. Huang, Y.-Y.; Wintner, A.; Seed, P. C.; Brauns, T.; Gelfand, J. A.; Hamblin, M. R. *Sci. Rep.* **2018**, *8* (1), 1–9.
135. Shen, X.; Dong, L.; He, X.; Zhao, C.; Zhang, W.; Li, X.; Lu, Y. *Photodiagn. Photodyn. Ther.* **2020**, *32*, 102051.
136. de Aguiar Coletti, T. M. S. F.; De Freitas, L. M.; Almeida, A. M. F.; Fontana, C. R. *Photomed. Laser Surg.* **2017**, *35* (7), 378–385.
137. Sułek, A.; Pucelik, B.; Kobielusz, M.; Barzowska, A.; Dąbrowski, J. M. *Int. J. Mol. Sci.* **2020**, *21* (22), 8716.
138. Sułek, A.; Pucelik, B.; Kuncewicz, J.; Dubin, G.; Dąbrowski, J. M. *Catal. Today* **2019**, *335*, 538–549.
139. Sułek, A.; Pucelik, B.; Kobielusz, M.; Łabuz, P.; Dubin, G.; Dąbrowski, J. M. *Catalysts* **2019**, *9* (10), 821.
140. França, M. D.; Santos, L. M.; Silva, T. A.; Borges, K. A.; Silva, V. M.; Patrocinio, A. O.; Trovó, A. G.; Machado, A. E. *J. Braz. Chem. Soc.* **2016**, *27*, 1094–1102.
141. Flak, D.; Yate, L.; Nowaczyk, G.; Jurga, S. *Mater. Sci. Eng. C* **2017**, *78*, 1072–1085.
142. Dąbrowski, J. M.; Pucelik, B.; Pereira, M. M.; Arnaut, L. G.; Macyk, W.; Stochel, G. *RSC Adv.* **2015**, *5* (113), 93252–93261.
143. Lobo, A. C.; Silva, A. D.; Tome, V. A.; Pinto, S. M.; Silva, E. F.; Calvete, M. J.; Gomes, C. M.; Pereira, M. M.; Arnaut, L. G. *J. Med. Chem.* **2016**, *59* (10), 4688–4696.
144. Iqbal, Z.; Chen, J.; Chen, Z.; Huang, M. *Curr. Drug Metab.* **2015**, *16* (9), 816–832.
145. Aslam, B.; Wang, W.; Arshad, M. I.; Khurshid, M.; Muzammil, S.; Rasool, M. H.; Nisar, M. A.; Alvi, R. F.; Aslam, M. A.; Qamar, M. U. *Infect. Drug Resist.* **2018**, *11*, 1645.
146. Mehellou, Y.; Willcox, B. E., A Two-Pronged Attack on Antibiotic-Resistant Microbes, Secondary Nature Publishing Group: 2021.
147. Bush, K.; Courvalin, P.; Dantas, G.; Davies, J.; Eisenstein, B.; Huovinen, P.; Jacoby, G. A.; Kishony, R.; Kreiswirth, B. N.; Kutter, E. *Nat. Rev. Microbiol.* **2011**, *9* (12), 894–896.
148. Stewart, P. S. *Int. J. Med. Microbiol.* **2002**, *292* (2), 107–113.
149. Hamblin, M. R. *Curr. Opin. Microbiol.* **2016**, *33*, 67–73.
150. Wainwright, M.; Maisch, T.; Nonell, S.; Plaetzer, K.; Almeida, A.; Tegos, G. P.; Hamblin, M. R. *Lancet Infect. Dis.* **2017**, *17* (2), e49–e55.
151. Demidova, T.; Hamblin, M. *Int. J. Immunopathol. Pharmacol.* **2004**, *17* (3), 245–254.
152. Mikula, P.; Kalhotka, L.; Jancula, D.; Zezulka, S.; Korinkova, R.; Cerny, J.; Marsalek, B.; Toman, P. *J. Photochem. Photobiol. B Biol.* **2014**, *138*, 230–239.

CHAPTER FOUR

The secret life of oligopyridines: Complexes of group 1 elements

Edwin C. Constable*
University of Basel, Basel, Switzerland
*Corresponding author: e-mail address: edwin.constable@unibas.ch

Contents

Abstract

This chapter describes the coordination chemistry of the group 1 metals with the oligopyridines. The oligopyridines are generally thought of as the archetypal ligands for the d-block metals. Although complexes of group 1 metals with 1,10-phenanthroline are relatively well known, this is not the case for the parent oligopyridines. The review attempts to illustrate the structural diversity encountered in these interesting and unusual complexes.

Abbreviations

18C6 18-crown-6
2-py pyridin-2-yl
B.M. Bohr Magneton
DME 1,2-dimethoxyethane
DMF *N*,*N*-dimethylformamide
DMSO methanesulfinylmethane, dimethyl sulfoxide

Advances in Inorganic Chemistry, Volume 79
ISSN 0898-8838
https://doi.org/10.1016/bs.adioch.2021.12.004

EHMO	extended Hückel molecular orbital
emim	1-ethyl-3-methylimidazolium
En	ethane-1,2-diamine
EPR	electron paramagnetic resonance
Et	ethyl
IR	infrared
IUPAC	International Union of Pure and Applied Chemistry
LCAO	linear combination of atomic orbitals
Me	methyl, CH_3
Mes	2,4,6-trimethylphenyl, mesityl
Mes*	2,4,6-tris(tert-butyl)phenyl
MO	molecular orbital
MTHF	2-methyltetrahydrofuran
NMR	nuclear magnetic resonance
OLED	organic light emitting diode
Ph	phenyl
PIN	preferred IUPAC name
PMDTA	N^1-(2-(dimethylamino)ethyl)-N^1,N^2,N^2-trimethylethane-1,2-diamine
THF	tetrahydrofuran
TMEDA	N^1,N^1,N^2,N^2-tetramethylethane-1,2-diamine

1. Introduction

If you ask any undergraduate student of chemistry what ligands the group 1 metal cations prefer, the answer is almost certain to be "crown ethers." While this is certainly true, it hides the fact that the coordination chemistry of the group 1 metals is rich and often surprising. This article concentrates upon the coordination behavior of group 1 metals with a particular class of ligands—those containing oligopyridine metal-binding domains.

The oligopyridines are the classical ligands of the 20th Century CE and are generally associated with the transition metals (Fig. 1).[1–17] Their ubiquity is associated with a combination of forming thermodynamically stable complexes due to the operation of the chelate effect and the low-lying π^* levels of the ligand making them excellent π-acceptors. Thermodynamic stability says nothing about kinetic stability, and oligopyridine complexes may be kinetically inert or labile. Complexes of **3** are typically more stable than those of **1** because the nitrogen donors in **3** are pre-organized for

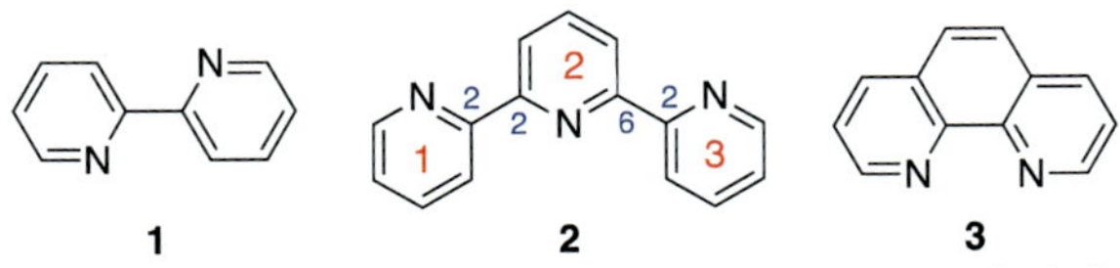

Fig. 1 The archetypal oligopyridines 2,2′-bipyridine (**1**) and $1^1,2^2{:}2^6,3^2$-terpyridine (**2**) and the closely related ligand 1,10-phenanthroline (**3**). The ligands **1** and **2** are represented in their equilibrium geometries—in order to coordinate in a chelating manner it is necessary to rotate a ring about the inter-annular C—C bond, thus making complexes of **3** typically more stable than those of **1**. The structure of **2** also identifies the ring numbering (red) and positions (blue) used to construct the IUPAC PIN of $1^1,2^2{:}2^6,3^2$-terpyridine.

coordination, in contrast to the equilibrium geometry of **1** which has the rings orthogonal or with a *trans* conformation of the donors.

Although compounds **1** and **3** have been known for almost 150 years, their coordination chemistry continues to surprise and delight us.[18] On the one hand, these chelating metal-binding domains have proved invaluable in supramolecular chemistry, providing building blocks with (relatively) predictable and robust coordination properties and on the other have opened up new reaction pathways and possibilities for the chemistry of simple molecules such as water.[19] The typical bonding modes involve the coordination of all of the potential nitrogen donors, but a variety of other behaviors are well-established, the most important of which is hypodentate binding, in which fewer than the maximum number of nitrogen atoms are coordinated.[20]

Ligands **1** and **2** are commonly known in the community as 2,2′-bipyridine and 2,2′:6′,2″-terpyridine with well-established abbreviations bpy (or bipy) and tpy (or terpy). Recently IUPAC has revised the PINs (Preferred IUPAC Names) for the higher oligopyridines to avoid the use of multiple primes, but retains the name 2,2′-bipyridine for **1** (Rule P-28.3.1).[21] The new numbering system for ring assemblies consisting of more than two rings is composed of primary and composite locants. Composite locants denote the positions in each ring as superscripts to locants indicating the position of a cyclic system in an assembly. Locants indicating points of attachment are placed in ascending order with locants denoting junctions separated by a comma and sets of junction locants separated by a colon. Thus, compound **2** is named $1^1,2^2{:}2^6,3^2$-terpyridine—the commas denote the connections or rings 1 and 2, and 2 and 3, respectively, and the superscripts indicate the positions by the pyridine rings are linked. I will use this nomenclature in the text.

Despite the emphasis on transition metal coordination chemistry, the binding of these ligands to group 1 metals has been discovered, forgotten and rediscovered multiple times. This chapter attempts to provide an overview of the topic and highlights just how common such complexes are and identifies opportunities for further development of this rich area of chemistry. The review only considers the oligopyridines and excludes 1,10-phenanthroline, **3**; this decision was made on the basis of the very well established group 1 chemistry of ligand **3** together with the observation that the structural chemistry is often dominated by aromatic-aromatic interactions. The emphasis of the review is on compounds in which interactions between the group 1 metal cation and the ligand nitrogen donors is established through spectroscopic or structural methods—with that caveat, I apologize for the errors and omissions that will inevitably occur.

2. Ligand-binding parameters

It is useful to summarize the parameters which may be used to characterize either an individual metal-ligand interaction or the properties of an entire coordination entity.

For polydentate ligands, the denticity—the number and type of atoms coordinated—is probably the primary parameter. This is most conveniently described using the IUPAC kappa notation and is illustrated for $1^1,2^2:2^6,3^2$-terpyridine in Fig. 2.[22] The "typical" chelating mode with all three nitrogen atoms coordinated to a single metal center is denoted κ^3N,

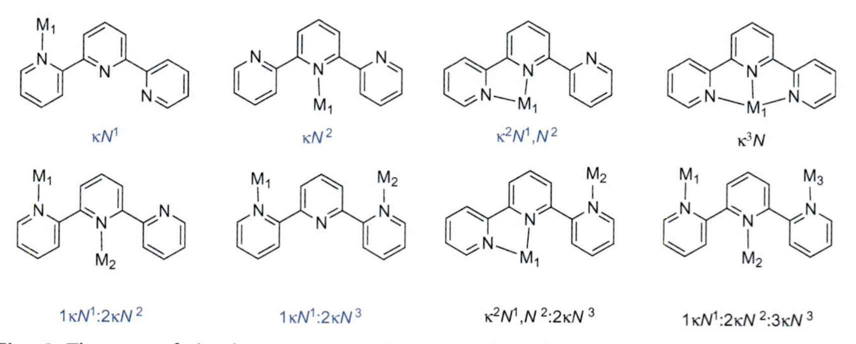

Fig. 2 The use of the kappa nomenclature to describe the coordination modes of $1^1,2^2:2^6,3^2$-terpyridine. The commonly encountered chelating mode in which all three nitrogen donors are coordinated to a single metal center is denoted κ^3N. The metals M_1, M_2 and M_3 may be the same or different. The terms in blue refer to hypodentate modes of coordination.

where the superscript three indicates the number of coordinated nitrogen atoms. Two other modes are possible in which all three nitrogen atoms are coordinated, in the binuclear and trinuclear species denoted $1\kappa^2N^1,2\kappa N^3$ and $1\kappa N^1,2\kappa N^2,3\kappa N^3$, respectively. The numeral before each kappa denotes the metal center. Finally, there are a number of coordination modes in which not all of the nitrogen donors bind metal centers; these bonding modes are described as hypodentate and may be mononuclear monodentate (κN^1 or κN^2), mononuclear bidentate (κ^2N^1,N^2), dinuclear ($1\kappa N^1$:$2\kappa N^2$, $1\kappa N^1$:$2\kappa N^3$ or $1\kappa^2N^1$:$2\kappa N^3$) or trinuclear ($1\kappa N^1$:$2\kappa N^2$:$3\kappa N^3$).[20]

The nature of the metal–ligand bond can be quantified in terms of the bond length, which is usually derived from solid state structural data, and the bond strength obtained from thermochemical measurements or spectroscopic analysis. For chelating oligopyridine ligands the ∠N–M–N bond angles is useful: although often referred to as the bite angle, this should not be viewed as a fundamental property of the ligand, as the angle is dependent upon the M–N bond lengths. Finally, for oligopyridine ligands, the torsion angle ∠N–C–C–N can provide useful information about the bonding mode (Fig. 3).

Finally, there are a number of quantitative values which can be applied to a consideration of the coordination entity as a whole. The first of these relates to the thermodynamic stability of the complex and are the stability constants for the formation of the compound. In the case of an ML_3 complex, three different stepwise stability constants K_1, K_2 and K_3 can be defined, related to three separate equilibria (Eqs. 1–3). The square brackets denote the equilibrium concentration of the species that they enclose.

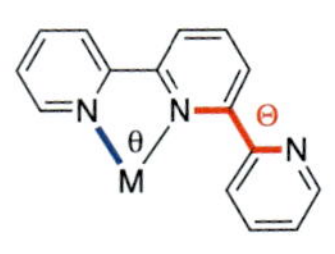

Fig. 3 The metrics that may be used to quantify the nature of metal-ligand bonding. The bond length (for example, the M–N distance in blue) is usually obtained from crystallographic data and reported in Å, nm or pm. The ∠N–M–N bond angle q (often erroneously called the bite angle) is also obtained from crystallographic data and reported in degrees. Finally, the torsion angle Q (for example, the ∠N–C–C–N angle in red) provides valuable information about the bonding. A torsion angle in the range $\pm 30°$ is typical for a κ^3N ligand, whereas larger values are associated with the other bonding modes indicated in Fig. 2.

$$M + L \rightleftharpoons ML \qquad K_1 = \frac{[ML]}{[M][L]} \tag{1}$$

$$ML + L \rightleftharpoons ML_2 \qquad K_2 = \frac{[ML_2]}{[ML][L]} \tag{2}$$

$$ML_2 + L \rightleftharpoons ML_3 \qquad K_3 = \frac{[ML_3]}{[ML_2][L]} \tag{3}$$

Often overall stability constants β_n are cited rather than the stepwise stability constants, and these are defined in Eqs. (4)–(6).

$$M + L \rightleftharpoons ML \qquad \beta_1 = \frac{[ML]}{[M][L]} \tag{4}$$

$$M + 2L \rightleftharpoons ML_2 \qquad \beta_2 = \frac{[ML_2]}{[M][L]^2} \tag{5}$$

$$M + 3L \rightleftharpoons ML_3 \qquad \beta_3 = \frac{[ML_3]}{[M][L]^3} \tag{6}$$

As the stability constants are often large numbers, it is convenient to report them as their log values, with the relationships between the stepwise and overall stability constants being given by

$$\log \beta_1 = \log K_1 \tag{7}$$

$$\log \beta_2 = \log K_1 + \log K_2 \tag{8}$$

$$\log \beta_3 = \log K_1 + \log K_2 + \log K_3 \tag{9}$$

Additional stability is associated with the formation of chelate rings (the chelate effect), binding to macrocyclic ligands (the macrocyclic effect) or encapsulating ligands (the cryptate effect). These values all relate to the thermodynamic stability of a complex.

It is also common to report kinetic data relating to the rate of formation or displacement of ligands from complexes, and these are typically reported as the rate constants or exchange constants for the reactions of interest.

3. Hardness and softness: A ubiquitous concept

We return to our starting point. Why did our chemistry student reply "crown ethers" to the question about group 1 coordination chemistry?

3.1 Class (a) and class (b) metals

By the 1950s, various research groups were measuring the stability constants for various metals with various ligands. In 1954, it was noted that most metal ions formed complexes for which the stability constants of complexes with halide ligands were in the sequence $F^- > Cl^- > Br^- > I^-$ but that for a few metals the sequence was reversed.[23] In a survey of all available stability constant data from 1958, this observation was further generalized and reformulated "There are two classes of acceptor: (a) those which form their most stable complexes with the first ligand atom of each Group, i.e., with N, O, and F, and (b) those which form their most stable complexes with the second or a subsequent ligand atom."[24] Metal ions in group 1 belonged to class (a), metal ions such as Hg^{2+} and Cu^+ to class (b) and the majority of the transition metal ions to a borderline class between the two. This classification is retained by IUPAC where a class(a) metal ion is "A metal ion that combines preferentially with ligands containing ligating atoms that are the lightest of their Periodic Group" and a class (b) ion one "that combines preferentially with ligands containing ligating atoms other than the lightest of their Periodic Group."[25]

In 1963, Pearson extended and redefined the concept, defining class (b) metal ions as soft and class (a) metal ions as either hard or intermediate.[26] He similarly defined ligands as either hard (fluoride, amines) borderline (bromide) or soft (iodide, sulfur). In keeping with the donor-acceptor description of the coordination bond, he described the metal centers as (Lewis) acids and the ligands as (Lewis) bases. A modern compilation of hard and soft acids and bases is presented in Table 1.

Table 1 A compilation of hard and soft acids and bases.

Ligands (Lewis bases)	
Hard; class (a)	F^-, Cl^-, H_2O, ROH, R_2O, OH^-, RO^-, RCO_2^-, CO_3^{2-}, NO_3^-, PO_4^{3-}, SO_4^{2-}, ClO_4^-, NH_3, RNH_2
Soft; class (b)	I^-, H^-, R^-, $(CN\text{-}\kappa C)^-$, CO-κC, RNC, RSH, R_2S, RS^-, $(SCN\text{-}\kappa S)^-$, R_3P, R_3As, R_3Sb, alkenes, arenes
Intermediate	Br^-, N_3^-, py, $(SCN\text{-}\kappa N)^-$, $ArNH_2$, NO_2^-, SO_3^{2-}
Metal centers (Lewis acids)	
Hard; class (a)	Li^+, Na^+, K^+, Rb^+, Cs^+, Mn^{2+}, Zn^{2+}, Al^{3+}, Ga^{3+}, In^{3+}, Sc^{3+}, Cr^{3+}, Fe^{3+}, Co^{3+}, Y^{3+}, VO^{2+}, VO_2^+
Soft; class (b)	M(0), Tl^+, Cu^+, Ag^+, Au^+, Hg_2^{2+}, Pb^{2+}, Hg^{2+}, Cd^{2+}, Pd^{2+}, Pt^{2+}, Tl^{3+}
Intermediate	Sn^{2+}, Fe^{2+}, Co^{2+}, Ni^{2+}, Cu^{2+}, Zn^{2+}, Ru^{3+}, Rh^{3+}, Ir^{3+}

The ligands and metal ions of particular interest to this review have been highlighted in red.

Pearson then made the proposal that hard acids prefer to coordinate with hard bases, and soft acids to soft bases. In the intervening 60 years, numerous attempts have been made to quantify scales of hardness and softness, with varying degrees of success. We will rely on a qualitative description of the phenomenon. Soft acids are weakly polarizing and hard acids are strongly polarizing, whereas soft bases are easily polarized while hard bases are poorly polarized. This can be further elaborated into the observation that soft acid–soft base interactions tend toward covalent bonding and hard acid-hard base interactions toward ionic bonding. We now see the group 1 metal cation interaction with an ether is a hard acid-hard base interaction and favored. This is the origin of the answer "crown ethers"—the hard acid–hard base interaction is further enhanced by the macrocyclic effect when using a crown ether as a macrocyclic ligand.

Pyridines are seen as intermediate ligands, so why are we interested in their interaction with group 1 metal ions. The description "intermediate" is actually given with the prejudice of a transition metal coordination chemist—pyridine ligands act as σ-donors using the lone pair on the nitrogen but also as π-acceptors through back bonding to the π*-levels of the ligand. In pyridine coordination compounds with group 1 metal ions, back-bonding is irrelevant and the electronegativity of the donor can be used as a measure of base strength. The Pauling electronegativity of nitrogen (3.04) is similar to that of oxygen (3.44) and we might expect to find group 1 pyridine complexes (albeit not so stable as those with ethers).

The Cambridge Structural Database of the Cambridge Crystallographic Data Center (CCDC version 2021.2.0) contains 15,824 entries for compounds containing a C–O–C fragment coordinated to a group 1 metal ion and 1244 entries for compounds with a pyridine ligand coordinated to group 1 through nitrogen.[27] There are certainly more oxygen donor ligands than pyridine donors, but the number of the latter is far from insignificant. We shall now look at these compounds in detail.

4. Bonding in group 1 complexes

Following from the recognition that group 1 metal cation coordination chemistry is dominated by hard base–hard acid interactions, it follows that the bonding in these compounds is *primarily* electrostatic. For the transition metal coordination chemist, accustomed to rationalizing structure upon interactions of d-orbitals with ligand orbitals, this has some interesting consequences.[28]

We can approximate the complexes to a positively charged cations and negative point charges to describe the ligand donor atoms. The electrostatic force between two charged species possessing charges q_1 and q_2 is given by Coulomb's equation (Eq. 10):

$$\text{Force} = \frac{q_1 q_{21}}{4\pi\varepsilon_0 r^2} \tag{10}$$

where ε_0 is the permittivity of free space, 8.85×10^{-12} F m^{-1} and r is the distance between the charged particles. Dimensional analysis confirms that the force has units of energy (joules in the S.I.) and that the energy is only dependent upon the distance between the charged particles and the magnitude of the charges. This means that if we only consider the electrostatic interactions between the positive and negatively charged particles, the spatial distribution of the charges (for example, tetrahedral or square-planar distribution of four donors *at the same distance)* has no influence on the total energy of the cation–ligand interactions. This, in turn, suggests that the spatial arrangement of the donors will be dictated by their mutual electrostatic repulsion modified by any structural constraints imposed by the connectivity between donor atoms.

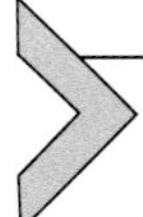

5. The compounds

5.1 2,2′-Bipyridine derivatives

5.1.1 The 2,2′-bipyridine anions

Herzog described the reaction of **1** with either sodium or lithium metal in THF to give initially red and eventually olive-green solutions containing the species {Li(**1**)} (red) and {Li_2(**1**)} (green),[29,30] which have been used to prepare low valent compounds of titanium,[29] manganese,[31] yttrium,[32] zirconium,[33] niobium,[34,35] iron,[36,37], beryllium,[30,38] zinc,[38,39] boron,[38] aluminum,[38] magnesium,[38] scandium[40] and gallium.[38] Herzog obtained crystalline {Li(**1**)(THF)} from the red solution and {Li_2(**1**)$(THF)_2$} from the green solution.[41] If the reaction with lithium is conducted in dioxan, only {Li(**1**)} is formed which can be isolated as the solid species {Li(**1**)(dioxan)}.[41] Coates subsequently showed that the reaction of **1** with lithium in diethyl ether yields only red {Li(**1**)} which was subsequently used to prepare [Be$(\mathbf{1})_2$].[30] The reaction of {Li_2(**1**)} in THF with PhNC is reported to generate {Li(**1**)}[42] One of the best solvents for the reaction of lithium with **1** was liquid ammonia, in which reaction was complete in seconds and allowed the preparation of either {Li(**1**)} or {Li_2(**1**)} simply by varying the stoichiometry of the reaction.[41] In a later publication in the series, Herzog describes the preparation of {Li(**1**)$(THF)_{1.84}$}, {Li$(\mathbf{1})_2$} and {Li_2(**1**)$(THF)_4$} as solid state species.[43]

Herzog reported that similar reactions were observed between **1** and sodium or potassium, although the tendency for the formation of {M_2L} species was greatest for lithium.[41] The solid state species {Na(**1**)(dioxan)} was described.[41]

The {M(**1**)} compounds were thought of as {$M^+(\mathbf{1}^-)$} species with the formulation rationalized by a comparison of the E° potentials of Li and Na (3.02 and 2.41 V, respectively) with the first reduction potential of **1** (1.46 V).[44] The radical anion structures for {Li(**1**)} and {Na(**1**)} were established when the EPR spectra in dioxan solution[45] and the solid state (for {Li(**1**)(dioxan)} and {Na(**1**)(dioxan)}) as well as the magnetic moments of {Li(**1**)(dioxan)} (μ_{eff} 1.5 B.M.) and {Na(**1**)(dioxan)} (μ_{eff} 1.65 B.M.) were reported.[44,46] The compound {Li(**1**)(Et_2O)} was prepared by the initial synthesis of {Li_2(**1**)} in Et_2O followed by reaction with a stoichiometric amount of **1**: magnetic measurements between 4.2 and 300 K were interpreted in terms of a strong antiferromagnetic interaction between adjacent radical anions.[47] Variable temperature magnetic data were reported for compounds {Li$(\mathbf{1})_2$} and {Li(**1**)$(dioxan)_2$}.[48]

Indications that the dianion $\mathbf{1}^{2-}$ might not be a simple diradical came when {Li_2(**1**)$(THF)_2$} was reported to be diamagnetic[46] and the observation that {Li_2(**1**)$(THF)_2$} in the solid state and {Li_2(**1**)} or {Na_2(**1**)} in solution were EPR silent.[44,45]

These first fully analyzed EPR studies were reported on red solutions of {Na(**1**)} in DME in 1962[49] although later workers were unable to reproduce the spectrum.[50] The 134 line spectrum showed a clear quartet splitting due to coupling to the spin 3/2 ^{23}Na nucleus and a quintet splitting due to two equivalent ^{14}N nuclei with spin 1, both of which are compatible with a symmetrical structure with significant interaction between the sodium and the organic radical. Fluid solution EPR spectra of {Li(**1**)} in dioxan were interpreted in terms of dominant {Li(**1**)} species whereas those of {K(**1**)} in MTHF contained free $\mathbf{1}^-$ radical anions.[51] The EPR spectra of 77 K frozen solutions of the product from the reaction of potassium with **1** or **4** in MTHF were interpreted in terms of dimeric biradical structures K[K$(\mathbf{1})_2$] or K[K$(\mathbf{4})_2$] with two orthogonal ligands on the central potassium and a K^+ counterion sufficiently closely associated with the anion to desymmetrize it (Fig. 4).[52] These studies were extended to an analysis of the EPR spectra of frozen solutions of the products of the reaction of **1** with

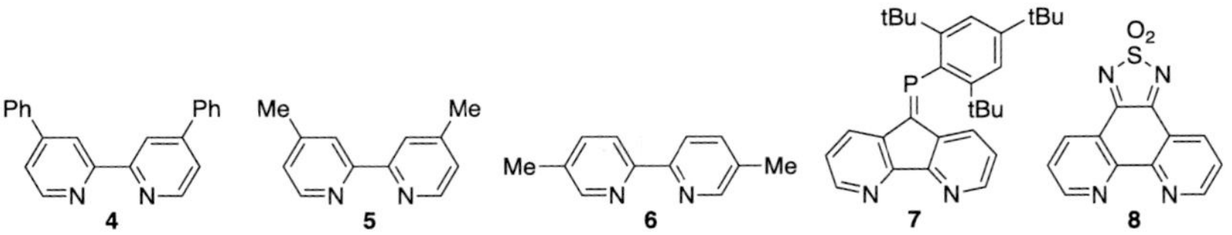

Fig. 4 Some simple 2,2′-bipyridine derivatives which have been converted to radical anions by reaction with group 1 metals.

sodium and cesium in THF, DME or MTHF.[53] The spectra were interpreted in terms of the presence of free radical anions or {M(**1**)} and biradical structures $M[M(\mathbf{1})_2]$ with the concentration independence of the ratio in the K–**1** system in MTHF being interpreted in terms of the equilibrium:

$$2\{K(1)\} \rightleftharpoons K^+ + \{K(1)_2\}^-$$

A variable temperature EPR study of {Na(**1**)} in DME over the range 188–323 K suggested, on the basis of the hyperfine coupling constants, that the compound should be regarded as dynamic with the sodium not localized in the chelate plane but rather exhibiting vibrations out of that plane.[54] The hyperfine coupling to ^{39}K in {K(**1**)} and {K(**4**)} in THF and other solvents is very small or non-detectable suggesting that these species might best be regarded as tight ion pairs rather than complexes.[55]

Combined EPR, ^{7}Li and ^{23}Na NMR studies of {M(**1**)} compounds in DME, DMF, THF or MTHF provided further insight into the solution structures.[56] In particular, the NMR studies were conducted at higher concentrations and provided evidence for aggregation, possibly of the $M[M(\mathbf{1})_2]$ type. A re-examination of the EPR spectra of {Li(**1**)}, {Na(**1**)} and {K(**1**)} in DME, DMF, THF or MTHF combined with calculations confirm the chelating nature of the metal-**1** interaction.[57,58] Dynamic nuclear polarization studies on {Li(**1**)} were relatively inconclusive; the ^{7}Li polarization is small and independent of temperature and solvent. The coupling constant a/h is negative but much smaller than might be expected for proposed chelated structures, presumably due to quadrupole relaxation induced by the Li - N bond.[59]

Computational chemistry was invoked relatively early in the search to understand the electronic structures of these neutral compounds, with the first LCAO-MO results being reported in 1962.[44] A subsequent LCAO-MO study confirmed the results and calculated the charge on the metal in {M(**1**)} (M = Li, Na or K) to be +0.912, +0.932 or + 0.970, respectively, and in accord with their electronegativities.[60] Gustav subsequently reported revised calculations at the EHMO level which indicated that the Christoffersen and Baker approach gave more realistic charge densities that the Mulliken model.[61,62] Numerous other levels of calculation have been reported for the anion and dianion.[63–65]

In view of the highly colored nature of the {M(**1**)} compounds, it is not surprising that their spectra have been investigated. The electronic spectra allow a facile distinction between (red) {Li(**1**)} with absorptions at 383, 516, 545, 745 and 826 nm and (green) {Li_2(**1**)} with maxima at 382, 618, 672, 740 and 820 nm.[37,63,64,66–68] The absorption spectra of {Na(**1**)}, {Na(**5**)}, {Na(**6**)} and {K(**1**)} in THF exhibit four $\pi^* \leftarrow \pi$ bands, of which the one close to 556 nm is most sensitive to the metal ion (Fig. 4). The broad details of the spectra of {Li(**1**)}, {Na(**1**)} and {K(**1**)} in THF were confirmed in subsequent studies together with the reporting of broad absorptions between 320 and 450 nm for aged solutions.[69] Solutions of {Li(**1**)} in Et_2O are luminescent with an emission close to 630 nm.[63]

Another diagnostic spectroscopic technique is vibrational (IR) spectroscopy and the $\mathbf{1}^-$ radical anion in {Li(**1**)$(S)_n$} (S = THF or Et_2O) exhibits characteristic ligand-based absorptions in the 1625–400 cm^{-1} region.[63,70] A very complete study of the resonance Raman spectra of {Li(**1**)} using ^2H and ^{15}N labeled ligands has allowed a complete analysis of the fundamental vibrational modes.[66]

To summarize the state of knowledge by the 1990s: (i) compounds with various group 1 metal ions were known with stoichiometry {M_2(**1**)}, {M(**1**)} and {M$(\mathbf{1})_2$}, (ii) {M_2(**1**)} compounds were diamagnetic and contained $\mathbf{1}^{2-}$ anions, (iii) {M(**1**)} species were paramagnetic and contained $\mathbf{1}^-$ radical anions, (iv) {M_2(**1**)}, {M(**1**)} tended to have solvent molecules associated with them, (v) spectroscopic and calculational methods established the nature of the bonding within the various **1** species, (vi) EPR studies gave conflicting information about the nature and symmetry of the metal-ligand interactions in {M(**1**)}, (vii) the metal-ligand interactions varied with the metal present, (viii) spectroscopically similar derivatives of **1** could be generated by electrochemical reduction and (ix) no structural data had been reported.

To date (2021), no solid-state structures of the archetypical lithium compounds, which are expected to exhibit the strongest N—Li bonding have been reported, but the overall picture changed between 1999 and 2009, when X-ray structures of sodium, potassium and rubidium complexes of $\mathbf{1}^-$ and $\mathbf{1}^{2-}$ species, albeit with additional donor ligands, were reported.

The reaction of sodium with **1** in DME-PhMe gave the deep red one-dimensional polymer $[(\{Na(DME\text{-}1\kappa O^1{:}2\kappa O^2)\}_2(\mathbf{1}\text{-}1\kappa^2 N{:}2'\kappa^2 N))_\infty]\cdot$PhMe (CSD Refcode JAXQUU, Fig. 5A) containing the $\mathbf{1}^{2-}$ anion.[71] Each $\mathbf{1}^{2-}$ anion binds both nitrogen atoms to two sodium cations (Na—N, 2.395, 2.369 Å) and additional Na—C interactions to the two carbons comprising the inter-annular C—C bond (Na...C, 2.872, 2.884 Å). When the reaction was repeated in PMDTA, the discrete deep red species $[[Na(PMDTA\text{-}1\kappa^3 N)]_2(1\text{-}1\kappa^2 N{:}2\kappa^2 N)]$ was obtained, exhibiting a similar coordination mode for the $\mathbf{1}^{2-}$ dianion (Na—N, 2.367, 2.385, 2.431, 2.439 Å; Na...C, 2.736, 2.775, 2.780, 2.787 Å) (CSD Refcode JAXRIJ, Fig. 5B). A third structure resulted when the reaction was performed in TMEDA-benzene which gave the black (or possibly deep red!) compound $[(Na_8O)Na_6(\mathbf{1})_6(TMEDA)_6].4C_6H_6$ containing a central $Na_{14}O$ cluster (the oxygen presumably arising from adventitious oxygen) which is

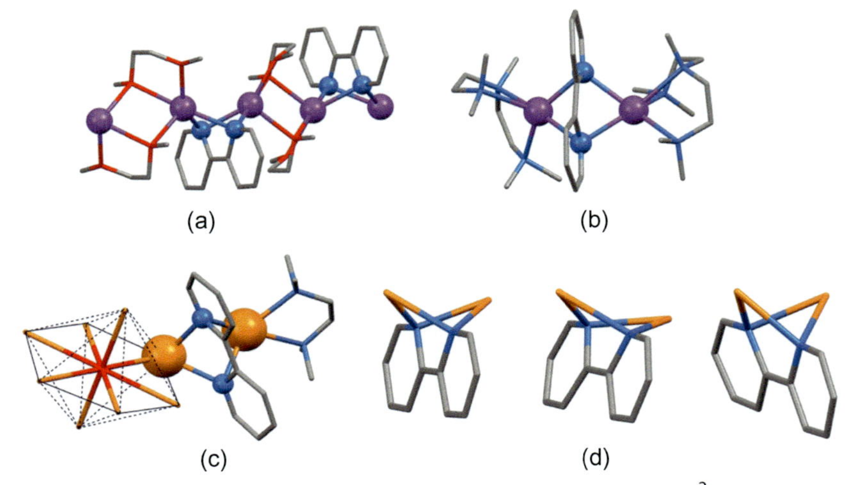

Fig. 5 The solid state structures of sodium complexes containing $\mathbf{1}^{2-}$ anions, all of which exhibit interactions between the anion and two sodium ions (A) part of the one-dimensional coordination polymer $[\{Na(DME\text{-}1\kappa O^1{:}2\kappa O^2)_2(\mathbf{1}\text{-}1\kappa^2 N{:}2'\kappa^2 N)\}_\infty]$ (CSD Refcode JAXQUU)[71]; (B) the discrete dinuclear compound species $[[Na(PMDTA\text{-}1\kappa^3 N)]_2(\mathbf{1}\text{-}1\kappa^2 N{:}2\kappa^2 N)]$ (CSD Refcode JAXRIJ)[71]; (C) the cluster complex $[(Na_8O)Na_6(\mathbf{1})_6(TMEDA)_6].4C_6H_6$ (CSD Refcode JAXROP)[71];in which only one of the six outer $\{Na_2(\mathbf{1})(TMEDA)\}$ units is shown and the hexagonal bipyramidal Na_8O cluster is indicated with dotted lines; (D) comparison of the coordination environment about the $\mathbf{1}^{2-}$ ligand in the three preceding compounds. Metal cations and $\mathbf{1}^{2-}$ donor atoms are shown in ball and stick representation and hydrogen atoms as well as lattice solvent molecules have been omitted for clarity.

encapsulated by six $\mathbf{1}^{2-}$ dianions and six TMEDA-κ^2N ligands. The central core comprises a hexagonal bipyramidal Na_8O cluster and the $\mathbf{1}^{2-}$ anion forms short contacts to the sodium ions of the hexagonal plane (Na—N, 2.357–2.406 Å) and longer contacts to the six outer sodium ions (Na... N, 2.438–2.539 Å), each of which also interacts with a TMEDA ligand (Na—N, 2.467–2.688 Å) (CSD Refcode JAXROP, Fig. 5C). There is a large number of other sodium-aromatic ring and longer Na...N interactions within this complex structure. The similarity of the three $\{Na_2(\mathbf{1})\}$ motifs is presented in Fig. 5D.

More recently the reaction of **1** with potassium or rubidium in en has been investigated and compounds containing either $\mathbf{1}^{-}$ or $\mathbf{1}^{2-}$ have been isolated.[72] Two dark purple polymorphs of a species exhibiting a 1:1:1 K-**1**-en stoichiometry were structurally characterized, $[\{K(\mathbf{1}\text{-}1\kappa^2N\text{:}2\kappa^2N)(\text{en-}1\kappa N^1\text{:}2\kappa N^2)\}_\infty]$ (CSD Refcode QUPGAK, Fig. 6A) and $[K_4(\mathbf{1})_4(\text{en})_4]$ (CSD Refcode QUPGEO, Fig. 6B). The one-dimensional polymer $[\{K(\mathbf{1}\text{-}1\kappa^2N\text{:}2\kappa^2N)(\text{en-}1\kappa N^1\text{:}2\kappa N^2)\}_\infty]$ comprises $\{K_2(\mathbf{1})_2\}$ units bridged by en ligands, with K—N distances to the in plane chelated **1** ligand of 2.820–2.841 Å and in the range 2.852–2.892 Å for the en ligands. In addition, each **1** ligand interacts with an orthogonal potassium ion at a distance of 2.960 Å. The core building block in the two-dimensional network is a $\{K_4(\mathbf{1})_4\}$ with short K—N bonds in the plane of the ligand and longer orthogonal contacts to the next potassium center (CSD Refcode QUPGEO, Fig. 6C). In contrast, the reaction of rubidium with **1** in en gave the two-dimensional coordination network $[\{Rb_2(\mathbf{1})(\text{en})_2\}_\infty]$ containing $\mathbf{1}^{2-}$ anions (CSD Refcode QUPGIS, Fig. 6D).

An example of a related compound is found in the structurally characterized compound $[K(\mathbf{7})(18C6)]\cdot 2THF$ (CSD Refcode DOMLOJ) in which the sterically demanding ligand **7**, containing the Mes* substituent on phosphorus, is converted to its radical anion by reaction with metallic potassium; the two nitrogen donors of ligand **7** are asymmetrically coordinated (K...N, 2.804, 3.266 Å) above the metal, which in turn lies above the plane of the O_6 donor set of the crown ether.[73] Detailed magnetic studies of derivatives of **8** and its radical anions have been reported, including solid state structures of potassium (CSD Refcodes MINQEH, MINSAF), rubidium (CSD Refcode MINQIL) and cesium (CSD Refcode MINQOR) derivatives.[74]

Very recently, it has been shown that doping **1** with sodium-potassium alloy close to room temperature generates a superconducting material with a critical temperature close to 7 K.[75]

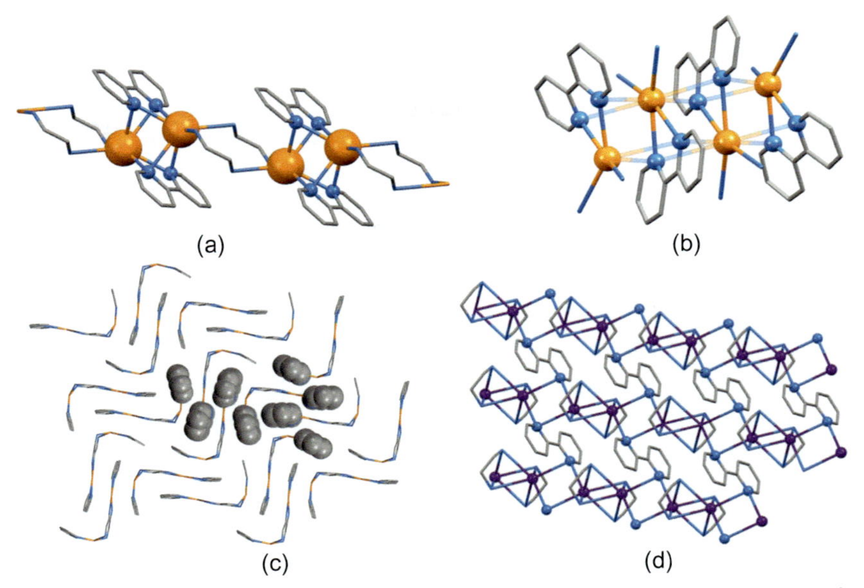

Fig. 6 The solid state structures of potassium complexes containing $\mathbf{1}^-$ anions and a rubidium compound containing $\mathbf{1}^{2-}$ anions (A) part of the one-dimensional coordination polymer $\mathbf{1}^{2-}$ anions (CSD Refcode QUPGAK) [72]; (B) part of the two-dimensional coordination network $[K_4(\mathbf{1})_4(en)_4]$ showing the $\{K_4(\mathbf{1})_4\}$ package with the longer, orthogonal, K–M bonds translucent and only the nitrogen donor atoms of the en ligands shown (CSD Refcode QUPGEO) [72]; (C) the extended two-dimensional network in $[K_4(\mathbf{1})_4(en)_4]$ (CSD Refcode QUPGEO) showing how the $\{K_4(\mathbf{1})_4\}$ package extends to create the sheet and (D) the two-dimensional network found in $[\{Rb_2(\mathbf{1})(en)_2\}_\infty]$ (CSD Refcode QUPGIS) Metal cations and $\mathbf{1}^{2-}$ donor atoms are shown in ball and stick representation and hydrogen atoms have been omitted for clarity.

5.1.2 Simple 2,2′-bipyridines

This section places the emphasis upon structurally characterized complexes of **1** and its derivatives with group 1 metals. Coordination networks and polymers of oligopyridines with group 1 metal ions have been recently reviewed and will not be covered in detail here. Herzog could not prepare simple homoleptic complexes of **1** with group 1 metal salts analogous to those described for ligand **3** in Section 5.2.2[76] although Vögtle subsequently reported the synthesis of $[Li(\mathbf{1})(SCN)(H_2O)_{0.5}]$ from the reaction of **1** with LiSCN in $MeCO_2H$ or $MeCO_2Me$.[77]

A number of studies of solutions containing **1** or derivatives and group 1 cations have been reported. In THF, DMF or MeOH, ^7Li and ^{13}C NMR spectroscopic studies suggested that **1** did not coordinate to Li^+, but in propylene carbonate or $MeNO_2$ complexes were formed with $Li(ClO_4)$, with good evidence for the formation of $\{Li(\mathbf{1})_2\}^+$ in $MeNO_2$.[78,79] More extensive ^7Li and ^{23}Na NMR spectroscopic studies with **1** and ($M(ClO_4)$ M = Li or Na) in a range of solvents established that both $[M(\mathbf{1})]^+$ and $[M(\mathbf{1})_2]^+$ (M = Li or Na) species could be formed in various conditions. In general, the higher the solvating ability of the solvents, as quantified by the Gutmann donor number, the less stable the complex with lithium compounds typically being more stable than those with sodium (Table 2).[80] Computational studies at the DFT level predicted a tetrahedral $[Li(\mathbf{1})_2]^+$ cation, with Li—N distances of 2.05 Å, slightly shorter than is typically found in the solid state.[79] In THF, MeCN and propylene carbonate (but not DMF, DMSO or MeOH), ^{23}Na and ^{13}C NMR spectroscopic studies show clear evidence for the interaction of **1** with $Na(BPh_4)$ although the stoichiometry was not fully established.[77,78] Similar ^{39}K NMR spectroscopic studies on solutions of **1** and $K(AsF_6)$ indicated the formation of weak complexes, but ^{133}Cs studies provided no evidence for the coordination of **1** to Cs^+ in solution.[78] The use of the "non-coordinating" ionic liquids (emim)X ($X = ClO_4$, $F_3CSO_2NSO_2CF_3$ or $EtOSO_3$) has also been investigated by ^7Li NMR spectroscopy; in the case of (emim)($F_3CSO_2NSO_2CF_3$) and (emim)(ClO_4) clear evidence for $[Li(\mathbf{1})_2]^+$ was found, although in the case of the perchlorate, the possibility of coordinated anion could not be excluded (Table 3).[81]

Table 2 Stability constants (log *K*) values for the formation of 1:1 and 1:2 complexes of **1** with $LiClO_4$ and $NaClO_4$ in various solvents.[80]

	$LiClO_4$		$NaClO_4$	
Solvent	**log K_1**	**log K_2**	**log K_1**	**log K_2**
Me_2CO	<0.5	<0.5	1.22 ± 0.02	0.43 ± 0.02
MeCN	2.11 ± 0.07	0.86 ± 0.02	1.31 ± 0.07	0.76 ± 0.07
$PhNO_2$			3.2 ± 0.03	
$MeNO_2$	2.44 ± 0.07	2.29 ± 0.07	2.39 ± 0.07	0.90 ± 0.02

Table 3 Stability constants (log β_2) values for the formation of 1:1 complexes of **1** with LiBr and $NaClO_4$ in various solvents.[79,81]

Solvent	**$Li(F_3CSO_2NSO_2CF_3)$**	**$Li(ClO_4)$**	**Reference**
(emim)(ClO_4)		4.45	81
(emim)($F_3CSO_2NSO_2CF_3$)	4.46		81
$MeNO_2$		4.76, 4.73	79

Table 4 Stability constants (log *K*) values for 1:1 complexes of **1** with LiBr and $NaClO_4$ in various solvents.[82,83]

Solvent	LiBr	$Na(ClO_4)$	Reference
95% EtOH	1.61 ± 0.05	1.30 ± 0.06	82
MeOH	0.45 ± 0.04	0.31 ± 0.01	83
MeCN	1.87 ± 0.01	1.67 ± 0.02	83
Me_2CO	1.85 ± 0.05	1.19 ± 0.04	83

Fluorometric measurements using competitive binding of murexide or direct observation of ligand emission have provided insight into the interaction of **1** with LiBr and $Na(ClO_4)$ in various solvents (Table 4).[82,83]

Finally, mass spectrometric methods have also been used to investigate group 1 cation interactions with **1** and its derivatives. The electrospray mass spectra of aqueous solutions containing Li^+ or Na^+, **1** and 5-methylphenazin-1(5*H*)-one (P), homoleptic $[M(\mathbf{1})]^+$ and heteroleptic $[M(\mathbf{1})(P)]^+$ species (M = Li or Na) were obtained; the collisionally activated dissociation of $[M(\mathbf{1})(P)]^+$ generated both $[M(P)]^+$ and $[M(\mathbf{1})]^+$.[84] Collisionally activated dissociation studies of methanol solutions containing **1** or **5** and MCl (M = Li, Na, K, Rb or Cs) established the formation of $[M(\mathbf{1})_2]^+$ (M = Li or Na) and $[M(\mathbf{1})_2]^+$ (M = Li, Na or K), but no complexes with rubidium or cesium.[85]

Structurally characterized homoleptic complexes of 2,2′-bipyridines with group 1 metal cations are relatively rare, with only two examples reported as of September 2021. The compound $[Li(\mathbf{1})_3]I \cdot 1.5\mathbf{1}$ is obtained from LiI-**1** melts and contains discrete octahedral $[Li(\mathbf{1})_3]^+$ cations with Li—N distances in the range 2.18–2.25 Å, and bite angles of 73.6–74.7° (CSD Refcode REXVOF, Fig. 7A).[86] In contrast, the cation $[Na(\mathbf{1})_3]^+$ found in $[Na(\mathbf{1})_3]I$ obtained from the crystallization of NaI and **1** solutions is close to trigonal biprismatic (CSD Refcode VEPBOI, Fig. 7B) with Na—N distances 2.477 and 2.554 Å and bite angles 66.39–66.40°.[87] The **1** ligands in the cation are helically distorted (Fig. 7B) and the cation is chiral, with both enantiomers present in the crystal lattice. The LiI–**1** system is complex and the phase diagram has been studied and the solid state species including $\{Li(\mathbf{1})I\}$ and $\{Li(\mathbf{1})_2I\}$ isolated, and one solid state species with a 1:1 stoichiometry, is the one-dimensional species shown in Fig. 7C, comprising a ladder-like $\{LiI\}_n$ core capped with **1**-κ^2N ligands (CSD Refcode REXVUL).[86] One-dimensional structures are found in other apparently simple species, and $[\{Na(\mathbf{1})(ClO_4)\}]$ has a similar structure to the previously

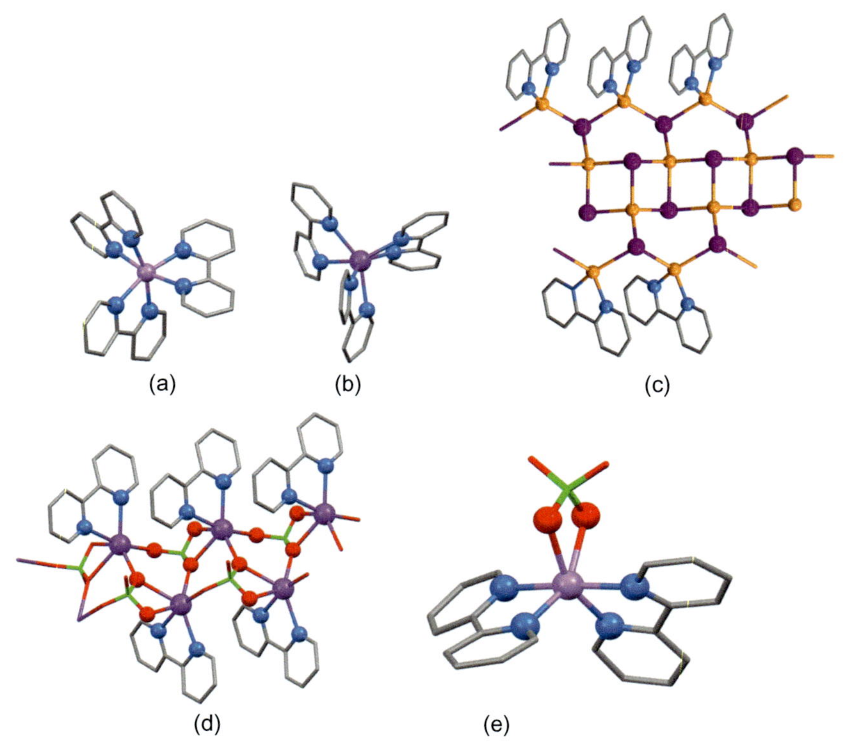

Fig. 7 (A) The homoleptic $[Li(\mathbf{1})_3]^+$ cation found in $[Li(\mathbf{1})_3]I\cdot 1.5\mathbf{1}$ is close to octahedral (CSD Refcode REXVOF)[86]; (B) in contrast to the trigonal biprismatic $[Na(\mathbf{1})_3]^+$ cation in $[Na(\mathbf{1})_3]I$ (CSD Refcode VEPBOI). [87] One-dimensional structures are found in (C) $[\{Li_2I_2(\mathbf{1})\}_n]$ (CSD Refcode REXVUL)[86] and (D) $[\{Na(\mathbf{1})(ClO_4)\}]$ (CSD Refcode DEMMIS).[88] (E) A return to a discrete mononuclear structure is observed in the compound $[Li(\mathbf{1}\text{-}\kappa^2N)(ClO_4\text{-}\kappa^2O)]\cdot\mathbf{1}$ (CSD Refcode UCORUB). [81] Metal cations and donor atoms are shown in ball and stick representation and hydrogen atoms have been omitted for clarity.

discussed iodide, with a central one-dimensional $\{Na(ClO_4)\}$ ladder, containing perchlorate ligands bridging three sodium centers, capped with $\mathbf{1}\text{-}\kappa^2N$ ligands (CSD Refcode DEMMIS, Fig. 7D). The sodium ions are in six-coordinate N_2O_4 environments.[87,88]

A number of other mononuclear complexes containing **1** or related ligands have been structurally characterized and compounds of the type $[M(\mathbf{1})_nX]^{z+}$ ($n = 1$ or 2) are better established. The compound $[Li(\mathbf{1}\text{-}\kappa^2N)(ClO_4\text{-}\kappa^2O)]\cdot\mathbf{1}$ (CSD Refcode UCORUB, Fig. 7E) was obtained from the reaction of LiI and **1** in a 1:3 ratio in (emim)(ClO_4)[81] Other mononuclear complexes which have been structurally characterized include [Li(**1**-κ^2*N*)Br(iPrOH-κ*O*)] (CSD Refcode KENREA), obtained from the reaction of LiI with **1** in ether followed by treatment with iPrOH,[89] [Li(**1**-κ^2*N*){SP(Ph$_2$)CH(Ph$_2$)PS-κ^2*S*}]·C$_6$H$_6$ (CSD Refcode MEXGON),[90] [Li(**1**-κ^2*N*)(BH$_4$-κ^2*H*)(THF)] (CSD Refcode SUSPOM), obtained from the reaction of Li(BH$_4$) with **1** in THF,[91] and [Li(**1**-κ^2*N*){2-pyP(*Me*)NSiMe$_3$-κ^2*N*}] (CSD Refcode SUPSUR).[92] This latter compound is of interest as the **1** ligand was formed *in situ* from the reaction of (2-py)$_3$P=NSiMe$_3$ with LiMe in ether.[92] In the mononuclear complex [Na(**1**-κ^2*N*)(L)(Et$_2$O-κ*O*) (HL= 1,3-bis(trimethylsilyl)indene, CSD Refcode MIGBUB)], the deprotonated indene acts as an η^5-donor to the metal.[93] The reaction of K(BPh$_4$) with **1** in acetone gives crystals of {K(**1**)(BPh$_4$)} which are moderately water stable, but nothing further is known regarding the coordination mode.[94]

Metalloligands are metal complexes which have additional peripheral coordination capacity and a number of group 1 complexes containing metalloligands have been reported; in [(**5**–κ^2*N*)Li(CH$_2$Cl$_2$–κ*Cl*){(MesNCONR–1κ*O*:2κ^2*N*)U(**5**–κ^{2}N)Cl$_3$}]·CH$_2$Cl$_2$ (*R* = 2,6-bis(isopropyl) benzene), one chlorine atom of a dichloromethane molecule occupies the fourth coordination site of the lithium (Li … Cl=2.751 Å, CSD Refcode FUDJEV), {Jilek et al., 2014, #130344} whereas the lithium has a more conventional N$_2$SCl environment in [(**1**–κ^2*N*)Li(S–1κ*S*:2κ*S*)(Cl–1κ*S*:2κ*S*){Ti(OR)(cp)}]·CH$_2$Cl$_2$ (R= 2,6-bis(isopropyl)benzene, CSD Refcode RAKCUB).[95] In [NaY(**9**)$_4$]·4MeOH a {Na(**9**-κ^3*N*^{2}*O*)} metalloligand acts as a bidentate *O*$_2$ donor to a {Y(**9**-κ^3*N*^{2}*O*)} acceptor, resulting in a discrete dinuclear species with a six-coordinate sodium center and an eight-coordinate yttrium (CSD Refcode XAGQIJ),[96] while in [Na(**10**-κ^2*N*){Zr(OAr)$_3$(**11**)}]·PhMe (Ar= 2,6-dimethylphenyl, CSD Refcode JASYUY) the {Na(**10**-κ^2*N*)} entity is embraced by the two deprotonated η^5-pyrroles.[97] Metalloligands can also form network structures, a good example of which is found in the one-dimensional system in which a {Mn(**1**-κ^2*N*)(CO-κ*C*)$_2$} unit acts as a 1κ*O*^{2}:2κ*O*^{2} bridging ligand connecting {Na(**1**-κ^2*N*)(Et$_2$O-κ*O*)} coordination entities.[98] The ligand H$_2$**12** has a tendency to form one-dimensional polymers based on metalloligands, and examples have been structurally characterized in which {(**12**-κ^2*N*)Na(Me$_2$SO-1κ*O*:2κ*O*)$_3$Na(**12**-κ^2*N*)} entities act as bridging ligands to form a one-dimensional chain with M^{2+} centers (M=Ni, CSD Refcode LEYWIW; M=Pd, CSD Refcode LEYWOC),[99] Related one-dimensional chains are found in the compounds (NEt$_4$)$_n$[(Hg{(**12**-κ^2*N*)$_2$Na(Me$_2$SO-κ*O*)})$_n$]·*n*H$_2$O (CSD Refcode LEYWUI),[99] and

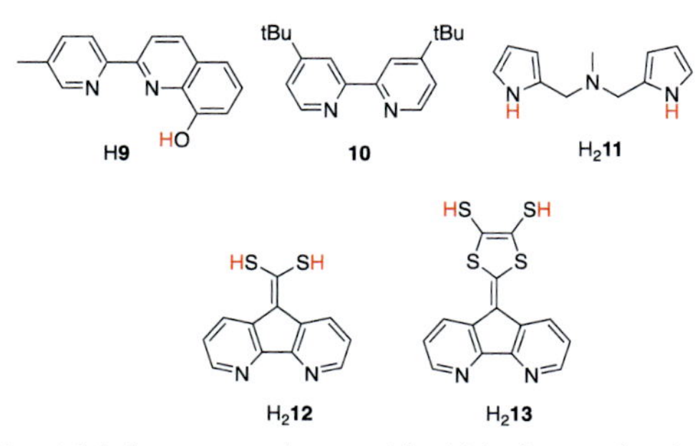

Fig. 8 Some ligands which form complexes with d-block metals which can act as metalloligands to group 1 cations.

[(Hg{(**13**-κ²*N*)₂Na₂(Me₂SO-κ*O*)₂(Me₂SO-1κ*O*:2κ*O*)₃})*ₙ*] (CSD Refcode REJPAY) containing the metalloligand {(**13**-κ²*N*)Na(Me₂SO-κ*O*)(Me₂SO-1κ*O*:2κ*O*)₃Na(Me₂SO-κ*O*)(**13**-κ²*N*)} (Fig. 8).[100]

One of the other common structural motifs is the formation of dinuclear Li_2 complexes, and structurally characterized examples include $[Li_2(\mathbf{14}{-}\kappa^2N)_2(I{-}1\kappa I{:}2\kappa I)_2]$ (CSD Refcode MORGIL, Fig. 9A),[101] $[Li_2(\mathbf{5}{-}\kappa^2N)_2(BH_4{-}1\kappa^2H^1H^2{:}2\kappa^2H^1H^3)_2]$ (CSD Refcode PIVVOF, Fig. 9B),[102] $[Li_2(\mathbf{1}{-}\kappa^2N)_2\{(ArO)_2PO_2{-}1\kappa O^1{:}2\kappa O^2\}_2]\cdot 2PhMe$ (Ar = 2,6-bis(isopropyl)phenyl, CSD Refcode SOHRIT),[103] $[Li_2(\mathbf{1}{-}\kappa^2N)_2(CF_3SO_2NSO_2CF_3{-}1\kappa^2O{:}2\kappa O)_2]$ (CSD Refcode UCOROV),[81] $[K_2(\mathbf{1}{-}\kappa^2N)_2(tBuOSiMe_2NSiMe_3{-}1\kappa N{:}2\kappa^2NO)_2]$ (CSD Refcode SARHUO),[104] $[Na_2(\mathbf{1}{-}\kappa^2N)(tBuOSiMe_2NSiMe_3{-}1\kappa N{:}2\kappa O)_2]$ (CSD Refcode SARHOI),[104] $[Li_2(\mathbf{1}{-}\kappa^2N)_2(2\text{-}pySe{-}1\kappa N{:}2\kappa Se)_2]$ (CSD Refcode YIMBAX),[105] $[Li_2(\mathbf{1}{-}\kappa^2N)_2(PhSe{-}1\kappa Se{:}2\kappa Se)_2]$ (CSD Refcode YILZUO),[105] and $[Li_2(\mathbf{1}{-}\kappa^2N)_2(L{-}1\kappa^2O^1O^2{:}2\kappa O^1)_2]$ (HL = 2,4,6-trinitrophenol, CSD Refcode VINDOL).[106]

Another motif which is surprising to the transition metal chemist is the $1\kappa N^1{:}2\kappa N^2$ bridging mode. This is often associated with the presence of anionic substituents on the 2,2′-bipyridine, typified in the structurally characterized compounds $[Li_2(\mathbf{15}{-}1\kappa^2N^1O{:}2\kappa^2N^2O)(THF)_4]$ (CSD Refcode JOQYUJ),[107] and $[Li_2(\mathbf{16}{-}1\kappa^2N^1C{:}2\kappa^2N^2C)(en)_2]\cdot 2C_6H_6$ (CSD Refcode TOHLAD, Fig. 10A)[110] The reaction of 4,5-diazafluorene, H**17**, with NaH in THF generates {Na(**17**)}, which was partially protonated upon crystallization to give $[Na_2(\mathbf{17}{-}1\kappa^2N^1N^2{:}2\kappa N^2)_2(H\mathbf{17}{-}\kappa^2N)_2]$ (CSD Refcode OFUPAH, Fig. 10B).[108] Another interesting example is found in the two-dimensional structure of $[\{Li_2(\mathbf{1}{-}1\kappa N^1{:}2\kappa N^2)Mo(\mathbf{1}{-}\kappa^2N)Cl_4\}_n]$ in which one 2,2′-bipyridine ligand is chelated to a molybdenum center and the second is coordinated to lithium centers in adjacent

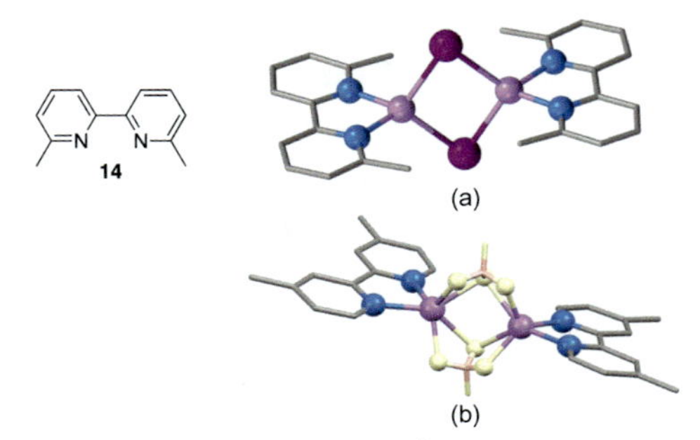

Fig. 9 Dilithium complexes include (A) $[Li_2(\mathbf{14}\text{-}\kappa^2N)_2(I{-}1\kappa I{:}2\kappa I)_2]$ (CSD Refcode MORGIL), [101] and (B) $[Li_2(\mathbf{5}\text{-}\kappa^2N)_2(BH_4{-}1\kappa^2H^1H^2{:}2\kappa^2H^1H^3)_2]$ (CSD Refcode PIVVOF).[102] Metal cations and donor atoms are shown in ball and stick representation and hydrogen atoms, other than those of the tetrahydridoborate, have been omitted for clarity.

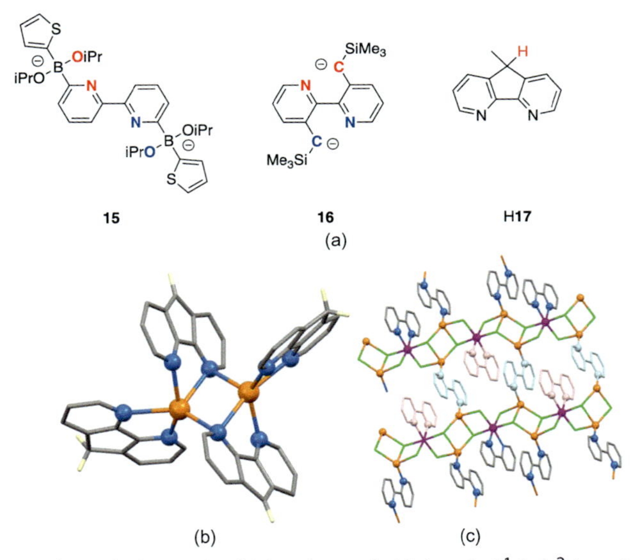

Fig. 10 (A) Two ligands (**15**, **16**) which adopt a bridging $1\kappa N^1{:}2\kappa N^2$ bonding mode to form dilithium complexes. The coordinating atoms are shown in bold and colored to indicate which of the two group 1 metal ions they are coordinated to; (B) the compound $[Na_2(\mathbf{17}{-}1\kappa^2N^1N^2{:}2\kappa N^2)_2(H\mathbf{17}{-}\kappa^2N)_2]$ contains both neutral and deprotonated 4,5-diazafluorene ligands (CSD Refcode OFUPAH)[108] and; (C) the two-dimensional structure of $[\{Li_2(\mathbf{1}{-}1\kappa N^1{:}2\kappa N^2)Mo(\mathbf{1}{-}\kappa^2N)Cl_4\}n]$ (CSD Refcode BOZGUV).[109] Metal cations and donor atoms are shown in ball and stick representation and hydrogen atoms, other than those at C9 of H**17**, have been omitted for clarity, Li in orange and bridging **1** shown in pale blue, chelating **1** in pink.

1-dimensional $\{Li_2(\mu\text{-}Cl)2Mo(\mu\text{-}Cl)_2\}$ chains (CSD Refcode BOZGUV, Fig. 10C).[109] Another two-dimensional structure is found in the compound $[(K(\mathbf{1})\{Au(CN)_2\})_n]$ in which each 2,2′-bipyridine ligand is coordinated two three potassium ions with K...N distances in the range 2.778–3.017 Å. These $\{K(\mathbf{1})\}_n$ one-dimensional chains are then bridged by linear $Au(CN)_2$ units to generate the two-dimensional structure (CSD Refcode PCAUPY01).[111]

A number of compounds which can loosely be described a supramolecular involving group 1 cations and ligands containing 2,2′-bipyridine metal-binding domains have been described. DFT calculations suggest that ligand **18**$^-$ should act as a chelating N_2O donor to Li^+ and the complex $\{Li(\mathbf{18})\}$ has been utilized as an electron injection layer in OLEDs with a driving voltage dependence almost independent of the film thickness.[112,113] The salt H(**19**) is the product of the reaction of 6-(pyrazol-3-yl)-2,2′-bipyridine with $K(BH_4)$ and has been structurally characterized. In the solid state, the double helical $[K_2(\mathbf{19})_2]$ complex (CSD Refcode RIBMAQ01, Fig. 11) is present containing two six-coordinate potassium centers, each coordinated to a chelating terdentate N_3 metal-binding domain from each ligand and with the BH_2 unit acting as the helical "hinge."[114,115]

5.1.3 Cyclic and encapsulating ligands with 2,2′-bipyridine metal-binding domains

A variety of macrocyclic ligands incorporating two 2,2′-bipyridine metal-binding domains have been investigated, both experimentally and *in silico*. The first examples were the macrocycles **20–23** (Fig. 12, macrocycles drawn as the tautomers found in the complexes) which were shown to bind Li^+ (but not Na^+ or K^+) with a color change from red to colorless

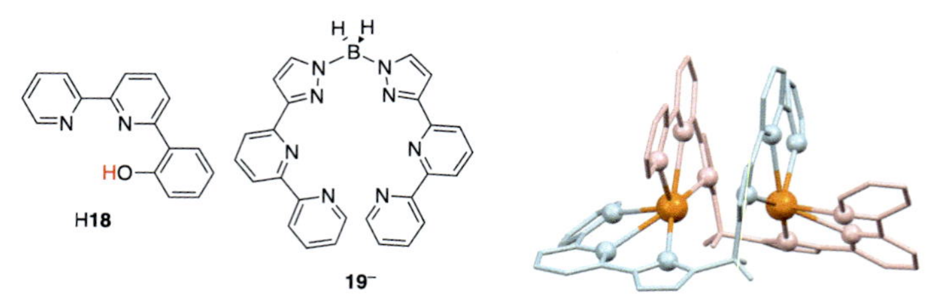

Fig. 11 The solid state structure of the double-helical complex $[K_2(\mathbf{19})_2]$ with the two threads colored to emphasize the structural features (CSD Refcode RIBMAQ01).[114] Metal cations and donor atoms are shown in ball and stick representation and hydrogen atoms, other than those of the BH_2, have been omitted for clarity.

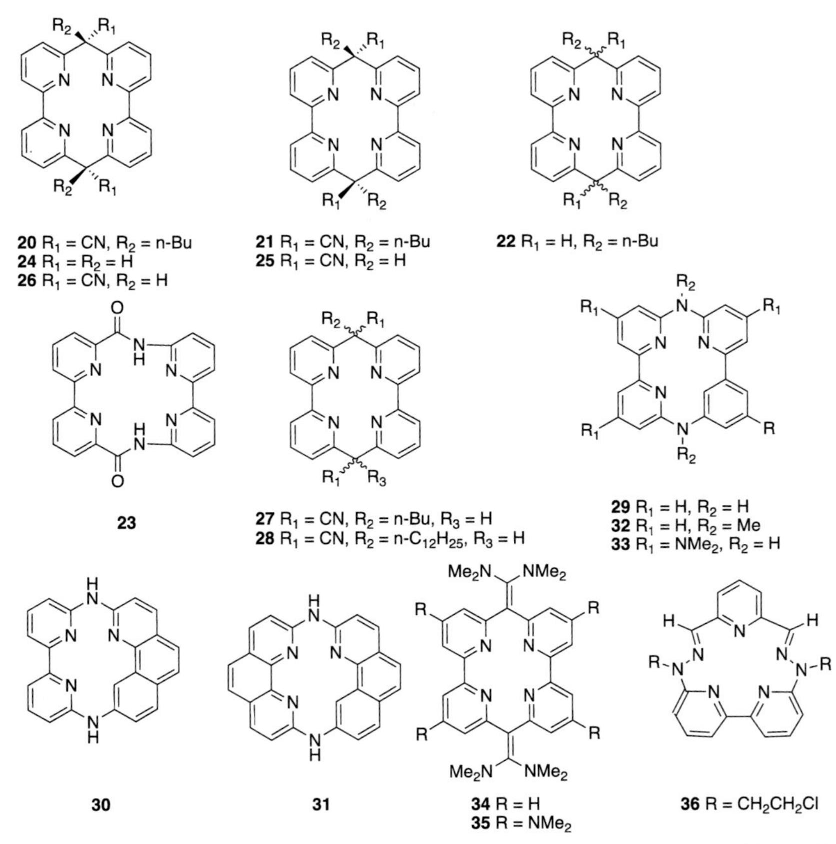

Fig. 12 The structures of macrocyclic ligands incorporating 2,2′-bipyridine metal-binding domains which have been shown to bind group 1 metal ions.

in CH_2Cl_2 or $CHCl_3$.[116,117] The ^{1}H NMR spectra of $CDCl_3$ solutions containing **20** or **21** and LiCl were interpreted in terms of a lithium cation lying above the N_4 plane of a bowed macrocycle. In the case of the complex with **21** a dynamic inversion was observed whereas no such behavior was observed for **20**. In methanol, complexes with **20** (log *K* LiCl 6.30; NaCl 2.60) are considerably more stable than those with **21** (log *K* LiCl 3.89; NaCl 2.11).[117] The fluorescence of macrocycle **21** was enhanced by 3 orders of magnitude in either CH_2Cl_2 or MeCN.[118] Interestingly, the related macrocycle **24** only shows a very modest fluorescence enhancement in the presence of lithium salts.[118] Calculational studies on **20** or **22** (CNDO/S CI with RHF/3-21G optimized geometries) confirmed the color changes associated with lithium binding by **22** arise from a major conformational change. It was suggested that fluorescence enhancement on binding Li^+ might be due to intersystem crossing between energetically close singlet and triplet states.[119] More detailed studies on the model macrocycles **24**, **25** and **26** and their lithium complexes found two local minima for the complex with **24** of C_{2h} and C_{2v} symmetry, respectively. The computational studies on the complexes of **25** and **26** are in accord with the ^{1}H NMR studies.[120] Macrocycle **27** has been evaluated and shown to be effective for the separation of lithium from seawater.[121] The related compound **28** is also a very effective lithium transport species, although the rate of uptake and intimate structure of the complex is anion dependent.[122]

Computational studies on macrocycles **29** and **30** and the related 1,10-phenanthroline analog **31** confirm their lithium binding potential,[123] while **32–35** are expected to be good ligands for Li^+, Na^+ and K^+.[124]

An unusual environment for a group 1 metal ion is found in the complex [Li(**36**)(MeOH)](PF_6) (Fig. 13A, CSD Refcode FAHJAZ) where the lithium adopts a pentagonal based pyramidal coordination geometry, with the five nitrogen donors of **36** providing the basal plane and a methanol the axial donor. The complex is obtained from methanol and is stable in oxygen donor solvents. [125]

In addition to macrocycles, encapsulating ligands related to cryptands have also been developed for binding group 1 metals. The earliest examples were **37** and **38** (Fig. 14) which were shown to give sodium complexes in solution and form solid state species [Li(**37**)ClO_4] and [Na(**38**)ClO_4].[128] The solid state structure of the complex [Na(**38**)]Br·$2H_2O$ confirmed that the sodium cation is located in the cryptand in an eight-coordinate environment (Fig. 13B, CSD Refcode DEKQIV; Na...N_{bridge} 2.706–2.828 Å, Na...N_{bpy} 2.568–2.637 Å, Na...O, 2.475–2.692 Å).[126] The solid-state structure of the salt in $[Na(\mathbf{38})]_2[W_6I_{14}]$ (CSD Refcodes TOMBIJ, TOMBIJ01) has also been recently reported.[129]

Subsequently the all-nitrogen donor systems **39** and **40** (Fig. 15) were obtained directly as the complexes [NaL]Br (L = **39** and **40**) from the synthesis suggesting the operation of a template effect. The ^{1}H NMR spectra of the complexes indicated a fluctional process related to twisting of the strands about the bridgehead N...N axis.[130] These compounds are members of an extensive series of encapsulating ligands incorporating 2,2′-bipyridine and related metal-binding domains in which the NaBr complexes are typically obtained directly from the synthesis (**41–54**).[131–133] Subsequently the synthetic procedure was extended to the preparation of

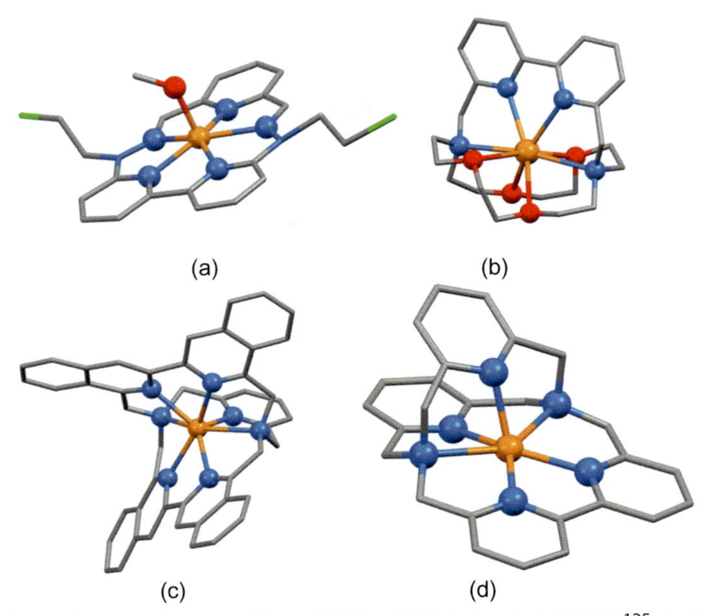

Fig. 13 (A) In the complex [Li(**36**)(MeOH)](PF_6) (CSD Refcode FAHJAZ),[125] the lithium is in a pentagonal pyramidal environment in which the N_5 plane is capped with a methanol ligand; (B) the structure of the cation in [Na(**38**)]Br·$2H_2O$ (CSD Refcode DEKQIV)[126] (C) the $[Li(\mathbf{48})]^+$ cation in [Li(**48**)]Br·$2CH_2Cl_2$ (CSD Refcode SOLLUA)[127] and (D) the cation present in [Li(**41**)]Br (CSD Refcode SOLMAH).[127] Metal cations and donor atoms are shown in ball and stick representation and hydrogen atoms have been omitted for clarity.

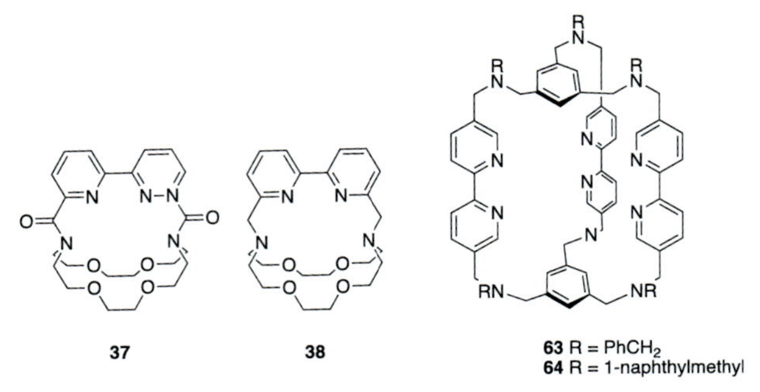

Fig. 14 The structures of encapsulating ligands incorporating 2,2′-bipyridine metal-binding domains which have been shown to bind group 1 metal ions.

[NaL]Br (L = **55–59**). [133,134] and complexes of the ligands **60–62**.[134] The solid state structures of [Li(**41**)]Br and [Li(**48**)]Br·$2CH_2Cl_2$ (CSD Refcode SOLLUA, Fig. 13C).[127] In [Li(**48**)]Br·$2CH_2Cl_2$ the lithium is in an asymmetric seven-coordinate environment with a short Li—N bond

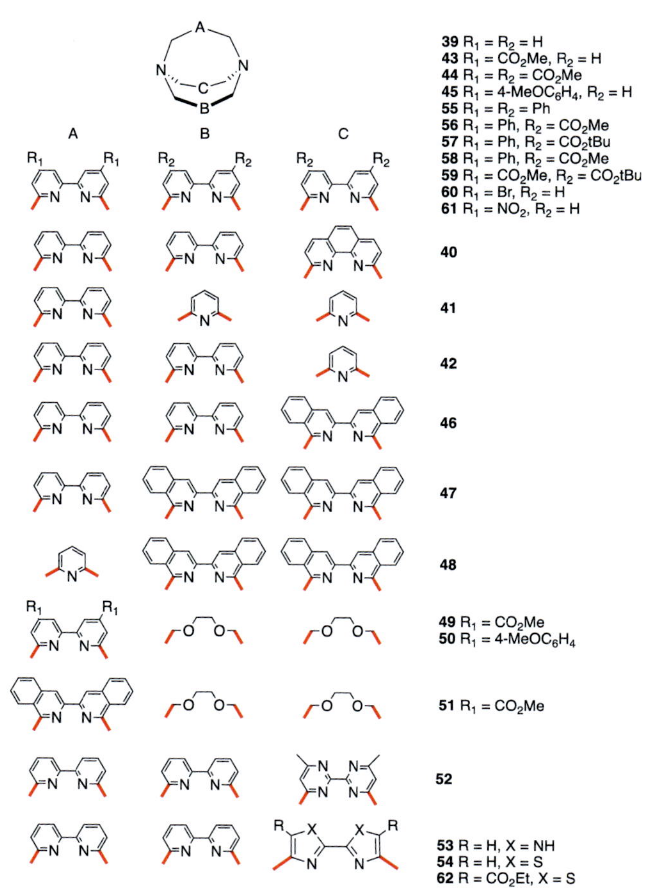

Fig. 15 The structures of encapsulating ligands incorporating 2,2′-bipyridine metal-binding domains which typically give NaBr complexes directly from the macrocyclization reactions.

to the pyridine at 2.069 Å, longer Li—N bonds to the 2,2′-biisoquinoline donors (Li—N, 2.448, 2.332 Å) and two additional long bonds to the bridgehead nitrogen atoms (Li—N, 2.555, 2.556 Å). In contrast, the lithium in [Li(**41**)]Br is best described as being in a pentagonal pyramidal coordination environment with a very different ligand conformation and short Li—N bonds to the axial pyridine (Li—N, 2.126 Å), and the equatorial pyridine (Li—N, 2.224 Å), 2,2′-bipyridine (Li—N, 2.176 Å), and longer contacts to the equatorial bridgehead nitrogen atoms (Li—N, 2.426 Å) (CSD Refcode SOLMAH, Fig. 13D). Calculational studies of the complexation of group 1 and other metal cations with **39** indicated that K^+ would form the most stable complex of the group 1 metals.[135] The complexes [NaL]Br (L = **63** or **64**) are both obtained directly from the macrocyclization reactions of 5,5′-bis(bromomethyl)-2,2′-bipyridine with capping triammines in the presence of Na_2CO_3.[136]

The electrosynthesis of the compound [Na(**39**)], described as a cryptatium, is of particular interest as the compound is obtained as blue-violet very air sensitive crystals by the electrochemical reduction of [Na(**39**)]Br in DMF. This compound is related to the species discussed in Section 5.1.1 and the crystal structure (CSD Refcode VOBSIO, Fig. 16A) shows that the Na—N distances to one of the 2,2′-bipyridine subunits (2.583 Å) are considerably shorter than those to the other two subunits (Na—N 2.817–2.827 Å). This implies that the radical anion is localized on one, and only 2,2′-bipyridine subunits.[137] The detailed electrochemical behavior of $[Na(L)]^+$ (L = **39, 52–54**) confirmed that the formation of cryptatium species is quite general when a reducible heterocycle is incorporated in the cryptand.[140,141]

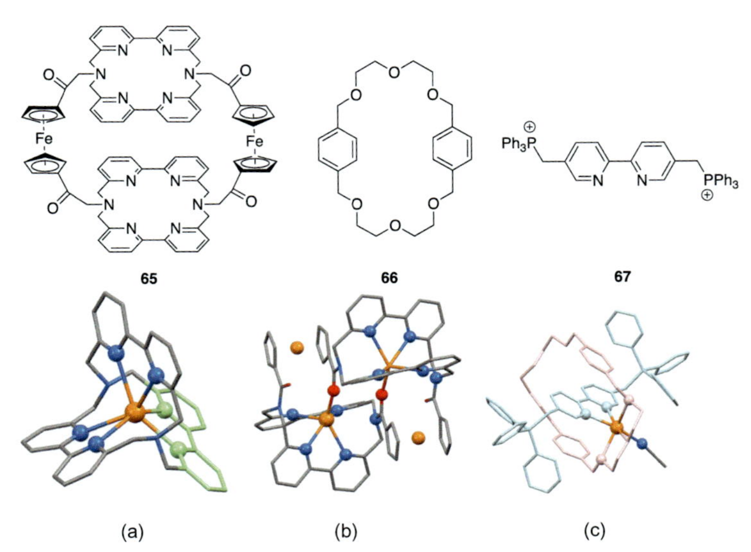

Fig. 16 (A) In the cryptatium [Na(**39**)] a sodium cation is located in a cryptand in which one of the 2,2′-bipyridine units (colored green) (CSD Refcode VOBSIO);[137] (B) the structure of the disodium cation in $[Na_2(\textbf{65})](ClO_4)_2 \cdot 4MeCN$ (CSD Refcode RUHRAN)[138] and (C) the rotaxane $[Na(\textbf{66})(\textbf{67})(MeCN)](PF_6)_3 \cdot 0.5H_2O \cdot 0.5MeOH$ in which a sodium cation anchors the ring (in pink) on the dumbbell (in blue) (CSD Refcode JINLAU).[139] Metal cations and donor atoms are shown in ball and stick representation and hydrogen atoms have been omitted for clarity.

Ligand **65** binds two sodium cations in the complex $[Na_2(\mathbf{65})](ClO_4)_2 \cdot 4MeCN$, with each in a square pyramidal N_4O coordination environment with Na—N distances in the range 2.373–2.579 Å, Na—O, 2.192 Å and an Na...Na distance of 5.567 Å (CSD Refcode RUHRAN, Fig. 16B).[138] Finally, a rotaxane $[Na(\mathbf{66})(\mathbf{67})(MeCN)](PF_6)_3 \cdot 0.5H_2O \cdot 0.5MeOH$ has been described in which a sodium cation anchors the ring (**66**) on the dumbbell (**67**) (CSD Refcode JINLAU, Fig. 16C).[139]

5.2 $1^1,2^2:2^6,3^2$-Terpyridine derivatives

Two types of terpyridine have been studied as ligands for group 1 metals—open chain derivatives of $1^1,2^2:2^6,3^2$-terpyridine and those in which the terpyridine metal-binding domain is incorporated into a macrocyclic or encapsulating structure.

5.2.1 Open-chain $1^1,2^2:2^6,3^2$-terpyridines

$1^1,2^2:2^6,3^2$-Terpyridine (**68**) derivatives are readily available, both commercially and through a variety of versatile and adaptable synthetic methodologies. The first group of complexes to be reported were not with **68** itself, but rather with the radical anion $(\mathbf{68})^-$. Nakamura studied the radical species formed from the reaction of **68** with Li, Na, Rb and Cs in the solvents DME, DMF, THF and MTHF and observed three characteristic absorptions in the visible spectrum. He assigned a species with absorption maximum 573 nm (obtained as the only species in DMF) to the free radical anion $(\mathbf{68})^-$, and species with absorption maxima close to 610 and 560 nm to **69** and **70,** respectively.[142] It was suggested, without any supporting evidence, that the metal had different numbers of solvent ligands in **69** and **70**. This paper is also one of the earliest examples postulating a hypodentate $1^1,2^2:2^6,3^2$-terpyridine ligand (Fig. 17).

ESMS experiments have shown that the threshold activation voltages for $[M(\mathbf{5.3.1\_1})_2]$ (M = Li, Na) are rather low indicating that the 1:2 complexes

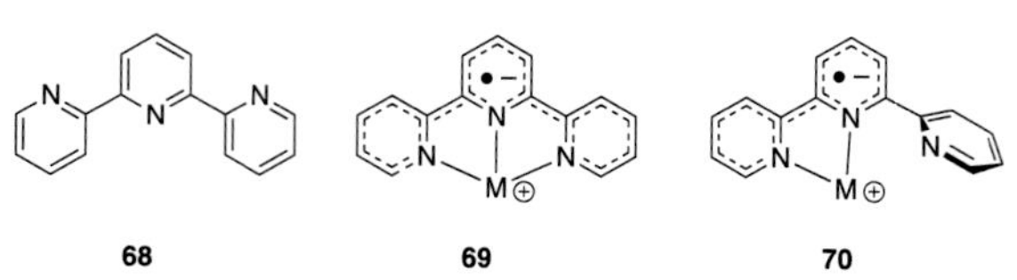

Fig. 17 The compound $1^1,2^2:2^6,3^2$-terpyridine (**68**) and the two forms of coordinated $(\mathbf{68})^-$ radical anion proposed by Nakamura (**69** and **70**). Structure **70** represents a very early example of a hypodentate oligopyridine.

are strongly destabilized with respect to the 1:1 species.[85] This observation was rationalized in terms of the first **5.3.1_1** ligand binding strongly and effectively delocalizing the positive charge.

The first examples of complexes with neutral **68** ligands were the compounds {Li(**68**)(SCN)} and {Na(**68**)(SCN)} obtained from the reaction of the metal salt with **68** in EtOAc or EtOAc-MeOH. Nothing is known about the structures of these species.[77]

The stability of complexes with **68** ligands is expected to increase in poorly coordinating solvents with low donor numbers such as the ionic liquids (emim)(NTf_2) and (emim)(ClO_4). The crystalline complexes {Li(**68**)($CF_3SO_2NSO_2CF_3$)} and {Li(**68**)(ClO_4)} were obtained by the direct reaction of **68** with Li($CF_3SO_2NSO_2CF_3$) in (emim)($CF_3SO_2NSO_2CF_3$) or Li(ClO_4) in (emim)(ClO_4), respectively.[143] Even though the crystals were obtained in the presence of an excess of **68**, no solid state species with a higher than a 1:1 M:L ratio were obtained. Both {Li(**68**)($CF_3SO_2NSO_2CF_3$)} and {Li(**68**)(ClO_4)} have been structurally characterized with the former being described as a 1D coordination polymer in solid state with five-coordinate lithium centers coordinated to a terdentate **68** ligand and two bridging $CF_3SO_2NSO_2CF_3$ ligands (Fig. 18A). Unfortunately, the data for {Li(**68**)($CF_3SO_2NSO_2CF_3$)} do not appear to have been deposited with the CCDC. In contrast, the perchlorate salt is a discrete dimer [(**68**)Li(μ-ClO_4)$_2$Li(**68**)] (CSD Refcode PIRWEU, Fig. 18B), again containing a terdentate **68** ligand. In both cases the lithium is five-coordinate in an N_3O_2 environment with both anions coordinated.

In CH_3NO_2 solution, ESMS measurements provided evidence for the formation of $[Li(\mathbf{68})_2]^+$ and prompted ^7Li NMR studies in (emim)($CF_3SO_2NSO_2CF_3$) and (emim)(ClO_4). In (emim)($CF_3SO_2NSO_2CF_3$), at low concentrations of **68** a variety of species assigned to mono, bi- and

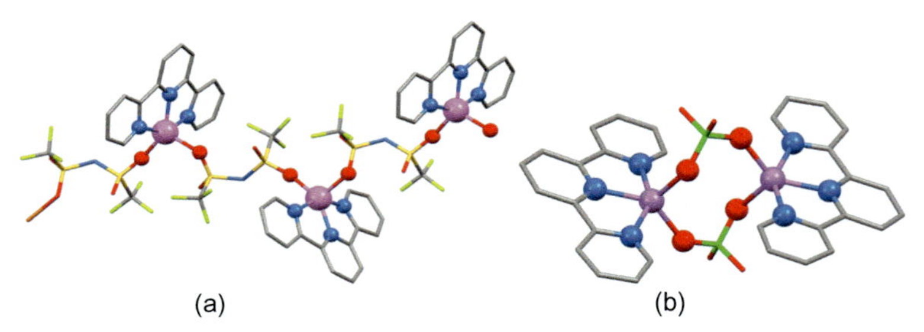

Fig. 18 The solid state structures of (A) {Li(**68**)($CF_3SO_2NSO_2CF_3$)}$_n$ and (B) [(**68**)Li(m-ClO_4)$_2$Li(**68**)] (CSD Refcode PIRWEU).[143]

terdentate **68** ligands were observed but at higher concentrations the data were consistent with the formation of $[Li(\mathbf{68})]^+$ and $[Li(\mathbf{68})_2]^+$. DFT calculations at the B3LYP/ LANL2DZp level showed a clear minimum for a six-coordinate $[Li(\mathbf{68})_2]^+$ ion with two terdentate **68** ligands. At low concentrations of **68** in (emim)(ClO_4) the behavior was slightly different with faster exchange between **68** and the smaller, weakly bound, perchlorate anion. The measured log β_2 values for the formation of $[Li(\mathbf{68})_2]^+$ in (emim)(NTf_2), (emim)(ClO_4) and CH_3NO_2 solution were 5.19, 4.68 and 5.80, respectively.

Support for the formation of monodentate **68** complexes with group 1 metal ions is found in the solid state structure of the complex $K_2[Ru(\mathbf{68}–\kappa^2N^1,N^2)(CN)_4]\cdot MeOH\cdot 0.5\ H_2O$ (CSD Refcode KAZHOJ) which builds a 1D coordination network. The ruthenium only binds to two of the three nitrogen donors of the **68** ligand, and the remaining pyridine is coordinated to the potassium, with the remainder of the coordination sphere of the potassium completed by four side-on cyanido ligands, and a methanol, giving a total coordination number of 10 (Fig. 19).[144]

The three compounds **71**, **72** and **73** (Fig. 20) have been prepared and their ability to bind group 1 picrate salts in water-saturated $CHCl_3$ compared (Table 5).[145] The preorganization of the metal-binding domain

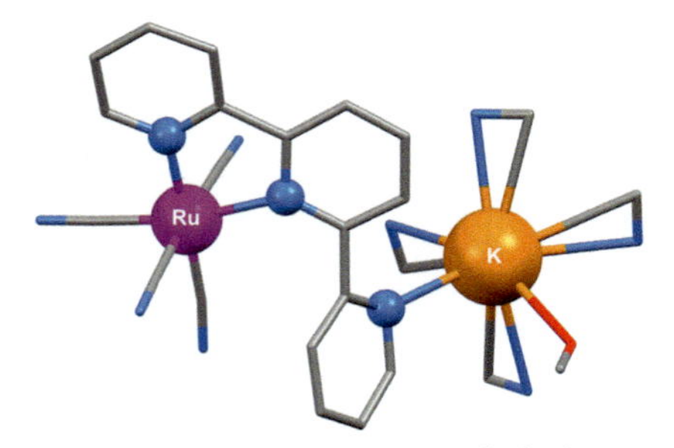

Fig. 19 The $RuK(\mathbf{68})(CN)_4$ unit present in $K_2[Ru(\mathbf{68}–\kappa^2N^1,N^2)(CN)_4]\cdot MeOH\cdot 0.5\ H_2O$ (CSD Refcode KAZHOJ).[144]

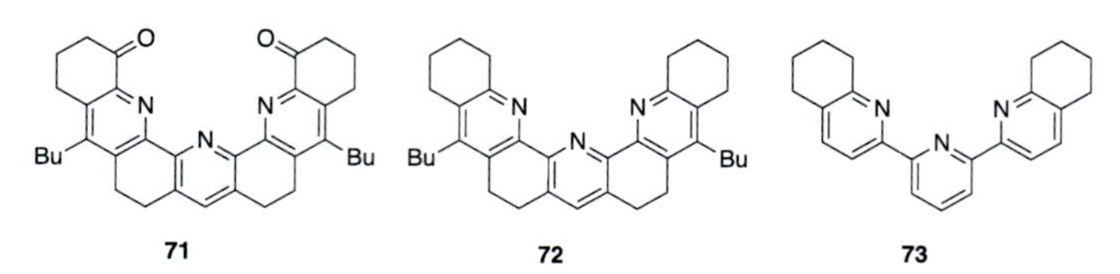

Fig. 20 Three compounds which have been shown to have selective interactions with group 1 metal picrates.

Table 5 Comparison of the binding (log K_S) of group 1 picrate salts in water-saturated $CHCl_3$ by compounds **71**, **72** and **73**.

Ligand	Li	Na	K	Rb	Cs
71	8.3	9.7	9.5	8.9	8.1
72	5.2	4.7	4.7	4.3	4.1
73	<3	<8	<3	<3	<3

in **71** and **72** dramatically enhances the stability of the complexes with respect to **73**. Comparison of the log K_S values for **71** and **72** suggest involvement of the ketone oxygen atom as a donor.

5.2.2 Cyclic ligands with $1^1,2^2:2^6,3^2$-terpyridine metal-binding domains

Relatively few macrocyclic ligands containing $1^1,2^2:2^6,3^2$-terpyridine metal-binding domains and which also bind group 1 metal ions have been reported. One example, and the compound that first attracted my interest to this area of chemistry, is found with the macroycle **74**, which forms a planar pentagonal complex [Li(**74**)](PF_6) in which the cations form a slipped stacking one-dimensional system with the interplanar distances of 3.261 and 3.401 Å (CSD Refcode GEWVEJ, Fig. 21).[146]

5.3 $1^1,2^2:2^6,3^2:3^6,4^2$-Quaterpyridine derivatives

No complexes of group 1 metals with to $1^1,2^2:2^6,3^2:3^6,4^2$-quaterpyridine derivatives have been reported.

5.4 $1^1,2^2:2^6,3^2:3^6,4^2:4^6,5^2$-Quinquepyridine derivatives

Group 1 complexes of ligands related to $1^1,2^2:2^6,3^2:3^6,4^2:4^6,5^2$-quinquepyridine (**75**) are relatively rare (Fig. 22). In 1979, Vögtle reported the isolation of the complex {Li(**75**)(ClO_4)(H_2O)$_2$}, but nothing further is known regarding the structure of this compound, although the lithium is presumably coordinated to all five nitrogen donors.[147] Only one example of a $1^1,2^2:2^6,3^2:3^6,4^2:4^6,5^2$-quinquepyridine derivative which forms group 1 complexes has been reported; compound **76** comprises a pair of rapidly interconverting atropisomers the rate of atropisomerization is reduced and the pitch of the helix changed upon coordination to sodium cations.[148]

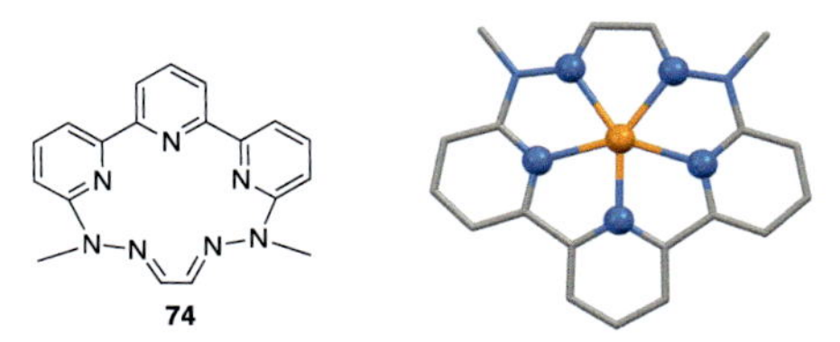

Fig. 21 A macrocycle containing a $1^1,2^2:2^6,3^2$-terpyridine metal-binding domain which forms a planar pentadentate lithium complex (CSD Refcode GEWVEJ).[146]

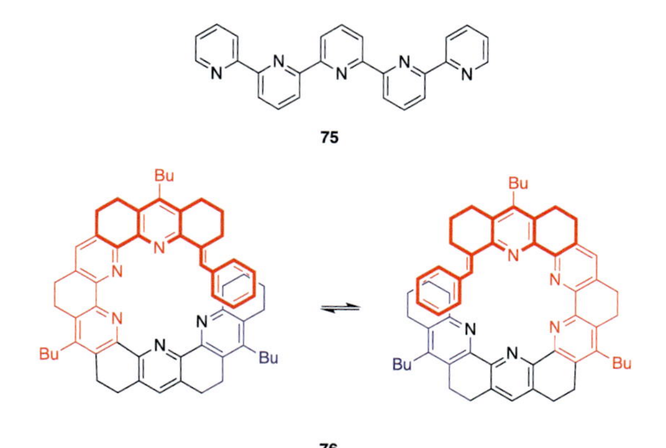

76

Fig. 22 The compound $1^1,2^2:2^6,3^2:3^6,4^2:4^6,5^2$-quinquepyridine (**75**) forms a 1:1 complex with $LiClO_4$ of unknown structure. The helical quinquepyridine derivative **76** exists as a pair of rapidly interconverting atropisomers; coloration has no significance other than to emphasize the stereochemistry. Coordination to Na^+ slows the rate of racemization and alters the helical pitch of the heterohelicene.

5.5 $1^1,2^2:2^6, 3^2:3^6, 4^2:4^6, 5^2:5^6, 6^2$-Sexipyridine and higher oligopyridine derivatives

Two types of sexipyridine have been studied in the context of binding group 1 metals—derivatives of linear $1^1,2^2:2^6$, $3^2:3^6$, $4^2:4^6$, $5^2:5^6$, 6^2-sexipyridine and cyclic sexipyridines, (IUPAC PIN, 1,2,3,4,5,6(2,6)-hexapyridinacyclohexaphanes).

5.5.1 *Linear $1^1,2^2:2^6, 3^2:3^6, 4^2:4^6, 5^2:5^6, 6^2$-sexipyridines*

The parent linear $1^1,2^2:2^6,3^2:3^6,4^2:4^6,5^2:5^6,6^2$-sexipyridine (**77**) is shown in Fig. 23. To date, no complexes of **77** have been isolated, but DFT calculations on the anticipated helical complexes $[M_1M_2(\mathbf{77})]$ (M_1, M_2 = Li, Na) have been carried out. In all cases, helical configurations were found as the minimum energy structures, with Li—N distances being shorter than

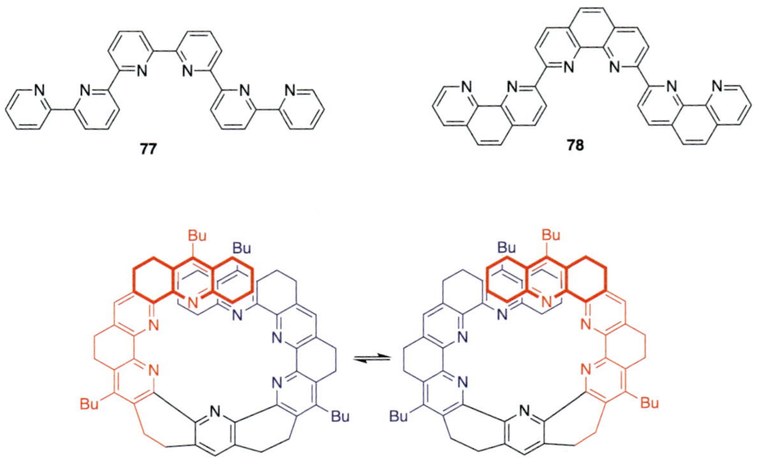

Fig. 23 (A) The structure of the parent linear $1^1,2^2:2^6,3^2:3^6,4^2:4^6,5^2:5^6,6^2$-sexipyridine (**77**) together with that of the closely related 1,10-phenanthroline derivative **78** and (B) the $1^1,2^2:2^6,3^2:3^6,4^2:4^6,5^2:5^6,6^2:6^6,7^2$-septipyridine derivative **79** exists as a pair of rapidly interconverting atropisomers; the coloration has no significance other than to emphasize the stereochemistry. Coordination to Na^+ slows the rate of racemization and alters the helical pitch of the heterohelicene.

Na—N, with consequences for the helical pitch. The shortest M–M distance was observed in [LiNa(**77**)]. The natural charge analyses indicated positive charges between 0.6 and 0.68 on lithium and *ca.* 0.8 on sodium in the three complexes, leading to a formulation approaching the formal description of **77** diradicals coordinated to group 1 metal cations. The work was motivated by the development of new strategies for design of non-linear optical materials and indicated that group 1 metal doping of oligopyridines might prove an effective strategy.[149] This report built upon an earlier computational studies on the winding and unwinding of **77** double helices upon doping with lithium or sodium.[150] The closely related compound **78** (Fig. 23) has been prepared from a double Friedländer reaction of 2,9-bis(acetyl)-1,10-phenanthroline. When the condensation is carried out using KOH as the base, a salt containing the $[K(\mathbf{78})]^+$ cation was isolated.[151] Although group 1 complexes of **77** have not been experimentally observed, the heterohelicene **79**, a $1^1,2^2:2^6,3^2:3^6$, $4^2:4^6,5^2:5^6,6^2:6^6,7^2$-septipyridine derivative, has been prepared as a pair of rapidly interconverting atropisomers (Fig. 23B) and it was shown that the rate of atropisomerization is reduced and the pitch of the helix changed upon coordination to sodium cations.[148]

5.5.2 Cyclic sexipyridines

The compound "cyclosexipyridine" (**80**, Fig. 24) with the IUPAC PIN of 1,2,3,4,5,6(2,6)-hexapyridinacyclohexaphane is an attractive synthetic target and the first report of synthetic approaches appeared in 1983.[152,153] Toner described the preparation of the C_2 symmetric $1^4,4^4$-bis(4-methylphenyl)- (**81**) and $1^4,4^4$-bis(4-ethylphenyl)- (**82**) derivatives as the complexes {Na(L)(OAc)}; although no sodium salts were used in the synthetic procedure, the presence of sodium in the product was inferred from field-desorption mass spectrometric data. Somewhat better characterized are the C_3 symmetric ligands (**84–86**), which could be obtained as mixtures of sodium and potassium complexes, with the sodium probably coming from the glassware. The complexes with **83–85** were very insoluble, but those with **86** were significantly more soluble and solution NMR studies showed that the mixture of sodium and potassium complexes did not possess threefold symmetry, presumably as a result of the out-of-plane twisting of the pyridine rings.[154] DFT calculations at the HF/3-21G and HF/6-31G* levels on the coordination of **86** to group 1 metal cations provide some insight to the binding, with sodium and potassium cations forming lowest energy structures with the ligand in D_3 symmetry (as suggested by Potvin), whereas lithium forms a complex of lower symmetry.[155] The calculated binding energies for Li^+, Na^+ and K^+ (counterpoise corrected at the B3-LYP/Basis1//HF/6-31G* level) are 485.0, 449.6 and 426.2 kJ mol^{-1}, respectively.

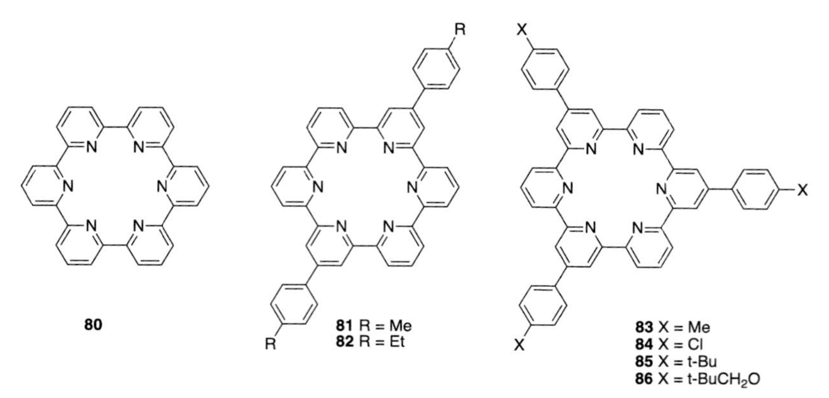

Fig. 24 The parent "cyclosexipyridine" (1,2,3,4,5,6(2,6)-hexapyridinacyclohexaphane), **5.6.2_1**, synthetic examples exhibiting C_2 (**81** and **82**) and C_3 (**83–86**) symmetry.

Better investigated is a second series of compounds which are more soluble and can be seen as partially reduced derivatives of "hexaazakekulene" (**87**, Fig. 25). This class of molecules bind group 1 metal ions exceptionally well and are collectively known as "torands" with the first example **88** being described by Bell in 1989.[157] The stability constants for the 1:1 adducts of **88** with sodium and potassium picrate are lg K_S 14.7 and 14.3, respectively, significantly stronger than the values for binding to the crown ether **89** of 6.4 and 8.3, respectively. The 1:1 potassium and rubidium picrate complexes of **88** have both been structurally characterized (CSD Refcode VOVTIJ and VOVTOP, Fig. 25A and B), with the potassium complex exhibiting a hexagonal planar coordination N_6 geometry and the larger rubidium cation adopting a hexagonal pyramidal structure. The torands adopt the same conformation, although the differing coordination of the picrate (potassium, monodentate K...ONO

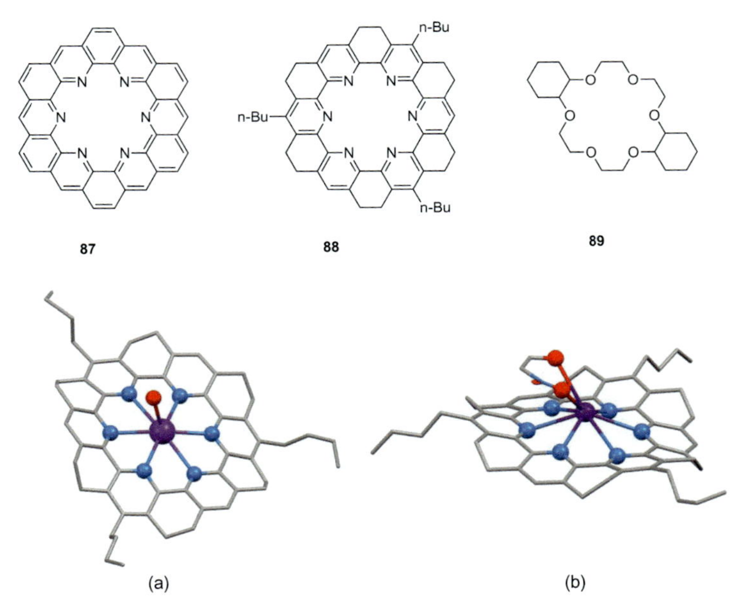

Fig. 25 The compound "hexaazakekulene" (**87**) and the structure of the torand **88**. The crown ether **89** binds to sodium or potassium picrate significantly less strongly than of **88**. (A) The structure of the complex [K(**88**)(picrate)] (CSD Refcode VOVTIJ),[156] hydrogen atoms omitted for clarity, donor atoms and metal center in ball-and-stick representation, only the oxygen atom of the coordinated picrate nitro group shown and (B) the structure of the complex [Rb(**88**)(picrate)]·$0.5CH_2Cl_2$ (CSD Refcode VOVTOP),[156] hydrogen atoms omitted for clarity, donor atoms and metal center in ball-and-stick representation, only the oxygen atom of the coordinated picrate nitro group is shown.

2.836 Å; rubidium, bidentate Rb...ONO 3.091 Å, Rb...OC 3.055 Å) makes a detailed conformational analysis complex.[156] Nevertheless, in both cases the macrocycle adopts a conformation in which the nitrogen atoms lie alternately above and below its mean plane.

The computational studies on the complex of **80** with Li^+ did not yield a single minimum and structural information on complexes of **88** with lithium cations, which are expected to be too small to coordinate to all six nitrogen donors of the ligand, are revealing. The reaction of **88** with a mixture of lithium picrate and hydroxide gives a compound of stoichiometry {Li(**88**)$(H_2O)_{1.5}$(picrate)} which has been structurally characterized and shown to be $[Li_2(\mathbf{88})_2(OH_2)_2(\mu\text{-}OH_2)](picrate)_2$ containing two {Li(**88**)(OH_2)} cations bridged by a water molecule (CSD Refcode VUHLIT, Fig. 26).[158] The lithium ions are displaced from the centroid of the macrocycle and the coordination geometry is best described as four-coordinate (Li—N, 2.03–2.30 Å, Li-OH_2, 1.87–1.90 Å, Li-μ-OH_2, 1.90–1.95 Å) lithium center exhibiting an additional longer interaction to a third nitrogen donor in each macrocycle (Li...N, 2.62–2.64 Å) and, as a consequence, the conformation is different to that observed in the mononuclear potassium and rubidium complexes. Langmuir-Blodgett studies of monolayers of **88** over aqueous sub-phases containing lithium, potassium or cesium salts showed a transition from horizontal to vertically oriented bilayers at higher pressure.[159]

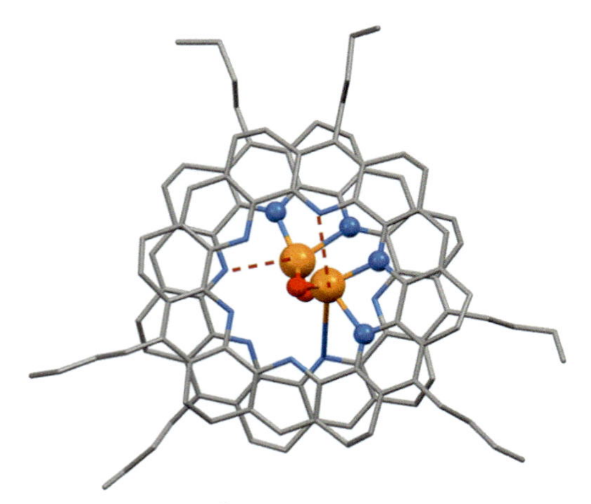

Fig. 26 The $[Li_2(\mathbf{88})_2(OH_2)_2(m\text{-}OH_2)]^{2+}$ dication (CSD Refcode VUHLIT)[158] contains two {Li(**88**)(OH_2)} cations bridged by a water molecule. Each lithium cation is four-coordinate with an additional, longer, contact to a third nitrogen donor from the macrocycle. Metal cations and donor atoms are shown in ball and stick representations, hydrogen atoms are omitted for clarity and the weaker interactions to the third nitrogen of each macrocycle are depicted with maroon dashed lines.

6. Conclusions and future perspectives

This review has attempted to give an overview of the general occurrence of group 1 metal complexes with the nitrogen donor oligopyridine ligands which are typically thought of as partners for d-block metal ions. Although I am aware that this review is, inevitably, not comprehensive, the disparity and beauty of the diverse coordination modes exhibited in complexes with group 1 metal ions makes the behavior with d-block metals look positively mundane. I hope that these observations might inspire others to further investigate the potential of these ligands outside the d-block.

Acknowledgments

As always, I thank the generations of talented colleagues and co-workers who have allowed me to play with oligopyridine ligands throughout the past 40 years. I acknowledge the long term support of the Swiss National Science Foundation for our research in this and related areas for the past 30 years.

References

1. Accorsi, G.; Listorti, A.; Yoosaf, K.; Armaroli, N. *Chem. Soc. Rev.* **2009**, *38*, 1690.
2. Bencini, A.; Lippolis, V. *Coord. Chem. Rev.* **2010**, *254*, 2096.
3. Brandt, W. W.; Dwyer, F. P.; Gyarfas, E. D. *Chem. Rev.* **1954**, *54*, 959.
4. Constable, E. C. *Adv. Inorg. Chem.* **1989**, *34*, 1.
5. Constable, E. C.; Housecroft, C. E. *Molecules* **2019**, *24*, E3951.
6. Kaes, C.; Katz, A.; Hosseini, M. W. *Chem. Rev.* **2000**, *100*, 3553.
7. Lindoy, L. F.; Livingstone, S. E. *Coord. Chem. Rev.* **1967**, *2*, 173.
8. McWhinnie, W. R.; Miller, J. D. *Adv. Inorg. Chem. Radiochem.* **1970**, *13*, 135.
9. Summers, L. A. *Adv. Heterocycl. Chem.* **1978**, *24*, 1.
10. Summers, L. A. *Adv. Heterocycl. Chem.* **1984**, *35*, 281.
11. Luman, C. R.; Castellano, F. N. In *Comprehensive Coordination Chemistry II*; McCleverty, J. A., Meyer, T. J., Eds.; Vol. 1; Elsevier: Oxford, 2003; p. 25.
12. Smith, A. P.; Fraser, C. L. In *Comprehensive Coordination Chemistry II*; McCleverty, J. A., Meyer, T. J., Eds.; Vol. 1; Elsevier: Oxford, 2003; p. 1.
13. Thummel, R. P. In *Comprehensive Coordination Chemistry II*; McCleverty, J. A., Meyer, T. J., Eds.; Vol. 1; Elsevier: Oxford, 2003; p. 41.
14. Bui, A. T.; Castellano, F. N. In *Comprehensive Coordination Chemistry II*; Constable, E. C., Parkin, G., Que, L., Eds.; Vol. 1; Elsevier: Oxford, 2021; p. 78.
15. Grice, K. A.; Nganga, J. K.; Naing, M. D.; Angeles-Boza, A. M. In *Comprehensive Coordination Chemistry III*; Constable, E. C., Parkin, G., Que, L., Eds.; Vol. 1; Elsevier: Oxford, 2021; p. 60.
16. Schilt, A. A. *Analytical Applications of 1,10-Phenanthroline and Related Compounds*; Pergamon: Oxford, 1969; p. 10.
17. McKenzie, E. D. *Coord. Chem. Rev.* **1971**, *6*, 187.
18. Constable, E. C.; Housecroft, C. E. *Aust. J. Chem.* **2020**, *73*, 390.
19. Constable, E. C. *Polyhedron* **2016**, *103*, 295.
20. Constable, E. C.; Housecroft, C. E. *Coord. Chem. Rev.* **2017**, *350*, 84.
21. Favre, H. A.; Powell, W. H. *Nomenclature of Organic Chemistry*; The Royal Society of Chemistry, 2014.

22. Connelly, N. G.; Damhus, T.; Hartshorn, R. M.; Hutton, A. T. *Nomenclature of Inorganic Chemistry*; The Royal Society of Chemistry, 2005.
23. Carleson, B. G. F.; Irving, H. *J. Chem. Soc.* **1954**, 4390.
24. Ahrland, S.; Chatt, J.; Davies, N. R. *Quart. Rev. Chem. Soc.* **1958**, *12*, 265.
25. IUPAC. Compendium of Chemical Terminology. In *The "Gold Book"*; 2nd ed.; Blackwell Scientific Publications: Oxford, UK, 1997.
26. Pearson, R. G. *J. Am. Chem. Soc.* **1963**, *85*, 3533.
27. Groom, C. R.; Bruno, I. J.; Lightfoot, M. P.; Ward, S. C. *Acta Crystallogr. B Struct. Sci. Cryst. Eng. Mater.* **2016**, *72*, 171.
28. Sarad, S. K.; Bhagwat, V. W.; Poonia, N. S. *Univ. Indore Res. J. Sci.* **1977**, *4*, 27.
29. Herzog, S.; Taube, R. *Z. Anorg. Allg. Chem.* **1960**, *306*, 159.
30. Coates, G. E.; Green, S. I. E. *J. Chem. Soc.* **1962**, 3340.
31. Herzog, S.; Schmidt, M. *Z. Chem.* **1963**, *3*, 392.
32. Herzog, S.; Gustav, K. *Z. Anorg. Allg. Chem.* **1966**, *346*, 150.
33. Herzog, S.; Zuehlke, H. *Z. Chem.* **1966**, *6*, 382.
34. Herzog, S.; Wulf, E. *Z. Chem.* **1966**, *6*, 434.
35. Patrikeeva, N. B.; Suvorova, O. N.; Domrachev, D. A. *Metalloorg. Khim.* **1991**, *4*, 684.
36. Herzog, S.; Weber, A. *Z. Chem.* **1968**, *8*, 66.
37. Mahon, C.; Reynolds, W. L. *Inorg. Chem.* **1927**, *1967*, 6.
38. Kaim, W. *Chem. Ber.* **1981**, *114*, 3789.
39. Boersma, J.; Mackor, A.; Noltes, J. G. *J. Organomet. Chem.* **1975**, *99*, 337.
40. Tupper, K. A.; Tilley, T. D. *J. Organomet. Chem.* **2005**, *690*, 1689.
41. Herzog, S.; Taube, R. *Z. Chem.* **1962**, *2*, 208.
42. Herzog, S.; Gutsche, E. *Z. Chem.* **1963**, *3*, 393.
43. Herzog, S.; Lühder, K. *Z. Chem.* **1966**, *6*, 475.
44. Taube, R.; Herzog, S. *Z. Chem.* **1962**, *2*, 225.
45. Elschner, B.; Herzog, S. *Arch. Sci.* **1958**, *11*, 160.
46. Perthel, R. *Z. Phys. Chem.* **1959**, *211O*, 74.
47. Inoue, M.; Hara, K.; Horiba, T.; Kubo, M. *Chem. Lett.* **1972**, 1055.
48. Wulf, E.; Herzog, S. *Z. Anorg. Allg. Chem.* **1972**, *387*, 81.
49. Zahlan, A.; Heineken, F. W.; Bruin, M.; Bruin, F. *J. Chem. Phys.* **1962**, *37*, 683.
50. dos Santos-Veiga, J.; Reynolds, W. L.; Bolton, J. R. *J. Chem. Phys.* **1966**, *44*, 2214.
51. König, E.; Fischer, H. Z. *Naturforsch. A* **1962**, *17*, 1063.
52. Brown, I. M.; Weissman, S. I.; Snyder, L. C. *J. Chem. Phys.* **1965**, *42*, 1105.
53. Van Voorst, J. D. W.; Zijlstra, W. G.; Sitters, R. *Chem. Phys. Lett.* **1967**, *1*, 321.
54. Gooijer, C.; Velthorst, N. H.; MacLean, C. *Mol. Phys.* **1972**, *24*, 1361.
55. Kaim, W. *J. Am. Chem. Soc.* **1982**, *104*, 3833.
56. Takeshita, T.; Hirota, N. *Chem. Phys. Lett.* **1969**, *4*, 369.
57. Takeshita, T.; Hirota, N. J. *Amer. Chem. Soc.* **1971**, *93*, 6421.
58. Takeshita, T.; Hirota, N. *J. Chem. Phys.* **1973**, *58*, 3745.
59. Meise-Gresch, K.; Mueller-Warmuth, W. *Ber. Bunsenges. Phys. Chem.* **1979**, *83*, 586.
60. Gustav, K. *Z. Chem.* **1968**, *8*, 193.
61. Gustav, K. *Z. Chem.* **1975**, *15*, 39.
62. Gustav, K. *Z. Phys. Chem.* **1976**, *257*, 28.
63. Noble, B. C.; Peacock, R. D. *Spectrochim. Acta, Part A* **1990**, *46A*, 407.
64. König, E.; Kremer, S. *Chem. Phys. Lett.* **1970**, *5*, 87.
65. Nakamura, T.; Soma, M.; Onishi, T.; Tamaru, K. *Z. Phys. Chem.* **1974**, *89*, 122.
66. Danzer, G. D.; Golus, J. A.; Strommen, D. P.; Kincaid, J. R. *J. Raman Spectrosc.* **1990**, *21*, 3.
67. Torii, Y.; Yazaki, T.; Kaizu, Y.; Murasato, S.; Kobayashi, H. *Bull. Chem. Soc. Jap.* **1969**, *42*, 2264.
68. Pappalardo, R. *Inorg. Chim. Acta* **1968**, *2*, 209.

69. Kalyanaraman, V.; Rao, C. N. R.; George, M. V. *J. Chem. Soc. B* **1971**, 2406.
70. Saito, Y.; Takemoto, J.; Hutchinson, B.; Nakamoto, K. *Inorg. Chem.* **2003**, *1972*, 11.
71. Bock, H.; Lehn, J. M.; Pauls, J.; Holl, S.; Krenzel, V. *Angew. Chem. Int. Ed Engl.* **1999**, *38*, 952.
72. Gore-Randall, E.; Irwin, M.; Denning, M. S.; Goicoechea, J. M. *Inorg. Chem.* **2009**, *48*, 8304.
73. Chen, C.; Hu, Z.; Li, J.; Ruan, H.; Zhao, Y.; Tan, G.; Song, Y.; Wang, X. *Inorg. Chem.* **2020**, *59*, 2111.
74. Shuku, Y.; Suizu, R.; Domingo, A.; Calzado, C. J.; Robert, V.; Awaga, K. *Inorg. Chem.* **2013**, *52*, 9921.
75. Peng, D.; Wang, R.-S.; Chen, X.-J. *J. Phys. Chem. C* **2020**, *124*, 906–912.
76. Herzog, S. *Die Naturwissenschaften* **1956**, *43*, 105.
77. Voegtle, F.; Mueller, W. M.; Rasshofer, W. *Isr. J. Chem.* **1980**, *18*, 246.
78. Schmidt, E.; Hourdakis, A.; Popov, A. I. *Inorg. Chim. Acta* **1981**, *52*, 91.
79. Schmeisser, M.; Zahl, A.; Scheurer, A.; Puchta, R.; van Eldik, R. Z. *Naturforsch. B: J. Chem. Sci.* **2010**, *65*, 405.
80. Madrakian, T.; Afkhami, A.; Ghasemi, J.; Shamsipur, M. *Polyhedron* **1996**, *15*, 3647.
81. Schmeisser, M.; Heinemann, F. W.; Illner, P.; Puchta, R.; Zahl, A.; van Eldik, R. *Inorg. Chem.* **2011**, *50*, 6685.
82. Ghasemi, J.; Shamsipur, M. *J. Coord. Chem.* **1993**, *28*, 231.
83. Ghasemi, J.; Shamsipur, M. *J. Coord. Chem.* **1992**, *26*, 337.
84. Stone, J. A.; Vukomanovic, D. *Int. J. Mass Spectrom.* **2001**, *210–211*, 341.
85. Satterfield, M.; Brodbelt, J. S. *Inorg. Chem.* **2001**, *40*, 5393.
86. Fischer, E.; Hummel, H.-U. *Z. Anorg. Allg. Chem.* **1997**, *623*, 483.
87. Buttery, J. H. N.; Effendy; Mutrofin, S.; Plackett, N. C.; Skelton, B. W.; Whitaker, C. R.; White, A. H. *Z. Anorg. Allg. Chem.* **2006**, *632*, 1851.
88. Buttery, J. H. N.; Effendy; Mutrofin, S.; Plackett, N. C.; Skelton, B. W.; Somers, N.; Whitaker, C. R.; White, A. H. Z. Anorg. Allg. Chem. **2006**, *632*, 1839.
89. Skelton, B. W.; Whitaker, C. R.; White, A. H. *Aust. J. Chem.* **1990**, *43*, 755.
90. Ren, W.; Chantrapromma, S. *Acta Crystallogr. Sect. E: Struct. Rep.* **2013**, *69*, m160.
91. Aguilar-Martinez, M.; Felix-Baez, G.; Perez-Martinez, C.; Noeth, H.; Flores-Parra, A.; Colorado, R.; Galvez-Ruiz, J. C. *Eur. J. Inorg. Chem.* **1973**, *2010.*
92. Steiner, A.; Stalke, D. *Angew. Chem., Int. Ed.* **1995**, *34*, 1752.
93. Hung-Low, F.; Bradley, C. A. *Inorg. Chem.* **2013**, *52*, 2446.
94. Grillone, M. D.; Nocilla, M. A. *Inorg. Nucl. Chem. Lett.* **1978**, *14*, 49.
95. Witt, E.; Stephan, D. W. *Inorg. Chem.* **2001**, *40*, 3824–3826.
96. Chong, B. S. K.; Rajah, D.; Allen, M. F.; Galán, L. A.; Massi, M.; Ogden, M.; Moore, E. G. *Inorg. Chem.* **2020**, *59*, 16194–16204.
97. Li, Y.; Banerjee, S.; Odom, A. L. *Organometallics* **2005**, *24*, 3272–3278.
98. Hartl, F.; Rosa, P.; Ricard, L.; Le Floch, P.; Záliš, S. *Coord. Chem. Rev.* **2007**, *251*, 557–576.
99. Baudron, S. A.; Hosseini, M. W.; Kyritsakas, N.; Kurmoo, M. *Dalton Trans.* **2007**, 1129–1139.
100. Baudron, S. A.; Hosseini, M. W. *Inorg. Chem.* **2006**, *45*, 5260–5262.
101. García-Rodríguez, R.; Simmonds, H. R.; Wright, D. S. *Organometallics* **2014**, *33*, 7113–7117.
102. Reger, D. L.; Collins, J. E.; Matthews, M. A.; Rheingold, A. L.; Liable-Sands, L. M.; Guzei, L. A. *Inorg. Chem.* **1997**, *36*, 6266–6269.
103. Kalugin, A. E.; Komarov, P. D.; Minyaev, M. E.; Lyssenko, K. A.; Roitershtein, D. M.; Nifant'ev, I. E. *Acta Crystallogr. Sect. E Crystallogr. Commun.* **2019**, *75*, 848–853.
104. Veith, M.; Böhnlein, J.; Huch, V. *Chem. Ber.* **1989**, *122*, 841–849.

105. Khasnis, D. V.; Buretea, M.; Emge, T. J.; Brennan, J. G. *J. Chem. Soc., Dalton Trans.* **1995**, 45.
106. Hundal, M. S.; Sood, G.; Kapoor, P.; Poonia, N. S. *J. Crystallogr. Spectrosc. Res.* **1991**, *21*, 201.
107. Weis, N.; Pritzkow, H.; Siebert, W. *Eur. J. Inorg. Chem.* **1999**, *1999*, 393–398.
108. Jiang, H.; Song, D. *Organometallics* **2008**, *27*, 3587–3592.
109. Helland, S. D.; Chang, A. S.; Lee, K. W.; Hutchison, P. S.; Brennessel, W. W.; Eckenhoff, W. T. *Inorg. Chem.* **2020**, *59*, 705–716.
110. Leung, W.-P.; Poon, K. S. M.; Mak, T. C. W.; Zhang, Z.-Y. *Organometallics* **1996**, *15*, 3262–3266.
111. Döring, C.; Strey, M.; Jones, P. G. *Acta Crystallogr. C Struct. Chem.* **2017**, *73*, 1104–1108.
112. Ohisa, S.; Karasawa, T.; Watanabe, Y.; Ohsawa, T.; Pu, Y.-J.; Koganezawa, T.; Sasabe, H.; Kido, J. *ACS Appl. Mater. Interfaces* **2017**, *9*, 40541–40548.
113. Pu, Y.-J.; Miyamoto, M.; Nakayama, K.-i.; Oyama, T.; Masaaki, Y.; Kido, J. *Org. Electron.* **2009**, *10*, 228–232.
114. Fleming, J. S.; Psillakis, E.; Couchman, S. M.; Jeffery, J. C.; McCleverty, J. A.; Ward, M. D. *J. Chem. Soc. Dalton Trans.* **1998**, 537–543.
115. Psillakis, E.; Jeffery, J. C.; McCleverty, J. A.; Ward, M. D. *Chem. Commun.* **1997**, 479–480.
116. Ogawa, S.; Narushima, R.; Arai, Y. *J. Am. Chem. Soc.* **1984**, *106*, 5760.
117. Ogawa, S.; Uchida, T.; Uchiya, T.; Hirano, T.; Saburi, M.; Uchida, Y. *J. Chem. Soc., Perkin Trans.* **1990**, *1*, 1649.
118. Ogawa, S.; Tsuchiya, S. *Chem. Lett.* **1996**, 709–710.
119. Takano, K.; Furuhama, A.; Ogawa, S.; Tsuchiya, S. *J. Chem. Soc., Perkin Trans.* **1999**, *2*, 1063–1068.
120. Furuhama, A.; Takano, K.; Ogawa, S.; Tsuchiya, S. *Bull. Chem. Soc. Jpn.* **2001**, *74*, 1241–1249.
121. Tsuchiya, S.; Nakatani, Y.; Ibrahim, R.; Ogawa, S. *J. Am. Chem. Soc.* **2002**, *124*, 4936–4937.
122. Morita, J.; Tsuchiya, S.; Yoshida, N.; Nakayama, N.; Ogawa, S. *Heterocycles* **2010**, *80*, 1103–1123.
123. Furuhama, A.; Takano, K.; Ogawa, S.; Tsuchiya, S. *J. Mol. Struct.* THEOCHEM. **2003**, *620*, 49–63.
124. Despotovic, I. *New J. Chem.* **2015**, *39*, 6151–6162.
125. Constable, E. C.; Chung, L.-Y.; Lewis, J.; Raithby, P. R. J. Chem. Soc., Chem. Commun. 1986, 1719.
126. Alzakhem, N.; Bischof, C.; Seitz, M. *Inorg. Chem.* **2012**, *51*, 9343–9349.
127. Cesario, M.; Guilhem, J.; Pascard, C.; Anklam, E.; Lehn, J. M.; Pietraszkiewicz, M. *Helv. Chim. Acta* **1991**, *74*, 1157.
128. Buhleier, E.; Wehner, W.; Voegtle, F. *Chem. Ber.* **1978**, *111*, 200.
129. Hummel, T.; Leis, W.; Eckhardt, A.; Ströbele, M.; Enseling, D.; Jüstel, T.; Meyer, H.-J. *Dalton Trans.* **2020**, *49*, 9795–9803.
130. Rodriguz-Ubis, J.-C.; Alpha, B.; Plancherel, D.; Lehn, J.-M. *Helv. Chim. Acta* **1984**, *67*, 2264–2269.
131. Alpha, B.; Anklam, E.; Deschenaux, R.; Lehn, J. M.; Pietraskiewicz, M. *Helv. Chim. Acta* **1988**, *71*, 1042.
132. Lehn, J. M.; Regnouf de Vains, J. B. *Tetrahedron Lett.* **1989**, *30*, 2209.
133. Chen, S.; Hong, Y.; Liu, Y.; Xue, M.; Zheng, Y.; Shen, Q. *Youji Huaxue* **2017**, *37*, 1198–1204.
134. Lehn, J. M.; Regnouf de Vains, J. B. *Helv. Chim. Acta* **1992**, *75*, 1221.
135. Puchta, R.; van Eldik, R. *Eur. J. Inorg. Chem.* **2007**, 1120–1127.

136. Ebmeyer, F.; Voegtle, F. *Chem. Ber.* **1989**, *122*, 1725.
137. Echegoyen, L.; DeCian, A.; Fischer, J.; Marie Lehn, J. *Angew. Chem., Int. Ed.* **1991**, *30*, 838–840.
138. Hall, C. D.; Truong, T.-K.-U.; Tucker, J. H. R. *Chem. Commun. (Cambridge)* **1997**, 2195–2196.
139. Chen, N. C.; Huang, P. Y.; Lai, C. C.; Liu, Y. H.; Wang, Y.; Peng, S. M.; Chiu, S. H. *Chem. Commun.* **2007**, 4122–4124.
140. Echegoyen, L.; Perez-Cordero, E.; Regnouf de Vains, J. B.; Roth, C.; Lehn, J. M. *Inorg. Chem.* **1993**, *32*, 572.
141. Echegoyen, L.; Xie, O.; Perez-Cordero, E. *Pure Appl. Chem.* **1993**, *65*, 441–446.
142. Nakamura, K. *Bull. Chem. Soc. Jap.* **1943**, *1972*, 45.
143. Pokorny, K.; Schmeisser, M.; Hampel, F.; Zahl, A.; Puchta, R.; van Eldik, R. *Inorg. Chem.* **2013**, *52*, 13167–13178.
144. Adams, H.; Alsindi, W. Z.; Davies, G. M.; Duriska, M. B.; Easun, T. L.; Fenton, H. E.; Herrera, J. M.; George, M. W.; Ronayne, K. L.; Sun, X. Z.; Towrie, M.; Ward, M. D. *Dalton Trans.* **2006**, 39–50.
145. Bell, T. W.; Cragg, P. J.; Firestone, A.; Kwok, A. D.-I.; Liu, J.; Ludwig, R.; Sodoma, A. *J. Org. Chem.* **1998**, *63*, 2232–2243.
146. Constable, E. C.; Doyle, M. J.; Healy, J.; Raithby, P. R. J. Chem. Soc., Chem. Commun. 1988, 1262.
147. Oepen, G.; Voegtle, F. *Liebigs Ann. Chem.* **1979**, 2114.
148. Bell, T. W.; Jousselin, H. *J. Am. Chem. Soc.* **1991**, *113*, 6283.
149. Zhang, F.-Y.; Xu, H.-L.; Su, Z.-M. *Int. J. Quantum Chem.* **2021**, *121*, e26478.
150. Zhang, F.-Y.; Xu, H.-L.; Su, Z.-M. *Dyes Pigm.* **2020**, *176*, 108203.
151. Zong, R.; Thummel, R. P. *Inorg. Chem.* **2005**, *44*, 5984–5986.
152. Toner, J. L. *Tetrahedron Lett.* **1983**, *24*, 2707.
153. Newkome, G. R.; Lee, H. W. *J. Am. Chem. Soc.* **1983**, *105*, 5956.
154. Masciello, L.; Potvin, P. G. *Can. J. Chem.* **2003**, *81*, 209–218.
155. Howard, S. T.; Fallis, I. A. *J. Chem. Soc. Perkin Trans.* **1999**, *2*, 2501–2506.
156. Bell, T. W.; Cragg, P. J.; Drew, M. G. B.; Firestone, A.; Kwok, D. I. A. *Angew. Chem. Int. Ed. Engl.* **1992**, *31*, 345–347.
157. Bell, T. W.; Firestone, A.; Ludwig, R. *J. Chem. Soc., Chem. Commun.* **1989**, 1902.
158. Bell, T. W.; Cragg, P. J.; Drew, M. G. B.; Firestone, A.; Kwok, D. I. A. *Angew. Chem. Int. Ed. Engl.* **1992**, *31*, 348–350.
159. Boguslavsky, L.; Bell, T. W. *Langmuir* **1994**, *10*, 991.

CHAPTER FIVE

Advanced characterization techniques for electrochemical capacitors

Elżbieta Frąckowiak[a,*], Anetta Płatek-Mielczarek[b], Justyna Piwek[a], and Krzysztof Fic[a]

[a]Institute of Chemistry and Technical Electrochemistry, Poznan University of Technology, Poznan, Poland
[b]Laboratory for Multiphase Thermofluidics and Surface Nanoengineering, Department of Mechanical and Process Engineering, ETH Zurich, Sonneggstrasse 3, Zurich, Switzerland
*Corresponding author: e-mail address: elzbieta.frackowiak@put.poznan.pl

Contents

Abstract

This chapter is a comprehensive overview of the recent advances in electrochemical capacitor characterization. Various modes, including *in-situ/operando* and *ex-situ/postmortem* techniques, are described and compared. All the advantages resulting from each approach are highlighted. Special attention is given to the current limits of different modes (*in-situ* vs *ex-situ*), and the need to verify fundamental research where full device testing is discussed.

Advances in Inorganic Chemistry, Volume 79
ISSN 0898-8838
https://doi.org/10.1016/bs.adioch.2021.12.006

Abbreviations

2D-NLDFT	2D non-local density functional theory
AC	activated carbon
AFM	atomic force microscopy
Al^{3+}	aluminum cation
AN	acetonitrile
BMI BF_4	1-butyl-3-methylimidazolium tetrafluoroborate
$BMIM^+$	1-butyl-3-methylimidazolium
C	total system capacitance [F]
C	carbon
C_+, C_-	capacitance of the positive and negative electrodes, respectively [F]
CB	carbon black
CDC	carbide-derived carbon
Ce	electrode capacitance [F]
C_{EDL}	capacitance [F]
CNT	carbon nanotube
C_{QCM}	mass sensitivity constant [ng cm^{-3} Hz^{-1}]
Cs^+	cesium cation
CuO	copper oxide
CVD	chemical vapor deposition
d	EDL thickness [m]
DEME TFSI	diethylmethyl(2-methoxyethyl)ammonium bis(trifluoromethylsulfonyl)imide
DFT	density functional theory
E	energy [Wh]
ε_0	vacuum permittivity [F m^{-1}]
EA	elemental analysis
EC	electrochemical capacitor
EDL	electrical double layer
EDLC	electric double layer capacitor
EIS	electrochemical impedance spectroscopy
EMIM BF_4	1-ethyl-3-methylimidazolium tetrafluoroborate
EMIM Cl	1-ethyl-3-methylimidazolium chloride
EMIM FSI	1-ethyl-3-methylimidazolium bis(fluorosulfonyl)imide
EMIM OTF	1-ethyl-3-methylimidazolium trifluoromethanesulfonate
EMIM SCN	1-ethyl-3-methyl imidazolium thiocyanate
EMIM TFSI	1-ethyl-3-methylimidazolium bis(trifluoromethylsulfonyl)imide
$EMIM^+$	1-ethyl-3-methylimidazolium cation
EQCM	electrochemical quartz crystal microbalance
EQCM-D	electrochemical quartz crystal microbalance with dissipation monitoring
ε_r	relative electrolyte permittivity [−]
ESR	equivalent series resistance
f_0	the fundamental resonance frequency of the crystal [Hz]
FTIR	Fourier transform infrared spectroscopy
GC-MS	gas chromatography-mass spectrometry
H	hydrogen
$H_2PO_4^-$	dihydrophosphate anion
H_2SO_4	sulfuric acid

HF	hydrofluoric acid
HNO_3	nitric acid
HPO_4^{2-}	hydrophosphate anion
IL	ionic liquid
IR	infrared
L	inductor
Li^+	lithium cation
Li_2SO_4	lithium sulfate
$LiNO_3$	lithium nitrate
M	the molar mass of adsorbed/desorbed species [g mol^{-1}]
MAS	magic angle spinning
MC	Monte Carlo
Mg^{2+}	magnesium cation
MnO_2	manganese(IV) oxide
MWCNT	multi-walled carbon nanotube
n	the number concentration of large electrode particles [−]
Na^+	sodium cation
NaF	sodium fluoride
NiO	nickel(II) oxide
NiOOH	nickel oxide hydroxide
NMR	nuclear magnetic resonance
NPT	isothermal–isobaric MD simulations
N_{QCM}	the frequency constant for the quartz crystal resonator [Hz cm]
N_xO_y	nitrogen oxides
OEMS	online electrochemical mass spectrometry
OLC	onion-like carbon
OMI PF_6	octylmethylimidazolium hexafluorophosphate
OPLS	optimized potentials for liquid simulations
PANI	polyaniline
P_c	maximal power [W]
PC	propylene carbonate
PEDOT	poly(3,4-ethylenedioxythiophene)
PEt_4 BF_4	tetraethylphosphonium tetrafluoroborate
PPy	polypyrrole
PTFE	polytetrafluoroethylene
PVDF	polyvinylidene fluoride
Pyr_{13} TFSI	N-propyl-N-methylpyrrolidinium bis(trifluoromethanesulfonyl)imide
PYR_{13}^+	N-propyl-N-methylpyrrolidinium cation
pzc	point of zero charge
R	system resistance [Ω]
r	the radius of large electrode particles [cm]
RFB	redox flow battery
RuO_2	ruthenium oxide
S	surface area of electrodes [m^2]
SCN^-	thiocyanate anion
SO_2	sulfur dioxide
SO_3	sulfur trioxide

SPECS	step potential electrochemical spectroscopy
SSA	specific surface area [$m^2\ g^{-1}$]
SWCNT	single-walled carbon nanotube
TAA^+	tetraalkyl ammonium-based cations
TCD	thermally conductive detector
TEA BF_4	tetraethylammonium tetrafluoroborate
$TFSI^-$	bis(trifluoromethanesulfonyl)imide anion
Ti	titanium
TPD-MS	temperature-programmed desorption with mass spectrometry
U	applied voltage [V]
WO_3	tungsten (VI) oxide
XPS	X-ray photoelectron spectroscopy
XRD	X-ray diffraction
z	the number of electrons exchanged = the valence number of adsorbed/desorbed species [−]
Δf	the frequency change = the experimental frequency shift [Hz]
Δf_{PVDF}	the frequency change of a pseudo-uniform electrode layer (depending on its thickness, permeability length, density and wetting properties) [Hz]
Δm	the mass change [g]
Δm_{QCM}	the areal mass change [$g\ cm^{-2}$]
ΔQ	the charge exchanged [C]
ΔR	resistance change
η	the dynamic viscosity of electrolyte [$g\ cm^{-1}\ s^{-1}$]
μ_q	the shear elastic modulus of quartz crystal [$g\ cm^{-1}\ s^{-2}$]
ρ	the bulk density of electrolyte [$g\ cm^{-3}$]
ρ_b	the bulk density of electrode particles [$g\ cm^{-3}$]
ρ_q	the density of quartz crystal [$g\ cm^{-3}$]

1. Introduction

Our daily energy consumption is continuously growing, so is the number of electronic devices we possess. Each, even small electronic device requires energy/power supply in the form of batteries or electrochemical capacitors. This situation is known to all of us; however, many other applications are associated with the energy storage market. Here, one should consider emerging renewable energy sources, socio-economic progress, an energy policy of countries, and technological development impacting our energy consumption and its further consequences for the environment.

New energy storage/conversion trends focus on regionalization, where local production and sustainable storage are on demand. Moreover, environmental pollution should be reduced due to the low-carbon emission program/economy by 2050. The efficiency of energy production is suggested to be improved utilizing advanced materials and technologies. Considering the increasing population (even 9.5 billion by 2050), the energy storage market expands and attracts more attention each year. The utilization of carbon-based fuels causes deterioration of the air all over the world. Therefore, scientists' attention is focused on renewable sources, such as solar, wind, hydro- or thermal energy (Fig. 1). According to the latest reports, fusion energy is also on target.[1] All energy sources exhibit a high potential to replace conventional energy conversion systems. Thus, fossil fuels may be mostly consigned to history in due course.

At the same time, the amount of energy from renewable sources has been at a constant level since the end of the 20th century. This might be related to the fluctuations in the power delivery, especially for technologies based on photovoltaics, wind energy or tides. Therefore, various energy conversion and storage systems attract scientific attention. The systems could be compared in the form of a Ragone plot (Fig. 2), presenting specific energy (Wh kg^{-1}) vs specific power (W kg^{-1}).

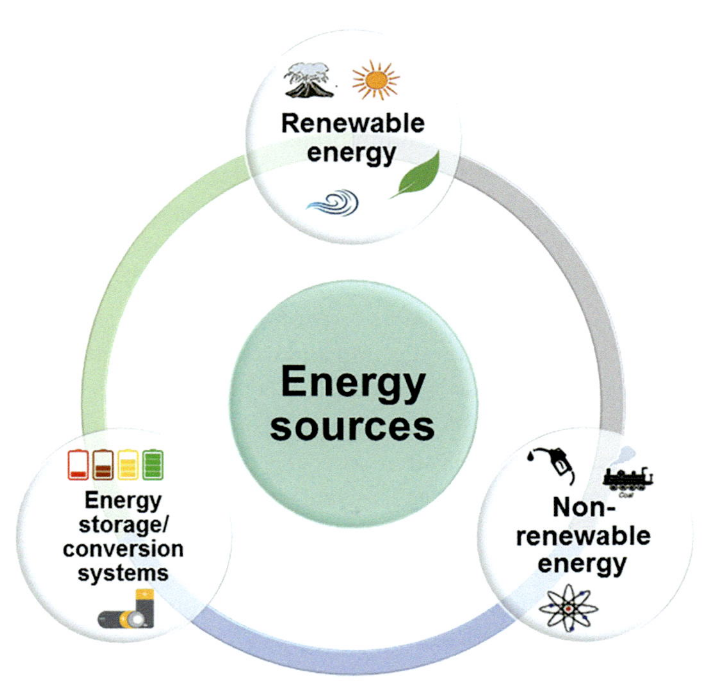

Fig. 1 Cross-dependence between energy sources, harvesting and storage systems.

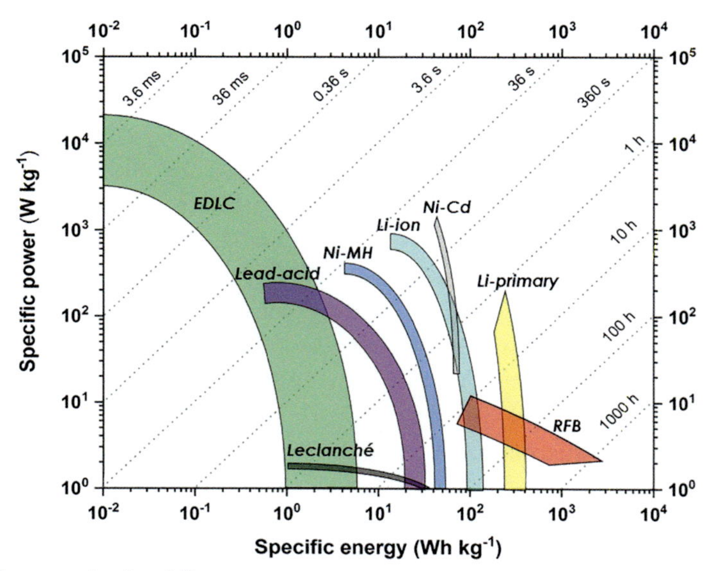

Fig. 2 Ragone plot for different energy storage/conversion systems.

It could be contemplated that both Li-ion and redox flow batteries (RFB) are the best choice for high energy applications. However, their power is quite limited. That is why redox flow batteries are mainly applied in energy storage of renewable energy sources, when the size and the time of delivering the energy are not that important as electrical efficiency (approx. 80%). The energy of a RFB directly depends on the volume of the battery and electrolyte concentration. Therefore, RFB are used for stationary applications. Fuel cells, with electrical efficiency of a level of 40–60%, are much more efficient than conventional combustion engines (20–30%). Therefore, their application is directed at transportation—they are successfully utilized in garbage trucks, vehicles for cleaning and cars. The second group of very energetic storage systems consists of Li-ion and Li-polymer batteries. Although their electrical efficiency is above 90%, their long-term cyclability is rather moderate. Moreover, their recycling process is still not fully developed, and lithium depletion triggers scientists to create other energy storage systems based on sodium or potassium metals. A sodium-sulfur battery is characterized by approx. 85% electrical efficiency. Such a relatively new solution had already dangerous consequences, e.g., fire. Safer systems, based on nickel, such as NiOOH/MH or NiOOH-Cd, with 70–90% efficiency, are commonly used in portable electronics. However, one of the most reliable and almost fully recyclable energy storage systems is based on redox reactions of Pb in a moderately concentrated H_2SO_4 medium (called lead-acid batteries). Their energy and power are moderate, but other features make them the most widespread source of energy/power in the world (recycling rate, cost and maintenance). Besides the battery systems, electrochemical capacitors (EC_S) emerged (ca. 1950) as energy storage systems filling the energy gap between conventional dielectric capacitors and rechargeable batteries. Their electrical efficiency is usually very high (>90%), and together with their cyclability (>1,000,000) have favorable properties. Moreover, their energy is two orders of magnitude higher than their dielectric relatives, and power is up to four times higher than the fastest energy storage battery system. Interestingly, both batteries and electrochemical capacitors may reciprocally support their performance. Therefore, continuous scientific attention is directed toward the optimization and improvement of these systems. Generally, scientific interest is focused on new materials. However, the performance of the device is tightly connected with the mechanism of processes at the electrode/electrolyte interface. Detailed assessment of these phenomena as well as learning of failure reasons are crucial for EC development. This chapter focuses on exploring advanced techniques for electrochemical capacitors characterization; techniques which are also used in other energy storage/conversion systems diagnoses.

2. Electrochemical capacitors: State-of-the-art

Electrochemical capacitors, are energy storage devices characterized by high power density (up to 10 kW kg^{-1}) with short charging/discharging time between 1 ms and 10 s. This makes ECs well-suited for peak current applications, e.g., memory back-ups, burst-mode power delivery, short-term energy storage or regenerative braking.[2]

An EC consists of two electrodes separated by an insulator (ionic conductive membrane), soaked in the electrolytic solution, as shown in Fig. 3. The operating principle of ECs is based on surface-confined electrostatic or Faradaic processes. Therefore, the electrodes of ECs should have a well-developed specific surface area since the amount of charge accumulated strongly depends on the electrode micro-texture. During the charging process, the electrolyte ions are adsorbed on the electrode surface, forming an electrical double-layer (EDL) in the potential-driven process. The ions desorption from the electrode to the electrolyte bulk takes place during the discharge process.[3,4]

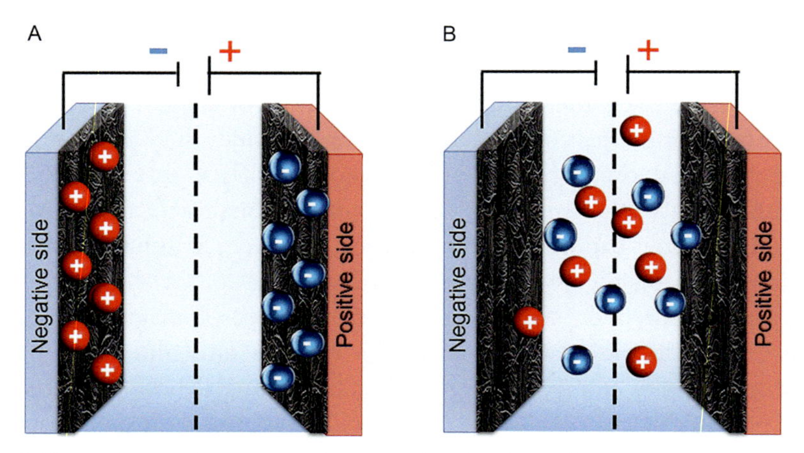

Fig. 3 Electrochemical capacitor scheme in the: (A) charged; (B) discharged mode.

As shown in Fig. 3, electrodes are connected in series and each electrode is equivalent to a single capacitor giving capacitance (C_{EDL}) according to the formula:

$$C_{EDL} = \frac{S\varepsilon_r\varepsilon_0}{d} \tag{1}$$

where

ε_r—relative electrolyte permittivity [−]

ε_0—vacuum permittivity [F m^{-1}]

d—EDL thickness [m]

S—surface area of electrodes [m^2]

Each EC system consists of two electrodes connected in series. Therefore, capacitance of the capacitor system (C) may be calculated from the given formula:

$$\frac{1}{C} = \frac{1}{C_+} + \frac{1}{C_-} \tag{2}$$

where

C_+, C_-—capacitance of the positive and negative electrodes, respectively [F]

Assuming that in the symmetric systems $C_+ = C_- = C_e$, it can be stated that the capacitance of the systems stands for the half of the electrode capacitance (C_e):

$$C = \frac{C_e}{2} \quad (3)$$

where

C—total system capacitance [F]

$C_e = C_+ = C_-$—electrode capacitance [F]

The energy and power are strongly related to the applied voltage. Therefore, this parameter is crucial when the ECs final parameters are considered as shown in the equations:

$$E = \frac{1}{2} CU^2 \quad (4)$$

where

E—energy [Wh]

C—total capacitance [F]

U—applied voltage [V]

$$P_C = \frac{U^2}{4R} \quad (5)$$

where

P_C—maximal power [W]

U—voltage [V]

R—system resistance [Ω]

2.1 Electrode materials

Carbon is well known as the material in various energy-related applications. It is a fundamental element of matter. All the polymers, called colloquially "plastics," are synthetic carbon-based networks. Variation of carbon allotropic forms is a prerequisite for so many features that allow us to find carbon almost everywhere. This common rationale is also the reason for the variety of electrode materials used in electrochemical capacitors. Specific coupling of carbon-based materials with electrolytic solutions (aqueous, organic or ionic liquids) is possible due to the advanced and tunable properties of the carbon matrix. Good electrical conductivity, low cost and high specific surface area (up to 2000 $m^2\,g^{-1}$) make the porous carbons preferable for ECs application with the specific capacitance ranging from 100 to 200 $F\,g^{-1}$. Various forms of carbons can be applied for ECs, e.g., activated carbons (ACs),[5] carbon nanotubes (CNTs),[6–8] onion-like carbons (OLC),[9] carbon blacks (CB),[10] carbide-derived carbons (CDCs)[11] and templated carbons.[12]

2.1.1 Activated carbons

Activated carbon is produced in the form of powder, granules or self-standing cloth/fibers. The morphology determines whether there is a need to use a polymer binder or not; this also affects the final properties of electrode material: its flexibility, durability, conductivity and specific surface area. Mostly, carbon materials are in powder form; therefore, the presence of a binder (ca. 5–10 wt.%) is inevitable for electrode formation. Additionally, a percolator is usually added to activated carbon electrode slurry (CBs, CNTs or OLC) to enhance conductivity.[10,13]

AC should be prepared only from carbon-rich natural or synthetic precursors (e.g., coconut shell, plant stem, seaweeds, fruit stones, and adenine).[14–16] It is not recommended to use edible products or non-sustainable production processes, especially as various eco-friendly solutions have been proposed. Moreover, the diversity of natural origin precursors maintains constant research attention towards preparation protocols of different carbon forms and their further activation processes.

One should also consider precursor abundance, elemental content (heteroatom presence) or carbonization yield. The price of AC is also of great importance, and it depends on the quality of the material. Hence, compared with the other electrode materials (metal oxides, electrically conducting polymers, silicon-based structures and others), ACs are relatively inexpensive.

ACs are very often enriched with various heteroatoms. Usually, oxygen or/and nitrogen groups are introduced to the carbon surface. To modify the carbon surface, the source of the desired element is necessary. O-rich carbons can be realized by oxidation in air,[17] acids[18] or electrochemical methods,[19] while N-rich carbons may be obtained via ammonia,[20] urea, melamine or polyaniline chemical treatment.[21–22] It should be highlighted that functional groups are the source of pseudo-capacitance, but they may also enhance carbon wettability; however, only a moderate amount of nitrogen and oxygen should be present. Nitrogen content should not exceed 8 wt.%.[23] Similarly, for oxygen—too high oxygen content decreases electrode conductivity and thus, reduces capacitor power output.

It should also be mentioned that the presence of functional groups has its limitations. Therefore, functional groups have their beneficial limits, which should not be exceeded with respect to their % contribution and/or type of doped species. Usually, shorter ECs lifetime and higher self-discharge rates are obtained for functionalized carbons. Besides electrochemical measurements of assembled ECs cells, carbon physicochemical properties should be monitored in-depth to understand ongoing processes fully. It can be realized by X-ray photoelectron spectroscopy (XPS), temperature-programmed desorption with mass spectrometry (TPD-MS), Fourier Transform Infrared Spectroscopy (FTIR), Raman spectroscopy or elemental analysis. Few of the techniques mentioned above can also be realized in the *operando/in-situ* mode. This aspect will be discussed further in detail.

2.1.2 Graphene-related materials

Graphite is an allotropic form of carbon. A single layer of graphitic structure is called graphene and application of it may be advantageous relative to other carbon systems used in ECs, such as carbon nanotubes or onion-like carbon. These materials represent a different dimensional structure from 0D for OLC, through 1D for CNTs up to 2D for pure graphene. Each of them has been implemented in electrochemical capacitors as electrode material. OLCs production cost is very high, as they may be synthesized in the ideal spherical form by arc-discharge or tilted form by a direct diamond annealing process. Both processes require high temperatures >1000 °C. It might be listed as the main obstacle for OLC in becoming prevalent in ECs. However, they can be successfully applied as conductive agents in the electrode slurry. It needs to be noted that OLC tends to agglomerate. It results not only from their small single particle size, but also especially in aqueous solution, from their hydrophobic character.

Fig. 4 presents the process steps to obtain a stable water dispersion of OLCs, ready to be used in the electrode material. The oxidation step is performed in the presence of acids, i.e., HNO_3 or H_2SO_4. The final product contains approx. 10% of oxygen, 90% of carbon, and is characterized by a specific surface area at the level of 300 $m^2\ g^{-1}$. It should be highlighted that OLCs are characterized by external surface structure, where almost no micropores are present. Their electrochemical response is very satisfactory; however, the performance was studied in the 3-electrode setup. ECs based on the OLCs easily manage the scan rate up to 1000 $mV\ s^{-1}$. The equivalent series resistance (ESR) of the system is lower than 0.5 Ohm under open circuit conditions. Unfortunately, only a few two-electrode studies can be found in the literature to date.

Two-electrode tests of OLC-based electric double layer capacitor (EDLC) with organic electrolyte (1.5 $mol\ L^{-1}$ TEA BF_4 in acetonitrile) have been reported.[25] Various annealing temperatures, up to 2000 °C, allows the specific surface area up to 500 $m^2\ g^{-1}$ to be increased. The specific capacitance reached by OLC-based electrode materials is on the same

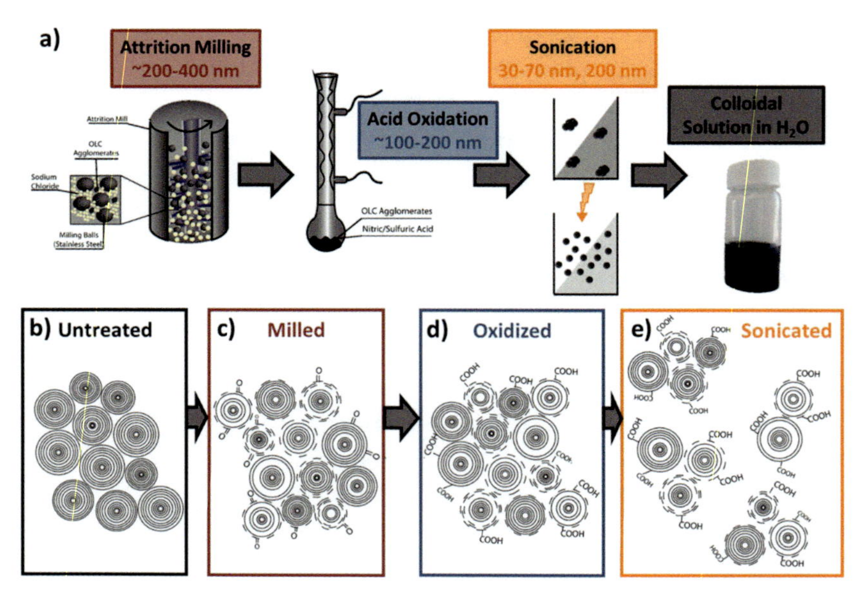

Fig. 4 Scheme taken from Ref. 24 presenting various processing steps to obtain a stable aqueous OLC dispersion.

level as for CNTs with similar specific surface area (SSA)—max. 40 F g^{-1}. However, this value is far from the ones reported for AC. The specific capacitance value of OLCs is not extraordinary, but the systems are capable of working efficiently up to high current densities. The resulting capacitance retention was plotted up to a current load of 200 mA cm^{-2}. For different annealing temperatures applied during OLC processing, it has been shown that the time constant is approx. 1 s, i.e., much shorter than in the case of most ACs. Therefore, for high power application, nanostructured electrodes are in favor as presented in Fig. 5.

Another successful OLCs application in EC was also made in an organic medium, i.e., 1 mol L^{-1} TEA BF_4 in propylene carbonate.[26] High rate performance is highlighted, as the time constant is equal to 26 ms. The report is focused on microcapacitor application where OLCs are found to be very suitable for this application. The energy of the device is smaller compared to the corresponding one with AC, but its power is much higher, i.e., 300 W cm^{-3}.

CNTs might be used directly or as a component of the electrode material. Their properties are preserved in the single-walled (SWCNTs) or multi-walled (MWCNTs) form. The main advantage of CNTs is the possibility of their direct growth on the current collector. It assures good

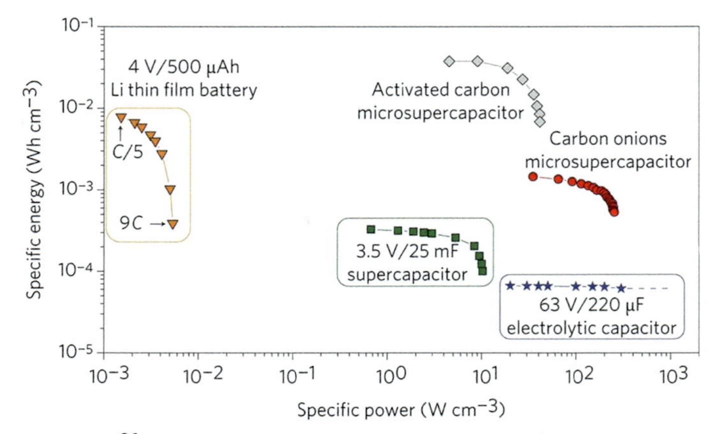

Fig. 5 Ragone plot[26] locating energy and power response of microsupercapacitors.

electrical contact and no resistance increase due to an electrode/current collector improved interface. Moreover, their specific 1D tubular form allows a subsequent modification of the external graphene layer to be carried out, e.g., by covering a thin layer of electro-conducting polymers or changing the surface chemistry. CNTs may be chemically activated—increasing their SSA from $500\,m^2\,g^{-1}$ for MWCNTs up to $1000\,m^2\,g^{-1}$ for SWCNTs. However, both the utilization of a catalyst for synthesis as well as activation and purification processes require the presence of hazardous compounds, such as strong alkaline and acid solutions, e.g., HF, HNO_3 or KOH. The application of the chemical vapor deposition (CVD) process within the template matrix seems to be less harmful, but its energetic cost is very high, as the temperatures used during synthesis are above 800 °C. Furthermore, the alumina template needs to be removed afterwards. In considering features of CNTs not only is the single fiber of high importance—straight or entangled, but also its structure—an open or a closed tube. Therefore, the variety of CNTs properties and their application in ECs has been meticulously studied over the last two decades.

2.2 Electrolytes

One may distinguish three main types of electrolytes applied in electrochemical capacitors, i.e., organic, aqueous electrolytes and ionic liquids.[27,28] Since the energy output of ECs strongly depends on the applied voltage, organic electrolytes are preferred for commercial use. The wide electrochemical window of 2.5–2.7 V can be easily reached for $1\,mol\,L^{-1}$

tetraethylammonium tetrafluoroborate (TEA BF_4) salt dissolved in acetonitrile (AN) or propylene carbonate (PC) solvent.[29,30] On the other hand, these kinds of electrolytes require an inert atmosphere for the system assembly, which significantly increases the initial costs of the capacitors. Moreover, they are characterized by lower conductivity than in the case of water-based electrolytes what strongly affects the ECs power output (Table 1). Therefore, aqueous electrolytes have become used recently, with ca. 1.23 V of the operating voltage due to the thermodynamic stability of water. It seems that neutral salts may slightly improve the operational window since the water decomposition potential is shifted. The stability of aqueous electrolytes strongly depends on pH and decomposition potentials are described by the following equations:

$$E_{ox} = 1.23 - 0.059\,pH \tag{6}$$

$$E_{red} = -0.059\,pH \tag{7}$$

As shown in Table 1, each electrolyte type has its positive and negative aspects. To improve the energy density of electrochemical capacitors, usually, two different approaches are used. One is based on capacitance increase by the introduction of redox active species.[31–33] They can be placed in the electrolyte or electrode material, giving the source of redox reaction or pseudo-capacitance. The second strategy for energy enhancement exploits the operating voltage widening (e.g., by using an organic electrolyte or by manipulating the pH of a water-based medium).[20] For the power rating, the selected electrolyte should be characterized by relatively high conductivity to obtain the high power of the ECs device. Recently, some studies confirmed that "water in salt" electrolytes can be good candidates for this purpose.[34–36]

Table 1 Compared parameters of ECs systems with various electrolytes.

Property	Organic electrolytes	Ionic liquids	Aqueous electrolytes
Conductivity	ca. 2–20 mS cm^{-1}	ca. 2–10 mS cm^{-1}	ca. 100 mS cm^{-1}
Operating voltage	2.5–2.7 V	3.0–3.5 V	0.8–1 V for H_2SO_4 & KOH medium up to 1.8 V for neutral medium
Technical aspects	Flammable, the inert atmosphere for system assembly	Inert atmosphere for system assembly	Environmental friendly, easy construction

In addition, "neat" ionic liquids are very often applied when the basic measurements are performed as only the ions (with well-known dimensions) are present in the system; thus, data interpretation due to lack of standard solvent molecules is expected to be easier.[37]

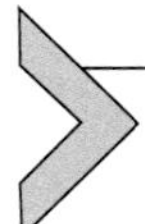

3. Advanced techniques for interfaces characterization in electrochemical capacitors

Besides the advanced technologies already known, there is still room for improvement of EC systems to ensure sustainable, commonly available and durable devices. Therefore, insightful studies are necessary to understand fully the ongoing processes, which take place during ECs operation. Not only is the complete device always characterized, but also the capacitor components or single processes separately. Hence, current characterization techniques include electrochemical measurements coupled with physicochemical property determination. This can be realized in two different modes:

(i) *ex-situ*,
(ii) *in-situ*.

In-situ is a Latin phrase that can be translated as: "on site," "locally" or "in place." It might have a different context in biology, chemistry or medicine. In electrochemistry, it refers to performing the electrochemical measurement coupled with another technique that gives the information about ongoing processes simultaneously or it determines the physicochemical properties of the studied material (e.g., electrolyte, electrode). One may also differentiate *in-situ* from *operando* mode. *Operando* measurements of physicochemical parameters control at the same time (online) electrochemical capacitor cell testing, while by *in-situ* mode, physicochemical parameters can be controlled periodically during the regular pauses of electrochemical tests. Very often both terms are used in a confused way in the literature. Unquestionably, *in-situ* or *operando* measurements ensure an insightful view into capacitor charging/discharging mechanisms. The opposite mode to *in-situ* is *ex-situ*. *Ex-situ* measurements concern separate electrochemical and physicochemical analyses separated by time and place. Sometimes such type of an experimental procedure is named *postmortem* analysis, where the pristine properties of analyzed material are compared to the ones after the electrochemical tests.

Fig. 6 presents the difference between *in-situ*, *ex-situ* (postmortem) and *operando* measurements. One should also note the construction differences

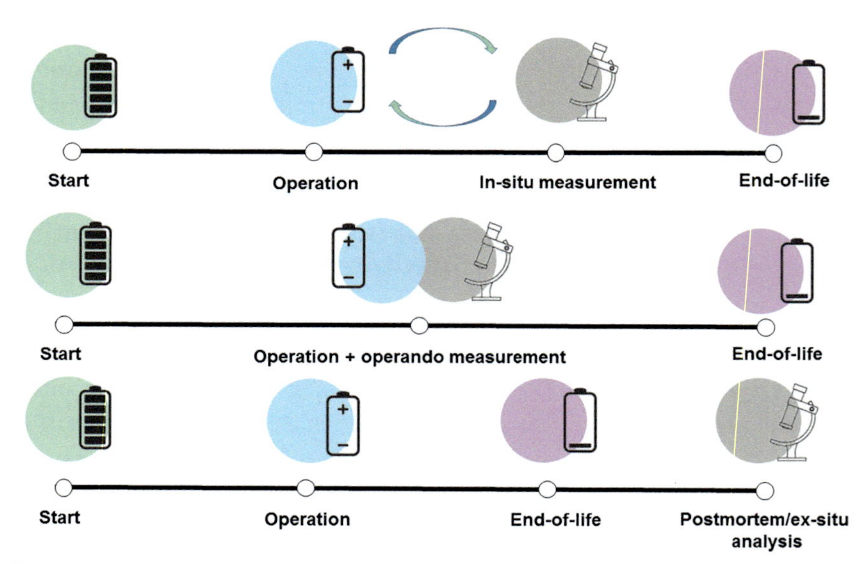

Fig. 6 Comparison between various measurement modes: (A) *in-situ*; (B) operando; (C) *ex-situ*/postmortem.

between various approaches. The *ex-situ*/postmortem approach concerns full device testing. This can be realized in a 2-electrode setup with/without the reference electrode. Simply, after the electrochemical test, the component is comprehensively analyzed. This allows both electrodes from the same system to be compared. For *in-situ/operando* techniques, usually a 3-electrode cell setup is implemented and only one electrode can be considered at a time. It means that only one electrode is analyzed *in-situ* and such measurements only simulate the real device performance. Separate cells should be built to determine the processes on positive and negative electrodes.[38,39] Nevertheless, one cannot clearly distinguish which analysis method is the better one. Each one has its distinctive features, and the best knowledge can be obtained when the information from both approaches is determined. Obviously, not all the measurements can be carried out in *operando* mode. Usually, the measurements include techniques which can monitor the deterioration of the sample (e.g., elemental analysis) or the measurements that require specific conditions (e.g., pressure changes in N_2 sorption at 77 K). Obviously, one cannot predict if in future, some novel solutions are invented. Undoubtedly, *operando/in-situ* techniques usually allow the processes at atomic and molecular levels in electrolyte/electrodes to be fully scanned. These fundamental studies should be later verified (as presented in Fig. 7) in the real capacitor system since not always a 3-electrode setup can be directly translated into a full device.

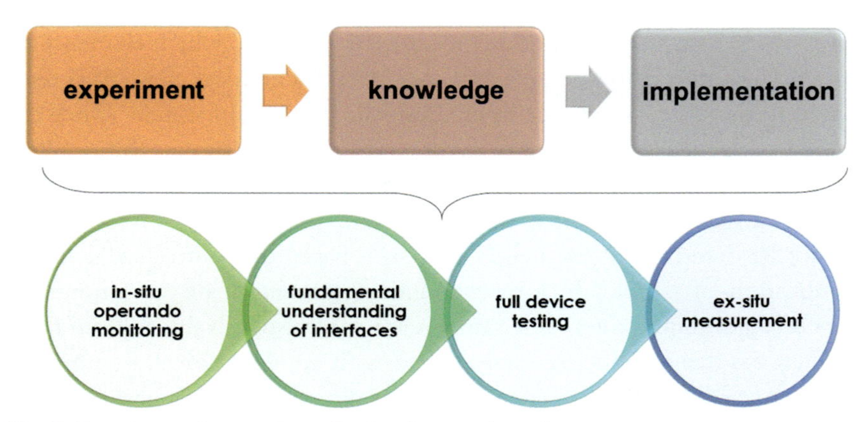

Fig. 7 Experimental procedures for fundamental studies.

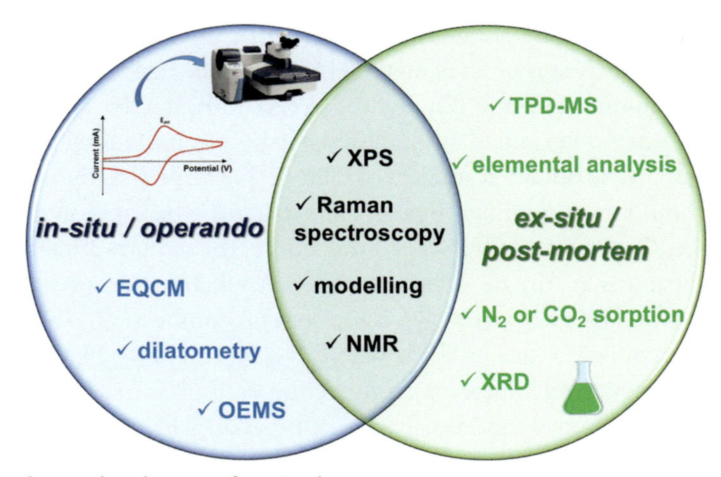

Fig. 8 Advanced techniques for ECs characterization.

In the following section, we would like to introduce contemporary state-of-the-art *in-situ* and *ex-situ* techniques used for ECs characterization. Fig. 8 shows the advanced techniques utilized, which will be included in this chapter. As seen, a few techniques (e.g., Raman spectroscopy, X-ray photoelectron spectroscopy, gas chromatography with mass spectroscopy) can be carried out both in *operando*/*in-situ* as well as in *ex-situ*/postmortem mode. The chapter will also summarize all the advantages and limitations of the methods used with a comprehensive comparison.

3.1 Elemental analysis

Elemental analysis (EA) is an analytical technique applied in chemistry to determine the elemental composition of chemical compounds and their composites. Through this method, it can be determined which elements are present and how many percent by mass of each chemical element is contained in the tested substance to establish the empirical formula. Once the overall molecular weight of the tested compound is known, the result of the analysis enables the molecular formula to be established. Nevertheless, the structure, or the way how atoms are interconnected by chemical bonds, cannot be determined. Elemental analysis is based on controlled, dynamic combustion of samples in a reactor and measurement of the amount of the corresponding oxides. The samples are combusted in the column filled with an oxidizing-reducing catalytic bed with an electronically-controlled temperature. The evolved gases are usually directed to the analyzer, which is usually a simplified version of the gas chromatograph. All gases (H_2O, CO, CO_2 and N_xO_y) are separated by chromatographic columns and then detected on a thermally conductive detector (TCD).

Elemental analysis is often named CHNS or CHN analysis since it determines the amount of carbon (C), hydrogen (H), nitrogen (N) and sulfur (S). Oxygen is often determined as the difference between the total percentage and the sum of the obtained amount of CHNS. Nevertheless, for good practices, it should be determined in a separate analysis where pyrolysis process takes place. Generally, in CHNS + O procedure, the samples are introduced to the apparatus in tin or silver crucibles. An inert atmosphere is used (He, Ar); for CHN analysis oxygen is injected for combustion. The catalytic bed usually contains CuO or WO_3 where N_2, N_xO_y, CO_2, H_2O, SO_2 and SO_3 gases are evolved. Therefore, a reduction column is necessary to eliminate the oxygen excess and finally obtain N_2, CO_2, H_2O and SO_2, afterwards the N, C, H, S, are respectively calculated. The oxygen amount is obtained from the pyrolysis process based on the CO peak. Some elemental analyzer configurations also enable the chlorine content to be determined. In elemental analysis a high temperature is used. Locally, the temperature in the range of 1800 °C is reached; therefore, many different samples could be measured by the EA technique. The elemental analysis ensures high accuracy. Elemental analysis was found to be very useful in electrochemistry. It is worth highlighting that EA enables the samples with fluorine content to be analyzed. Since fluorine causes quartz glass recrystallization, silver crucibles should be used. Such possibility simplifies electrode properties determination since usually PTFE or PVDF binders are used for electrode formulation. EA is a typical *ex-situ*/postmortem technique since the sample is completely combusted during EA and therefore, it cannot be further used. Usually, the electrode materials are analyzed to determine oxygen and nitrogen content. It was found that the comparison of elemental analysis results with other surface-related techniques (e.g., TPD-MS or XPS) is beneficial. On this basis, it can be determined whether the oxygen/nitrogen presence is rather related to the surface groups or sample structure. Kleszyk *et al.* used the EA technique to compare the oxygen content measured by elemental analysis and TPD-MS for Burley carbons. For both methods, oxygen decreases with carbonization temperature rise. Since the EA detects oxygen in the total electrode mass (whole electrode framework), while TPD only surface functionalities, the values estimated from EA are higher compared to the ones obtained by TPD.[40] Nevertheless, it is not always obvious. An inversed situation was detected for the negative electrodes after the ageing test with lithium nitrate as an electrolyte. Herein, the oxygen data obtained from XPS are higher than those from composition analysis, suggesting that there are only surface-related ageing causes connected with the lithium carbonate solid-state deposit.[41]

3.2 Electrochemical quartz crystal microbalance

An electrochemical quartz crystal microbalance (EQCM) allows measuring the mass change of the electrode material and its correlation with the potential-driven ions adsorption/desorption process. In principle, the electrode material is deposited onto a piezoelectric crystal (called resonator, when covered with a metallic current collector) that vibrates with a constant frequency. Changing the mass at the electrode/electrolyte interface either by ions adsorption, their desorption or reorganization, deposit formation, *in-situ* polymerization and other processes directly influences the frequency amplitude and can be correlated to the corresponding mass change using the Sauerbrey equation. Knowledge of the movement of the ions such as cations adsorption when potential lower than *pzc* is applied, and *vice versa*, one may evaluate specific ion fluxes responsible for electric double-layer formation for electrochemical capacitor studies. It has been accepted in the scientific community that one might distinguish mainly three different phases in a charge storage mechanism, i.e., counter-ion adsorption, co-ion desorption and anion-cation exchange, called ion reorganization depending on the potential applied. However, only in a small number of cases can such clearly differentiated processes be described. Usually, ongoing processes result from all these movements, which overlap with each other depending on the potential applied, especially close to the porous electrode material and in its pore volume. One cannot exclude the competitive behavior of the ions within the electrolyte. Therefore, it is of particular interest to discriminate the dominant mechanism in the energy storage systems in order to propose some improvements.

Designing of an EQCM experiment is the same as standard electrochemical tests. The exact correlation of the mass and frequency is possible due to the application of the Sauerbrey equation. It is valid for rigid, solid material coatings. This equation can be presented at different levels of sophistication, depending on user experience, experimental approach and EQCM-device (Eqs. 8–11):

$$\Delta f = -\Delta m_{QCM}/C_{QCM} \tag{8}$$

$$\Delta f = -f_0^2 \cdot \Delta m_{QCM}/N_{QCM} \cdot \rho_q \tag{9}$$

$$\Delta f = -f_0^2 \cdot \Delta m_{QCM}/\sqrt{\mu_q \cdot \rho_q} \tag{10}$$

$$\Delta f = \Delta f_{PVDF} \cdot \left(1 - n \cdot \pi \cdot r^2\right) - \frac{3}{2} \cdot f_0^{\frac{3}{2}} \cdot \sqrt{\rho \cdot \eta} \cdot \sqrt{\pi \cdot \rho_q \cdot \mu_q}^{-1} \cdot n \cdot r^2 \cdot \frac{\pi^2}{2} - \frac{4}{3} \cdot f_0^2 \cdot \pi \cdot r^3 \cdot \sqrt{\rho_q \cdot \mu_q}^{-1} \cdot n \cdot (\rho_b + \rho) \tag{11}$$

where

Δf—the frequency change = the experimental frequency shift [Hz]
C_{QCM}—mass sensitivity constant [ng cm^{-3} Hz^{-1}]
Δm_{QCM}—the areal mass change [g cm^{-2}]
μ_q—the shear elastic modulus of quartz crystal [g cm^{-1} s^{-2}]
ρ_q—the density of quartz crystal [g cm^{-3}]
f_0—the fundamental resonance frequency of the crystal [Hz]
N_{QCM}—the frequency constant for the quartz crystal resonator [Hz cm]
n—the number concentration of large electrode particles [–]
r—the radius of large electrode particles [cm]
ρ_b—the bulk density of electrode particles [g cm^{-3}]
ρ—the bulk density of electrolyte [g cm^{-3}]
η—the dynamic viscosity of electrolyte [g cm^{-1} s^{-1}]
Δf_{PVDF}—the frequency change of a pseudo-uniform electrode layer (depending on its thickness, permeability length, density and wetting properties) [Hz]

The Sauerbrey equation allows combining the measured frequency change with a mass change detected on the piezoelectric crystal surface. Eq. (11) is one of the most comprehensive, considering its application in the electrochemical system (tested under potentiodynamic conditions), as it considers real carbon coating features. It assumes a non-uniform carbon layer composed of spherical particles mixed with a polymer binder (usually for EQCM application, PVDF is used). Thus, one may say that this equation resembles the factual conditions at the electrode/electrolyte interface when porous carbon is applied. To discuss quantitatively which species are responsible for EDL creation and recognize ion fluxes at the molecular level in the system, it is useful to implement the Faraday equation (Eq. 12). It combines mass change with charge exchanged (also called the quantity of electric charge, charge density, charge capacity, charge state) into a mass/charge ratio plot:

$$\Delta m = \frac{\Delta Q \cdot M}{F \cdot z} \tag{12}$$

where

Δm—the mass change [g]

ΔQ—the charge exchanged [C]

M—the molar mass of adsorbed/desorbed species [$g\,mol^{-1}$]

z—the number of electrons exchanged = the valence number of adsorbed/desorbed species [−]

All the above-mentioned calculation methodologies assume that the carbon coating does not differ in its viscoelastic properties from the piezoelectric crystal. The principle of EQCM application in energy storage research assumes that a thin coating layer behaves similarly to a piezoelectric crystal, so all the piezoelectric features can be transferred into coating behavior as carbon acts as an integral part of a crystal. Thus, electrode material, i.e., carbon coating resembles rigid, solid, thin-film behavior. Indeed, one may consider hand-made coating as uniform and easily reproducible electrode film, but on a molecular level, minor defects can cause an enormous difference in the ion fluxes during the charge/discharge process. Thus it is essential to verify the quality of a carbon coating prior to and during the electrochemical tests. First, resistance change (ΔR) is controlled over the whole potential range and it provides direct information about the coating state in the electrochemical system. Capacitive processes occur at the electrode/electrolyte interface when only a slight change in resistance is measured, i.e., the frequency shift of resonator results only from ions adsorption/desorption onto its surface. Peeling or detachment of the carbon coating from a piezoelectric crystal is easily recognizable, since the resistance value decreases quickly and significantly. Nevertheless, it is crucial to monitor this factor simultaneously with current and frequency response during operation. Furthermore, ΔR gives information about the electrochemical stability of the studied coating in terms of its fabrication method and in the function of potential applied (electrochemical stability). Second, another advantageous approach, which does not exclude resistance control, is to verify qualitatively the frequency change measured as a function of time. As with each electrochemical test, EQCM studies cannot be established on the basis of a single experiment or a single repetition of a charge/discharge process. Thus, controlling and evaluating a frequency wavelength and amplitude are crucial in order to discuss its reproducibility during several cycles of ions adsorption and desorption. A specific and unambiguous answer about the coating state and its viscoelastic properties is given by dissipation monitoring, introduced in the EQCM-D apparatus. Dissipation provides information about energy loss of carbon material (the acoustic wave dampening by carbon coating) after removal or pause of the externally forced oscillation of the piezoelectric resonator. Especially if viscoelastic properties modeling of the material coating is addressed in the studies, knowledge of the dissipation factor is essential. A low dissipation factor (D) suggests that oscillation will persist for a longer time than in the case of a high dissipation factor when no external power is supplied. Dissipation is an inverse parameter to quality EQCM, which might be calculated from basic wave characteristics: bandwidth and center frequency value of the wave (length and amplitude). Dissipation can also be correlated with a wave decay time to e^{-1} value. Considering acoustic or viscoelastic properties of a piezoelectric crystal, the latter resembles the mass connected to a spring (elastic modulus of a crystal) and damper (dissipation factor). From the electrical point of view, one has a parallel connection of an LCR model: inductor, capacitor and resistor circuit and static capacitance, that represents an oscillating quartz crystal (L—inductor: displaced mass, C—capacitor: stored energy, R—resistance: energy losses) and surface capacitance of a crystal or coating. As a consequence, EQCM can be considered as an electronic or acoustic system, so energy losses can be measured using various techniques and approaches. Thus, to study quantitatively conducted experiments, one should measure admittance, dissipation and/or resistance. All of these features provide insight into the viscoelastic properties of a coating deposited on a piezoelectric crystal and allow for application of the Sauerbrey equation.

Application of EQCM in electrochemical studies gives interesting information about the gravimetric, morphological and mechanical probe of the carbon coating in the vicinity of the electrolyte, as summarized in Fig. 9.[42]

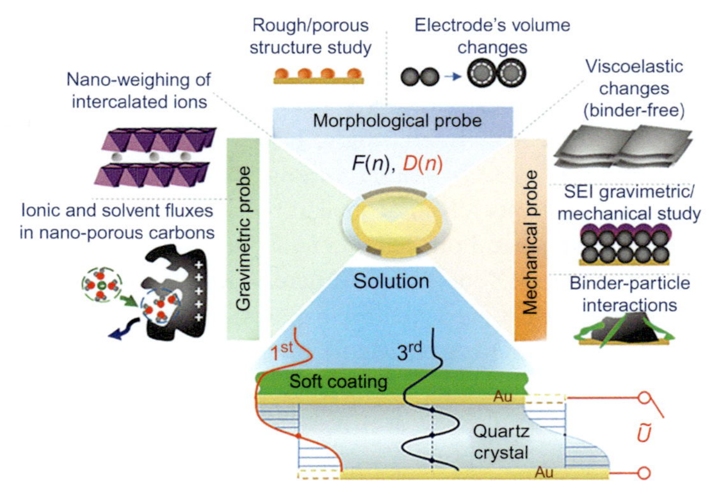

Fig. 9 Application of EQCM measurements in energy storage/conversion systems.[42]

Knowing the limitations and principles of EQCM, this technique is a powerful tool for the description of the charging mechanism.[43] Studies carried out with ionic liquids as electrolytes and their mixtures with organic solvents combined with CDC carbon materials gave information that dynamic changes of ions occur according to:

(i) concentration of ionic species,
(ii) type of solvent used,
(iii) cation-anion interactions,
(iv) textural properties of carbon (pore size and pore volume).[44,45]

It proves that electrode materials should match the ions from the electrolytic solution, and *vice versa*, in terms of their effective adsorption/desorption. The smaller the average pore size, the higher the desolvation degree of ions and larger spatial hindrance for reversible ions adsorption. Moreover, in the polarization zone close to the point of zero charge (*pzc*) ca. ± 200 mV, one can observe mostly an ion mixing-zone also called reorganization,[44] which results from simultaneously co-ions desorption and counter-ions adsorption, even though the process has been artificially limited by the potential range only to counter-ions adsorption and desorption.[46] In aqueous-based systems it is believed that this region is much wider, as owing to various electrolyte pH values and their variation in the electrode bulk, not only electrolyte-based ionic species take an active part in EDL formation, but also H^+ and OH^- from the solvent.[38,47–49] Contrarily, organic solvent molecules usually passively participate in the EDL formation by being pulled into a solvation shell of an ion.[50] Because the EQCM study helps to unveil the charge storage mechanism, it is widely accepted for a novel, advanced carbonaceous materials characterization, such as MXenes[51] or graphene.[52] Interestingly, it was shown that changing the alkali metal from Li^+ to Cs^+ causes various gravimetric responses.[51] Li^+ cation was found with its solvation shell at the electrode/electrolyte interface, which means that both: counter-ion and solvent molecules are intercalated into the MXene phase. Whereas, for Cs^+ the recorded mass change was smaller than predicted for bare Cs^+ cation..[45,53] Such studies, also applied for alkaline-earth (Me^{2+}) and tetraalkyl ammonium-based (TAA^+) cations, show that larger cations are responsible for electrode expansion. Unfortunately, in the case of MXenes the EQCM studies are usually limited only to cation adsorption,[54] as this material can play a role of negative electrode only in electrochemical capacitors or the anode in batteries.[55] Thus, these investigations are fundamental in terms of the optimization of MXene electrode charging by proper cation matching from the electrolyte, but do not provide a direct insight into full cell operation.[56] A different energy storage mechanism was observed for graphene-based materials compared with MXenes or ACs.[50,52] The first main difference is that single-layer graphene does not undergo volume changes during the charging and discharging process. Moreover, in the potential range higher than *pzc*, ions are expelled from the graphene surface. Whereas in the lower potential ranges than *pzc*, ions are reorganizing at the electrode/electrolyte interface which leads to dense EDL.[52] For graphene, EQCM application can shed light on the coating properties, i.e., thickness, mass and quality.[57] It has been proven that the dry transfer method leads to around 10 times thicker graphene layer than predicted.[57–59] In addition, the semi-permeable properties of graphene as a membrane must be considered, as ions can diffuse causing a change of ions, the population of ions and potential gradient in the graphene vicinity. In addition, for graphene-based materials, strong solvation-cation interactions were observed for small alkali metal cations (Li^+).[60] EQCM studies, when combined with EIS also disclose ion transport kinetics.[49] Faster ions take part as a dominant energy storage mechanism.[61] It has been shown that EDL formation in aqueous electrolytic solutions is not a one-step ion adsorption, but a continuous, multi-step process that also involves various ion fluxes depending on the amount of charge flowing through the resonator. Such an observation is valid for CNTs and ACs.[38,49] Thus, summarizing all the above-mentioned observations EQCM allows specific charge/discharge mechanism descriptions in energy storage devices.

It ensures the fundamental study of ion fluxes and their interactions as a potential-driven process. It detects concentration of ions, the population at the electrode/electrolyte interface that can lead to the improvement of the charging process in terms of electrode material-electrolyte matching. One should note that the gravimetric probe is a resultant from movement of all ions. Thus, detailed electrochemical characterization is crucial to be complementary with EQCM data. In addition, EQCM discloses if ions adsorbed are in a solvated or desolvated state approaching the surface of the polarized electrode (not including the desolvation process ongoing in the pore volume). This information is anion-, cation-type, solvent type and concentration-dependent. Thus, optimization of charging process efficiency for both negative and positive electrodes should be carried considering both: electrolyte and electrode material features.

3.3 *In-situ* dilatometry measurements

Dilatometry studies correlate electrochemical performance with charging mechanisms for various carbonaceous materials.[62] The possibility of using a normal size working electrode (usually with a diameter ca. 10 mm) is a big advantage of this technique compared with the other *in-situ* analysis.[63,64] Therefore, even though the volume changes/expansion of electrodes are measured directly during electrochemical charging/discharging, the changes almost resemble the conditions of the full cell operation. Temperature control is crucial for dilatometry studies, as it directly influences the viscosity of the electrolytic solution and the wettability of carbon material. It has been observed that ions adsorption from 1 mol L^{-1} TEA BF_4 in AN into AC porosity does not particularly lead to high electrode expansion, contrary to the mesopore SWCNTs.[62] Thus, change in the volume results from intercalation of ions into crystalline domains—observed mostly for battery-type materials. On the other hand, expansion of a microporous AC-based electrode can be related to a decrease of a coulombic efficiency of charging/discharging process, as in principle this should not occur to a great extent.[65] Interestingly, for all materials studied with 1 mol L^{-1} TEA BF_4 in AN or PC, the asymmetry of positive and negative electrode profiles was observed in terms of the electrochemical response, as well as the electrode thickness change. Solvent oxidation and further carbon material oxidation and corrosion trigger moderate or even a small volume alteration in the potentials higher than *pzc*.[66] Contrarily, potentials lower than *pzc* induce strong cation insertion, intercalation, and/or adsorption into carbon texture, which mechanically pushes the electrode composite towards its boundary limits. As already discussed, small alkali metal cations in aqueous solutions can attract a high number of solvent molecules into their hydration shell causing a larger electrode swelling effect.

Various carbon-based electrodes with different textural properties (from microporous to mesoporous ones) have been tested in the same salt, 1 mol L^{-1} TEA BF_4, using two solvents: AN and PC (Fig. 10). The plot (in Fig. 10) shows that each carbon is characterized by unique potential-driven expansion, depending on the solvent used. Thus, the charge storage mechanism is one-of-a-kind for each electrode-electrolyte pair. Therefore, it is necessary to continue the *in-situ* characterization of well-known systems and as well, the new advanced ones. The structure and composition of the EDL in a solvent-free electrolyte (ILs) is fundamentally different from the one observed in organic or aqueous solutions.[46] The recorded maximum strain for a carbon electrode was on the level of 2%, which is found to be high for ECs system. The pressure of the dilatometer sensor on the working electrode during the experiment is also an important experimental parameter.[63] Usually, the force is in the range of 1–2 N.[46,65] It has been shown also that different cations cause different expansion hysteresis, even though the applied carbon material was the same.[46] It is important, as such delay (electrode volume returns to the state before adsorption of ions) can influence the time constant of the system and may

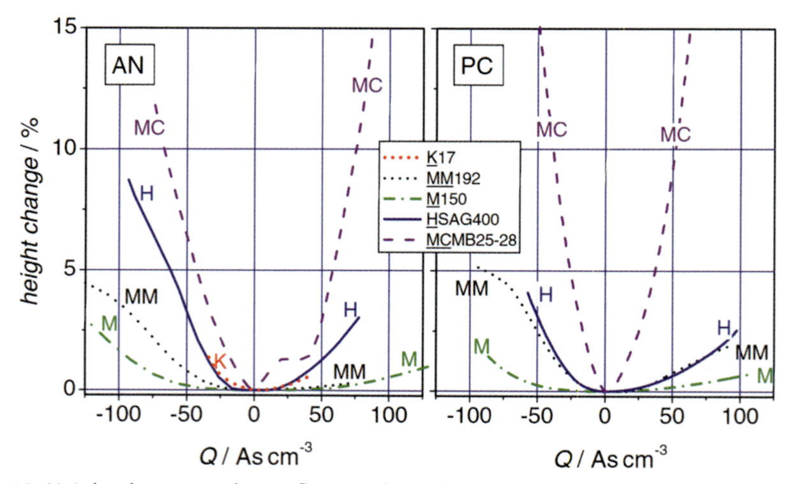

Fig. 10 Height change vs charge flowing through the electrode in the potentials higher and lower than pzc for five studies of carbon materials with different textural properties. Used electrolyte: 1 mol L^{-1} TEA BF_4 in AN (left) or PC (right).[65]

induce irreversible electrode changes when used in too harsh load conditions. It proves that the size of adsorbed cation (desolvated and/or solvated) is crucial for electrode volume expansion. $BMIM^+$ adsorption into microporous activated carbon is characterized by higher strain—1.1%, than recorded for $EMIM^+$—0.7% (for EMIM TFSI) or 0.25% (for EMIM BF_4).[46] Thus, not only electrode strain but also hysteresis can be the result of complex ion-ion interactions (depending on the anion applied) and the volume of each ionic species present in the electrolytic solution. The strain recorded for anion adsorption is usually much smaller, which is in accordance with the EQCM test in EMIM TFSI, where in the positive potential ranges (higher than *pzc*) ion mixing/reorganization is the main charge storage mechanism.[44,46] Interestingly, the electrode expansion in organic electrolytes starts at the relatively high overpotential (both cathodic and anodic), i.e., $\geq \pm 1$ V.[64,65,67] However, if the electrochemical capacitor operates in the safe voltage range—electrode expansion is equilibrated by the free electrolyte volume shrinkage—and the overall cell volume change should be negligible, especially in the pouch cell construction.[65] Currently, dilatometry studies have been implemented in redox-active electrolytic solution studies. In the system where (1-ethyl-3-methylimidazolium ferrocenylsulfonyl-(trifluoromethylsulfonyl)-imide EMIM FcNTf redox ionic liquid in AN was coupled with AC YP50-F, charge storage on the positive and negative electrode differs from each other.[67] On the positive electrode, first, co-ion desorption occurs and is followed at high voltage by a combination of faradaic reaction with counter-ion adsorption. Contrarily, on the negative electrode, counter-ion adsorption is a main charge storage mechanism in the entire potential range studied. Such a phenomenon explains a self-discharge of a redox-based system, as the co-ion desorption is one of the charging mechanism steps and, thus, is inevitable.[67] A detailed description of the charge storage mechanism is possible due to *in-situ* dilatometry studies implementation combined with the SPECS technique.[68] Interestingly, when two carbon electrode materials differ from the textural point of view, but not in the structure—their anodic and cathodic charge storage mechanisms are similar.[68,69] Cations ($EMIM^+$) follow the ion exchange and co-ions adsorption mechanism in the anodic potential scan (potentials higher than *pzc*). Anions ($TFSI^-$) are also adsorbed via an ion-exchange mechanism, replaced by a perm-selective one at higher potentials.[68] Therefore, the ion-exchange region, called reorganization, is a unique property taking place when moderate voltage is applied. This potential difference is not powerful enough to overcome strong carbon-ion and ion-ion interactions. Interestingly, when carbon electrode expansion with an aqueous solution containing HPO_4^{2-}, $H_2PO_4^-$ and Na^+ ions was studied, the difference between the behavior of the ions based on their valence state was shown.[70] It can be related to the electrostatic attraction between ions and the carbon surface, competitive to their potential driven adsorption. Taking into account the size of ionic species, the equal electrode volume changes in the case of the electrolyte containing HPO_4^{2-} and Na^+ suggest possible hydrophosphate anion confinement into carbon porosity. That leads to the disproportionation of the charge delivered to the charge released and lowers the charging/discharging process efficiency. When the same valency of cation and anion was used, it was shown that both can easily exchange themselves in the fast ion-exchange process triggering uneven electrode height change.[70]

3.4 Modeling

Molecular simulations and modelling studies are also focused on the fundamental aspects of the charge storage mechanism in electrochemical capacitors. There is a variety of possibilities to model the ion fluxes behavior, inside and outside carbon porosity, considering of electrode structure, texture, applied potential and other factors. To show the powerful abilities of ion dynamic simulations, especially when combined with *in-situ* characterization techniques, a few examples will be discussed herein. However, the broad diversity of molecular simulation studies applied in the energy storage/conversion field is too large to be included herein. It needs also to be highlighted that limiting conditions to make a reliable simulation run, are usually experimentally determined. Therefore, modelling studies without considering the processes taking place in the real system are without merit. They must be performed in close cooperation with each other (real experiments and modeled ones) to push optimally the boundaries of data interpretation and reveal insight into the charge storage mechanism.

One of the recent publications discussed reorientation of ions, fluxes of ions and the effect of electrode expansion for five ionic liquids (EMIM Cl, EMIM BF_4, EMIM OTF, EMIM FSI, EMIM TFSI) combined with graphene or MXenes electrode material ($Ti_3C_2F_2$, $Ti_3C_2O_2$, and $Ti_3C_2(OH)_2$)—presented in Fig. 11.[71] To perform reliable molecular simulations, many aspects, including the properties of materials studied, must be considered. In this study, carbon atoms building graphene layers were neutral, and carbon atoms in MXenes acquired specific atomic charges.[72]

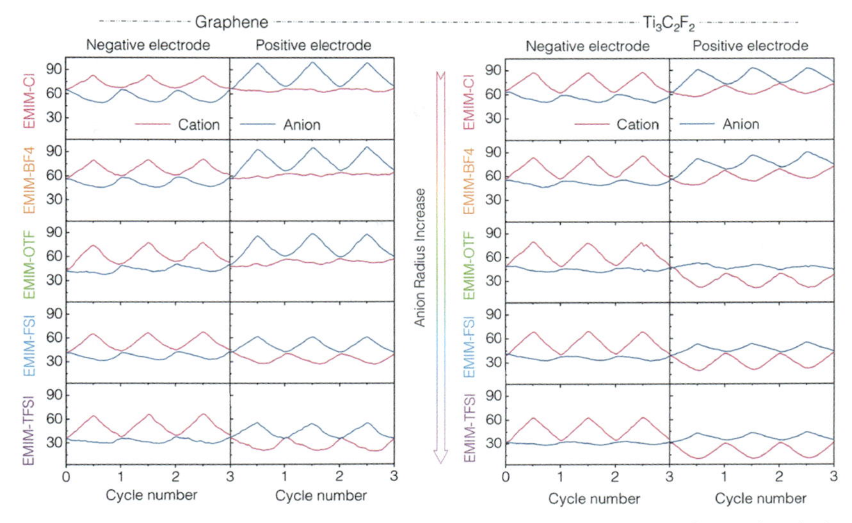

Fig. 11 Evolution of the ion numbers in the graphene and $Ti_3C_2F_2$ electrodes during charging and discharging.[71]

Overall electrode charge was neutral in both cases. For ions—the OPLSAA force field was used in the simulation process.[73,74] The temperature was set at 450 K with a time-step of 1 fs. These parameters are directly correlated with ion dynamics, and effective charge screening that can be present in the real ionic liquid-based system and are crucial for obtaining a reliable outcome.[75–78] Selected ionic liquids are stable up to higher temperatures than the selected one, i.e., 450 K.[79–81] Electrostatic interactions in the long-range distance were calculated using the particle-particle, particle-mesh Ewald scheme (Fourier-based method). Van der Waals interactions were cut off at 1 nm. For the charging mechanism, the galvanostatic charge/discharge process was imitated. The scale of such experiments is much smaller (in nanoscale instead of micrometer real experiments or even larger ones).[68,82] However, carefully selected parameters allow obtaining precise information, insightful for the full device optimization. For the graphene material, almost similar electrode expansion was observed for the anodic and cathodic polarization as previously discussed in Section 3.3. It is worth highlighting that on single-layer graphene, expansion should not occur, as the electrode surface is flat and uniform. This result indicates the few-layer graphene material or, how it should be termed, graphene-like material was considered. Moreover, for selected interlayered materials—electrode expansion in the potentials lower than *pzc* is independent of the anion present in the system. Such observation is logical, as no solvent is present in the systems, and thus, no cation solvation shells are observed—bare $EMIM^+$ takes part in EDL formation for each ionic liquid studied. For potentials higher than *pzc* the electrode expansion/contraction is both electrode and anion type dependent. Interestingly, not only an increase of electrode thickness but also its contraction was observed. Fig. 11 summarizes the ion population during the charging process in anodic and cathodic scans for graphene and $Ti_3C_2F_2$ MXene electrode materials.

By comparing simulated results with those obtained in the laboratory on a real device (with EMIM BF_4), it can be shown that properly conducted simulations are in accordance with experimental data.[37,83] For MXene-electrodes, simulated data resemble data determined using *in-situ* characterization.[84,85] Such a statement indicates that modelling data can be used to predict the behavior of a real system, even if not yet experimentally studied, and highlighted their advantages.

A recent report on the identification of failure mechanisms in electrochemical capacitors using Monte Carlo (MC) simulations reveals a very interesting and realistic performance degradation cause.[86] One needs to consider the energy, mass and charge transfer simultaneously with electrochemical reactions in EC. Thus, in other research, usually lumped or circuit models were implemented.[87–89] However, although being relatively easy to apply, these models do not fully consider the local distribution of the dependent values within the modeled electrochemical system. An improved approach includes mechanistic models that combined either 14 or 17 input data.[90,91] Furthermore, machine learning techniques were successfully implemented to predict specific capacitance of carbon-based ECs.[92,93] To fulfil the gap in the modeling of the parameter interactions of various ECs and their influence on the performance, Monte Carlo simulations were used. They allow to consider in isothermal conditions, charge conservation phenomenon in the macro- to micro-scale and double-layer charging together with Faradaic contribution of RuO_2-based electrodes. Macroscale is understood in this context as ions and electrons transport between the electrodes. Contrarily, H^+ diffusion into the RuO_2 structure is a micro-scale process. Therefore, Monte Carlo simulations capture charges and species distribution across the electrode thickness. This study includes 13 stochastic parameters that are treated as input data, that can be divided into five subgroups: (1) design and/or geometry: electrode, current collector and separator thickness; (2) electrochemistry: geometric double-layer capacitance ($F\ cm^{-2}$), current density; (3) operation: temperature; (4) materials: electrodes and current collectors conductivity and ionic conductivity of the electrolyte; (5) physical properties: electrodes and separator porosity and diameter of the RuO_2 particle.[86] MC simulations run at low and high charging loads. Variability of input data was quantified by sensitivity analysis. It was possible to determine critical parameters influencing the performance of EC (exactly on the specific capacitance value).

Fig. 12 summarizes the work using MC simulations for selected electrochemical systems, composed of RuO_2-based electrode material with an H_2SO_4 aqueous electrolytic solution (the system already studied in modelling studies).[94,95] In such a composition, ionic charges are carried out in the bulk of the electrolyte, electronic charges in the bulk of the electrode and various forms of ionic species in the liquid phase (both in the bulk of electrolyte and electrode texture). The system studied was divided into 1D processes: charge and ionic species transport, and 2D process: diffusion. All of the equations used for the data simulation are summarized in the supplementary information file to the publication[86]—if an individual would like to repeat or follow this procedure, it is necessary to become familiar with all the boundary assumptions that are taken into account. Porous electrode theory has been implemented and verified on the basis of laboratory results.[96] MC simulations consider input data fluctuations, which are characterized by a normal distribution with a mean value, standard deviation and random variation.[97] However, the error bars were set to 0.01 (between the population and the mean value) and the standard deviation to 10%. Such assumptions are in accordance with the industrial approach and concur with manufacturing requirements.[98] Furthermore, temperature fluctuations have also been considered. The representativeness of input data was verified using the Kolmogorov-Smirnov test, which sets the confidence level higher than 0.05. Next, input data were correlated with the response variable using Pearson's coefficient that lies in the range -1 to $+1$.[99] Subsequently, verification was performed using the rank correlation coefficient of Spearman's.[100] A Pearson coefficient close to $+1$ indicates a strong positive correlation, contrarily to -1 value—indicative of a strong negative correlation. Therefore, to prove the correlation of simulation with experimentally obtained data, the artificial values of specific capacitance on the level of 765 F g^{-1} with a specific energy density of 27 Wh kg^{-1} at 9 W kg^{-1} were reported for a 5 mA charging/discharging process.[97] Higher current (5 A) resulted in a much lower specific capacitance, on the level of 14 F g^{-1} which reduces also specific energy and power values. Calculated standard deviations were on the level of 72 F g^{-1} for 5 mA

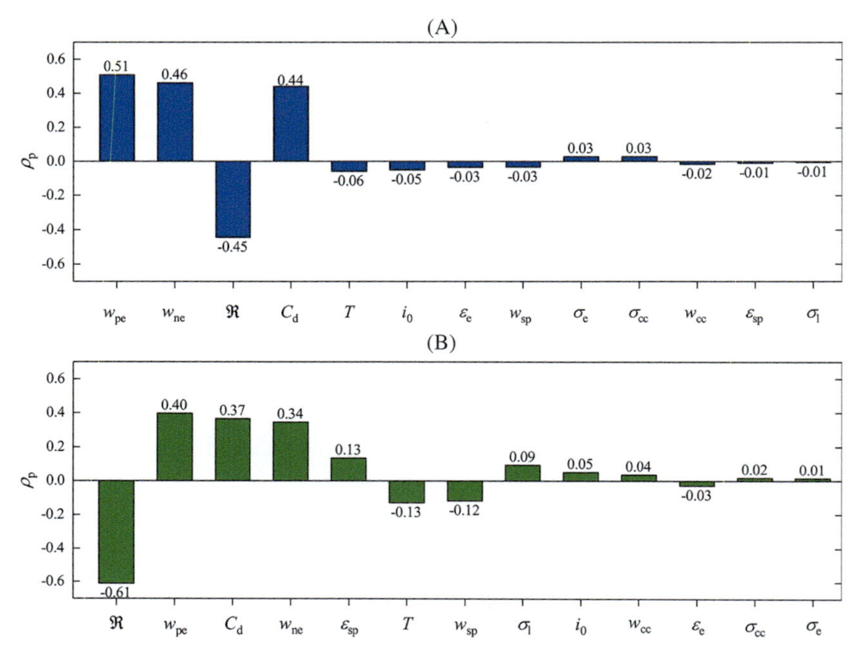

Fig. 12 Ranking of stochastic parameters when all of them were varied simultaneously at charging/discharging currents of: (A) 5 mA, (B) 5000 mA.[86]

charging/discharging process and 0.5 F g^{-1} for 5 A. They represent ±9% and ±4%, respectively. Referring to Fig. 12, representing Pearson's correlation coefficients for input data, it can be noticed readily which selected parameter affects the specific capacitance. The 0X axis is presented based on the absolute values of the coefficients, indicating the strength of the influence and moreover its direction (positive or negative). Parameters that resulted in Pearson's correlation coefficient lower than 0.05 were not considered relevant. It can be observed that for two currents studied: 5 mA and 5 A, different parameters are responsible as the main factor affecting the overall performance of the cell (measured in specific capacitance). At low current load, it seems that electrode thickness—more relevant for a positive than for a negative electrode, is the most crucial factor. When the charging/discharging process was simulated with a high current, i.e., 5 A, the particle size of active material (RuO_2) turned out to be the most limiting factor. Such observation could be expected as higher electrode thickness leads to higher resistance of ion transport into and from active material. Particle size has a tremendous effect when the charging/discharging process is conducted at high current loads—when the diffusion limitations are critical.[101] The time scale of the energy storage process is important, and not all improvements can be beneficial to the same extent for various working modes. Therefore, optimization of electrode thickness depending on the active material used is a crucial step for the advancement of electrochemical capacitors.[102] However, the real case never looks so straightforward, as the optimization process is somehow a compromise between a few approaches that can be contradictory to each other.[12,103,104] This conclusion is shown to be valid also for different electrode materials such as nickel oxide or carbon.[105,106] Diffusion limitations do not concern only RuO_2-based electrodes but are present in all composite electrodes. Diffusion of ions is current-related and is very sensitive to a time-scale of the charging/discharging process. It prevailed that electrode-material electronic conductivity or type of current collectors are marginal parameters influencing the specific capacitance of the system. However, to the best of our knowledge, that can be only stated for current collectors that are not degrading faster than electrode material,[107] which depends on the coating properties and mechanical quality of the current collector itself.[108] Furthermore, for the fast charging/discharging process higher number of parameters seem to have a real impact on the performance. Not only the electrode thickness or particle size but also the porosity of a separator has an important role. Using MC simulation, it was possible to determine also the nature of the correlation between the input data selected and model-specific capacitance. Thus, it shows a direct correlation of how the parameters are affecting one other and influence the overall outcome—capacitance value. The overall discussion examines observed correlations with the experimental data and well-known electrochemical principles. For example, electrode porosity (ε_e) has a lower limit to not inhibit ionic diffusion inside the bulk and deteriorate performance.[109]

Other information that can be gained from molecular simulations/modelling work is a better understanding of the charging dynamics[110]—especially when highly viscous electrolytic solutions like ionic liquids are used. For this purpose, 1-ethyl-3-methyl imidazolium thiocyanate EMIM SCN (with 256 ionic pairs) was coupled with two planar graphene electrodes (with 448 carbon atoms each). Fixed separation of electrodes was established at 6 nm—that the bulk phase of electrolyte volume is present in the system. Such separation distance lies far away from the real cell construction, where usually thin separators are used from 260 μm to maximally a reported value of 520 μm. System size was determined by NPT MD simulations, ions—using OPLS all-atom force fields. All physicochemical parameters of either anion or cation were established based on literature data.[73,111] Carbon atoms in the graphene layer could be characterized as Lennard-Jones particles with a diameter of 0.34 nm and a depth of the potential on the level of 0.360 kJ mol^{-1}.[112] Atomic species interactions were based on a Lorentz-Berthelot law. For the simulation, the LAMMPS program was employed with 3D-periodic Ewald summation and Coulombic interactions assumptions.[99] A Velocity Verlet algorithm was used for the velocity calculations, with a cut off of 1 fs. Equilibration of the model system was conducted for 10 ns at 350 K isothermal conditions (considered annealing temperatures range from 350 to 800 K). Electrochemical charging was conducted using a constant potential hold. Electrode atoms were modeled using a Gaussian function, with a function parameter of 19.97 nm^{-1}—the same for graphene or graphite-based materials.[113] Importantly, the charge of each graphene electrode atom is not fixed; this results from the local electrostatic environment and potential difference applied between the electrodes. Interestingly, the probability of charge distribution shows a non-Gaussian behavior at all potential differences applied (from 0 to $\pm$4 V). For ionic liquid-based systems, the charge distribution profile is caused by the high viscosity of the electrolyte medium and large shape asymmetry of anions and cations. The first layer of ions in EDL at polarized graphene material is displayed at 0.35 nm owing to the size of electrolyte ions. For the potential difference higher than 3 V, sandwich-type EDL formation is predicted, i.e., counter-ion layer, co-ion layer, counter-ion layer. It has been found that ions rearrangement in the vicinity of polarized electrode results from maximization of Coulomb interaction between ionic species and the electrode surface.[114–116] It is important that simulated data exhibit a similar trend as observed in the real EC, and that high anodic potentials do not directly lead to the anions adsorption as it is preceded by a stepwise cation repulsion and ion mixing process.[117] It seems that behavior of ions is not that much dependent on the charging method applied (potentiodynamic or potentiostatic conditions) but more on the electrode-electrolyte interactions. For a more thorough EDL understanding, orientation of ions should also be considered. When the potential difference equals 0 V, both ionic species at the electrode/electrolyte interface are arranged parallel to the electrode surface. $EMIM^+$, as an ion-containing aromatic ring, creates strong π-π interaction to the carbon structure. It seems that the SCN^- anion reflects the $EMIM^+$ arrangement, according to the Coulombic interaction maximization law. Such composition is maintained within three subsequent ionic layers in the EDL structure. At high potential difference (3 V) $EMIM^+$ rotates and desorbs from the first EDL layer, making a void space for SCN^- anions. Their concentration in the vicinity of positively charged electrode material increases. A similar mechanism is observed on the cathode surface (electrode polarized toward lower potential values); however, all anions are repelled from the first ionic layer. In this study, differential capacitance has also been studied. Nevertheless, currently, this parameter can also be determined experimentally which leads to the verification of the stated hypothesis.[68,118] The point of zero charge for the EMIM SCN ionic liquid in the vicinity of the graphene electrode has been calculated to be 0.2 V. Moreover, a vacating effect in the charge storage mechanism has been described and it is believed that it occurs in all planar electrode-based systems and for those with high viscosity electrolytic solutions. Observations drawn from an EMIM SCN experiment were confirmed using the EMIM BF_4 ionic liquid. Therefore, considering the tunable properties of ionic liquids, the charge dynamics are more crucial than the chemical structure or density of ions. Furthermore, from MD simulation, it has been revealed that ca. 50% of the population of ions in the entire electrolyte volume takes part in EDL formation on the planar graphene-based electrode material. It shows that applied potential difference does not screen the ionic charges clearly, but a mixed ionic composition in the various electric double layers can be found. Both charging and discharging dynamics are characterized by a short time constant, on the level of 2 ns. When the electrode texture is enhanced, the charge and discharge process become asymmetric.[119] A planar electrode surface does not face spatial hindrance obstacles, when EDL is formed. Upon comparing the simulated outcome of the charging/discharging mechanism to the experimental data, it can be observed that modeled IL are characterized by a high activity/facile to move in the bulk of electrolyte. On the contrary, in all IL-based EC studies, the high viscosity of the electrolytic solution medium is the main drawback for high-power applications. A possible explanation of such difference is the electrode distance applied in MD simulations. Therefore, the non-realistic distance between electrodes and lack of a separator induce a fast charging dynamic, even in the relatively "slow," "bulky" electrolytic medium, as the ions have more space to rearrange themselves in the bulk of electrolyte. To increase the power abilities of IL-based ECs, one needs to consider thin separator design and application. Moreover, graphene production technology does not allow full implementation of these features and limits full commercialization. Thus, increasing the pore volume, roughness of electrode and pore size—one introduces several spatial hindrances that can affect ion dynamics during EDL formation.[120,121] Therefore, the proper electrode-electrolyte matching is crucial for advanced energy storage devices, as the best optimum between physicochemical and electrochemical features need to be balanced and established.[12]

One can conclude from the papers discussed here that this type of electrochemical capacitor investigation (*via* molecular dynamic simulations) is very advanced and unique. It is, however, crucial to verify observed phenomena with a real experiment to extrapolate other results (impossible to be conducted in the real laboratory scale) and obtain an insight into the operational performance of the EC system. The multiplicity of possible electrode-electrolyte combinations leaves this approach open for further studies.

3.5 Nuclear magnetic resonance (NMR)

In-situ NMR spectroscopy allows distinguishing ionic and molecular species adsorbed at the electrode/electrolyte interface from those present in the electrolyte bulk.[122] An applied potential difference triggers the chemical shift of the in-pore resonance. It originates from two phenomena: reorganization of ions and local change of electronic density of carbon material.[123] The latter seems to be the dominant contribution, contrary to the previous findings concerning EDLC charging/discharging.[124] Interestingly, only *in-situ* NMR (modelled or real-time measured) combined with experimental data leads to new insight into the charge storage mechanism. Both provide information based on the different parameters and are complementary to each other. In modelling data, as previously discussed, mostly planar or highly organized carbon structures are considered. In the real application, micro-mesoporous carbons are applied with various structural and textural properties. It has been observed for the coupled NMR studies with modeled data that, usually, the chemical shift is moved toward higher values irrespective of the potential applied (either anodic or cathodic scans) in organic electrolytic solutions. Therefore, this provides information about reorganization of ions in the pore volume, not about the charge type itself. Contrarily, for an aqueous-based system (with an aqueous solution of NaF), the chemical shift is changed towards lower values, probably due to nuclei formation at the electrode/electrolyte interface.[125] However, the richness of aqueous-based electrolytic solutions in the redox contribution, together with the active participation of the hydration shell in EDL formation, makes this study more complex and therefore usually is beyond current scientific scope. The novel methodology assumes the interplay of real experiments, together with MD simulations and theoretical studies of density functional theory (DFT) in order to obtain a high quality and reliable NMR spectrum.[126–129] For this purpose, many calculation steps were required, such as a model of carbon particles based on the isothermal nitrogen adsorption experiments or in-depth analysis of experimental NMR spectra (conducted with one activated carbon and various electrolytic solutions) to verify free energy of ionic species. It was possible to distinguish between ions behavior in slit pores and in real wide pore size carbon material combining real data and modelled ones. The inaccessible sites, free energy and resonant frequency of ions and solvent molecules were assigned prior to the simulation run.[127,130,131] Interestingly, for a realistic carbon particle model, the lattice $20 \times 20 \times 20$ was used, where each lattice site has a slit pore texture. However, the whole lattice cube is then characterized by a wide pore size distribution resembling one of YP-50F activated carbon.[132] Moreover, pore surface/distribution has also been assigned. In order to increase the applicability of the proposed model, simultaneously with its verification, several electrolytic solutions were used: 1.5 mol L^{-1} BMI BF_4 in AC,[133,134] 1.5 mol L^{-1} PEt_4 BF_4 in AC[73,135] and EMIM TFSI.[135] Even though chemical shifts determined experimentally vary from those modelled, the same trend can be observed: negative potentials are characterized by a higher difference in chemical shifts than positive potentials. It is important to note that the reorganization contribution of ions and ring currents are not additive values. The average chemical shift is a function of the ion distance from the carbon surface. Reorganization of ions leads to the negative chemical shift of the positively charged electrode. Observations for the slit pore model in comparison to the realistic lattice model show the same trend: for negatively charged carbon particles; the chemical shift is much higher than for the positive one. However, to a different extent: +5.1 ppm (positive electrode) and +8.0 ppm (negative electrode) vs +2.1 ppm (positive slit pore) and +3.6 ppm (negative slit pore). This could be assumed, as a lattice model contains a wide range of slit pores—therefore, ion-pore interactions are larger. Realistic experimental data give results in between the simulated data either for slit pores or a realistic model.[136] Agreement between experimental and simulated data usually face some obstacles in the negative potential ranges, where significant differences in the chemical shift values can be observed. It results in inaccuracy of the cation adsorption model, especially when compared to the anion potential-driven adsorption. The ion mixing zone is present in the real system, as well as in the modelled one. However, this zone is hard to be ideally mapped. Furthermore, all simulated data indicate that in the pores smaller than 1 nm, highly effective ion adsorption occurs; this is observed as high specific capacitance is recorded for such uniform microporous carbon materials.

NMR studies were also applied in the structure and dynamic study of ions confined in the EDL[137,138] and were successfully carried out for ionic liquids (Pyr_{13} TFSI and EMIM TFSI, with or without organic solvent—AN) and a porous carbon electrode (95 wt.%—YP-50F and 5 wt.%—PTFE). Beside the high viscosity of the IL studied, it was shown that this electrolyte spontaneously penetrates the entire electrode pore volume—which results in a satisfactory electrochemical performance. In this particular system, the anion potential driven adsorption/desorption plays a crucial role in the charging mechanism contrary to that of the cations. Both carbon and selected electrolyte (5 μL) were placed into double-resonance zirconium dioxide magic angle spinning (MAS) rotors with an outer diameter of 2.5 mm. This construction allows *in-situ* NMR data to be recorded, simultaneously preserving the electrolyte from evaporation during measurement.[138] The strength of the magnetic field used has been reported—7.1 and 9.4 T, which corresponds to 300.2 and 400.4 MHz frequency. The probe background was recorded for ^{1}H spectra and then subtracted from the data obtained using a simple pulse – acquired sequence of 5 kHz MAS. Each spectrum collected (^{19}F, ^{1}H, ^{2}H) was referenced to known chemical substances. To study the influence of the temperature, the MAS sample was calibrated with ^{207}Pb on lead nitrate.[139,140] 5 kHz frequency corresponds to the temperature change lower than 1 °C; for this reason, a delay time was introduced between measuring sequences to obtain a quantitative data set. For NMR spectra analysis, as well as for XPS[141–143] or Raman spectroscopy,[144,145] the deconvolution process is very important and often requires the experience of the user. The intensities of the spectra must to be normalized in order to compare the samples between themselves. In contrast to *in-situ* NMR studies, the *ex-situ* approach allows for standard coin cell utilization—the processes taking place inside the cell are the resultant of all ongoing phenomena. Therefore methodology using two approaches in parallel (*ex-situ*, *in-situ* and/or modeling) is recommended, so the outcome of both procedures is verified. It is worth highlighting the fact that the *ex-situ* approach requires the sample to run and stop at the desired moment, determined by the potential, time or state (as also valid for other *ex-situ* techniques like specific surface area determination SSA,[41,146] X-ray diffraction (XRD),[147] elemental analysis[41,148,149] and other). Therefore, the very precise electrochemical protocol is required with minimal statistical repetition (minimum three samples) and should be followed to minimize the error of the *ex-situ* analysis as the preceding test could be time- or cost-consuming.

It has been demonstrated comparing *ex-situ* with *in-situ* NMR data that charge storage is related to chemical shift changes of confined ions into carbon porosity.[132] However, this results from the properties of carbon more than the type of ion itself. The anion-cation interplay is observed in the ion dynamics in the electrode porosity.[150] The same anion $TFSI^-$ reveals different mobility when combined with different cations: $PYR_{13}{}^+$ or $EMIM^+$. Organic solvent presence (in this case—AN), as expected, increased ions mobility when compared to the pure ionic liquid. Depending on the potential applied a different desolvation state has been found for ions, and in-pore to out-pore ionic population varies.[150] As can be seen, all the NMR studies are aimed at describing the energy storage mechanism on the molecular level and successfully brings a stepwise solution to each puzzle presented, in order to obtain a full energy storage mechanism picture.[151] However, observations emanating from the studies are very specific and depend on many components of the systems and parameters applied during operation.

3.6 X-ray photoelectron spectroscopy (XPS)

To study any chemical composition change, the application of XPS analysis is very appropriate. The analysis method is not sensitive to the electrode bulk compositional variation (contrary to TPD-MS or EA measurements (described separately)), but it can detect with high precision, functionalities present on the carbon surface. Therefore, the application of XPS *ex-situ* is a very common approach to study electrode aging upon electrochemical operation.[41,47,146]

For the specifically conducted aging test, one can follow the fade mechanism for EC with an aqueous-based electrolyte. This has been carried out for nitrate-based ECs with lithium nitrate as the electrolyte with concentrations: 0.2, 0.5, 1.0 and 5.0 mol L^{-1} and activated carbon cloth as the electrode material (Fig. 13). Comparison of electrode surface functionalities (for both positive and negative electrodes) after electrochemical fade of the system (specific capacitance drop by 20%) allow different mechanisms

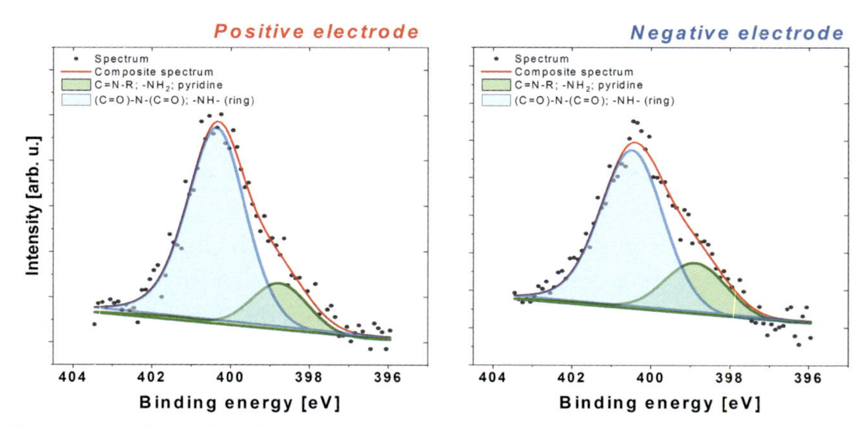

Fig. 13 Specific surface functionalities measured by XPS (*ex-situ*): N1s deconvoluted spectra for the positive (A) and negative (B) electrode of ECs with 0.5 mol L^{-1} $LiNO_3$ after alternate floating.[41]

occurring on both electrodes to be observed. The positive electrode has been mostly oxidized, and some of the nitrogen has been introduced via oxygen-bonds (C=O)—N—(C=O) to the carbon structure (400.4 eV). The amount of nitrogen decorating carbon electrodes after electrochemical aging of EC varies but does not exceed 3.5%. Chemical bonding via oxygen to carbon structure indicates that during electrochemical operation nitrogen more favorably reacts with oxygen, than incorporates itself to the carbon matrix (which also results from the energy level barrier of both processes that is higher for nitrogen binding directly into the benzene structure). Even from *ex-situ* conducted XPS analysis, one can explain many ongoing phenomena at the electrode/electrolyte interface during electrochemical operation.

The only experimental limitation of this technique is the high vacuum condition. Thus, if the chemisorbed species are weakly bonded to the carbon, they will not be detected in the XPS analysis. XPS is a very sensitive measurement tool, it can trace elements in the ppm concentration level. The only not-traceable elements are hydrogen and helium, owing to their ideal electron configuration. *In-situ* XPS was successfully applied in the graphene potential-driven perm-selectivity studies with DEME TFSI.[152] Arising from this, evidence for the finite potential drop across double layers formed at the graphene/electrolyte and electrolyte/metallic current collector interfaces, was obtained. The imbalance of ion population at the solid-liquid interface has some consequences in the voltage development across the cell. The observation was accomplished based on the verification of F1s, N1s and C1s signals. Two nitrogen peak positions in N1s spectra results from the IL cation ($DEME^+$—with one centrally located, positively charged nitrogen atom) and anion ($TFSI^-$—with centrally located, negatively charged nitrogen atom). The imbalance of cation and anion population is reversible dependent on the bias voltage. The system was tested in the voltage range from −2.5 up to +2.5 V. Successful quantification of ion movement has been achieved *in-situ* for multilayer graphene-based system (the counter electrode was gold foil). Thermal ($T > 50\,^{\circ}C$) and electrochemical ($U > 3.0$ V) stability of the EMIM BF_4 ionic liquid was studied in contact with model carbon material CDC using *in-situ* XPS. Interestingly, in this study, the presence of intermediate and final chemical compounds resulting from the decomposition/degradation of ionic liquid (oxidation and reduction reactions) was investigated. In accordance with modelling data, the formation of ionic liquid ion dimers was proven.[153,154] Especially for the $EMIM^+$ cation, where carbon atoms undergo a reduction reaction leading to radical formation. At higher voltages ($U > 3.2$ V) a subsequent reduction reaction occurs of the nitrogen atom in the $EMIM^+$ cation. It results in the appearance of the two new forms: $N\text{-}CH_3$ and $N\text{-}C_2H_5$ bonds.[155] It seems that the $EMIM^+$ cation is less stable than the BF_4^- anion, which does not undergo a reduction reaction. Considering the observed phenomena, stable EDL is formed only in the limited voltage (up to 3.0 V). Exceeding this value, even if not observed on the electrochemical curves, some redox reactions already take place and change locally the chemistry at the electrode surface, leading to a faster degradation process. For ionic liquids, one can conclude that a dimerization phenomenon occurs since dimeric forms are created. This process can also be observed in the macroscale as the color of the ionic liquid solution changes from transparent to brownish/yellowish.[155] However, the application of a surface-sensitive analytical technique does not reveal whether any carbon corrosion process occurs as well, or only electrolyte redox reactions.

3.7 *In-situ* gas analysis

A very interesting insight into EC operation is proposed by coupling electrochemical measurements with *in-situ* gas analysis, i.e., *online* mass spectrometry (OEMS) or *in-situ* gas analysis (GC-MS).[66] From the experimental point of view, these techniques are very similar. In the customized cell, the full device is subjected to standard electrochemical analysis. The key parameter that makes the difference is the flow of gas that flushes the cell and transports volatile products to the detector. Such procedure requires knowledge of possible gaseous species that can be formed—for specific molecules, the m/e signals can be the same for totally two different substances, such as $m/z_{CO} = 12 + 16 = 28 \text{ g mol}^{-1}$ and $m/z_{N2} = 14 \times 2 = 28 \text{ g mol}^{-1}$. Nevertheless, OEMS allows to detect gases continuously and combines them with electrochemical performance. Even though the approach with mass spectroscopy is not new in research practice, its *in-situ* application is very challenging and needs to be performed with caution. The flow rates of flushing gas, the temperature, size of the cell, the sum of the detected gases, localization of the gas collection outlets and many others—are the main engineering issues that must be solved to obtain representative, reliable and reproducible data. For commercial energy storage/conversion devices, an increase in internal pressure due to the formation of the gaseous by-product is one of the main reasons for failure. Therefore, OEMS studies seem to be of considerable importance and should be carefully conducted to find and propose a safe (without any explosion risk) device. Although the gases released in water-based systems (CO, CO_2, O_2, H_2 and N_xO_y or SO_x) are moderately safe when appropriately confined, especially when compared to volatile organic ones—the pressure built up leads to mechanical damage of the separator and, in effect, causes short-circuiting with local overheating spots. Additionally, the type of the evolved gases depends on electrolyte origin. A thorough understanding of the gas formation and its chemical composition is crucial in order to propose EC systems with decreased decomposition rate or at least inhibited gas evolution. It has been proven that during short term cycling, gases like CO and CO_2[156] can be generated. They can be further "consumed" during the discharge process, giving an impression of a *cell-breathing* process. Predominantly, this process seems to be harmless for the long-term operation, since any deterioration of electrochemical parameters has not been observed so far. Additionally, different electrochemical conditions were compared and cycling was shown to be more harmful from a structural and mechanical point of view to the system operation.[156] Electrode integrity faces substantial problems when a dynamic gas formation occurs at the electrode/electrolyte interface. Finally, asymmetric aging of the positive and negative electrode was confirmed by testing the gases evolved in the vicinity of the electrodes.[39,157] Usually, in aqueous EC positive electrode undergoes oxidation and further corrosion processes. Unfortunately, at moderate and high voltages, this process seems to be irreversible, leading to the specific surface area drop and capacitance fade. The negative electrode, owing to hydrogen electro-sorption, operates as if to balance the deterioration of the positive electrode, but with time it ages as well.

Fig. 14 presents the concept of *in-situ* gas analysis together with complementary advanced analytical techniques. It is important to plan carefully the experimental methodology to prove or provide evidence for the hypothesis from primary electrochemical tests. Interestingly, in organic-based EC the construction of the device was evaluated based on the amount of gas evolved during electrochemical operation.[158] This aspect is highly important from an industrial point of view and indicates that cell construction itself can also affect electrode/electrolyte stability. Moreover, the inhibition of gas evolution in improved cell design varies to a different extent for each type of gas that is evolved.[158] Evolution trends of different gases depend strongly on the voltage applied. Pre-charging seems to be another experimental method, based on the adjustment of the operational conditions, to inhibit gas evolution, *ergo* aging of the cell.[159–161]

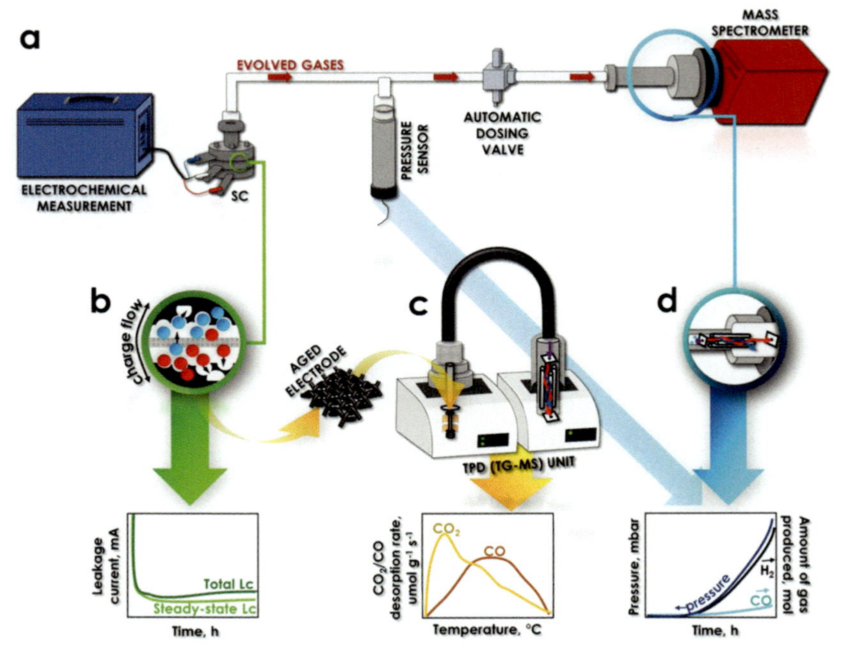

Fig. 14 Schematic concept of the approach used for assigning the charge passing through Li_2SO_4-based EC under floating at 1.5 V: (A) Simplified scheme of the OEMS system; (B–D) Illustration of the analyses realized for determining the processes taking place during high voltage performance of the EC taken from Ref. 39.

In-situ gas analysis is successfully implemented in energy storage/ conversion devices—proving the aging phenomenon and managing the undesirable pressure increase during electrochemical operation.[162,163]

3.8 Other analytical techniques

This subgroup of the analytical techniques successfully applied in electrochemical capacitors study is based on battery research (both *in-situ* and *ex-situ*). Until now, there is no extensive usage of these techniques in EC, but promising trials have already been carried out. In the authors' opinion, they deserve to be highlighted as a new characterization approach. Moreover, some of the analyses cannot be conducted *in-situ*.

3.8.1 *Infrared (IR) spectroscopy*

IR spectroscopy especially focuses on the carbon surface chemistry in contact with specific electrolytes. One of the reports elucidates the stability of ions (EMIM BF_4) based on the carbon functionalities change after being in contact with electrolyte at boundary potential values, i.e., $E < -2.0$ V and $E > +1.6$ V vs Ag wire in AgCl/IL.[164] At the low potential value, the $EMIM^+$ cation undergoes a dimerization process. At the high potential values, BF_4^- is destabilized and undergoes a decomposition reaction. It proves that the stability limit of electrolyte designated at planar, metallic, low surface area materials cannot be directly translated to porous carbon-based systems.[165] Notably, IR spectra cannot be recorded satisfactorily if any gas evolution occurs. Thus, the potential window is limited for a 3-electrode set-up on which only the working electrode is observed and faces the IR source. To simplify the optical system, highly conductive carbon material can be used, and there is no necessity for an Au thin layer as the current collector. Another electrolyte that was studied, EMIM Tf was in contact with a RuO_2 electrode or EMIM TFSI with CDC, onion-like carbons and carbon nanofibers.[166–168] Knowing the exact ion structure, it is possible to follow its behavior during charge and discharge. Potential-driven adsorption of ions was established as the primary charge storage mechanism. Moreover, the population of ions was described according to the electrolyte composition, and potential difference applied. However, this methodology is limited to outer surface processes, as the pore volume is not reachable by an IR source.

3.8.2 Temperature programmed desorption (TPD)

This technique is also used for carbon surface chemical composition evaluation.[169,170] Unfortunately, it cannot be conducted *in-situ* as the characterization technique is based on the thermal destruction of the sample. Therefore, electrochemical tests are performed before cell disassembly and carbon electrodes are tested *ex-situ* without the possibility of being introduced into a new system afterwards. Usually, a constant temperature rate is applied, and gas products are transferred either in a vacuum or by a neutral gas (such as Ar) to the mass spectrometer to quantify the signal and correlate it with an exact gas composition.[170–172] From the gas evolution detected at a particular temperature, specific surface functionalities can be determined.[173] It seems that not only for graphitic carbon materials but also for microporous ones, the contribution of surface defects in the specific capacitance is relatively high. Therefore, the structure of electrode material is important as well as its texture.[174] Such analysis can also inform about the aging mechanism behind the electrochemical fade of the device—as gas evolved from the heated sample results from the exact chemical composition, and the electrolyte decomposition or electrolyte-electrode interaction upon the high voltage or current loads.

3.8.3 Raman spectroscopy

Detailed structure analysis can also be conducted using Raman spectroscopy.[175] Such analysis can be realized *ex-situ* and *in-situ*, giving either information about the electrode/composite structure or its change upon electrochemical operation. If the analysis is done correctly, it is a non-destructive method. However, the user should be aware of the laser power, proper laser length (usually used 622 and 532 nm) and exposition time in order to obtain reliable results.[145,176,177] Moreover, besides recording the material spectra, one needs to consider their deconvolution for detailed output data. *In-situ* Raman studies of CNTs electrodes revealed that metallic or semiconducting character of the material results in various activities upon the potential load.[178] At low polarizations (close to the *pzc* value) only metallic CNTs shows specific capacitance. Studies also indicate that C-C bond length is a unique parameter and its changes during the charging/discharging process are very susceptible to the electrolyte applied. In addition, the charging/discharging process has been shown to be dynamic, as all structural changes are reversible to some extent, while the system has not yet been aged.[179] Redox processes can be clearly identified using Raman mapping. Thus, hydrogen sorption,[47,176] iodine adsorption[38,180] or other species can be detected. Intensity changes of the specific band can be correlated with specific redox reactions ongoing at the electrode/electrolyte interface, e.g., nitrate anion reduction.[176]

3.8.4 Texture characterization: Specific surface area, pore size distribution and pore volume

Analysis of the texture of carbon used as an electrode material is crucial for describing the charge storage mechanism. However, such analysis cannot be conducted *in-situ*. Nevertheless, it gives such a considerable insight into possible ions behavior, ions fluxes and charge storage accumulation in the EDL that should not be neglected. Advancements of specific surface area determination with pore size distribution and pore volume estimation are based on two approaches: (1) utilization of various adsorbent gases (N_2, CO_2, Ar, Kr) and/or their combination[12,181]; (2) update, novel mathematical models implementation in order to obtain more realistic carbon texture representation (2D-NLDFT, methods for new hard carbon materials,[182] etc.). Specific surface area determination is based on the isothermal adsorption of gas molecules in the specific relative pressures that result in the isotherm curve. These data can be translated/recalculated to pore size distribution, pore volume and specific surface area value using various models and equations. Thus, the experience of the user is crucial in order to make a reliable data evaluation of collected isotherms.[183] Up to date, there is still a need to find a universal correlation between carbon texture and the electrochemical behavior of EC. There are several research projects in progress that are trying to address this issue.[184,185] However, as the interplay of various variables in the electrochemical system is very complex, for such a study, usually model electrolytes are used (such as ionic liquids).[186] For low operational loadings (current or voltage)—usually small micropores (<0.8 nm) are the determining specific capacitance of the system.[184,187] For high power applications, a small mesopore fraction does not guarantee satisfactory rate handling, as this parameter is also related to carbon structural properties.[12,184] Owing to a non-linear correlation, there is a constant need for carbon texture optimization and characterization.[188,189]

3.8.5 Atomic force microscopy (AFM)

This characterization approach has been utilized to study nucleation of the redox active material (MnO_2) during EC electrochemical operation that boosts the energy output of the device.[190] MnO_2 was evaluated from a mechanical point of view when the expansion of the material is foreseen as a result of ions intercalation into a layered structure. As a result of the AFM measurement profile, thickness is obtained from the screened electrode surface that can be followed according to the charged state of the electrode. AFM in the conductive mode allows the nanometer resolution detection of the non-conductive or less conductive sites in the carbon/electrode matrix.[191,192] Such an approach is advantageous when the polymer-binder carbon electrode material is used in EC. Not only the content but also the arrangement (distribution) of polymer fillers (PEDOT, PANI, PPy) can be determined. Such analysis is conducted in the Tapping Mode with the small, nondestructive force of 5 nN. A topography and conductivity map is registered in parallel, which is a significant advantage of this mapping method. To provide input data for conductivity calculations, the voltage applied was 100 mV, with a resistance of 100 kOhm in order to minimize the current not to exceed 1 μA. It has been shown that mechanical deposition of NiO at the electrode surface results in a more homogenous structure than the chemical process. In the case of polymer additives—the output data addresses the coverage of carbon matrix surface by the polymer matrix. It also provides information about the polymer-carbon adhesion that can lead to the electrode preparation method optimization. AFM enables many different operation modes to be applied in research practice.[193] One of the approaches is based on laser deflection; however, it requires transparent liquid media. The deflection signal is calculated from volts to nanometers, and further to force in nN. The AFM tip in this mode is moving in the x-y-direction; height changes cause its movement into the z-direction which can be observed on the recorded mapping profile. The second used mode is contactless, essential for nondestructive soft-matter investigations.[194,195] Then, attractive forces between the sample surface and the AFM tip are used. The tip is vibrating with known frequency and amplitude and any changes between its interaction with the surface studied, triggers frequency and/or amplitude change. This mode is applicable mostly in the gas phase, as for the liquid phase, there are too many oscillating movement/damping phenomena occurring that are not necessarily related to surface topography. When the AFM tip is made of sufficiently stiff material, AFM studies can also reveal mechanical properties of the probe (Young modulus).[196] This approach can be used in both media: gaseous and liquid. Interestingly, employing AFM studies in EC investigations (graphite substrate with the $EMIM^+$ $TFSI^-$ ionic liquid) leads to complementary data to EQCM and dilatometry studies being obtained.[197–199] More charge is passing through the carbon electrode. Hence, more mass (EQCM), volume (dilatometry) and thickness (AFM) change are recorded. Various materials were tested: gold, silicon, graphite together with either an aqueous or IL electrolyte.[200,201] It is shown that ion separation (for $EMIM^+$ $TFSI^-$) starts in the range 3–5 nm from the electrode/electrolyte interface, from which leads to consideration of "electrolyte bulk."[199] AFM measurements are sensitive to the external potential applied, thus, most of the tests are done under open circuit potential conditions or studied at only a single location upon the charge.[202] Additionally, various electrolyte compositions (PYR_{14} TFSI,[203] OMI PF_6,[204] H_2O additive[205]) were verified and it was shown that chemical composition of liquid electrolyte (ions type, solvent used) drastically influences the structure of the layer of ions upon polarization.[203,206] AFM was successfully employed in the *in-situ* MXenes characterization—its susceptibility to different alkali and earth alkali metal intercalation.[207] It allows the mechanical properties of MXene with the type of intercalated cation (Li^+, Na^+, K^+, Cs^+, Mg^{2+}) to be correlated. It seems that the MXene electrode uses only a part of its entire volume in the charge storage mechanism, and it is mostly related to the size of the intercalated cation. In the case of Mg^{2+} intercalation, more homogenous mechanical changes in the MXene structure are observed than in the case of K^+ [197]. Interestingly, a similar approach to AFM is implemented in electrochemical force microscopy to probe local ion dynamics at the solid-liquid interface.[208] As a result, one can obtain a response profile vs applied voltage. It shows the complexity of studied dependences on the voltage applied, especially when discussing various salt concentrations. Charge screening phenomena, depend nonlinearly on the voltage applied, which results from the diffusion limitations (even at the planar, metallic electrode surface, i.e., gold). High AFM probe resolution finds its application in other energy storage/conversion devices studies.[209–211]

3.8.6 X-ray diffraction (XRD)

X-ray diffraction plays a crucial role in the crystal structure determination and/or chemical composition based on the crystalline structure. This technique can be applied in battery research. The principle of the XRD technique is based on the diffraction into various directions of X-ray radiation that is caused by the properties of the crystalline structure. Based on the angle of this diffraction, the crystal structure responsible for the diffraction pattern can be described—the position of the atoms in the crystal structure, chemical bonds or disorder values. For the material characterization, specific sample preparation is required (grinding) to fill the sample holder for diffraction measurement. Nonetheless, the XRD pattern can also be detected for thin-film materials. Therefore, it is a very powerful technique for different electrode material characterization. Few studies have been reported on its application for EC characterization, especially when based on advanced electrode materials. One of the studies proves the intercalation of Al^{3+} cations into the structure of the 3D nanowire network/fluorine-doped tin oxide.[212] XRD was used for pseudo-capacitance contribution detection at different voltage intervals. Interestingly, here the electrochemical data and XRD spectra were collected simultaneously, whereas in the case of the battery system it is usually done *ex-situ*. The electrochemical system is charged up to the desired state of charge, then disassembled and the sample is characterized in terms of its structure/composition. Even in non-graphitic carbon materials, the volume changes, therefore expansion and increase of the interlayer distance can be observed and recorded using XRD.[212] In Li-ion capacitors, XRD analysis is employed in order to study the Li ion storage mechanism—thus, a battery-type electrode.[213] 2Θ peak positions—(101) and (141)—are correlated with electrode potential and reorganization during the charge/discharge process. It demonstrates new phase formation related to Li^+ insertion. Peak positions are reversible, indicating a reversible lithiation/delithiation process into the $NaNbO_3$ electrode material.

4. Summary and future perspectives

All the *in-situ* advanced characterization techniques described herein aim to understand the energy storage mechanism and lead to proposals for some system optimization. The assumption that electrochemical capacitor operation is only based on potential-driven adsorption of oppositely charges ions onto electrode surface is not consistent with the molecular phenomena occurring at the electrode/electrolyte interface. *In-situ* analytical techniques allow the realistic mechanism for individual systems to be described. This unveils that the energy storage mechanism is complex and results from interactions between ions (anions and cations), solvent molecules (if present) and carbon properties (structure, texture and chemical composition). Furthermore, electrochemical operation conditions (potential, current, temperature, time) also play a crucial role in the charging/discharging process. With all this information, one can predict the exact charging mechanism and/or possible failure of the individual system.

Various techniques are useful if the user is cognizant and can evaluate the difference between experimental error/device shift and observed phenomena. Application of the *in-situ* techniques described can help in answering the questions related to the charge storage mechanism. However, due to cell construction limitations, these experiments usually cannot be considered as a faithful, representative probe of ongoing phenomena in the real system. To reconstruct the interfacial processes, their origin must be well-known and adequate experimental conditions must be ensured. In many cases, cell construction (very often made in-house) is the most difficult part before implementing *in-situ* characterization techniques. Therefore, interdisciplinary studies are of tremendous importance; that is a sharing of experiences from various fields to propose novel techniques, and application of known techniques, or giving rise to a new insight into measuring possibilities. Additionally, such studies are usually limited to a three-electrode setup. This is another obstacle in the way of reliable *in-situ* characterization. Highlighting again, basic knowledge and standard electrochemical characterization need to be obtained/done *a priori* to *in-situ* investigations. *In-situ* measurements are very sensitive to any variable, and input parameter variations. Currently, there is no limit to required measurements of components and their properties of each device. The issues are as follows: how to interpret the data? Are the data reliable? Are the data representative? What information is sourced from the data set? An experienced and highly skilled individual can be capable of answering these questions. To reproduce any advanced *in-situ* measurements, all parameters should be reported, as the change of one variable can significantly change the processes ongoing at the electrode/electrolyte interface. The current state-of-the-art knowledge is scarce regarding the universal model for electrochemical capacitor operation. Up to date, there is no ubiquitous mechanism description that can be used for all: aqueous-, organic- or ionic liquid-based electrochemical capacitors. Therefore, there is still room for advanced characterization, and efforts to propose a realistic charging principle on the molecular scale are needed. Each discovery and each observation bring us closer to a better understanding of energy storage devices and imply optimization direction for future systems.

Acknowledgment

The authors acknowledge the Polish Ministry of Education and Science for the financial support received as 0911/SBAD/2101 project.

References

1. https://news.mit.edu/2021/MIT-CFS-major-advance-toward-fusion-energy-0908?fbclid=IwAR14X9wxSXKDEdNMmkZZU6F1yru2XBN0fDe87tV9c2jnQmn9VvgrLvoVGrc.
2. Winter, M.; Brodd, R. J. *Chem. Rev.* **2004**, *104*, 4245–4269.
3. Beguin, F.; Presser, V.; Balducci, A.; Frackowiak, E. *Adv. Mater.* **2014**, *26*. 2219-51, 2283.
4. Lu, M.; Beguin, F.; Frackowiak, E. *Supercapacitors: Materials, Systems, and Applications*; Weinheim: John Wiley & Sons Incorporated: Weinheim, 2013.
5. Sulaiman, K. S.; Mat, A.; Arof, A. K. *Ionics* **2016**, *22*, 911–918.
6. Skunik, M.; Chojak, M.; Rutkowska, I. A.; Kulesza, P. J. *Electrochim. Acta* **2008**, *53*, 3862–3869.
7. Frackowiak, E.; Foroutan Koudahi, M.; Tobis, M. *Small* **2021**, *17*, 1–10. e2006821.
8. Frackowiak, E.; Metenier, K.; Bertagna, V.; Beguin, F. *Appl. Phys. Lett.* **2000**, *77*, 2421–2423.
9. McDonough, J. K.; Frolov, A. I.; Presser, V.; Niu, J. J.; Miller, C. H.; Ubieto, T.; Fedorov, M. V.; Gogotsi, Y. *Carbon* **2012**, *50*, 3298–3309.
10. Wang, Y.; Du Pasquier, A.; Li, D.; Atanassova, P.; Sawrey, S.; Oljaca, M. *Carbon* **2018**, *133*, 1–5.
11. Beidaghi, M.; Gogotsi, Y. *Energy Environ. Sci.* **2014**, *7*, 867–884.
12. Platek-Mielczarek, A.; Nita, C.; Matei Ghimbeu, C.; Frackowiak, E.; Fic, K. *ACS Appl. Mater. Interfaces* **2021**, *13*, 2584–2599.
13. Zhu, Z. T.; Tang, S. H.; Yuan, J. W.; Qin, X. L.; Deng, Y. X.; Qu, R. J.; Haarberg, G. M. *Int. J. Electrochem. Soc.* **2016**, *11*, 8270–8279.
14. Marsh, H. *Activated carbon*, 1st ed.; Elsevier: Amsterdam; London, Oxford, 2006.
15. Inagaki, M. *Materials Science and Engineering of Carbon: Fundamentals*, 2nd ed.; Butterworth-Heinemann: Waltham, MA, 2014.
16. Fic, K.; Platek, A.; Piwek, J.; Frackowiak, E. *Mater. Today* **2018**, *21*, 437–454.
17. Hsieh, C. T.; Teng, H. *Carbon* **2002**, *40*, 667–674.
18. Nian, Y. R.; Teng, H. S. *J. Electrochem. Soc.* **2002**, *149*, A1008–A1014.
19. Hu, C. C.; Wang, C. C. *J. Power Sources* **2004**, *125*, 299–308.
20. Slesinski, A.; Matei-Ghimbeu, C.; Fic, K.; Beguin, F.; Frackowiak, E. *Carbon* **2018**, *129*, 758–765.
21. Salinas-Torres, D.; Shiraishi, S.; Morallon, E.; Cazorla-Amoros, D. *Carbon* **2015**, *82*, 205–213.
22. Marsh, H.; Rodríguez Reinoso, F.; Rodriguez-Reinoso, F. *Activated Carbon*; Oxford: Elsevier Science & Technology: Oxford, 2006.
23. Frackowiak, E. *Phys. Chem. Chem. Phys.* **2007**, *9*, 1774–1785.
24. Van Aken, K. L.; Maleski, K.; Mathis, T. S.; Breslin, J. P.; Gogotsi, Y. *ECS J. Solid State Sci. Technol.* **2017**, *6*, M3103–M3108.
25. Portet, C.; Yushin, G.; Gogotsi, Y. *Carbon* **2007**, *45*, 2511–2518.
26. Pech, D.; Brunet, M.; Durou, H.; Huang, P.; Mochalin, V.; Gogotsi, Y.; Taberna, P. L.; Simon, P. *Nat. Nanotechnol.* **2010**, *5*, 651–654.
27. Krause, A.; Balducci, A. *Electrochem. Commun.* **2011**, *13*, 814–817.
28. Haque, M.; Li, Q.; Rigato, C.; Rajaras, A.; Smith, A. D.; Lundgren, P.; Enoksson, P. *J. Power Sources* **2021**, 485.
29. Ruch, P. W.; Cericola, D.; Foelske, A.; Kotz, R.; Wokaun, A. *Electrochim. Acta* **2010**, *55*, 2352–2357.
30. Ruther, R. E.; Sun, C.-N.; Holliday, A.; Cheng, S.; Delnick, F. M.; Zawodzinski, T. A.; Nanda, J. *J. Electrochem. Soc.* **2016**, *164*, A277–A283.
31. Gorska, B.; Bujewska, P.; Fic, K. *Phys. Chem. Chem. Phys.* **2017**, *19*, 7923–7935.
32. Bujewska, P.; Gorska, B.; Fic, K. *Synth. Met.* **2019**, *253*, 62–72.

33. Chun, S. E.; Evanko, B.; Wang, X.; Vonlanthen, D.; Ji, X.; Stucky, G. D.; Boettcher, S. W. *Nat. Commun.* **2015**, *6*, 7818.
34. Martins, V. L.; Mantovi, P. S.; Torresi, R. M. *Electrochim. Acta* **2021**, *372*, 137854.
35. Lannelongue, P.; Bouchal, R.; Mourad, E.; Bodin, C.; Olarte, M.; le Vot, S.; Favier, F.; Fontaine, O. *J. Electrochem. Soc.* **2018**, *165*, A657–A663.
36. Galek, P.; Frackowiak, E.; Fic, K. *Electrochim. Acta* **2020**, 334.
37. Hantel, M. M.; Weingarth, D.; Kotz, R. *Carbon* **2014**, *69*, 275–286.
38. Platek-Mielczarek, A.; Frackowiak, E.; Fic, K. *Energy Environ. Sci.* **2021**, *14*, 2381–2393.
39. Przygocki, P.; Ratajczak, P.; Beguin, F. *Angew. Chem. Int. Ed. Engl.* **2019**, *58*, 17969–17977.
40. Kleszyk, P.; Ratajczak, P.; Skowron, P.; Jagiello, J.; Abbas, Q.; Frackowiak, E.; Beguin, F. *Carbon* **2015**, *81*, 148–157.
41. Piwek, J.; Platek, A.; Frackowiak, E.; Fic, K. *J. Power Sources* **2019**, 438.
42. Shpigel, N.; Levi, M. D.; Sigalov, S.; Daikhin, L.; Aurbach, D. *Acc. Chem. Res.* **2018**, *51*, 69–79.
43. Malka, D.; Attias, R.; Shpigel, N.; Melchick, F.; Levi, M. D.; Aurbach, D. *Isr. J. Chem.* **2021**, *61*, 11–25.
44. Tsai, W. Y.; Taberna, P. L.; Simon, P. *J. Am. Chem. Soc.* **2014**, *136*, 8722–8728.
45. Levi, M. D.; Sigalov, S.; Salitra, G.; Aurbach, D.; Maier, J. *ChemPhysChem* **2011**, *12*, 854–862.
46. Jäckel, N.; Patrick Emge, S.; Krüner, B.; Roling, B.; Presser, V. *J. Phys. Chem. C* **2017**, *121*, 19120–19128.
47. Fic, K.; Platek, A.; Piwek, J.; Menzel, J.; Slesinski, A.; Bujewska, P.; Galek, P.; Frackowiak, E. *Energy Storage Mater.* **2019**, *22*, 1–14.
48. Wu, Y. C.; Taberna, P. L.; Simon, P. *Electrochem. Commun.* **2018**, *93*, 119–122.
49. Escobar-Teran, F.; Arnau, A.; Garcia, J. V.; Jimenez, Y.; Perrot, H.; Sel, O. *Electrochem. Commun.* **2016**, *70*, 73–77.
50. Wu, Y. C.; Ye, J.; Jiang, G.; Ni, K.; Shu, N.; Taberna, P. L.; Zhu, Y.; Simon, P. *Angew. Chem. Int. Ed. Engl.* **2021**, *60*, 13317–13322.
51. Levi, M. D.; Lukatskaya, M. R.; Sigalov, S.; Beidaghi, M.; Shpigel, N.; Daikhin, L.; Aurbach, D.; Barsoum, M. W.; Gogotsi, Y. *Adv. Energy Mater.* **2015**, *5*, 1400815.
52. Ye, J.; Wu, Y. C.; Xu, K.; Ni, K.; Shu, N.; Taberna, P. L.; Zhu, Y.; Simon, P. *J. Am. Chem. Soc.* **2019**, *141*, 16559–16563.
53. Shpigel, N.; Levi, M. D.; Sigalov, S.; Mathis, T. S.; Gogotsi, Y.; Aurbach, D. *J. Am. Chem. Soc.* **2018**, *140*, 8910.
54. Shpigel, N.; Chakraborty, A.; Malchik, F.; Bergman, G.; Nimkar, A.; Gavriel, B.; Turgeman, M.; Hong, C. N.; Lukatskaya, M. R.; Levi, M. D.; Gogotsi, Y.; Major, D. T.; Aurbach, D. *J. Am. Chem. Soc.* **2021**, *143*, 12552–12559.
55. Gavriel, B.; Shpigel, N.; Malchik, F.; Bergman, G.; Turgeman, M.; Levi, M. D.; Aurbach, D. *Energy Storage Mater.* **2021**, *38*, 535–541.
56. Shao, H.; Xu, K.; Wu, Y. C.; Iadecola, A.; Liu, L. Y.; Ma, H. Y.; Qu, L. T.; Raymundo-Pinero, E.; Zhu, J. X.; Lin, Z. F.; Taberna, P. L.; Simon, P. *ACS Energy Lett.* **2020**, *5*, 2873–2880.
57. Dolleman, R. J.; Hsu, M.; Vollebregt, S.; Sader, J. E.; van der Zant, H. S. J.; Steeneken, P. G.; Ghatkesar, M. K. *Appl. Phys. Lett.* **2019**, *115*, 53102.
58. Fauzi, F.; Rianjanu, A.; Santoso, I.; Triyana, K. *Sens. Actuator A Phys.* **2021**, *330*, 112837.
59. Kakenov, N.; Balci, O.; Salihoglu, O.; Hur, S. H.; Balci, S.; Kocabas, C. *Appl. Phys. Lett.* **2016**, *109*, 53105.
60. Goubaa, H.; Escobar-Teran, F.; Ressam, I.; Gao, W. L.; El Kadib, A.; Lucas, I. T.; Raihane, M.; Lahcini, M.; Perrot, H.; Sel, O. *J. Phys. Chem. C* **2017**, *121*, 9370–9380.

61. Lé, T.; Aradilla, D.; Bidan, G.; Billon, F.; Delaunay, M.; Gérard, J. M.; Perrot, H.; Sel, O. *Electrochem. Commun.* **2018**, *93*, 5–9.
62. Ruch, P. W.; Kotz, R.; Wokaun, A. *Electrochim. Acta* **2009**, *54*, 4451–4458.
63. Michael, H.; Jervis, R.; Brett, D. J. L.; Shearing, P. R. *Batteries Supercaps* **2021**, *4*, 1378–1396.
64. Hahn, M.; Barbieri, O.; Campana, F. P.; Kotz, R.; Gallay, R. *Appl. Phys. A: Mater. Sci. Process.* **2006**, *82*, 633–638.
65. Hahn, M.; Barbieri, O.; Gallay, R.; Kotz, R. *Carbon* **2006**, *44*, 2523–2533.
66. Fic, K.; He, M. L.; Berg, E. J.; Novak, P.; Frackowiak, E. *Carbon* **2017**, *120*, 281–293.
67. Wang, Y.; Rochefort, D. *Meet. abstr. Electrochem. Soc.* **2020**, *6*, 530. MA2020-01.
68. Galek, P.; Bujewska, P.; Donne, S.; Fic, K.; Menzel, J. *Electrochim. Acta* **2021**, *377*, 138115.
69. Hantel, M. M.; Presser, V.; Kotz, R.; Gogotsi, Y. *Electrochem. Commun.* **2011**, *13*, 1221–1224.
70. Moreno, D.; Bootwala, Y.; Tsai, W. Y.; Gao, Q.; Shen, F. Y.; Balke, N.; Hatzell, K. B.; Hatzell, M. C. *Environ. Sci. Technol. Lett.* **2018**, *5*, 745–749.
71. Xu, K.; Merlet, C.; Lin, Z. F.; Shao, H.; Taberna, P. L.; Miao, L.; Jiang, J. J.; Zhu, J. X.; Simon, P. *Energy Storage Mater.* **2020**, *33*, 460–469.
72. Muckley, E. S.; Naguib, M.; Wang, H. W.; Vlcek, L.; Osti, N. C.; Sacci, R. L.; Sang, X.; Unocic, R. R.; Xie, Y.; Tyagi, M.; Mamontov, E.; Page, K. L.; Kent, P. R. C.; Nanda, J.; Ivanov, I. N. *ACS Nano* **2017**, *11*, 11118–11126.
73. Canongia Lopes, J. N.; Deschamps, J.; Pádua, A. A. H. *J. Phys. Chem. B* **2004**, *108*, 2038–2047.
74. Canongia Lopes, J. N.; Padua, A. A. *J. Phys. Chem. B* **2006**, *110*, 3330–3335.
75. Schroder, C. *Phys. Chem. Chem. Phys.* **2012**, *14*, 3089–3102.
76. Bowron, D. T.; D'Agostino, C.; Gladden, L. F.; Hardacre, C.; Holbrey, J. D.; Lagunas, M. C.; McGregor, J.; Mantle, M. D.; Mullan, C. L.; Youngs, T. G. *J. Phys. Chem. B* **2010**, *114*, 7760–7768.
77. Bhargava, B. L.; Balasubramanian, S. *J. Chem. Phys.* **2007**, *127*, 114510.
78. Chaban, V. *Phys. Chem. Chem. Phys.* **2011**, *13*, 16055–16062.
79. Efimova, A.; Pfutzner, L.; Schmidt, P. *Thermochim. Acta* **2015**, *604*, 129–136.
80. Cao, Y. Y.; Mu, T. C. *Ind. Eng. Chem. Res.* **2014**, *53*, 8651–8664.
81. Ishikawa, M.; Sugimoto, T.; Kikuta, M.; Ishiko, E.; Kono, M. *J. Power Sources* **2006**, *162*, 658–662.
82. Xu, K.; Ji, X.; Zhang, B.; Chen, C.; Ruan, Y. J.; Miao, L.; Jiang, J. J. *Electrochim. Acta* **2016**, *196*, 75–83.
83. Kaasik, F.; Tamm, T.; Hantel, M. M.; Perre, E.; Aabloo, A.; Lust, E.; Bazant, M. Z.; Presser, V. *Electrochem. Commun.* **2013**, *34*, 196–199.
84. Jackel, N.; Kruner, B.; Van Aken, K. L.; Alhabeb, M.; Anasori, B.; Kaasik, F.; Gogotsi, Y.; Presser, V. *ACS Appl. Mater. Interfaces* **2016**, *8*, 32089–32093.
85. Lin, Z. F.; Rozier, P.; Duployer, B.; Taberna, P. L.; Anasori, B.; Gogotsi, Y.; Simon, P. *Electrochem. Commun.* **2016**, *72*, 50–53.
86. Kannan, V.; Somasundaram, K.; Fisher, A.; Birgersson, E. *Int. J. Energy Res.* **2021**, *45*, 16947–16962.
87. Zhang, Y. H.; Du, G. P.; Lei, Y. X. *IET Power Electron.* **2020**, *13*, 3171–3179.
88. Quynh, N. V.; Ali, Z. M.; Alhaider, M. M.; Rezvani, A.; Suzuki, K. *Int. J. Energy Res.* **2020**, *45*, 5766–5780.
89. Inthamoussou, F. A.; Pegueroles-Queralt, J.; Bianchi, F. D. *IEEE Trans. Energy Convers.* **2013**, *23*, 690–697.
90. Tong, W.; Koh, W. Q.; Birgersson, E.; Mujumdar, A. S.; Yap, C. *Int. J. Energy Res.* **2015**, *39*, 778–788.
91. Kannan, V.; Fisher, A.; Birgersson, E. *J. Energy Storage* **2021**, *35*, 102269.

92. Su, H. P.; Lin, S.; Deng, S. W.; Lian, C.; Shang, Y. Z.; Liu, H. L. *Nanoscale Adv.* **2019**, *1*, 2162–2166.
93. Zhu, S.; Li, J. J.; Ma, L. Y.; He, C. N. A.; Liu, E. Z.; He, F.; Shi, C. S.; Zhao, N. Q. *Mater. Lett.* **2018**, *233*, 294–297.
94. Somasundaram, K.; Birgersson, E.; Mujumdar, A. S. *J. Electrochem. Soc.* **2011**, *158*, A1220–A1230.
95. Zheng, J. P. *J. Electrochem. Soc.* **2003**, *150*, A484–A492.
96. Newman, J.; Tiedemann, W. *AIChE J.* **1975**, *21*, 25–41.
97. Zheng, J. P.; Cygan, P. J.; Jow, T. R. *J. Electrochem. Soc.* **1995**, *142*, 2699–2703.
98. Schmidt, O.; Thomitzek, M.; Roder, F.; Thiede, S.; Herrmann, C.; Krewer, U. *J. Electrochem. Soc.* **2020**, *167*, 60501.
99. Sreenivasan, R. An Introduction to Error Analysis: The Study of Uncertainties in Physical Measurements. *Sigma Xi Scientific Research Society* **1983**, *71*, 1–430.
100. Tuffry, S.; Tufféry, S. *Data Mining and Statistics for Decision Making*, 1. Aufl. ed.; Wiley: New York, 2011.
101. Kim, H.; Popov, B. N. *J. Electrochem. Soc.* **2003**, *150*, A1153–A1160.
102. Tsay, K. C.; Zhang, L.; Zhang, J. J. *Electrochim. Acta* **2012**, *60*, 428–436.
103. ten Elshof, J. E.; Yuan, H.; Gonzalez Rodriguez, P. *Adv. Energy Mater.* **2016**, *6*, 1600355.
104. Lawes, S.; Riese, A.; Sun, Q.; Cheng, N. C.; Sun, X. L. *Carbon* **2015**, *92*, 150–176.
105. Pilban Jahromi, S.; Pandikumar, A.; Goh, B. T.; Lim, Y. S.; Basirun, W. J.; Lim, H. N.; Huang, N. M. *RSC Adv.* **2015**, *5*, 14010–14019.
106. Rennie, A. J.; Martins, V. L.; Smith, R. M.; Hall, P. J. *Sci. Rep.* **2016**, *6*, 22062.
107. Meller, M.; Fic, K.; Menzel, J.; Frackowiak, E. *Phys. Chem. Electrol.* **2014**, *61*, 1–8.
108. Jeżowski, P.; Nowicki, M.; Grzeszkowiak, M.; Czajka, R.; Béguin, F. *J. Power Sources* **2015**, *279*, 555–562.
109. Patake, V. D.; Lokhande, C. D.; Joo, O. S. *Appl. Surf. Sci.* **2009**, *255*, 4192–4196.
110. Noh, C.; Jung, Y. *Phys. Chem. Chem. Phys.* **2019**, *21*, 6790–6800.
111. Dhungana, K. B.; Faria, L. F.; Wu, B.; Liang, M.; Ribeiro, M. C.; Margulis, C. J.; Castner, E. W., Jr. *J. Chem. Phys.* **2016**, *145*, 024503.
112. Hummer, G.; Rasaiah, J. C.; Noworyta, J. P. *Nature* **2001**, *414*, 188–190.
113. Wang, Z.; Yang, Y.; Olmsted, D. L.; Asta, M.; Laird, B. B. *J. Chem. Phys.* **2014**, *141*, 184102.
114. Shim, Y.; Jung, Y.; Kim, H. J. *J. Phys. Chem. C* **2012**, *116*, 18574–18575.
115. Vatamanu, J.; Borodin, O.; Bedrov, D.; Smith, G. D. *J. Phys. Chem.C* **2012**, *116*, 7940–7951.
116. Singh, R.; Rajput, N. N.; He, X.; Monk, J.; Hung, F. R. *Phys. Chem. Chem. Phys.* **2013**, *15*, 16090–16103.
117. Jo, S.; Park, S. W.; Noh, C.; Jung, Y. *Electrochim. Acta* **2018**, *284*, 577–586.
118. Davey, S. B.; Cameron, A. P.; Latham, K. G.; Donne, S. W. *Electrochim. Acta* **2021**, 396.
119. Vatamanu, J.; Borodin, O.; Olguin, M.; Yushin, G.; Bedrov, D. *J. Mater. Chem. A* **2017**, *5*, 21049–21076.
120. Iacob, C.; Sangoro, J. R.; Kipnusu, W. K.; Valiullin, R.; Kärger, J.; Kremer, F. *Soft Matter* **2012**, *8*, 289–293.
121. Dong, D.; Vatamanu, J. P.; Wei, X.; Bedrov, D. *J. Chem. Phys.* **2018**, *148*, 193833.
122. Griffin, J. M.; Forse, A. C.; Tsai, W. Y.; Taberna, P. L.; Simon, P.; Grey, C. P. *Nat. Mater.* **2015**, *14*, 812–819.
123. Sasikumar, A.; Belhboub, A.; Bacon, C.; Forse, A. C.; Griffin, J. M.; Grey, C. P.; Simon, P.; Merlet, C. *Phys. Chem. Chem. Phys.* **2021**, *23*, 15925–15934.
124. Li, K.; Bo, Z.; Yan, J.; Cen, K. *Sci. Rep.* **2016**, *6*, 39689.

125. Luo, Z. X.; Xing, Y. Z.; Liu, S.; Ling, Y. C.; Kleinhammes, A.; Wu, Y. *J. Phys. Chem. Lett.* **2015**, *6*, 5022–5026.
126. Forse, A. C.; Griffin, J. M.; Presser, V.; Gogotsi, Y.; Grey, C. P. *J. Phys. Chem. C* **2014**, *118*, 7508–7514.
127. Forse, A. C.; Griffin, J. M.; Grey, C. P. *Solid State Nucl. Magn. Reson.* **2018**, *89*, 45–49.
128. Kilymis, D.; Bartok, A. P.; Pickard, C. J.; Forse, A. C.; Merlet, C. *Phys. Chem. Chem. Phys.* **2020**, *22*, 13746–13755.
129. Moran, D.; Stahl, F.; Bettinger, H. F.; Schaefer, H. F., 3rd; Schleyer, P. *J. Am. Chem. Soc.* **2003**, *125*, 6746–6752.
130. Deringer, V. L.; Merlet, C.; Hu, Y.; Lee, T. H.; Kattirtzi, J. A.; Pecher, O.; Csanyi, G.; Elliott, S. R.; Grey, C. P. *Chem. Commun.* **2018**, *54*, 5988–5991.
131. Merlet, C.; Pean, C.; Rotenberg, B.; Madden, P. A.; Daffos, B.; Taberna, P. L.; Simon, P.; Salanne, M. *Nat. Commun.* **2013**, *4*, 2701.
132. Forse, A. C.; Griffin, J. M.; Merlet, C.; Carretero-Gonzalez, J.; Raji, A. R. O.; Trease, N. M.; Grey, C. P. Nat. Energy. **2017**, *2*, 1–7. 16216.
133. Merlet, C.; Salanne, M.; Rotenberg, B.; Madden, P. A. *Electrochim. Acta* **2013**, *101*, 262–271.
134. Tsuzuki, S.; Shinoda, W.; Saito, H.; Mikami, M.; Tokuda, H.; Watanabe, M. *J. Phys. Chem. B* **2009**, *113*, 10641–10649.
135. Price, M. L. P.; Ostrovsky, D.; Jorgensen, W. L. *J. Comput. Chem.* **2001**, *22*, 1340–1352.
136. Griffin, J. M.; Forse, A. C.; Wang, H.; Trease, N. M.; Taberna, P. L.; Simon, P.; Grey, C. P. *Faraday Discuss.* **2014**, *176*, 49–68.
137. Forse, A. C.; Griffin, J. M.; Merlet, C.; Bayley, P. M.; Wang, H.; Simon, P.; Grey, C. P. *J. Am. Chem. Soc.* **2015**, *137*, 7231–7242.
138. Wang, H.; Forse, A. C.; Griffin, J. M.; Trease, N. M.; Trognko, L.; Taberna, P. L.; Simon, P.; Grey, C. P. *J. Am. Chem. Soc.* **2013**, *135*, 18968–18980.
139. Lucier, B. E. G.; Reidel, A. R.; Schurko, R. W. *Can. J. Chem.* **2011**, *89*, 919–937.
140. Beckmann, P. A.; Dybowski, C. *J. Magn. Reson.* **2000**, *146*, 379–380.
141. Sahoo, G.; Polaki, S. R.; Pazhedath, A.; Krishna, N. G.; Mathews, T.; Kamruddin, M. *ACS Appl. Energy Mater.* **2021**, *4*, 791–800.
142. Rajendiran, R.; Nallal, M.; Park, K. H.; Li, O. L.; Kim, H. J.; Prabakar, K. *Electrochim. Acta* **2019**, *317*, 1–9.
143. Ghimbeu, C. M.; Raymundo-Pinero, E.; Fioux, P.; Beguin, F.; Vix-Guterl, C. *J. Mater. Chem.* **2011**, *21*, 13268–13275.
144. Menzel, J.; Frackowiak, E.; Fic, K. *Electrochim. Acta* **2020**, *332*, 135435.
145. Platek, A.; Nita, C.; Ghimbeu, C. M.; Frackowiak, E.; Fic, K. *Electrochim. Acta* **2020**, *338*, 135788.
146. Piwek, J.; Platek-Mielczarek, A.; Frackowiak, E.; Fic, K. *J. Power Sources* **2021**, *506*, 230131.
147. Dong, R.; Song, Y.; Yang, D.; Shi, H. Y.; Qin, Z. M.; Zhang, M. Y.; Guo, D.; Sun, X. Q.; Liu, X. X. *J. Mater. Chem. A* **2020**, *8*, 1176–1183.
148. Kolanowski, L.; Gras, M.; Bartkowiak, M.; Doczekalska, B.; Lota, G. *Waste Biomass Valoriz.* **2020**, *11*, 3863–3871.
149. Le Comte, A.; Pognon, G.; Brousse, T.; Belanger, D. *Electrochem.* **2013**, *81*, 863–866.
150. Wang, H.; Koster, T. K.; Trease, N. M.; Segalini, J.; Taberna, P. L.; Simon, P.; Gogotsi, Y.; Grey, C. P. *J. Am. Chem. Soc.* **2011**, *133*, 19270–19273.
151. Ilott, A. J.; Trease, N. M.; Grey, C. P.; Jerschow, A. *Nat. Commun.* **2014**, *5*, 4536.
152. Camci, M. T.; Ulgut, B.; Kocabas, C.; Suzer, S. *J. Phys. Chem. C* **2018**, *122*, 11883–11889.
153. Xiao, L.; Johnson, K. E. *J. Electrochem. Soc.* **2003**, *150*, E307–E311.
154. Kroon, M. C.; Buijs, W.; Peters, C. J.; Witkamp, G. J. *Green Chem.* **2006**, *8*, 241–245.

155. Kruusma, J.; Tonisoo, A.; Parna, R.; Nommiste, E.; Lust, E. *J. Electrochem. Soc.* **2014**, *161*, A1266–A1277.
156. He, M. L.; Fic, K.; Frackowiak, E.; Novak, P.; Berg, E. J. *Energy Environ. Sci.* **2016**, *9*, 623–633.
157. He, M.; Fic, K.; Frackowiak, E.; Novak, P.; Berg, E. J. *ChemElectroChem* **2019**, *6*, 566–573.
158. Kim, Y. T.; Kim, K. B.; Jin, C. S.; Yoo, E. H. *Met. Mater. Int.* **2015**, *21*, 1123–1132.
159. Slesinski, A.; Frackowiak, E. *J. Solid State Electrochem.* **2018**, *22*, 2135–2139.
160. Slesinski, A.; Miller, J. R.; Beguin, F.; Frackowiak, E. *J. Electrochem. Soc.* **2017**, *164*, A2732–A2737.
161. He, L. Z.; Li, J. L.; Gao, F.; Wang, X. D.; Ye, F.; Yang, J. *Electrochem.* **2011**, *79*, 934–940.
162. Xie, J.; Liang, Z.; Lu, Y. C. *Nat. Mater.* **2020**, *19*, 1006–1011.
163. He, M.; Boulet-Roblin, L.; Borel, P.; Tessier, C.; Novák, P.; Villevieille, C.; Berg, E. J. *J. Electrochem. Soc.* **2015**, *163*, A83–A89.
164. Romann, T.; Oll, O.; Pikma, P.; Tamme, H.; Lust, E. *Electrochim. Acta* **2014**, *125*, 183–190.
165. Laheäär, A.; Przygocki, P.; Abbas, Q.; Béguin, F. *Electrochem. Commun.* **2015**, *60*, 21–25.
166. Pal, B.; Yasin, A.; Kaur, R.; Tebyetekerwa, M.; Zabihi, F.; Yang, S. Y.; Yang, C. C.; Sofer, Z.; Jose, R. *Renew. Sust. Energ. Rev.* **2021**, *149*, 111418.
167. Richey, F. W.; Dyatkin, B.; Gogotsi, Y.; Elabd, Y. A. *J. Am. Chem. Soc.* **2013**, *135*, 12818–12826.
168. Richey, F. W.; Tran, C.; Kalra, V.; Elabd, Y. A. *J. Phys. Chem. C* **2014**, *118*, 21846–21855.
169. Boethe, A.; Pourhosseini, S. E. M.; Ratajczak, P.; Beguin, F.; Balducci, A. *J. Power Sources* **2021**, *496*, 229841.
170. Pourhosseini, S. E. M.; Bothe, A.; Balducci, A.; Beguin, F.; Ratajczak, P. *Energy Storage Mater.* **2021**, *38*, 17–29.
171. Ghimbeu, C. M.; Gadiou, R.; Dentzer, J.; Vidal, L.; Vix-Guterl, C. *Adsorpt. Sci. Technol.* **2011**, *17*, 227–233.
172. Batisse, N.; Raymundo-Pinero, E. *ACS Appl. Mater. Interfaces* **2017**, *9*, 41224–41232.
173. Figueiredo, J. L.; Pereira, M. F. R.; Freitas, M. M. A.; Orfao, J. J. M. *Carbon* **1999**, *37*, 1379–1389.
174. Moussa, G.; Ghimbeu, C. M.; Taberna, P. L.; Simon, P.; Vix-Guterl, C. *Carbon* **2016**, *105*, 628–637.
175. Forse, A. C.; Merlet, C.; Allan, P. K.; Humphreys, E. K.; Griffin, J. M.; Aslan, M.; Zeiger, M.; Presser, V.; Gogotsi, Y.; Grey, C. P. *Chem.Mater.* **2015**, *27*, 6848–6857.
176. Menzel, J.; Frackowiak, E.; Fic, K. *J. Power Sources* **2019**, *414*, 183–191.
177. Gambou-Bosca, A.; Belanger, D. *J. Electrochem. Soc.* **2015**, *162*, A5115–A5123.
178. Ruch, P. W.; Hardwick, L. J.; Hahn, M.; Foelske, A.; Kotz, R.; Wokaun, A. *Carbon* **2009**, *47*, 38–52.
179. Nunes, W. G.; Pires, B. M.; De Oliveira, F. E. R.; de Marque, A. M. P.; Cremasco, L. F.; Vicentini, R.; Doubek, G.; Da Silva, L. M.; Zanin, H. *J. Energy Storage* **2020**, *28*, 101249.
180. Przygocki, P.; Abbas, Q.; Babuchowska, P.; Beguin, F. *Carbon* **2017**, *125*, 391–400.
181. Beda, A.; Vaulot, C.; Matei Ghimbeu, C. *J. Mater. Chem. A* **2021**, *9*, 937–943.
182. Jagiello, J.; Chojnacka, A.; Pourhosseini, S. E. M.; Wang, Z.; Beguin, F. *Carbon* **2021**, *178*, 113–124.
183. Ania, C. O.; Armstrong, P. A.; Bandosz, T. J.; Beguin, F.; Carvalho, A. P.; Celzard, A.; Frackowiak, E.; Gilarranz, M. A.; László, K.; Matos, J.; Pereira, M. F. R. *Carbon* **2020**, *164*, 69–84.
184. Suarez, L.; Barranco, V.; Centeno, T. A. *J. Colloid. Interface Sci.* **2021**, *588*, 705–712.
185. Moreno-Fernández, G.; Schütter, C.; Rojo, J. M.; Passerini, S.; Balducci, A.; Centeno, T. A. *J. Solid State Electrochem.* **2017**, *22*, 717–725.

186. Liu, Y. H.; Rety, B.; Ghimbeu, C. M.; Soucaze-Guillous, B.; Taberna, P. L.; Simon, P. *J. Power Sources* **2019**, *434*, 226734.
187. Suárez, L.; Centeno, T. A. *J. Power Sources* **2020**, *448*, 227413.
188. Choma, J.; Jagiello, J.; Jaroniec, M. *Carbon* **2021**, *183*, 150–157.
189. Pameté Yambou, E.; Gorska, B.; Pavlenko, V.; Beguin, F. *Electrochim. Acta* **2020**, *350*, 136416.
190. Tao, X. Y.; Du, J.; Sun, Y.; Zhou, S. L.; Xia, Y.; Huang, H.; Gan, Y. P.; Zhang, W. K.; Li, X. D. *Adv. Funct. Mater.* **2013**, *23*, 4745–4751.
191. Alekseev, A.; Efimov, A.; Lu, K.; Loos, J. *Adv. Mater. (Weinheim, Ger.)* **2009**, *21*, 4915–4919.
192. Majchrzycki, L.; Nowicki, M.; Czajka, R.; Lota, K. *Micro* & Nano Lett. **2014**, *9*, 69–72.
193. Gao, Q.; Tsai, W. Y.; Balke, N. *Electrochem. Sci. Adv.*; 2021; pp. 1–27. e2100038.
194. Kocun, M.; Labuda, A.; Gannepalli, A.; Proksch, R. *Rev. Sci. Instrum.* **2015**, *86*, 083706.
195. Labuda, A.; Hohlbauch, S.; Kocun, M.; Limpoco, F. T.; Kirchhofer, N.; Ohler, B.; Hurley, D. *MTO* **2018**, *26*, 12–17.
196. Butt, H. J.; Cappella, B.; Kappl, M. *Surf. Sci. Rep.* **2005**, *59*, 1–152.
197. Gao, Q.; Come, J.; Naguib, M.; Jesse, S.; Gogotsi, Y.; Balke, N. *Faraday Discuss.* **2017**, *199*, 393–403.
198. Black, J.; Strelcov, E.; Balke, N.; Kalinin, S. V. *Electrochem. Soc. Interface* **2014**, *23*, 53–59.
199. Black, J. M.; Walters, D.; Labuda, A.; Feng, G.; Hillesheim, P. C.; Dai, S.; Cummings, P. T.; Kalinin, S. V.; Proksch, R.; Balke, N. *Nano Lett.* **2013**, *13*, 5954–5960.
200. Atkin, R.; El Abedin, S. Z.; Hayes, R.; Gasparotto, L. H. S.; Borisenko, N.; Endres, F. *J. Phys. Chem. C* **2009**, *113*, 13266–13272.
201. Endres, F.; Borisenko, N.; El Abedin, S. Z.; Hayes, R.; Atkin, R. *Faraday Discuss.* **2012**, *154*, 221–233. discussion 313-33, 465-71.
202. Carstens, T.; Gustus, R.; Hofft, O.; Borisenko, N.; Endres, F.; Li, H.; Wood, R. J.; Page, A. J.; Atkin, R. *J. Phys. Chem. C* **2014**, *118*, 10833–10843.
203. Jitvisate, M.; Seddon, J. R. T. *J. Phys. Chem. Lett.* **2018**, *9*, 126–131.
204. Zhong, Y. X.; Yan, J. W.; Li, M. G.; Zhang, X.; He, D. W.; Mao, B. W. *J. Am. Chem. Soc.* **2014**, *136*, 14682–14685.
205. Zhong, Y. X.; Yan, J. W.; Li, M. G.; Chen, L.; Mao, B. W. *ChemElectroChem* **2016**, *3*, 2221–2226.
206. Cheng, H.-W.; Stock, P.; Moeremans, B.; Baimpos, T.; Banquy, X.; Renner, F. U.; Valtiner, M. *Adv. Mater. Interfaces* **2015**, *2*, 1500159.
207. Gao, Q.; Sun, W. W.; Ilani-Kashkouli, P.; Tselev, A.; Kent, P. R. C.; Kabengi, N.; Naguib, M.; Alhabeb, M.; Tsai, W. Y.; Baddorf, A. P.; Huang, J. S.; Jesse, S.; Gogotsi, Y.; Balke, N. *Energy Environ. Sci.* **2020**, *13*, 2549–2558.
208. Collins, L.; Jesse, S.; Kilpatrick, J. I.; Tselev, A.; Varenyk, O.; Okatan, M. B.; Weber, S. A.; Kumar, A.; Balke, N.; Kalinin, S. V.; Rodriguez, B. J. *Nat. Commun.* **2014**, *5*, 3871.
209. Yu, W.; Fu, H. J.; Mueller, T.; Brunschwig, B. S.; Lewis, N. S. *J. Chem. Phys.* **2020**, *153*, 020902.
210. Wang, S. W.; Liu, Q.; Zhao, C. L.; Lv, F. Z.; Qin, X. Y.; Du, H. D.; Kang, F. Y.; Li, B. H. *Energy Environ. Sci.* **2018**, *1*, 28–40.
211. Breitung, B.; Baumann, P.; Sommer, H.; Janek, J.; Brezesinski, T. *Nanoscale* **2016**, *8*, 14048–14056.
212. Wang, S.; Xu, H. B.; Hao, T. T.; Wang, P. Y.; Zhang, X.; Zhang, H. M.; Xue, J. Y.; Zhao, J. P.; Li, Y. NPG Asia Mater. **2021**, *13*, 1–11.
213. Zou, Y. L.; Liu, D. Y.; Meng, X. R.; Liu, Q. T.; Zhou, Y.; Li, J. M.; Zhao, Z. Y.; Chen, D.; Kuang, Y. B. *J. Energy Chem.* **2021**, *56*, 504–511.

CHAPTER SIX

Mesoporous silica-based catalysts for selective catalytic reduction of NO_x with ammonia—Recent advances

Lucjan Chmielarz* and Aleksandra Jankowska
Jagiellonian University in Kraków, Faculty of Chemistry, Kraków, Poland
*Corresponding author: e-mail address: chmielar@chemia.uj.edu.pl

Contents

Abstract

Mesoporous silicate materials due to their high specific surface area, porosity and well-defined pore structure, are very promising for catalytic applications, including selective catalytic reduction of nitrogen oxides with ammonia (NH_3-SCR), which is one of the most important technologies for the conversion of toxic nitrogen oxides, present in flue gases, to nitrogen. Deposition of catalytically active metal species into the surface of mesoporous silicates, should result in the generation of a large number of well-distributed catalytically active sites. Significant selectivity to desired reaction products depends on suitable tailoring of deposited metal species into specific needs of the catalytic process as well as on high homogeneity of these deposited catalytically active metal species. These aspects of the design of the NH_3-SCR catalyst based on mesoporous silicate materials are presented, analyzed, and discussed. The review presents recent advances of studies of the copper, iron and manganese-containing catalysts, also in combination with other components, for selective reduction of nitrogen oxides with ammonia. The influence of methods used for catalytically active components deposition into mesoporous silicates, and therefore types and distribution of introduced species, the activating role of the additional components as well as the role of the reactions associated with NH_3-SCR are presented and discussed.

Advances in Inorganic Chemistry, Volume 79
ISSN 0898-8838
https://doi.org/10.1016/bs.adioch.2021.12.007

Abbreviations

BET	Brunauer, Emmett and Teller theory
CH	conventional heating
EDX	energy dispersive X-ray
ESP	electrostatic precipitator
FTIR	Fourier transform infrared
IE	ion-exchange
IM	impregnation
KIT-6	Korea Advanced Institute of Science and Technology cubic ordered mesoporous silicate
M41S	family of mesoporous molecular sieves
MCM-41/-48/-50	Mobil Composition of Matter no. 41/48/50 (hexagonal, cubic, and lamellar ordered mesoporous silicate)
MDD	molecular designed dispersion
MW	microwave oven
NH_3-SCR	selective catalytic reduction of nitrogen oxides with ammonia
NO_x	nitrogen oxides (NO and NO_2)
P123	nonionic triblock copolymer surfactant
PILCs	pillared interlayered clays
SBA-15	Santa Barbara Amorphous type 15 (mesoporous silicate)
SEM	scanning electron microscopy
STEM	scanning transmission electron microscopy
TEOS	tetraethyl orthosilicate
TIE	template ion-exchange
TIP	titanium isopropoxide
TOF	turnover frequency

1. Introduction

Mesoporous silicate materials due to their uniform mesoporous structure, very high specific surface area and porosity, have been intensively studied as catalytic supports for nearly three decades. The first successful synthesis of the M41S series of mesoporous silicate materials was reported by researchers of Mobil Oil Company in 1992.[1] The syntheses of these mesoporous silicates, including MCM-41 with hexagonal pore ordering, MCM-48 with a cubic porous structure and MCM-50 with the lamellar porous structure, were based on the alkylammonium surfactant directed method.[2] Surfactant molecules are organized in solution resulting in micellar structures, which play the role of templates of mesopores in silicate materials.

The concept of MCM-41 synthesis by the surfactant directed method is schematically presented in Fig. 1.

As can be seen, the final step of the mesoporous silica synthesis is the removal of surfactants from pores by calcination or extraction with suitable solvents. The extraction method has been recently adapted not only for the removal of cationic alkylammonium surfactants from pores of as-prepared MCM-41 but also for simultaneous deposition of metal cations into silicates.[3]

Development of the synthesis methods of mesoporous silicates has been undoubtedly one of the most spectacular achievements in the field of material science of the last few decades.[1,2] Due to extremely high surface area and uniform porosity in the mesopore range, these materials have been considered very promising for various catalytic applications. Initially, there was a hope that mesoporous silicates, similarly to zeolites, could be used as catalysts or catalytic supports for selective conversion of various compounds. Due to larger pores in mesoporous silicates, there was some optimism of an extension of the catalytic function of zeolites as well for bulkier molecules that could be converted to products in mesopores of silicates. The concepts and application of them in mesoporous silica catalysis are nowadays regarded as significantly less promising of successful outcomes. It is a consequence of there being too many differences between mesoporous silicate materials and zeolites. First, walls of mesoporous silicate species, in contrast to zeolites, are composed of amorphous silicates, which results in their lower thermal, hydrothermal, and mechanical stability. Pure silicate mesoporous materials do not exhibit the presence of acidic sites. Such acidic sites can be generated by the incorporation of heteroatoms into silicate walls, (e.g., Al^{3+}), but in contrast to zeolites, such sites are characterized by heterogeneity of their acid strength, which limits the application of such modified mesoporous silicate in acid-catalyzed reactions. Pure silicate mesoporous materials do not exhibit

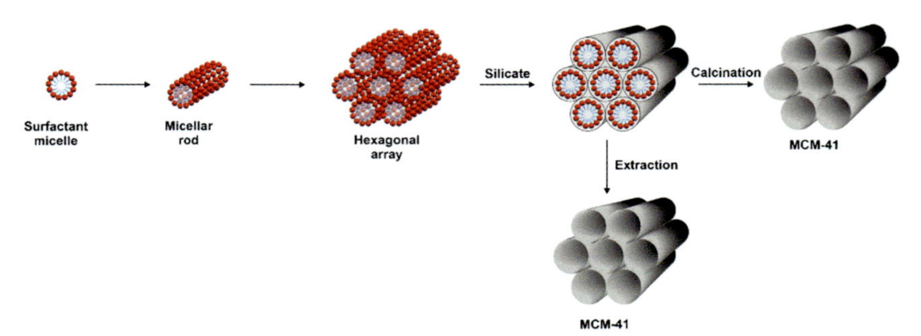

Fig. 1 Scheme for MCM-41 synthesis by surfactant directed method.

ion-exchange properties, which limits deposition of catalytically active metal species in uniform and highly dispersed form. Introduction of hetero-atoms, such as aluminum, into silicate walls of mesoporous materials, in contrast to the situation with zeolites, results only in limited ion-exchange properties[4] and therefore deposition of significant amounts of highly dispersed metal species by an ion-exchange method, in this case, is difficult or even impossible. Various methods were tested and implemented for deposition of catalytically active metal species into mesoporous silicate materials.[5] Impregnation methods result in deposition of metals species in various forms and aggregation, which may influence the selectivity of the catalytic processes. Moreover, the introduction of larger amounts of metal species may result in partial blocking of the pore system in mesoporous silicates. Co-condensation methods, based on the incorporation of metal species into mesoporous silicate walls during their synthesis, result in materials with decreased textural parameters in comparison to pure silicate mesoporous materials and moreover only part of the introduced metal species is exposed on the surface and is available for the catalytic processes. Grafting of metal species on the surface of mesoporous silicate materials by the reaction of surface silanol groups with organometallic compounds results in high and uniform dispersion of deposited metal species. However, the loading of such metal species is limited, and the relatively high costs of this method significantly limit their scale-up.[6] A very promising way of metal species deposition into mesoporous silicates, resulting in high and uniform dispersion of metal species that can be deposited in large amounts, seems to be the template ion-exchange (TIE) method.[3] The TIE method is based on an exchange of cationic surfactants in as-prepared mesoporous silicate (non-calcined) for metal cations from solution.[3] Such as-prepared mesoporous silicates exhibit ion-exchange properties related to the interaction of negatively charged surface $\equiv$Si—O$^-$ groups with cationic surfactants, which can be exchanged for metal cations. Such ion-exchange properties are irretrievably lost during calcination of mesoporous silicate materials due to the transformation of $\equiv$Si—O$^-$ groups to silanol groups ($\equiv$Si—OH), which do not exhibit ion-exchange properties. Thus, it seems that the application of the TIE method for catalytic functionalization of mesoporous silicates is very promising.

The problem of limited thermal and hydrothermal stability of mesoporous silicate materials was partially solved by the development of silicates with thicker walls (e.g., SBA-15)[7] or hydrophilization of the silicate surface by anchoring of hydrophobic functional groups.[8] However, in general, it seems that catalysts based on mesoporous silicates could be effectively used in low- and middle-temperature catalytic processes. Thus, there is some limitation in application of mesoporous silicate materials in catalysis but simultaneously there is still much optimism for their great role in the formulation of effective catalysts for various processes.

Selective catalytic reduction of nitrogen oxides, NO and NO_2, with ammonia (NH_3-SCR) is a technology used for the conversion of nitrogen oxides in flue gases emitted by stationary sources, such as electric power stations as well as industrial and municipal solid waste boilers. The NH_3-SCR process was patented in the United States by the Engelhard Corporation in 1957.[9] In the following years, optimisation of the NH_3-SCR process was continued, mainly in Japan and the United States, resulting in less expensive and more durable catalytic systems. The first large-scale NH_3-SCR installation was implemented by the IHI Corporation in 1978.[10] The NH_3-SCR process can be described by Eqs. 1 and 2:

$$4\,NO + 4\,NH_3 + O_2 \rightarrow 4\,N_2 + 6\,H_2O \quad (1)$$

$$2\,NO_2 + 4\,NH_3 + O_2 \rightarrow 3\,N_2 + 6\,H_2O \quad (2)$$

Ammonia and air are introduced into flue gases to convert nitrogen oxides according to reactions (1) and (2). Commercial catalysts of the NH_3-SCR process are based on V_2O_5-WO_3-TiO_2 and V_2O_5-MoO_3-TiO_2 metal oxide systems. Such catalysts effectively operate in the temperature range of 300–400 °C, with high selectivity to nitrogen. At temperatures below 300 °C, the efficiency of nitrogen oxides conversion is not satisfactory, while at temperatures above 400 °C, the side reactions of direct ammonia oxidation (reaction Eqs. 3–5) may decrease the efficiency of the NH_3-SCR process:

$$4\,NH_3 + 5\,O_2 \rightarrow 4\,NO + 6\,H_2O \quad (3)$$

$$2\,NH_3 + 2\,O_2 \rightarrow N_2O + 3\,H_2O \quad (4)$$

$$4\,NH_3 + 3\,O_2 \rightarrow 2\,N_2 + 6\,H_2O \quad (5)$$

Thus, NH_3-SCR is mature and well-optimized technology, relatively well fitted to exhaust installations currently used in electric power stations and industrial boilers. However, there are some weak points in this technology. Possibly the most important is related to operating of the NH_3-SCR monolithic converters with dusty gases, which may result in clogging of channels in catalytic monoliths by dust particles and finally decrease the efficiency of nitrogen oxides conversion. Currently, in the majority of exhaust gas installations the NH_3-SCR units are located upstream of an electrostatic precipitator (ESP) and therefore, such units operate satisfactorily with a dusty gas stream (high-dust NH_3-SCR). This problem could be solved by de-dusting of flue gases in the ESP unit prior to directing exhaust gas to the NH_3-SCR converter unit (low-dust NH_3-SCR). However, the ESP units operate with relatively cold flue gases of temperature about 250 °C or even lower. Thus, in such new installation the de-dusted gas, downstream of the ESP unit, must be heated to above 300 °C prior to directing to the NH_3-SCR converter.[11] Therefore, this solution requires additional operating costs caused by heating of the flue gases. Another solution could be an effective conversion of nitrogen oxides at temperature below 250 °C without the need to heat the exhaust gases between the EPS and NH_3-SCR units. Of course, to implement this proposed new solution, development of effective catalysts for the low-temperature NH_3-SCR process is necessary. The schematic presentations of exhaust gas treatment with high-dust and low-dust NH_3-SCR are shown in Fig. 2.

Mesoporous silicate materials, possessing high surface area and porosity, seem to be very promising supports of the catalysts for the low-temperature NH_3-SCR process. Because the process in question must be conducted at relatively low temperatures, the limited thermal and hydrothermal stability of mesoporous silicates should not give rise to problems.

This paper reviews the recent studies of mesoporous silicates modified with transition metals as catalysts for the NH_3-SCR process. A significant part of this review is based on studies carried out by the authors.

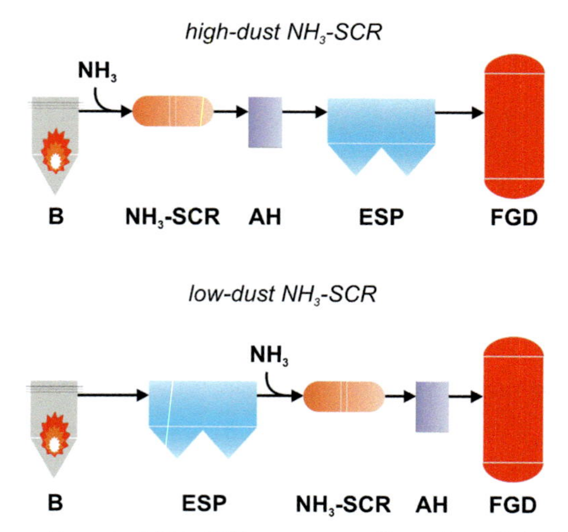

Fig. 2 High-dust and low-dust NH_3-SCR process configurations (B-boiler, AH-air heater, ESP-electrostatic precipitator, NH_3-SCR-monolithic converter, FGD-flue gas desulfurization).

2. Selective catalytic reduction of NO_x with ammonia—NH_3-SCR

Mesoporous silicate materials, due to their properties of high surface area and uniform porous structure, are considered very promising supports for NH_3-SCR catalysts. Deposition of catalytically active metals in the form of well-defined and uniform species should result in generation of a large number of catalytically active sites for nitrogen oxide conversion. Uniform metal species should guarantee high selectivity to desired reaction products, while a high density of such metal species deposited on large surface area mesoporous silicates should result in their high catalytic activity. Thus, both the selection of metal species, catalytically active in the NH_3-SCR process, as well as methods used for their deposition into mesoporous silicate materials are very important issues in the formulation of effective catalysts for reduction of nitrogen oxides with ammonia. Among transition metals, the most widely studied as catalytically active components of the NH_3-SCR catalysts are copper,[12,13] manganese[14] and iron.[15,16] The form and aggregation of deposited metal species are very important factors, since they may determine not only their activity and selectivity but also the temperature window of their effective operation.[17,18] Thus, the methods used for the metal species deposition which control their form and loading, are vitally important in the synthesis of the catalysts with tailored properties.

2.1 Mesoporous silicates modified with copper

In general, copper is reported to be the catalytically active component of the catalysts for low-temperature NH_3-SCR operation.[19,20] Qui et al.[21] studied mesoporous silicate of MCM-41 type modified with copper by two methods. First, a series of the catalysts was prepared by a hydrothermal co-condensation method and resulted in copper species incorporation into silicate walls of MCM-41, while the second series of the samples was obtained by an impregnation method. The copper loading in the catalysts was in the range from 1.0 to 16.7 wt%. It was reported that an increase in copper loading in the samples obtained by the hydrothermal co-condensation method resulted in a gradual disappearance of the ordered porous structure of MCM-41. This effect was explained as resulting from different lengths of the Si—O bond (0.160 nm[22]) and Cu—O bond (0.193 nm[23]). Moreover, it was shown that dispersion of copper decreases with increasing loading of this metal in the silicate samples.[21] However, the size of copper oxide particles within the catalysts obtained by the co-condensation method was significantly lower (2–5 nm) compared with CuO particles in the samples prepared by the impregnation method (10–20 nm). The NO conversion, measured at 250 °C, increased with increasing copper loading in the catalysts. Comparison of the NO conversion obtained for the samples prepared by hydrothermal co-condensation and impregnation methods with similar copper loadings showed significantly higher activity when containing smaller copper oxide species were introduced into MCM-41 by the co-condensation method. Moreover, the turnover frequency (TOF) values of NO conversion increased with decreasing copper loading (e.g., for MCM-41 with Cu loadings of 1.0 and 16.7 wt%, the TOF values were $2.8 \cdot 10^{-3}$ and $0.9 \cdot 10^{-3}\ s^{-1}$, respectively at 250 °C).[21] Thus, it clearly shows that less aggregated copper oxide species yield better catalytic activity in the NH_3-SCR process than when larger CuO aggregates are introduced into MCM-41. Similar studies were reported by Jankowska et al.[24] for spherical MCM-41 modified with copper by the co-condensation method. The SEM micrographs of mesoporous silicate of MCM-41 type with classical (cylindrical) and spherical morphology are shown in Fig. 3.

The MCM-41 samples with a copper content of 2.1 and 5.6 wt% were prepared, analyzed, and tested as catalysts of the NH_3-SCR process.[24] Similarly, as presented in earlier studies,[21] incorporation of copper into silicate walls of MCM-41 resulted in a partial destruction of the ordered porous structure. In both samples, copper was present mainly in the form of monomeric cations and a lower contribution of oligomeric copper oxides species. The NO conversion in the NH_3-SCR tests started at about 125 °C and increased to the level of about 90% at 300 °C (Fig. 4, Cu-S-MCM-41-C). Decrease in efficiency of the NO conversion, observed at temperatures above 300 °C, was related to the secondary process of direct ammonia oxidation by oxygen present in the reaction mixture. The NO conversion

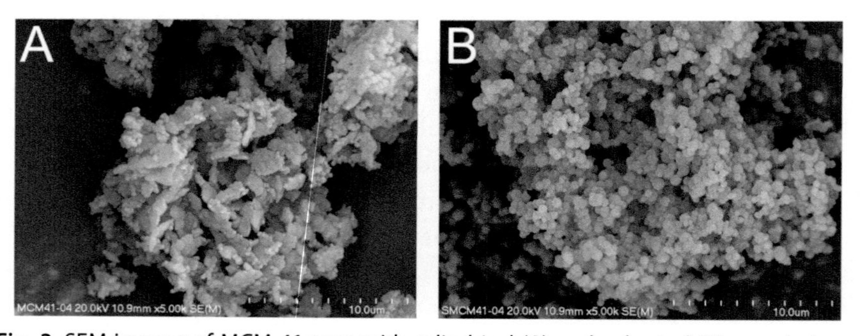

Fig. 3 SEM images of MCM-41 type with cylindrical (A) and spherical (B) morphology.

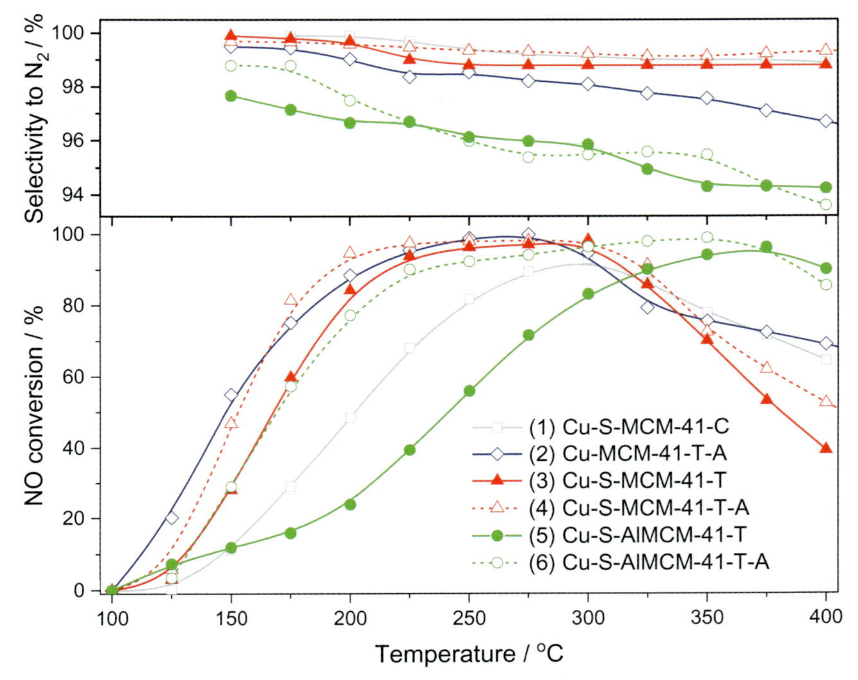

Fig. 4 Results of catalytic NH_3-SCR tests for MCM-41 modified with copper. Gas mixture: NO—0.25 vol%, NH_3—0.25 vol%, O_2—2.5 vol%, He—97 vol%, total flow rate—40 $cm^3\ min^{-1}$, VHSV—24,000 $cm^3\ h^{-1}\ g^{-1}$, catalyst 0.1 g. Catalysts: (1) Cu-S-MCM-41-C, spherical MCM-41 with 5.6 wt% Cu incorporated by the co-condensation method[24]; (2) Cu-MCM-41-T-A, cylindrical MCM-41 with 12.9 wt% Cu deposited by the TIE method with post-treatment with ammonia[25]; (3) Cu-S-MCM-41-T, spherical MCM-41 with 9.9 wt% Cu deposited by TIE method[26]; (4) Cu-S-MCM-41-T-A, spherical MCM-41 with 6.6 wt% Cu deposited by the TIE method with post-treatment with ammonia[26]; (5) Cu-S-AlMCM-41-T, spherical silicate-alumina MCM-41 with 6.0 wt% Cu deposited by the TIE method[27]; (6) Cu-S-AlMCM-41-T-A, spherical silicate-alumina MCM-41 with 7.9 wt% Cu deposited by the TIE method with post-treatment with ammonia.[27]

profiles of the catalysts with copper loading of 2.1 and 5.6 wt% are very similar, with only slightly higher efficiency of the catalysts with higher copper content. The TOF values determined for the NO conversion at 250 °C in the presence of the MCM-41 samples with copper loading of 2.1 and 5.6 wt % were $7.6 \bullet 10^{-3}$ and $4.9 \bullet 10^{-3}\ s^{-1}$, respectively. The higher TOF values measured for highly dispersed copper species[24] compared to those determined for the catalysts containing copper oxide aggregates[21] support the hypothesis that well-dispersed forms of this metal, preferably monomeric cations, are necessary to obtain active catalysts of the NH_3-SCR process.

Liu and Teng[28] compared catalytic activity of MCM-41 modified with copper deposited by different methods – template ion-exchange (TIE), impregnation (IM) and ion-exchange with silanol groups (IE). In all cases, a solution of $Cu(NO_3)_2$ was used for copper deposition. The TIE method is based on the ability to exchange cationic organic surfactants for metal cations in as-synthesized mesoporous silicates (non-calcined) obtained by using cationic surfactants as the structures directing agents. In such freshly prepared mesoporous silicates, cationic surfactants are attached to the surface $\equiv Si—O^-$ groups by Coulombic interactions and therefore relatively easily can be replaced for metal cations by the ion-exchange method. In general, the methods used for copper deposition result in aggregated copper oxide species containing both Cu^{2+} and Cu^+ cations. It was postulated that Cu^+ species possess higher activity than Cu^{2+} in catalytic reduction of NO.[28] Moreover, the bridging oxygen in the -Cu-O-Cu- species can be relatively easily transferred into reacting molecules and therefore exhibits oxidizing ability.[28] It was reported[28] that all methods used for copper deposition resulted mainly in aggregated copper oxide species; however, the size of such CuO aggregates was smaller in the case of the catalysts obtained by the IE method, than in the samples modified by the TIE procedure. The catalysts obtained by the IE method yield higher catalytic activity in the NH_3-SCR process than other samples, a result that can be explained by different sizes of deposited copper oxide aggregates. An increase in copper loadings in the catalysts resulted in a more significant decrease in NO conversion at higher temperatures due to the secondary process of direct ammonia oxidation. This is consistent with the hypothesis that the bridging oxygen in -Cu-O-Cu- can be readily transferred into the reacting molecules, e.g., ammonia, resulting in their oxidation. Studies focused on development of the TIE method were continued by Kowalczyk et al.[25] In the first step, the conditions of the TIE procedure were optimized. It was shown that replacement of water by methanol as the solvent for the copper salt resulted in improvement of copper species dispersion. Additionally, it was reported that deposition of copper nearly exclusively in the form of monomeric copper cations can be obtained by treatment of the samples directly after the TIE procedure with ammonia solution. Such high dispersion of copper was observed for the sample with such high copper loading as 12.9 wt%.[25] The results of the NH_3-SCR tests for these catalysts are very promising. The NO conversion started at about 100 °C and increased to the level of 90–100% at 250–275 °C. Efficiency of the NO reduction decreased at higher temperatures as a result of the secondary process of direct ammonia oxidation by oxygen present in the reaction mixture. Selectivity to nitrogen, which is a desired product of the NH_3-SCR process was above 96% at temperatures below 350 °C (Fig. 4, Cu-MCM-41-T-A). The activity of the catalysts studied increased with an increase in copper loading. Thus, these catalysts seem to be very promising for the low-temperature NH_3-SCR process. Jankowska et al.[26] extended these studies for MCM-41 with the spherical morphology. In the case of spherical MCM-41, treatment of the samples modified by copper directly after the TIE procedure with ammonia, significantly improved dispersion of the metal, which was deposited mainly in the form of monomeric copper cations. This is a similar outcome to results reported by Kowalczyk et al.[25] Moreover, the FTIR analysis of the MCM-41 samples modified and non-modified with copper, before and after calcination, proved that the mechanism of TIE is based on an exchange of the cationic alkylammonium surfactants (S^+) attached to surface $\equiv Si—O^-$ species for copper cations from the solution resulting in surface $(\equiv Si—O^-)_2Cu^{2+}$ and possibly also $\equiv Si—O^- Cu^{2+} OH^-$ and $\equiv Si—O^- Cu^+$ species. The NO conversion started at about 125 °C and in the range of 200–325 °C was above 90%. At higher temperatures decrease in the NO conversion was caused by the competing process of direct ammonia oxidation (Fig. 4, Cu-MCM-41-T-A). In the case of the catalysts non-modified with ammonia, containing more aggregated copper oxide species, the NO conversion above a level of 90% was obtained in the range of 225–300 °C only, for the most active catalyst (Fig. 4, Cu-MCM-41-T). Thus, it seems that more aggregated copper species are less active in the low-temperature NH_3-SCR process compared with the analogous process involving highly dispersed species, mainly monomeric copper cations. Moreover, aggregated copper species are more effective in the separate process of direct ammonia oxidation. The selectivity toward nitrogen is above the level of 97% at temperatures below 350 °C, both for the catalysts modified and non-modified with ammonia. Comparison of the catalysts based on spherical and cylindrical MCM-41 with similar copper loadings shows better activity for the spherical silicate-based catalysts. Possibly, this effect is related to shorter channels in spherical MCM-41 and therefore more effective internal diffusion of reactants inside pores of the catalyst.

Surface acidic sites play a very important role in the NH_3-SCR process.[29,30] Such sites can activate ammonia molecules for the NH_3-SCR reaction or accumulate ammonia molecules on the catalyst surface and protect them against oxidation. Acidic sites can be generated in mesoporous silicates by incorporation of heteroatoms, such as aluminum[27] or niobium,[31] into silicate walls, Jankowska et al.[27] synthesized spherical alumina-silicate

MCM-41 by the co-condensation method using tetraethyl orthosilicate (TEOS) as a silicate source and $NaAlO_2$ as an aluminum source. The molar Si/Al ratio in the spherical MCM-41 obtained was 27. It was shown that incorporation of aluminum into silicate walls resulted in the generation of surface acidic sites. In a further study, copper was deposited by the TIE method, with and without post-treatment with ammonia solutions. As reported in earlier studies,[25,26] in the case of alumina-silicate MCM-41, treatment of the copper modified samples with ammonia solution significantly improved dispersion of this metal.[27] For the most active catalyst obtained by the TIE method with ammonia post-treatment, the NO conversion above the level of 90% was obtained in the range of 225–375 °C. (Fig. 4, Cu-S-AlMCM-41-T-A). Thus, incorporation of alumina into silicate walls resulting in the generation of acidic sites, significantly extends an effective operation window from consideration of higher temperatures. This effect is possibly related to chemisorption of ammonia molecules on acidic sites, which in such form are protected against direct oxidation.[27] This hypothesis is supported by a relatively small efficiency of direct ammonia oxidation observed at temperatures above 375 °C. The catalysts obtained by the TIE method but without post-treatment with ammonia were catalytically active only at higher temperatures, but also in this case the efficiency of the separate process of direct ammonia oxidation in the high-temperature range was relatively low (Fig. 4, Cu-S-AlMCM-41-T). Thus, it could be concluded that surface acidic sites related to aluminum cations incorporated into silica walls improve catalytic performance at temperatures above 300 °C. On the other hand, significant differences in the efficiency of the low-temperature NH_3-SCR process for the catalysts treated and untreated with ammonia solutions show that highly dispersed copper species, preferably monomeric cations, are significantly more catalytically active in the low-temperature NO conversion, compared with activity when more aggregated copper oxide species are present.

Ziółek et al.[31] studied mesoporous silicates of MCM-41 type containing niobium incorporated into silicate walls as NH_3-SCR catalysts. The silica-niobium mesoporous samples were prepared by the co-condensation method. In the next step, copper was deposited by the ion-exchange method. It was shown that incorporation of niobium into the silicate resulted in the generation of acidic sites of Lewis type, which activate the catalyst in the NH_3-SCR reaction. A correlation between niobium loading and surface concentration of Lewis type acidic sites as well as catalytic activity in the NO conversion was reported. The comparison of Cu/Nb-MCM-41 and Cu/Al-MCM-41 showed that copper deposited on niobium modified silicate is much easier reduced than in the case of alumina doped mesoporous silica. Moreover, it was reported that copper in the form of Cu^+ is effectively stabilized on Nb-MCM-41 and can be relatively easy re-oxidized to Cu^{2+} by interaction with NO molecules. Thus, the introduction of niobium into MCM-41 not only resulted in the generation of Lewis acidic sites but also improved the redox properties of deposited copper species, which is important in the NH_3-SCR process.

Imyen et al.[32,33] studied the promoting effect of zinc deposited into Cu/MCM-41 catalysts of the NH_3-SCR process. The concept of these studies was based on reports indicating the activating role of zinc introduced into ZSM-5 zeolite in the NH_3-SCR process.[34] It was postulated that this promoting effect of zinc is possibly related to the formation of additional acidic sites and, what is more important, stabilization of Cu^+ cations, which has been suggested as the main active sites for nitrogen oxides reduction.[35,36] Various methods of copper and zinc deposition into MCM-41, including co-condensation, impregnation, and ion-exchange, were used.[32,33] In general, it was shown that the form of deposited copper species is very important. Moreover, the introduction of zinc into the Cu/Al-MCM-41 samples influenced their acidic properties and the nature of copper species in the catalysts. Both Lewis and Brønsted acidic sites were generated by deposition of Zn^{2+} cations. It was shown that the introduction of zinc to the Cu/Al-MCM-41 catalysts promoted the contribution of Cu^+ species in the reduced samples by stabilizing Cu^+ cations due to hindering these species from being completely reduced to Cu^0. Moreover, it was postulated that Zn^{2+} cations provided additional sites for NO adsorption in the form of nitrate species, which has been considered as one of the key intermediates for the NH_3-SCR process. Thus, it seems that this direction of studies is promising and may influence the formulation of effective catalysts for the low-temperature NH_3-SCR process.

Concluding, copper modified MCM-41 is a key material for active and selective catalysts of the low-temperature NH_3-SCR process. Copper loading and dispersion are very important in designing effective and selective catalysts. In general, catalytic activity increases with copper loading. Furthermore, dispersed copper species, mainly monomeric copper cations, increase the low-temperature activity of the NH_3-SCR catalysts. More aggregated copper species are more active in the secondary reaction of direct ammonia oxidation and therefore limit the efficiency of conversion of nitrogen oxides at higher temperatures. The high-temperature performance of the Cu-MCM-41 catalysts can be improved by incorporation of aluminum into silica walls of MCM-41, resulting in the formation of surface acidic sites.

Possibly, ammonia molecules chemisorbed on such sites are stabilized against their direct oxidation and therefore ammonia is available for the NH_3-SCR process also in the high-temperature range. However, it should be emphasized that an ultimate goal relates to the use of Cu-MCM-41 catalysts in the low-temperature NH_3-SCR process.

2.2 Mesoporous silicates modified with iron

Iron is among transition metals intensively studied as active components of the NH_3-SCR catalysts.[37,38] Jankowska et al.[24] studied iron modified mesoporous silicate of MCM-41 type with spherical morphology. Iron was introduced into silicate walls by the co-condensation method with the intended molar Si/Fe ratios of 10, 20, 50 and 100. It was shown that an increase in iron loading resulted in a gradual destruction of the ordered porous structure as well as a decrease of specific BET surface area. In all the samples iron was present in the form of highly dispersed iron species, mainly isolated cations. However, the contribution of oligomeric iron oxide species increased with iron loading on the samples. Moreover, a decrease in the size of silicate-iron spheres and an increase in disordering of their spherical shape were observed with the increasing iron loading in the samples. The NO conversion, depending on the iron loading, started at about 150–200 °C. In the case of the most active catalysts, with the intended Si/Fe molar ratio of 20, the NO conversion above the level of 90% was obtained in the range of 300–450 °C, while at higher temperatures the efficiency of NO reduction decreased due to the secondary process of direct ammonia oxidation by oxygen present in the reaction mixture (Fig. 5, Fe-S-MCM-41-C). Thus, this catalyst effectively operated in the high-temperature NH_3-SCR process. It was shown that the catalytic activity of the studied samples increased with an increase of iron content from the molar Si/Fe ratio of 100 to 20. However, the catalysts with the Si/Fe ratio of 10, presented the lowest catalytic activity in the whole series of the samples. This effect shows a very important role of the type of iron species in the NH_3-SCR process. In the sample with the highest iron loading (Si/Fe = 10), there are more aggregated iron oxide species that are possibly less catalytically active in the NO conversion compared to more dispersed species of this metal. It clearly shows in the comparison of TOF values, determined for the reaction at 250 °C. For the samples with the Si/Fe ratios of 20 and 10, TOF values were $4.5 \cdot 10^{-3}$ and $1.6 \cdot 10^{-3}\ s^{-1}$, respectively.[24] What is very promising, was the selectivity to nitrogen was on the level of about 99% for all catalysts of this series, also in the high-temperature range characteristic of direct ammonia oxidation. The studies of the reaction mechanism using temperature-programmed methods

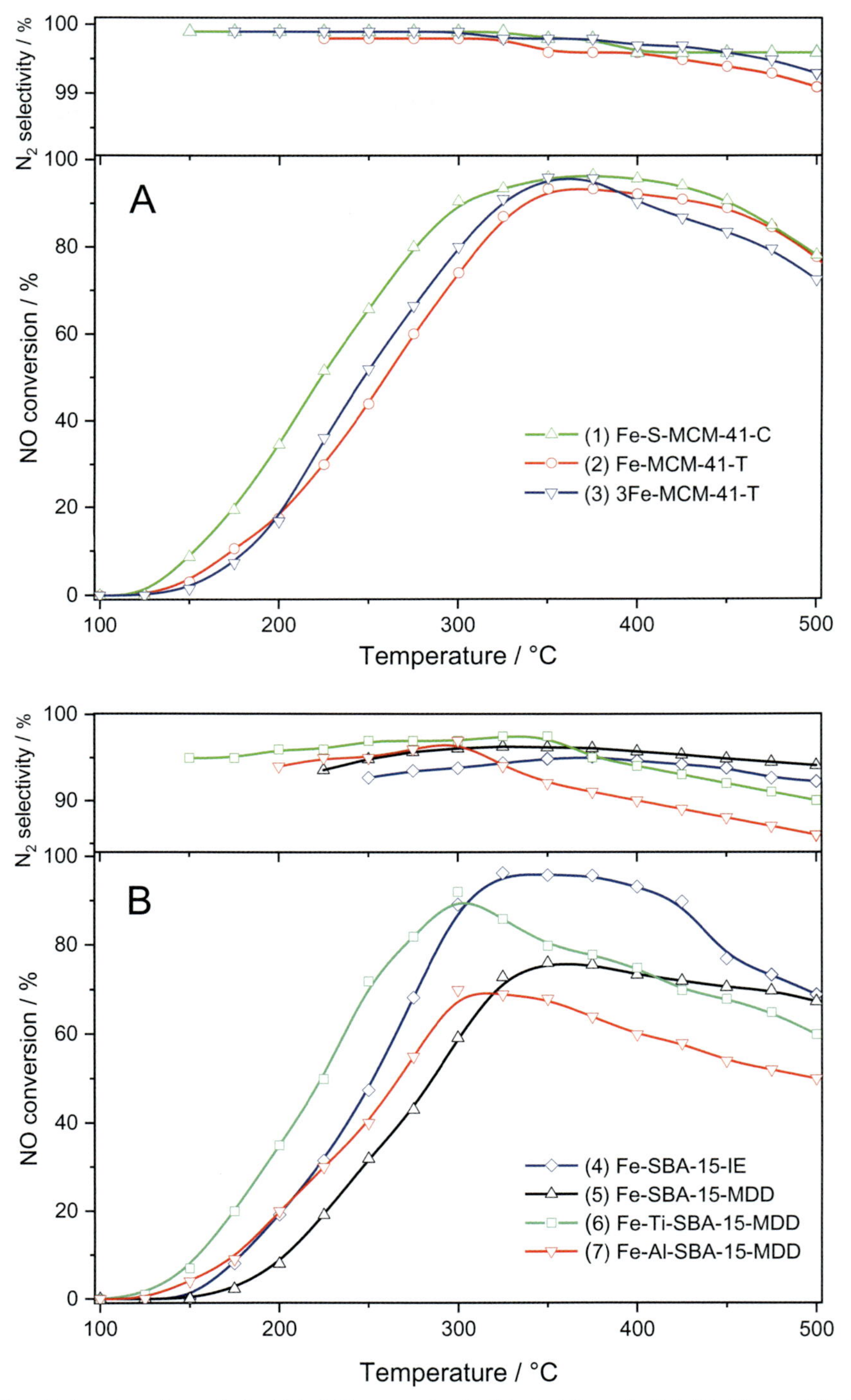

Fig. 5 Results of catalytic NH_3-SCR tests for MCM-41 (A) and SBA-15 (B) modified with iron. Gas mixture: NO—0.25 vol%, NH_3—0.25 vol%, O_2—2.5 vol%, He—97 vol%,

(Continued)

showed that the main reaction pathway possibly includes chemisorption and activation of ammonia molecules, which in the next step in reaction with NO.[24] However, this hypothesis does not exclude other possible mechanisms of NO conversion. Li et al.[41] studied iron modified mesoporous silica of the MCM-41 type with cylindrical morphology obtained from kaolin. The samples with different iron loading were obtained by the hydrothermal co-condensation method. The quality of the mesoporous materials obtained was relatively high and comparable with MCM-41 silicates synthesized by other methods. An increase in iron content in the samples resulted in a partial decrease of MCM-41 porous structure ordering. It was shown that the dosage of Fe^{3+} ions in the range of molar Si/Fe ratio of 50–20 increased the number and strength of catalytic sites. While for the samples with the lower molar Si/Fe ratio, 10 and 12.5, the formation of aggregated iron oxide species decreased the efficiency of the NH_3-SCR process. Thus, in general, the results presented by Li et al.[41] are very similar to those reported by Jankowska et al.[24]

Kowalczyk et al.[39] studied pure silica and silicate-titanium MCM-41 modified with iron by the template ion-exchange (TIE) method as catalysts of the NH_3-SCR process. Silicate-titanium MCM-41 was synthesized by the co-condensation method using tetraethyl orthosilicate (TEOS) and titanium isopropoxide (TIP) as silica and titanium sources, respectively. The iron sources were methanol solutions of $FeCl_2$ and $[Fe_3(OAc)_6O(H_2O)_3]NO_3$. The intention of these studies was deposition of iron in the form of highly dispersed species, mainly monomeric cations, in the case of using $FeCl_2$, and small oligomeric iron oxide aggregates in the case of the oligomeric $[Fe_3(OAc)_6O(H_2O)_3]NO_3$ precursor. It was reported that in the first case iron was deposited mainly in the highly dispersed form, while in the second case apart from highly dispersed iron species, Fe_2O_3 nanorods outside of the MCM-41 pores, were also formed (Fig. 6).

Fig. 5—Cont'd total flow rate—40 $cm^3\ min^{-1}$, VHSV—24,000 $cm^3\ h^{-1}\ g^{-1}$, catalyst 0.1 g. Catalysts: (1) Fe-S-MCM-41-C, cylindrical MCM-41 with 4.0 wt% Fe incorporated by the co-condensation method[24]; (2) Fe-MCM-41-T-, cylindrical MCM-41 with 5.8 wt % Fe deposited by the TIE method using $FeCl_2$ solution[39]; (3) 3Fe-MCM-41-T, cylindrical MCM-41 with 10.2 wt% Fe deposited by the TIE method using a solution of $[Fe_3(OAc)_6O(H_2O)_3]NO_3$.[39]; (4) Fe-SBA-15-IE, SBA-15 with 6.5 wt% Fe deposited by ion-exchange of $[Fe_3(OAc)_6O(H_2O)_3]^+$ with the $-SO_3H$ species grafted on the silica surface[40]; (5) Fe-SBA-15-MDD, SBA-15 with 2.2 wt% Fe by the molecular dispersion method (MDD) using iron acetylacetonate as a metal precursor[40]; (6) Fe-Ti-SBA-15-MDD, SBA-15 with 2.7 wt% Ti and 1.8 wt% Fe, both metals deposited by MDD method with using titanium and iron acetylacetonates[40]; (7) Fe-Al-SBA-15-MDD, SBA-15 with 0.9 wt% Al and 1.9 wt% Fe, both metals deposited by the MDD method using titanium and iron acetylacetonates.[40]

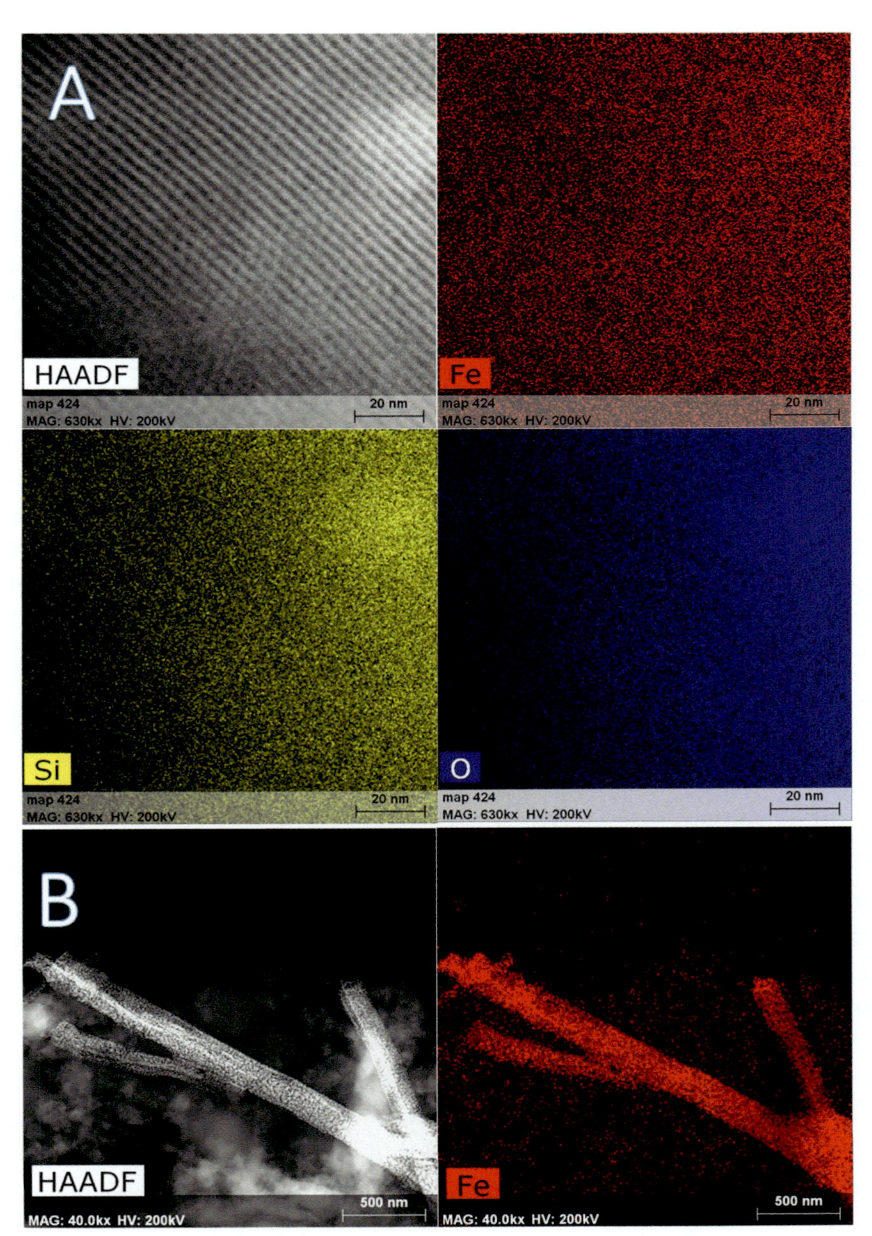

Fig. 6 STEM image and EDX element density maps of Fe, Si, O in the MCM-41 sample modified with iron oligomeric cations (A) and STEM image and Fe EDX density map of rod found in the material (B). *Reprinted with permission from Kowalczyk, A.; Piwowarska, Z.; Macina, D.; Kuśtrowski, P.; Rokicińska, A.; Michalik, M.; Chmielarz, L.* Chem. Eng. J. ***2016**, 295, 167–180. Copyright 2016 Elsevier.*

Silicate and silicate-titanium MCM-41 modified with a methanol solution of $FeCl_2$ yielded very similar profiles for the NO conversion and selectivity toward N_2,[39] indicating that additional acidic sites, generated by titanium, do not play a very important role in catalyst activation in the NH_3-SCR process. For the most active catalysts of these series, the NO conversion above a level of 90% was obtained in the range of 350–450 °C (Fig. 5A, Fe-MCM-41-T). At higher temperature, direct oxidation of ammonia decreased the efficiency of the NH_3-SCR process. On the other hand, the catalysts that were obtained using $[Fe_3(OAc)_6O(H_2O)_3]NO_3$ as an iron precursor, that is, the samples with significant contribution of aggregated iron oxide species were less effective in the NH_3-SCR process. For the most active catalysts of this series, the NO conversion above the level of 90% was obtained in a relatively narrow temperature range of 325–375 °C (Fig. 5A, 3Fe-MCM-41-T). This effect is related to intensive ammonia oxidation at higher temperature, which is more significant than for other series of the samples. Thus, it could be concluded that aggregated iron oxide species are more catalytically active in the separate process of ammonia oxidation than dispersed iron cations deposited on the surface of MCM-41 and therefore limit the efficiency of the NO reduction with ammonia at higher temperatures. The selectivity to nitrogen was above the level of 98% up to a temperature of 550 °C, thus not only the selectivity of the NH_3-SCR process was very pronounced, but also nitrogen was the main product of direct ammonia oxidation.

Macina et al.[40] studied the catalytic activity of various iron species deposited on mesoporous silicate of SBA-15 type in the NH_3-SCR process. Iron species were deposited by two different methods resulting in monomeric cations and small oligomeric iron oxide species. A first series of the catalysts was obtained by the molecular designed dispersion (MDD) method based on the selective reaction of iron acetylacetonate with surface silanol groups of silicate material. Calcination of silicates, modified by such a method, resulted in a deposition of iron in highly dispersed forms, mainly monomeric cations. The second series of the catalysts was obtained by grafting of 3-(mercaptopropyl) trimethoxysilane on SBA-15 followed by its oxidation to alkylsulfonic surface species by H_2O_2. Such species exhibit ion-exchange properties, and therefore in the next step trinuclear iron acetate oligocations, $[Fe_3(OAc)_6O(H_2O)_3]^+$, were introduced into modified SBA-15 by the ion-exchange method. Calcination of the samples resulted in the formation of small oligomeric iron oxide species. The catalysts containing such small oligomeric iron oxide species were found to be more catalytically active in the NH_3-SCR reaction compared with the catalysts containing nearly exclusively monomeric iron cations. For the most active catalysts, the NO conversion above the level of 90% with selectivity above 93% was achieved in the range of 300–425 °C (Fig. 5B, Fe-SBA-15-IE). The parallel reaction of direct ammonia oxidation decreased the efficiency of the NO conversion at higher temperatures. Although, the content of iron in the catalyst obtained by the MDD method is lower (2.2 wt%) than in silicates modified with aggregated iron oxide species (4.4 and 6.5 wt%), the comparison of TOF values showed slightly higher activity when small oligomeric iron oxide species are present. On the other hand, the catalyst obtained by the MDD method, containing mainly monomeric iron cations, was less active in the process of direct ammonia oxidation (Fig. 5B, Fe-SBA-15-MDD), which is in full agreement with the studies presented by Kowalczyk et al.[39]

Chmielarz et al.[42] reported that catalytic activity of SBA-15 modified with iron by the molecular designed dispersion (MDD) method in the NH_3-SCR reaction can be increased by additional deposition of aluminum or titanium. The catalysts were obtained by introduction of aluminum and iron or titanium and iron into SBA-15 by the MDD method using acetylacetonates of aluminum, titanium, and iron as precursors of deposited metal species. Deposition of iron on SBA-15 resulted nearly exclusively in highly dispersed monomeric Fe^{3+} cations. In the case of SBA-15 modified with aluminum or titanium, deposition of iron resulted not only monomeric iron cations, but also in small aggregated Fe_xO_y aggregates. Such iron species deposited on Al-SBA-15 and Ti-SBA-15 were reduced at temperatures significantly lower compared with those for monomeric Fe^{3+} cations introduced on the surface of pure silicate SBA-15. Modification of SBA-15 with alumina resulted in a generation of acidic sites of both Lewis and Brønsted types, while on the surface of SBA-15 doped with titanium, mainly Lewis type acidic sites were identified. The best catalytic results were obtained for the Fe-Ti-SBA-15 catalyst, which was more active than other Fe-containing catalysts, and additionally operated at lower temperatures. Thus, it could be concluded that small oligomeric iron oxide species are more catalytically active than monomeric iron cations, and possibly also Lewis acidic sites are more important in the NH_3-SCR process, in comparison with a catalyst possessing Brønsted acidic sites. The Fe-Al-SBA-15 catalyst was less active than Fe-Ti-SBA-15 but more active than Fe-SBA-15. Moreover, Fe-Ti-SBA-15 operated at lower temperatures than Fe-SBA-15 and Fe-Al-SBA-15 (Fig. 5B, Fe-SBA-15-MDD, Fe-Ti-SBA-15-MDD, and Fe-Al-SBA-15-MDD). Thus, introduction of titanium into Fe-SBA-15 not only increased its catalytic activity but also shifted the effective operation window into lower temperatures. The catalytic activity of the catalysts obtained by the MDD method is limited by relatively low iron loadings (1.8–2.2 wt%), which is related to the limited reactivity of silanol groups of silicate and precursors of deposited metallic species.

Studies focused on determination of the roles of various iron species in the NH_3-SCR process were also reported by Yan et al.[43] The Fe-Al-SBA-15 catalysts with different aluminum and iron content were prepared by co-condensation methods. In the case of the first series of the catalysts, the reaction mixture containing: tetramethyl orthosilicate, iron nitrate, aluminum isopropoxide, nonionic triblock copolymer surfactant (P123) and hydrochloric acid, was conventionally heated (CH). The second series of the samples was obtained by treatment of such a reaction mixture in a microwave oven (MW). The catalysts with the Al/Si molar ratio of 0.03, 0.05 and 0.07 were obtained. The Fe/Si molar ratio in the SBA-15 samples was 0.03, 0.05 or 0.07. Microwave treatment significantly influenced the properties of the materials obtained. First, the samples obtained by the MW assisted method were characterized by petal-like morphology, while materials obtained by the conventional method (CH) presented disk-like morphology. In general, for the samples of the MW series the specific surface area and pore diameter were higher compared to the catalysts of the CH series. Moreover, in the samples of the MW series, a significant contribution of oligomeric Fe_xO_y aggregates was detected. In contrast, the conventional synthesis method (CH) resulted mainly in highly dispersed iron cations incorporated into SBA-15 walls. Synthesis methods also influenced the form of aluminum in the samples. In the case of the MW series, for the samples with the Al/Si ratio of 0.03, aluminum was present only in tetrahedral coordination incorporated into silicate walls of SBA-15. An increase in aluminum content resulted in deposition of it in small amounts on the silicate surface in the form of octahedrally coordinated cations. In the case of the samples obtained by the conventional method (CH) in addition to the tetrahedrally coordinated aluminum cations incorporated into silica walls, the presence of octahedrally coordinated aluminum cations was detected, as well for the samples with the lowest aluminum loading (Al/Si = 0.03). In general, the catalysts obtained by the MW assisted method resulted in significantly better catalytic properties than the catalysts of the CH series. Surprisingly, the best catalytic results in the MW series were observed for the sample with the lowest aluminum (Al/Si = 0.03) and iron (Fe/Si = 0.03) loadings. In this case, the NO conversion above the level of 90% was obtained in the range of

350–480 °C (maximal temperature of the catalytic test). The other reaction, direct ammonia oxidation, was not observed up to 480 °C, and therefore these catalysts are very promising for the high-temperature NH_3-SCR process. An increase in iron loading resulted in activation of the catalysts in the reaction of direct ammonia oxidation. Thus, it seems that larger iron oxide aggregates formed in the samples with higher iron loading, in contrast to small oligomeric Fe_xO_y species dominating in the samples with the lower iron content, are catalytically active in the reaction of direct ammonia oxidation. Moreover, a very important role of small oligomeric Fe_xO_y species in NO to NO_2 oxidation was postulated. Aluminum tetrahedral coordination present in silicate walls generated acidic sites of Brønsted type, which was proven by identification of NH_4^+ cations in the samples pre-adsorbed with ammonia. The mechanism including NO to NO_2 oxidation on small oligomeric Fe_xO_y species, the reaction of NO_2 with NH_4^+ cation resulting in an intermediate surface $NO_2(NH_4^+)_2$ species, which in the next step reacts with NO resulting in N_2 and H_2O, was postulated. Thus, this mechanism is in full agreement with the general equation of a fast NH_3-SCR reaction (Eq. 6), which is postulated as one of the main pathways of nitrogen oxide reduction in the low-temperature range.[44]

$$2\,NH_3 + NO + NO_2 \rightarrow 2\,N_2 + 3\,H_2O \quad (6)$$

Promising catalytic results were reported for SBA-15 coated with TiO_2 and doped with iron species.[45] SBA-15 was modified with titanium by the reaction of surface silanol groups of silica with titanium isopropoxide, $Ti(OC_3H_7)_4$, resulting in the samples possessing the molar Ti/Si ratios of 0.075, 0.142 and 0.172. In the next step, iron was deposited on Ti-modified SBA-15 by the impregnation method. Titanium formed a pseudo-layer of TiO_2 on the silica surface, while iron was deposited, among other forms, as aggregates of Fe_2O_3 with the size in the range of 20–50 nm. It was shown that the majority of such aggregated iron oxide species is located directly on the TiO_2 layers. The maximum of pore diameters gradually decreased with increasing content of introduced titanium, indicating successful deposition of TiO_2 layers inside pores. Titanium introduced to SBA-15 resulted in the formation of acidic sites able to adsorb and activate ammonia molecules for the NH_3-SCR reaction. Moreover, a reduction of iron species deposited on TiO_2 layers covering the SBA-15 surface was observed at temperature lower by about 50 °C compared to iron introduced into commercial bulky TiO_2. As a result of these properties, the Fe/TiO_2-SBA-15 catalysts exhibit much better catalytic properties within the NH_3-SCR reaction relative to the reference Fe/TiO_2 catalyst. The catalytic activity of the catalysts based on mesoporous silicate increased with the increasing content of titanium. For the most active catalyst of this series, the maximum of the NO conversion occurred at 300 °C. Above this temperature, the efficiency of the NO conversion decreased due to the reaction of direct ammonia oxidation. However, it should be mentioned that this effect is relatively limited—for the most active catalyst, the NO conversion decreased to about 80% at 400 °C. The selectivity with respect to nitrogen was nearly 100% up to a temperature of 400 °C. Very interesting results were obtained after introduction of water vapor into the reaction mixture. It was shown that the presence of water vapor (5 vol%) activated the Fe/TiO_2-SBA-15 catalyst at lower temperatures and limited the reaction of direct ammonia oxidation at higher temperatures. It was shown that additional Brønsted acidic sites are created in the presence of water vapor, possibly by hydrolysis of ≡Ti—O—Ti≡ bridges with the formation of acidic ≡Ti—OH groups. Possibly, ammonia adsorbed on such species is activated to the NH_3-SCR reaction yet is protected against direct oxidation by oxygen present in the reaction mixture. Infrared studies of the catalyst samples pre-adsorbed with NO and NH_3 molecules showed the presence of surface -NO_2, -NH_2 and NH_4^+ species, which are possibly the intermediate products of the NH_3-SCR process. The formation of such nitrite species indicates the activity of the catalysts in NO to NO_2 oxidation, and therefore also possible conversion of nitrogen oxides according to the fast NH_3-SCR mechanism (Eq. 6). Possibly a very important role in this process is assigned to ≡Ti—OH groups, which were reported to bind NO_2 in the form of highly reactive nitrite species and therefore shift the thermodynamic equilibrium of NO to NO_2 oxidation.[37]

Zhang et al.[46] reported studies of SBA-15 modified with iron and molybdenum in the role of NH_3-SCR catalysts. Transition metals were introduced into mesoporous silicates during synthesis by the co-condensation method. Iron was introduced to the samples with the Si/Fe molar ratio in the range of 12–51, while the Si/Mo ratio was in the range of 41–547. It was shown that iron was mainly incorporated into silicate walls, while molybdenum was not incorporated into silicate walls but highly dispersed on the surface of the samples. The samples obtained yielded only limited catalytic activity at higher temperatures. Possibly low activity of these catalysts is related to the limited contribution of surface iron species available for catalysts. Due to relatively thick walls in SBA-15 part of the iron species was possibly occluded in such silicate walls and was not available for catalysis. Bimetallic catalysts afforded better activity than monometallic iron-containing catalysts. The activation role of molybdenum was attributed to the generation of additional acidic sites for sorption and catalytic activation of ammonia and possibly also nitrogen oxides.

Concluding, iron-modified mesoporous silicates effectively operate in the high-temperature NH_3-SCR process. The performance of such catalysts depends on iron loading as well as forms of deposited iron species. In general, the efficiency of the NO conversion increases with an increase in iron loading. However, the formation of aggregated iron oxide species, that occurs preferentially in the samples with high iron loadings, results in activation of the catalysts in the separate reaction of direct ammonia oxidation and therefore limits their performance in the high-temperature NH_3-SCR process.

2.3 Mesoporous silicates modified with manganese

Manganese containing catalysts were reported to be active catalysts of the NH_3-SCR process. Various catalytic systems, including bulky manganese oxides[47,48] as well as manganese deposited on different supports, such as zeolites,[49] TiO_2,[50] Al_2O_3,[51] activated carbon[52] or modified clay minerals,[53,54] were considered as potential catalysts of the NH_3-SCR process. However, the number of reports of studies of manganese deposited into mesoporous silicates is limited. Manganese may exist in different oxidations states, such as Mn^{2+}, Mn^{3+} and Mn^{4+}. It was reported that Mn^{4+} is the most active species in the NH_3-SCR reaction[55] because it can promote NO oxidation to NO_2 and therefore facilitate the fast NH_3-SCR reaction responsible for the low-temperature conversion of nitrogen oxides (Eq. 6). In the oxidation of NO to NO_2, the lattice oxygen is used and manganese oxide with the higher valence state, such as Mn^{4+}, loses oxygen and is converted to manganese oxides with the lower valence states, such as Mn^{3+} or Mn^{2+}. On the other hand, Mn^{2+} can adsorb oxygen from the gas phase and transfer it into the lattice oxygen. Thus, the oxygen cycle in such system can be formed. Mn^{3+} is believed to form surface $Mn^{3+}NO_3^-$ associates, while ammonia molecules can be chemisorbed on Mn^{4+} to form Mn^{4+}-NH_3 by the accommodation of a free electron pair of ammonia into unoccupied d-orbitals of Mn^{4+} (donor-acceptor bound).[56] Thus, Mn^{2+}, Mn^{3+} and Mn^{4+} play different roles in the NH_3-SCR process and the contribution of the manganese species present in different oxidation states determines the performance of the catalysts.

Li et al.[57] studied MCM-41, modified with alumina and manganese by the impregnation method, as catalysts of the NH_3-SCR process.

It was shown that Mn/MCM-4 operated effectively in the low-temperature range and 100% of the NO conversion was obtained in the range of 200–320 °C. At higher temperatures, the efficiency of the NH_3-SCR process was significantly decreased due to the process of direct ammonia oxidation. Moreover, the selectivity for nitrogen markedly decreased above 240 °C. The NO conversion, as well as selectivity for nitrogen, were very significantly improved at higher temperatures after deposition of aluminum. Deposition of aluminum provided acidic sites for ammonia and nitrogen oxide species adsorption and activation for the NH_3-SCR process. It was postulated that NO can be oxidized to NO_2 by the active lattice oxygen of MnO_x species. Subsequently, NO_2 can be adsorbed and therefore stabilized in the form of NO_3^- anions on aluminum sites. Concurrently, the cleavage of an N—H bond in ammonia molecules chemisorbed on the Lewis acidic sites, generated by aluminum deposition, may result in the formation of $-NH_2$ species. Moreover, the migration of H species, or NH_3 combining with the surface hydroxyls may result in the formation of NH_4^+ cations adsorbed on the Brønsted acidic sites. Both these types of surface species were postulated to be intermediates in the NH_3-SCR mechanisms.[58] It was shown that deposition of aluminum on the Mn/MCM-41 catalyst did not result in significant changes in the contribution of Mn^{2+}, Mn^{3+} and Mn^{4+} species, but modified its redox properties.

Water vapor and SO_2 are typical components of exhaust gases produced by fuel combustion, which may result in poisoning of the NH_3-SCR catalysts. Therefore, the analysis of catalyst stability in the presence of these components is a very important issue of effective catalyst design. Such studies were carried out for the Mn/MCM-41 catalysts. It was shown that aluminum introduction into the Mn/MCM-41 catalyst resulted in an improvement of its resistance for catalytic deactivation by water vapor. In the case of SO_2, this stabilizing effect of aluminum introduction was much less significant.

Liang et al.[59] compared the catalytic activity of manganese deposited on pure silica and silicate-alumina (Si/Al = 5) SBA-15. Manganese (about 15 wt%) was deposited on SBA-15 and Al-SBA-15 by the impregnation method. Incorporation of aluminum into the SBA-15 walls increased its surface acidity by the generation of both Lewis and Brønsted acidic sites. Deposition of molybdenum on SBA-15 and Al-SBA-15 also resulted in increased acidity of Lewis and Brønsted types. However, the stronger Lewis sites were found in Mn/Al-SBA-15 rather than in Mn/SBA-15. Both catalysts, Mn/SBA-15 and Mn/Al-SBA-15, produced very significant activity in the low-temperature NH_3-SCR process; however, the activation effect of the aluminum incorporation into SBA-15 was also observed. The NO conversion started at about 100 °C and reached 100% at about 185 °C for Mn/Al-SBA-15 and at 200 °C for Mn/SBA-15. A decrease in the NO conversion, due to the reaction of direct ammonia oxidation, was observed above 250 °C for Mn/Al-SBA-15 and at 200 °C for Mn/SBA-15. Thus, incorporation of aluminum in SBA-15 not only activated the Mn/Al-SBA-15 catalyst at lower temperatures but also limited its activity in the process of direct ammonia oxidation. The activation role of aluminum was related to the generation of stronger acidic sites of Lewis type, which can adsorb ammonia molecules and generate -NH and $-NH_2$ species that were postulated to be key intermediates in the NH_3-SCR process. Moreover, it was supposed that NH_4^+ species generated on Brønsted acidic sites acted only as spectators.

Ma et al.[60] recorded their studies of the Ce-Mn/MCM-41 catalysts of the NH_3-SCR process. The catalytic support, MCM-41, was produced starting from diatomite. The quality of the mesoporous silicate obtained was comparable to the MCM-41 materials produced by the "classical" method based on TEOS condensation. Manganese and cerium were deposited on the surface of the silica support by the impregnation method using an aqueous solution of $Mn(CH_3COO)_2·4H_2O$ and $Ce(NO_3)_3·6H_2O$, respectively. The catalysts gave rise to relatively good activity in the low-temperature NH_3-SCR process; however catalytic performance depended on manganese and cerium loading in the samples. An increase in manganese loading should generate more active sites for the NO conversion, but on the other hand, deposition of larger amounts of manganese may result in the formation of more aggregated MnO_x species with limited accessibility to surface Mn^{n+} sites. Cerium was reported to be catalytically active in NO to NO_2 oxidation and therefore should increase the low-temperature nitrogen oxide conversion by the fast NH_3-SCR mechanism (Eq. 6). Long and Yang[61] reported that an introduction of small amounts ($\leq$ 2 wt%) of cerium significantly improved the activity of Fe/Ti-PILC catalysts in NO to NO_2 oxidation and therefore, also activated them in the low-temperature NH_3-SCR process. However, an increase in CeO_2 loading to 5 wt% resulted in deactivation of the catalyst due to the formation of a cubic CeO_2 phase, which is less effective in NO to NO_2 oxidation than dispersed cerium oxide species. In the case of the Ce-Mn/MCM-41 catalysts the CeO_2 loading was in the range of 2.58–6.65 wt%[60]; thus the presence of cerium in the form of a less catalytically active cubic form could be expected. Thus, this direction of the studies, focused on development of the low-temperature NH_3-SCR catalysts based on bimetallic Ce—Mn systems, seems to be very interesting.

Huang et al.[62] studied mesoporous silicates modified with manganese or both manganese and iron, deposited by the impregnation method, as catalysts of the NH_3-SCR process. The role of manganese and iron loadings, their ratios and calcination temperature on the catalytic performance of modified silicates were analyzed. The total loadings of metals were in the range of 15 to 40 wt%, while the Fe/Mn ratio was from 0 to 1.4. The calcination of mesoporous silicate impregnated with manganese and iron precursors was carried out at 400, 500, 600 and 700 °C. It was reported that manganese was present in the form of well-dispersed metal oxide species, as well as in the case of the samples being calcined at high temperatures. This is in contrast to the situation for iron, which formed Fe_2O_3 crystallites after calcination of the samples at 600 °C. Moreover, an increase in calcination temperature significantly decreased the specific surface area and pore volume of the catalysts. For example, for the catalyst with 30% of total metals loading and the Fe/Mn ratio of 1, the specific surface area and pore volume decreased from 460 m^2/g and 0.47 cm^3/g to 100 m^2/g and 0.11 cm^3/g after increasing calcination temperature from 400 to 700 °C, respectively. Mesoporous silicate modified only with manganese yielded relatively poor catalytic activity in the low-temperature NH_3-SCR process. A very significant increase in catalytic activity was observed for the samples containing both manganese and iron. For the most active catalyst (30% metals loading, Fe/Mn ratio of 1, calcined at 400 °C) of this series, the NO conversion above the level of 90% was obtained at a temperature as low as 100 °C and did not decrease below this level at least up to 180 °C (maximal temperature of the catalytic test presented in[62]). It was shown that the catalysts containing both manganese and iron were significantly more catalytically active in NO to NO_2 oxidation compared with mesoporous silica modified only with manganese. Thus, it seems that in the case of silicates modified with both manganese and iron the main pathway of nitrogen oxides conversion in the low-temperature range is the fast NH_3-SCR reaction (Eq. 6). In general, the catalysts studied demonstrated relatively high resistance for poisoning by water vapor and SO_2. The NO conversion for the reaction conducted at 100 °C decreased by about 10% after introduction of 10 vol% water vapor into the reaction mixture; however, the NO conversion level at 150 °C was the same for the reaction conducted with the presence and absence of water vapor in the reaction mixture. Simultaneous addition of water vapor (10 vol%) and SO_2 (250 ppm) to the reaction mixture decreased the NO conversion by about 10%; however, after the withdrawal of water vapor and SO_2 the NO conversion was partially restored. Deposition of ammonium sulfate on the catalyst surface was postulated to be the main reason for its deactivation.

A very promising trend in preparation of porous metal oxide materials, as well as for catalysis, is the nano-casting method using porous silicates as hard templates. In general, nano-casting is the process in which a template with relevant structures on the length scale of nanometers is filled with another material, and the initial template is removed afterwards, resulting in porous material with the reproduced structure of pores in the initial template. Gao et al.[63] prepared a series of mesoporous manganese oxides by the nano-casting method using KIT-6, SBA-15, and MCM-41 as hard templates. A solution of $Mn(NO_3)_2$, used as a manganese oxide precursor, was introduced into the pores of hard templates. After drying and calcination of the samples, the hard templates were removed by their treatment in a solution of NaOH. Thus, it should be noted that in this case, in contrast to other studies reported, mesoporous silicates were used only as templates for the synthesis of porous metal oxide catalysts. In the materials obtained, mainly the MnO_2 phase was identified. The specific surface area and pore volume were in the ranges of 110–120 m^2/g and 0.25–0.39 cm^3/g, respectively. Thus, the textural parameters of these samples are lower than typical mesoporous silicates but higher than bulky manganese oxides. Catalytic studies showed a very favorable catalytic activity of such mesoporous manganese catalysts in the low-temperature NH_3-SCR process. The NO conversion was observed at temperatures as low as 50 °C, and conversion above the level of 90% was obtained in the range of 100–250 °C in the case of the catalysts obtained when using SBA-15 and KIT-6 as hard templates, and in the range of 100–200 °C for the catalyst prepared from using a MCM-41 template. For comparison, bulky MnO_2, which was tested as a reference catalyst, reached the maximum of the NO conversion at 200 °C at the level of 60%. Mesoporous manganese catalysts offered relatively high resistance for poisoning by water vapor and SO_2. In the case of the catalysts obtained using KIT-6, the NO conversion decreased from 100 to 90% after introduction of 100 ppm SO_2 into the reaction mixture at 150 °C and was on this level for the next 18 h (to the end of the stability test). Similarly, introduction of 5 vol% water vapor decreased the NO conversion from 100 to 90%, and this conversion level was stable to the end of the stability test (12 h, 150 °C). Thus, this direction of studies seems to be very promising for development of effective catalysts for the low-temperature NH_3-SCR process.

Concluding, manganese is a very promising component of the catalysts for the low-temperature NH_3-SCR process. The form of deposited manganese species is very important. The highly dispersed small metal oxide aggregates seem to be the most active in the low-temperature NH_3-SCR reaction. Due to the existence of Mn^{x+} cations in various oxidation states (Mn^{2+}, Mn^{3+}, and Mn^{4+}) and the relatively easy transitions between these oxidation states, the oxidation of NO to NO_2 with the re-oxidation of Mn^{x+} cations by oxygen from the gas phase is possible in the presence of Mn-containing catalysts. These reactions increase the role of the fast NH_3-SCR pathway in the low-temperature conversion of nitrogen oxides. The efficiency of NO to NO_2 oxidation can be improved by introduction of additional components, such as cerium. Moreover, the catalytic activity of silicates modified with manganese can be improved by the generation of acidic sites, able to adsorb and activate ammonia and possibly also nitrogen oxides for the NH_3-SCR reaction.

2.4 Mesoporous silicates modified with other metals

Apart from mesoporous silicates modified with copper, iron, and manganese, which seem to be very promising catalysts of the NH_3-SCR process, other metals deposited into silicates were reported to be active in the reactions studied.

Kwon et al.[64] studied vanadium modified MCM-41 to determine the role vanadium may serve in the NH_3-SCR process. Vanadium was introduced into mesoporous silica by the co-condensation method and its content in the samples was on the level of about 10 and 30 wt%. Vanadium was present mainly in the form of V_2O_5 aggregates. An increase in vanadium loading significantly reduced the specific surface area and porosity of the samples. Moreover, it was shown that vanadium species introduced into silicates generated relatively weak acidic sites. Results of the catalytic tests are surprising. In the case of the V-MCM-41 catalyst with about 10 wt% of vanadium loading, the NO conversion was at a constant level of about 45–48% in the broad temperature range of 100 to 400 °C. An increase in the vanadium loading to about 30 wt% resulted in a significant activation of the catalyst in the NH_3-SCR reaction. In this case, 100% of the NO conversion was achieved in the temperature range of 300–400 °C, thus very similar to the commercial V_2O_5-TiO_2 based catalysts.[58] Similar catalytic systems were studied by Segura et al.[65]; however, in this case vanadium and titanium were deposited on the surface of SBA-15 by the molecular designed dispersion (MDD) method. The MDD method is based on the selective reaction of organometallic precursors of metals with surface silanol groups of silicates. Calcination of the silicate samples modified by this method resulted in deposition of metal species in highly dispersed forms. Acetylacetonates of vanadium and titanium were used as precursors of deposited metal species. Experimental data showed that deposition of titanium and vanadium into SBA-15 by the MDD method resulted in highly dispersed TiO_x and VO_x species. The presence of the isolated V and Ti species on SBA-15, even for the samples with a high titanium and vanadium content, was verified. An increase in vanadium loading resulted also in the appearance of polymeric chains of VO_x. Modification of the silicate support with titanium and vanadium generated acidic sites. However, the acidic sites generated by deposition of titanium were stronger than those assigned to vanadium species. It was shown that the catalysts containing both vanadium and titanium were significantly more active than SBA-15 modified separately only with titanium or vanadium. For the most active catalysts doped with titanium, NO conversion above the level of 80% was obtained at 450 °C. In the case of the most active catalysts obtained by deposition of vanadium, the NO conversion above a level of 80% was achieved at 400 °C. Thus, these catalysts operated only in the high-temperature range. Simultaneous presence of titanium and vanadium in the catalysts resulted in their activation at lower temperatures. The activity of these catalysts increased with an increasing loading of vanadium and titanium. For the most active catalyst of this series, NO conversion close to 100% was obtained in the range of 300–350 °C. At higher temperature, the efficiency of the NO conversion significantly decreased due to the process of direct ammonia oxidation. The MDD method seems to be very promising for the preparation of the catalysts containing highly dispersed metal species. However, a significant drawback of this method is the limited content of deposited metal, as a result of the relatively low reactivity of the surface silanol groups of silicate with precursors of the catalytically active components. Moreover, it should be recalled that one of the main motivations for development of new catalysts, that is, alternatives for commercial V_2O_5-TiO_2 based systems, has been the relatively high volatility of V_2O_5, which on the one hand results in decreasing content of the catalytic component, yet on the other hand results in environmental pollution by vanadium oxide particles.[66]

Song et al.[67] analyzed the catalytic performance of MCM-41 modified with CeO_2 and SO_4^{2-} in the NH_3-SCR process. The CeO_2 and SO_4^{2-} modified silicates were prepared by the wet impregnation method. The first catalyst (Cat-A) was prepared by simultaneous introduction of cerium and SO_4^{2-} into MCM-41. The second (Cat—B) was prepared by sequential impregnation of silica with H_2SO_4 and then $Ce(NO_3)_3 \cdot 6H_2O$ solutions, followed by calcination. The third, (Cat—C) was prepared similarly but with the opposite order of deposited species – first cerium precursor and then H_2SO_4 solution. It is well known that the CeO_2-based catalysts provide outstanding reduction capability via the redox shift between Ce^{4+} and Ce^{3+}, which significantly improves the oxygen storage-release capacity[68] and therefore, are active in NO to NO_2 oxidation.[69] Thus, the CeO_2 based catalysts could be expected to be active in the low-temperature NO conversion by the fast NH_3-SCR reaction pathway (Eq. 6). In addition, sulfonation of the silicate surface should result in the formation of acidic sites for sorption and activation of ammonia, and possibly also NO molecules. In general, experimental results confirmed the expectations; however, it was also shown that the catalytic performance strongly depended on the method used for deposition of catalytically active components.[67] The best catalytic activity was achieved by Cat-A (obtained by simultaneous deposition of cerium and SO_4^{2-} into MCM-41). In this case, the NO conversion was above the level of 90%, with the selectivity for nitrogen, above 96%, was obtained in the range of 275–450 °C. For Cat—B, obtained by deposition of SO_4^{2-} and then cerium, the NO conversion above the level of 90% was observed in the range of 325–450 °C. In this case, the selectivity for nitrogen up to 450 °C, was above 96%. The lowest catalytic activity was recorded by Cat—C, obtained by deposition of cerium and then SO_4^{2-}. In this case, the NO conversion level of 90% was not achieved. It was of interest that the Cat-A and Cat-C samples were more catalytically active in NO to NO_2 oxidation than Cat-B was. Therefore, there is no simple correlation between activity in NO oxidation and the NH_3-SCR process. It was shown that the cerium introduced into Cat-B is present mainly in the form of cubic CeO_2 crystallites with the size larger in comparison to such species present in Cat-A and Cat—C. Long and Yang[61] postulated that the cubic CeO_2 phase is less effective in NO to NO_2 oxidation than dispersed cerium oxide species. Thus, it could be a reasonable explanation of the lower activity of Cat-B in NO to NO_2 oxidation, but still its higher activity in the NH_3-SCR process in comparison to that of Cat-C is a question. Probably, this effect could be explained by a very small concentration of deposited SO_4^{2-} in Cat-C comparing to Cat-A and Cat—B. It resulted in a lower concentration of acidic sites able to adsorb and activate ammonia and possibly also NO molecules in the NH_3-SCR reaction. Thus, it seems that the control of the type and size of deposited cerium species as well as surface acidity of the samples are very important for designing effective catalysts for the low-temperature

NH_3-SCR process. The most active catalyst of this series, Cat-A, yielded high stability in the reaction conducted in the flow of the reaction mixture containing water vapor; however, simultaneous introduction of both water vapor and SO_2 into the reaction mixture decreased the NO conversion by about 20%.

Moreno-Tost et al.[70] studied silicate-zirconium mesoporous silicates doped with cobalt as catalysts for the NH_3-SCR process. Mesoporous support was obtained by the co-condensation method, while cobalt was deposited by the impregnation method. The cobalt loading in the samples was in the range of 5–15 wt%. The catalysts obtained recorded high activity in the low-temperature NH_3-SCR process. For the most active catalyst of this series (with about 12 wt% of cobalt loading) the NO conversion reached nearly 90% with high selectivity for nitrogen at 200 °C. At higher temperatures, the reaction of direct ammonia oxidation significantly limited the efficiency of the NO conversion. Thus, the temperature window of the effective operation of this catalyst is very limited. The higher catalytic activity of dispersed cobalt species than metal oxide aggregates was postulated. Possibly highly dispersed cobalt species, mainly monomeric cobalt cations, due to relatively easy changes between Co^{3+} and Co^{2+} oxidation states may activate NO molecules for the NH_3-SCR reaction. It was postulated that Co^{2+} can transfer oxygen from the gas phase to the catalyst, thereby oxidation of Co^{2+} to Co^{3+}, where NO is adsorbed and oxidized to NO_2. As has already been mentioned NO_2 is necessary for the fast NH_3-SCR reaction (Eq. 6), which is one of the main low-temperature pathways of nitrogen oxides conversion. The continuation of these studies involved introduction of platinum to the catalysts by the impregnation method.[71] The molar ratios of Co/Pt in the samples were 10, 20 and 30. One purpose of platinum introduction to the catalysts was related to the potential improvement of Co^{3+} to Co^{2+} reduction. Moreover, cobalt ions should impede the reduction of Pt^{2+} to Pt^0. Modification of the catalysts with platinum species improved their activity in the low-temperature range. The maximum of the NO conversion was at about 200 °C, but similarly to the catalysts non-modified with platinum, above this temperature significant decrease in the efficiency of the NO conversion due to the reaction of direct ammonia oxidation, was observed. Moreover, deposition of platinum into catalysts decreased their selectivity to nitrogen. Interesting results were obtained for the catalytic tests conducted in the presence of water vapor and SO_2 introduced into the reaction mixture. Introduction of 10 vol% H_2O into the reaction mixture had consequences for the temperatures of the optimal NO conversion.

For the catalysts with high platinum content (Co/Pt molar ratio of 10), the maximum of the NO conversion was located at about 150 °C. A decrease in platinum loading resulted in the shift of this maximum to about 200 °C for the catalysts with the Co/Pt ratio of 20 and 300 °C for the sample with the Co/Pt ratio of 30. In the case of the catalyst without platinum, the maximum of the NO conversion was located at about 400 °C. This interesting effect was explained by the competitive adsorption of water and reactants of the NH_3-SCR process for these same adsorption sites. It seems that in the case of the catalysts without platinum, and with low platinum content, water molecules are adsorbed on the catalytically active sites. Desorption of water molecules from such sites probably makes these sites available for ammonia and possibly also NO. Thus, the NO conversion was shifted to higher temperatures. Another possible conclusion is related to the location of such active sites, which are probably on cobalt species. Introduction of platinum protected the catalysts against the adverse effects of water. The possible mechanism of such protection could be related to preferential water adsorption on platinum rather than on cobalt species. This hypothesis could be tested and possibly verified by additional studies. Deactivation of the catalysts by water was fully reversible, and after the removal of water from the reaction mixture the original levels of the NO conversion and selectivity for nitrogen were restored. The introduction of 100 ppm SO_2 into the reaction mixture decreased the efficiency of the NO conversion. However, in the case of the catalysts modified with platinum the maximum of the NO conversion the temperature was about 200 °C. For the catalyst not modified with platinum, the maximum temperature of the NO conversion was shifted to about 350 °C. Thus, platinum introduced into catalysts improved their resistance to poisoning by SO_2. Catalysts based on silicate-zirconium mesoporous materials modified with cobalt[71] or both cobalt and platinum[71] are interesting catalysts for the low-temperature NH_3-SCR process. However, a very serious disadvantage of such catalysts is their very narrow temperature window of effective operation, as well as sensitivity for the presence of water vapor and SO_2 in the reaction mixture. In the next stage, mesoporous silicate-zirconium supports were modified with cobalt and iridium.[72] The samples with the cobalt loading of 16 wt% and Co/Ir molar ratios of 10, 20, 30 and 60 were prepared by the impregnation method. Deposition of iridium into cobalt modified mesoporous supports increased their dispersion and improved redox properties. Cobalt was present in the samples in the form of isolated cobalt cations and Co_3O_4 oxide aggregates. Results of the catalytic studies clearly showed that modification of cobalt-containing catalysts with iridium significantly improved their catalytic performance in the NH_3-SCR process. The magnitude of the NO conversion obtained for the catalysts containing both cobalt and iridium was higher compared with that for the catalysts containing only cobalt or only iridium. Moreover, the temperature window of effective operation of such binary catalysts was significantly extended. For the most active catalysts, with the Co/Ir molar ratio of 30, the NO conversion above the level of 90% and selectivity above 95% was obtained in the range of 150–200 °C. The key role was assigned to the presence of monomeric cobalt cations, in the case of catalysis with cobalt and iridium present, which was also proposed in catalysis involving mesoporous silicate-zirconium doped with cobalt and platinum.[71,73] Oxygen from the gas phase possibly oxidizes Co^{2+} to Co^{3+}, where NO is adsorbed and oxidized to NO_2. In the next step, NO_2 and NO could be reduced to N_2 by ammonia according to the fast NH_3-SCR reaction (Eq. 6), which is one of the main pathways of the low-temperature conversion of nitrogen oxides. The catalytic activity of cobalt species in NO to NO_2 oxidation was reported and considered as a crucial step of the low-temperature NH_3-SCR, also by other authors.[72] The binary Co—Ir catalysts possess a very good tolerance for water vapor.[73] The addition of water vapor to the reaction mixture did not produce significant changes in their catalytic performance. However, the presence of 100 ppm of SO_2 in the feed resulted in a significant decrease of the catalytic activity probably due to an increase of the surface concentration of Co^{2+} originating from the reducing properties of SO_2.

3. Summary and perspectives

The development of catalysts for the low-temperature NH_3-SCR process is still one of the most demanding challenges necessary for retrofitting the existing process. A suitable design of a new system for more effective and less expensive purification of flue gases emitted by stationary sources is a critical requirement. Mesoporous silicates, due to their high surface area, porosity and uniform porous structure are excellent supports for such low-temperature NH_3-SCR catalysts. Deposition of uniform catalytically active species on such a large surface area of mesoporous silicate materials should result in active and selective catalysts. Among various metals deposited into mesoporous silicates, copper and manganese seem to be the most promising components of the low-temperature NH_3-SCR catalysts. Iron, which is also widely studied in this reaction, effectively operates in the middle- and high-temperature NH_3-SCR process. Various methods, including impregnation, co-condensation, grafting, ion-exchange with grafted organic species, template ion-exchange, were used for deposition of metal species into mesoporous silicates. However, the main goal of future studies should be development of a method resulting in high loading of the uniform metal species. This is possibly a key to make these materials more attractive for broader applications in catalysis, not only in laboratories but possibly in the future on an industrial scale. An additional important issue is the generation of acidic sites and increasing the efficiency of NO to NO_2 oxidation, as the reaction associated with the NH_3-SCR. Various examples of the generation of such properties were presented; however, studies in this direction are still necessary to design properly a suitable contribution of acidity and oxidation properties for the formulation of the optimal catalyst for the low-temperature NH_3-SCR process. Another important problem of the NH_3-SCR catalyst system operating in the low-temperature range, which is reported to a limited extent in the relevant literature, is the resistance of the catalysts for deposition under reaction conditions to solid by- products, such as NH_4NO_3, $(NH_4)_2SO_4$ or NH_4HSO_4, on the surface of the catalyst; these may limit the availability of the active sites for the reactants. This problem should be included in future studies of the catalysts for the low-temperature NH_3-SCR process.

Acknowledgments

Part of the studies presented in the paper was carried out in the framework of project 2018/31/B/ST5/00143 from the National Science Centre (Poland). AJ has been partly supported by the EU Project POWR.03.02.00-00-I004/16.

References

1. Beck, J. S.; Vartuli, J. C.; Roth, W. J.; Leonowicz, M. E.; Kresge, C. T.; Schmitt, K. D.; Chu, C. T. W.; Olson, D. H.; Sheppard, E. W.; McCullen, S. B.; Higgins, J. B.; Schlenker, J. L. *J. Am. Chem. Soc.* **1992**, *114*, 10834–10843.
2. Kresge, C. T.; Vartuli, J. C.; Roth, W. J.; Leonowicz, M. E. *Stud. Surf. Sci. Catal.* **2004**, *148*, 53–72.
3. Kowalczyk, A.; Borcuch, A.; Michalik, M.; Rutkowska, M.; Gil, B.; Sojka, Z.; Indyka, P.; Chmielarz, L. *Micropor. Mesopor. Mater.* **2017**, *240*, 9–21.
4. Luan, Z.; Hartmann, M.; Zhao, D.; Zhou, W.; Kevan, L. *Chem. Mater.* **1999**, *11*, 1621–1627.
5. Chmielarz, L.; Rutkowska, M.; Kowalczyk, A. *Adv. Inorg. Chem.* **2018**, *72*, 323–383.
6. Chmielarz, L.; Kuśtrowski, P.; Drozdek, M.; Rutkowska, M.; Dziembaj, R.; Michalik, M.; Cool, P.; Vansant, E. F. *J. Porous Mater.* **2011**, *18*, 483–491.
7. Chaudhary, V.; Sharma, S. *J. Porous Mater.* **2017**, *24*, 741–749.
8. Cavuoto, D.; Zaccheria, F.; Ravasio, N. *Catalysts* **2020**, *10*, 1337.
9. Cohn, J.G.E.; Steele, D.R.; Andersen, H.C. US patent US2975025A, 1961.

10. Surhone, L. M.; Timpledon, M. T.; Marseken, S. F. *Selective Catalytic Reduction*; Betascript Publishing, 2010.
11. Mladenović, M.; Paprika, M.; Marinković, A. *Renew. Sustain. Energy Rev.* **2018**, *82*, 3350–3364.
12. Święs, A.; Kowalczyk, A.; Michalik, M.; Díaz, U.; Palomares, A. E.; Chmielarz, L. *RSC Adv.* **2021**, *11*, 10847–10859.
13. Święs, A.; Kowalczyk, A.; Rutkowska, M.; Díaz, U.; Palomares, A. E.; Chmielarz, L. *Catalysts* **2020**, *10*, 734.
14. Zhang, S.; Zhang, B.; Liu, B.; Sun, S. *RSC Adv.* **2017**, 7, 26226–26242.
15. Rutkowska, M.; Borcuch, A.; Marzec, A.; Kowalczyk, A.; Samojeden, B.; Moreno, M. J.; Díaz, U.; Chmielarz, L. *Micropor. Mesopor. Mater.* **2020**, *304*, 109114.
16. Grzybek, J.; Gil, B.; Roth, W. J.; Skoczek, M.; Kowalczyk, A.; Chmielarz, L. *Spectrochim. Acta A Mol. Biomol. Spectrosc.* **2018**, *196*, 281–288.
17. Pang, C.; Zhuo, Y.; Weng, Q.; Zhu, Z. *RSC Adv.* **2018**, *8*, 6110–6119.
18. Chen, J.; Peng, G.; Zheng, W.; Zhang, W.; Guo, L.; Wu, X. *Catal. Sci. Technol.* **2020**, *10*, 6583–6598.
19. Jodłowski, P. J.; Kuterasiński, Ł.; Jędrzejczyk, R. J.; Chlebda, D.; Gancarczyk, A.; Basąg, S.; Chmielarz, L. *Catalysts* **2017**, 7, 205.
20. Basąg, S.; Kocoł, K.; Piwowarska, Z.; Rutkowska, M.; Baran, M.; Chmielarz, L. *Reac. Kinet. Mech. Catal.* **2017**, *121*, 225–240.
21. Qiu, J.; Zhuang, K.; Lu, M.; Xu, B.; Fan, Y. *Catal. Commun.* **2013**, *31*, 21–24.
22. Gomes, H. T.; Selvam, P.; Dapurkar, S. E.; Figueiredo, J. L.; Faria, J. L. *Micropor. Mesopor. Mater.* **2005**, *86*, 287–294.
23. Huang, Y. J.; Wang, H. P.; Lee, J.-F. *Chemosphere* **2003**, *50*, 1035–1041.
24. Jankowska, A.; Chłopek, A.; Kowalczyk, A.; Rutkowska, M.; Michalik, M.; Liu, S.; Chmielarz, L. *Molecules* **2020**, *25*, 5651.
25. Kowalczyk, A.; Święs, A.; Gil, B.; Rutkowska, M.; Piwowarska, Z.; Borcuch, A.; Michalik, M.; Chmielarz, L. *Appl. Catal. B: Environ.* **2018**, *237*, 927–937.
26. Jankowska, A.; Chłopek, A.; Kowalczyk, A.; Rutkowska, M.; Mozgawa, W.; Michalik, M.; Liu, S.; Chmielarz, L. *Micropor. Mesopor. Mater.* **2021**, *315*, 110920.
27. Jankowska, A.; Kowalczyk, A.; Rutkowska, M.; Michalik, M.; Chmielarz, L. *Molecules* **1807**, *2021*, 26.
28. Liu, C.-C.; Teng, H. *Appl. Catal. B: Environ.* **2005**, *58*, 69–77.
29. Liu, K.; Yan, Z.; Shan, W.; Shan, Y.; Shiad, X.; He, H. *Catal. Sci. Technol.* **2020**, *10*, 1135–1150.
30. Millan, R.; Cnudde, P.; Hoffman, A. E. J.; Lopes, C. W.; Concepción, P.; van Speybroeck, V.; Boronat, M. *J. Phys. Chem. Lett.* **2020**, *11*, 10060–10066.
31. Ziółek, M.; Sobczak, I.; Nowak, I.; Decyk, P.; Lewandowska, A.; Kujawa, J. *Micropor. Mesopor. Mater.* **2000**, *35-36*, 195–207.
32. Imyen, T.; Yigit, N.; Dittanet, P.; Barrabes, N.; Föttinger, K.; Rupprechter, G.; Kongkachuichay, P. *Ind. Eng. Chem. Res.* **2016**, *55*, 13050–13061.
33. Imyen, Y.; Yigit, N.; Poo-Arporn, Y.; Föttinger, K.; Rupprechter, G.; Kongkachuichay, P. *J. Nanosci. Nanotechnol.* **2019**, *19*, 743–757.
34. Yuan, E.; Han, W.; Zhang, G.; Zhao, K.; Mo, Z.; Lu, G.; Tang, Z. *Catal. Surv. Asia* **2016**, *20*, 41–52.
35. Intana, T.; Föttinger, K.; Rupprechter, G.; Kongkachuichay, P. *Colloids Surf., A* **2015**, *467*, 157–165.
36. Chamnankid, B.; Samanpratan, R.; Kongkachuichay, P. *J. Nanosci. Nanotechnol.* **2012**, *12*, 9325–9332.
37. Jankowska, A.; Kowalczyk, A.; Rutkowska, M.; Mozgawa, W.; Gil, B.; Chmielarz, L. *Catal. Sci. Technol* **2020**, *10*, 7940–7954.
38. Boroń, P.; Rutkowska, M.; Gil, B.; Marszałek, B.; Chmielarz, L.; Dźwigaj, S. *ChemSusChem* **2019**, *12*, 692–705.

39. Kowalczyk, A.; Piwowarska, Z.; Macina, D.; Kuśtrowski, P.; Rokicińska, A.; Michalik, M.; Chmielarz, L. *Chem. Eng. J.* **2016**, *295*, 167–180.
40. Macina, D.; Piwowarska, Z.; Góra-Marek, K.; Tarach, K.; Rutkowska, M.; Girman, V.; Błachowski, A.; Chmielarz, L. *Mater. Res. Bull.* **2016**, *78*, 72–82.
41. Li, S.; Wu, Q.; Lu, G.; Zhang, C.; Liu, X.; Cui, C.; Yan, Z. *J. Mater. Eng. Perform.* **2013**, *22*, 3762–3768.
42. Chmielarz, L.; Kuśtrowski, P.; Dziembaj, R.; Cool, P.; Vansant, E. F. *Micropor. Mesopor. Mater.* **2010**, *127*, 133–141.
43. Yan, H.; Qu, H.; Bai, H.; Zhong, Q. *J. Mol. Catal A: Chem.* **2015**, *403*, 1–9.
44. Chen, Q.; Guo, R.; Wang, Q.; Pan, W.; Wang, W.; Yang, N.; Lu, C.; Wang, S. *Fuel* **2016**, *181*, 852–858.
45. Guo, K.; Jiawei, J.; Osuga, R.; Zhu, Y.; Sun, J.; Tang, C.; Kondo, J. N.; Dong, L. *Appl. Catal. B: Environ.* **2021**, *287*, 119982.
46. Zhang, H.; Tang, C.; Sun, C.; Qi, L.; Gao, F.; Dong, L.; Chen, Y. *Micropor. Mesopor. Mater.* **2012**, *151*, 44–55.
47. Zhang, N.; Li, L.; Guo, Y.; He, J.; Wu, R.; Song, L.; Zhang, G.; Zhao, J.; Wang, D.; He, H. *Appl. Catal. B: Environ.* **2020**, *270*, 118860.
48. Fang, X.; Liu, Y.; Cheng, Y.; Cen, W. *ACS Catal.* **2021**, *11*, 4125–4135.
49. Chen, M.; Sun, Q.; Yang, X.; Yu, J. *Inorg. Chem. Comm.* **2019**, *105*, 203–207.
50. Huang, J.; Huang, H.; Jiang, H.; Liu, L. *Catal. Today* **2019**, *332*, 49–58.
51. Yang, G.; Zhao, H.; Luo, X.; Shi, K.; Zhao, H.; Wang, W.; Chen, Q.; Fan, H.; Wu, T. *Appl. Catal. B: Environ.* **2019**, *245*, 743–752.
52. Jiang, L.; Liu, Q.; Ran, R.; Kong, M.; Ren, S.; Yang, J.; Li, J. *Chem. Eng. J.* **2019**, *370*, 810–821.
53. Chmielarz, L.; Dziembaj, R.; Grzybek, T.; Klinik, J.; Łojewski, T.; Olszewska, D.; Papp, H. *Catal. Lett.* **2000**, *68*, 95–100.
54. Chmielarz, L.; Dziembaj, R.; Grzybek, T.; Klinik, J.; Łojewski, T.; Olszewska, D.; Węgrzyn, A. *Catal. Lett.* **2000**, *70*, 51–56.
55. Lu, X.; Song, C.; Jia, S.; Tong, Z.; Tang, X.; Teng, Y. *Chem. Eng. J.* **2015**, *260*, 776–784.
56. Xiang, J.; Wang, L.; Cao, F.; Qian, K.; Su, S.; Hu, S.; Wang, Y.; Liu, L. *Chem. Eng. J.* **2016**, *302*, 570–576.
57. Li, J.; Guo, J.; Shi, X.; Wen, X.; Chu, Y.; Yuan, S. *Appl. Surf. Sci.* **2020**, *534*, 147592.
58. Busca, B.; Lietti, L.; Ramis, G.; Berti, F. *Appl. Catal. B: Environ.* **1998**, *18*, 1–36.
59. Liang, X.; Li, J.; Lin, Q.; Sun, K. *Catal. Commun.* **2007**, *8*, 1901–1904.
60. Ma, M.; Ma, X.; Cui, S.; Liu, T.; Tian, J.; Wang, Y. *Materials* **2019**, *12*, 3654.
61. Long, R. Q.; Yang, R. T. *Appl. Catal. B: Environ.* **2000**, *27*, 87–95.
62. Huang, J.; Tong, Z.; Huang, Y.; Zhang, J. *Appl. Catal. B: Environ.* **2008**, *78*, 309–314.
63. Gao, J.; Han, Y.; Mu, J.; Wu, S.; Tan, F.; Shi, Y.; Li, X. *J. Colloid Interface Sci.* **2018**, *516*, 254–262.
64. Kwon, W. H.; Park, S. H.; Kim, J. M.; Park, S. B.; Jung, S.-C.; Ki, S. C.; Jeon, J.-K.; Park, Y.-K. *J. Nanosci. Nanotechnol.* **2016**, *16*, 1744–1747.
65. Segura, Y.; Chmielarz, L.; Kuśtrowski, P.; Cool, P.; Dziembaj, R.; Vansant, E. F. *Appl. Catal. B: Environ.* **2005**, *61*, 69–78.
66. Vomiero, A.; Mea, G. D.; Ferroni, M.; Martinelli, G.; Roncarati, G.; Guidi, V.; Comini, E.; Sberveglieri, G. *Mater. Sci. Eng., B* **2003**, *101*, 216–221.
67. Song, Z.; Wu, X.; Zhang, Q.; Ning, P.; Fan, J.; Liu, X.; Liu, Q.; Huang, Z. *RSC Adv.* **2016**, *6*, 69431–69441.
68. Gao, X.; Jiang, Y.; Zhong, Y.; Luo, Z. Y.; Cen, K. F. *J. Hazard. Mater.* **2010**, *174*, 734–739.
69. Lin, F.; He, Y.; Wang, Z.; Ma, O.; Whiddon, R.; Zhu, Y.; Liu, J. *RSC Adv.* **2016**, *6*, 31422–31430.

70. Moreno-Tost, R.; Santamaria-González, J.; Maireles-Torres, P.; Rodriguez-Castellón, E.; Jiménez-López, A. *Appl. Catal. B: Environ.* **2002**, *38*, 51–60.
71. Moreno-Tost, R.; Santamaria-González, J.; Rodriguez-Castellón, E.; Jiménez-López, A. *Appl. Catal. B: Environ.* **2004**, *52*, 241–249.
72. Stakheev, A. Y.; Lee, C. W.; Park, S. J.; Chong, P. J. *Catal. Lett.* **1996**, *38*, 271–278.
73. Moreno-Tost, R.; Rodriguez Castellon, E.; Jimenez-Lopez, A. *J. Mol. Catal. A: Chem.* **2006**, *248*, 126–134.

CHAPTER SEVEN

Neutral and charged group 13–16 homologs of carbones EL_2 ($E = B^-–In^-$; Si–Pb; $N^+–Bi^+$, $O^{2+}–Te^{2+}$)

Wolfgang Petz[a,*] **and Gernot Frenking**[a,b,*]
[a]Fachbereich Chemie, Philipps-Universität Marburg, Marburg, Germany
[b]Institute of Advanced Synthesis, School of Chemistry and Molecular Engineering, Nanjing Tech University, Nanjing, China
*Corresponding author: e-mail addresses: petz@staff.uni-marburg.de; frenking@chemie.uni-marburg.de

Contents

Advances in Inorganic Chemistry, Volume 79
ISSN 0898-8838
https://doi.org/10.1016/bs.adioch.2021.12.008

Abstract

The review covers experimental and theoretical work on neutral and charged compounds EL_2 ($E=B^-$–In^-; Si–Pb; N^+–Bi^+, O^{2+}–Te^{2+}) where the ligands L bind to an atom or ion with four valence electrons of the groups 13–16. The molecules are homologs of carbones CL_2, which have been identified as divalent carbon(0) compounds in 2006. Whereas bonding in classical carbones was identified in terms of dative interactions $L \rightarrow C \leftarrow L$, the homologs exhibit bonds that can be described with single and double electrons or with a mixture of the bonding categories. Much progress has been made in the recent past in this still rapidly developing field. The analysis of the bonding nature and the isolation of new examples of low-valent main-group compounds and their application in catalysis contributed to the fast expansion of knowledge in an exciting area of inorganic chemistry. The chemistry of the lower-valent main groups is a prolific field for theoretical and experimental chemists, constantly yielding surprising discoveries, and it remains a challenge for the inventive chemist.

1. Introduction

Carbones CL_2 are compounds where a carbon(0) atom is bonded to two σ-donor ligands L through dative interactions and in which the four valence electrons of carbon are retained as two lone electron pairs with σ and π symmetry. The carbon-ligand bonds may be described with the Dewar-Chatt-Duncanson (DCD) model [1] in terms of σ donation $L \rightarrow C \leftarrow L$ and π backdonation $L \leftarrow C \rightarrow L$ just like transition-metal complexes.[2] Carbones are often represented only by the former formula showing the σ-donor bonds, but the contribution of π-backdonation can become very strong if L is a good π-acceptor such as CO in carbon suboxide $C(CO)_2$.

Carbones have been introduced in chemistry in 2006,[3] when the electronic structure of hexaphenylcarbodiphosphorane $C(PPh_3)_2$, which was synthesized by Ramirez in 1961,[4] was analyzed with quantum chemical methods, wherein it was suggested to represent the bonding situation with dative bonds $(Ph_3P) \rightarrow C \leftarrow (PPh_3)$, which may alternatively be described as diylid with the classical Lewis-structure $(Ph_3P)^+$–$C^{2-}(PPh_3)^+$. It was said that the carbon–phosphorous bonds in $C(PPh_3)_2$ "come from $P \rightarrow C$ donor–acceptor interactions rather than electron-sharing $C-R$ bonds in carbenes CR_2. The chemistry of the former compounds exhibits novel features which distinguish them from carbenes."[3] Dative bonding in di-coordinated carbon compounds had already earlier been suggested by

Varshavskii in 1980,[5] but the subject received little attention. The possible involvement of two lone pairs at carbon to explain the reactivity of $C(PPh_3)_2$ was also suggested by Kaska in 1978,[6] but the prevailing model for the bonding situation exhibited electron-sharing single or double bonds.

At this point, it is useful to clarify the meaning of a bonding model to avoid misunderstandings. A bond model is an attempt to represent the complicated electronic structure of a molecule in a simple scheme that serves as an ordering system for understanding the structure and reactivity of a molecule. A bonding model is not right or wrong, it is more or less useful. This was discerningly expressed by Michael Dewar who wrote in 1984: "The only criterion of a model is usefulness, not its 'truth'."[7] Sometimes more than one description may seem appropriate for a molecule. For example, the EDA (Energy Decomposition Analysis) calculation of a carbene stabilized Si_2H_2 showed that the compound could equally be well described with dative bonds as with classical double bonds.[8] The bent equilibrium geometry of carbon suboxide $C(CO)_2$ can easily be explained with dative bonds instead of electron-sharing bonds that should lead to a linear form,[9] but a recent re-evaluation using valence-bond calculations suggest that the actual bonding, situation is best described by a resonance between several mesomeric structures.[10] This does not reject the dative bonding model but indicates that a single picture is sometimes not sufficient.

The suggestion of dative bonds using arrows for carbones CL_2 was not undisputed and it received some criticism, which initiated a healthy dispute[11,12] and led to further theoretical studies about the scope and limitation of the model.[13] A breakthrough came with the prediction that carbodicarbene $C(NHC)_2$ (NHC = N-heterocyclic carbene) are stable compounds with dative bonds $(NHC) \rightarrow C \leftarrow (NHC)$ and a comparable angular structure to hexaphenylcarbodiphosphorane $C(PPh_3)_2$.[14] The previously unknown compounds were synthesized shortly thereafter in the groups of Bertrand[15] and Fürstner.[16] Study of Carbodicarbenes[17] has become a very active field of experimental research due to the catalytic properties of the compounds.[18]

Shortly after the bonding model with dative bonds $L \rightarrow C \leftarrow L$ was proposed for carbones, it was suggested that there might be stable low-valent heavy group-14 compounds of atoms E (E = Si–Pb) termed tetrylones, which exhibit the same type of dative bonds $L \rightarrow E \leftarrow L$.[19] A bonding analysis of compounds that were isolated and introduced as the first examples of trisilaallenes $R_2Si=Si=SiR_2$ and trigermanallenes $R_2Si=Si=SiR_2$[20] suggested that the molecules should be considered as silylones $L\rightarrow Si\leftarrow L$ and germylones $L\rightarrow Ge\leftarrow L$, which easily explains the markedly bent equilibrium geometries.[19] The dative bonding model was then successfully employed for the isolation of new silylones[21] and germylones[22] and the first isolation of a stannylone $L\rightarrow Sn\leftarrow L$.[23] A review about the experimental work in the field of tetrylones was published in 2017.[24]

Further theoretical and experimental studies showed that complexes with dative bonds may also be found for low-valent atoms of the groups 13 and 15.[25] A spectacular result was the synthesis and structural characterization of the boron complex (cAAC)$\rightarrow$(BH)$\leftarrow$(cAAC) (cAAC = cyclic alkylaminocarbene) by Bertrand et al.[26] where the particular bonding properties of the cAAC ligand[27] are used. The reactivity of the complex showed that the boron atom in (BH)$(cAAC)_2$ reacts like a Lewis base due to its π lone pair that is partly delocalized rather than a Lewis acid. Another surprising result was the isolation of the dicarbonyl complex OC$\rightarrow$(BR)$\leftarrow$CO where R is a bulky terphenyl group by Braunschweig and co-workers.[28] The anionic complex $OC\rightarrow B^{-}\leftarrow CO$ was observed in a low-temperature matrix study by Zhou et al.[29]

Fig. 1 shows schematically the most common formulas for ylidone complexes; ylidone is the generic name for divalent main-group complexes EL_2. The graphic representation **I** shows the lone electron-pairs at atom E, which may engage in π backdonation $L\leftarrow E\rightarrow L$ whose contribution depends on the π acceptor strength of L and π donor strength of E. It has been shown that the bending angle is mainly determined by the π backdonation. The angle becomes more acute (wider) when the contribution of π backdonation is smaller (bigger).[12b] A linear geometry may be found in molecules with very strong π backdonation. The arrows in **I** identify the σ donor orbitals of L, which are sometimes explicitly shown as straight lines (formula **II**). Formula **III** shows EL_2 as classical diylids, which

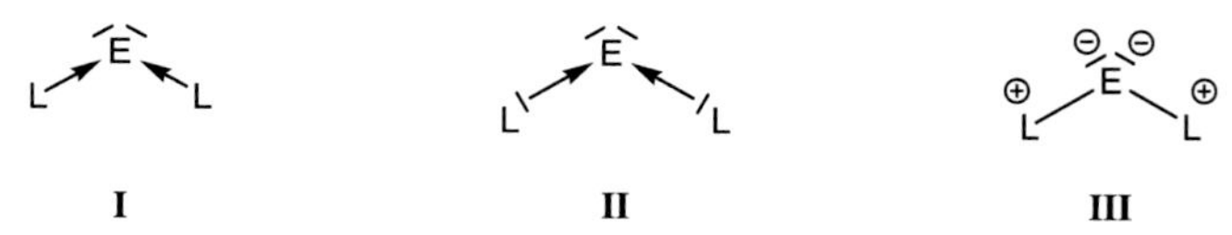

Fig. 1 Schematic representation of the most common graphical representations of ylidones EL_2.

Table 1 Overview of isoelectronic ylidone compounds EL_2 which are covered in this review.

Examples	Formula	Name	E NMR	Range (ppm)
Yes	$[BL_2]^-$	Borylone	^{11}B NMR	
Yes	$[AlL_2]^-$	Aluminone		
No	$[GaL_2]^-$	Gallylone		
No	$[InL_2]^-$	Indylone		
Yes	CL_2	Carbone	^{13}C NMR	
Yes	SiL_2	Silylone	^{29}Si NMR	
Yes	GeL_2	Germylone		
Yes	SnL_2	Stannylone	^{119}Sn NMR	
Yes	PbL_2	Plumbylone		
Yes	$[NL_2]^+$	Nitreone	^{15}N NMR	
Yes	$[PL_2]^+$	Phosphorone	^{31}P NMR	−150 to −260
Yes	$[AsL_2]^+$	Arseone		
Yes	$[SbL_2]^+$	Stibione		
Yes	$[BiL_2]^+$	Bismutone		
Yes	$[OL_2]^{2+}$	Oxygeone		
Yes	$[SL_2]^{2+}$	Sulfurone		
Yes	$[SeL_2]^{2+}$	Selenone	^{77}Se NMR	
Yes	$[TeL_2]^{2+}$	Telurone		

conforms with the Lewis model but the character of dative bonding is not obvious. The use of arrows for dative bonds was introduced by Sidgwick, who used it also for the description of chemical bonds in divalent carbon compounds.[30]

Theoretical and experimental studies of group 13 and group 14 complexes EL_2 have been reviewed some years ago.[24,31] Many new results were obtained in the interim in this rapidly growing field. The model of dative bonding was extended to group-15 cations $L \rightarrow E^+ \leftarrow L$ (E = N–Bi)[32] and group-16 dications $L \rightarrow E^{2+} \leftarrow L$ (E = O–Te) for which numerous examples are now experimentally known. Table 1 summarizes the type of compounds, which are covered in this review. It should be helpful as a guideline for future studies in the field, which is a very active area of theoretical and experimental research in main-group chemistry.

2. Group 13 compounds $[EL_2]^-$, E = B, Al

2.1 Borylone $[BL_2]^-$

2.1.1 Borylone with a C-B-C core

Boron carbonyl species were prepared by co-condensation of the reactive species generated by laser ablation of bulk boron target with CO/Ne mixtures onto a CsI window at 4 K. The anion $[B(CO)_2]^-$ (**B-1**) was identified by IR spectroscopy; one CO stretching and one B-C stretching vibration implies a linear C-B-C arrangement. *Ab initio* calculations at the CCSD(T)/aug-cc-pVTZ level of theory show that the $[B(CO)_2]^-$ anion exhibits a linear equilibrium geometry close to that of $C(CO)_2$ as depicted in Fig. 2.[29]

2.2 Aluminone $[AlL_2]^-$

2.2.1 Aluminone with a C-Al-C core

Laser ablated Al atoms were codeposited with CO and irradiated with a 400 nm Hg lamp. **Al-1** was studied by IR spectroscopy. The Al—C and C—O bond lengths were predicted to be 1.955 and 1.175 Å respectively; the C-Al-C angle was calculated to be 72°;[33] see also calculated values to **Al-1** in Ref [29]; the anion **Al-1** is shown in Fig. 3.

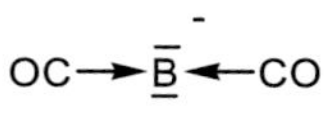

OC→B←CO (−)

B-1

Fig. 2 Experimental and calculated linear arrangement of $[B(CO)_2]^-$.

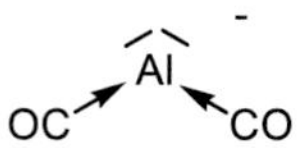

OC→Al←CO (−)

Al-1

Fig. 3 Graphical representation of the anion $[Al(CO)_2]^-$.

3. Group 14 compounds EL_2, E = C, Si, Ge, Sn, Pb

A theoretical study on $E(PPh_3)_2$ compounds (with E = C, Si, Ge, Sn, Pb) was performed by the group of Frenking.[34] First and second PAs were calculated at 279.4, 186.0 (Si), 276.0, 174.8 (Ge) 272.8, 164.0 (Sn), and 270.7, 147.1 (Pb) kcal/mol, showing a continuous decline of the second PA down this series; see also quantum chemical calculations using *DFT* and *ab initio* methods of EL_2 compounds with various ligands L[35] and studies of EL_2 compounds (E = C to Sn; L = cAAC and NHC) at the BP86/TZ2P level of theory.[36] A minireview on tetrylones (E = Si, Ge, Sn, Pb) is outlined in Ref. [37].

3.1 Carbones CL_2

Carbones, CL_2 compounds, will not be treated here. To this thematic, several review articles were reported previously. A general overview on species that bear two lone pairs of electrons at the same C-center is summarized in Ref. [38] transition metal adducts of carbones are described in Ref. [39], and those of main group fragments in Ref. [40]. Two contributions,[41,42], in the series Structure and Bonding (Springer Edition) also deal with carbone transition metal addition compounds.

3.2 Silylones SiL_2

The higher homologs of carbodicarbenes, CDPs, with a C-Si-C core, **Si-1** to **Si-4**, termed as silylone with a Si-Si-Si core, **Si-5** to **Si-11**, and related compounds with a Ge-Si-Ge core, **Si-12** and **Si-13** and those with Ge-Si-P (**Si-14**) and Si-Si-C core (**Si-15**, **Si-16**) are known to date. For asymmetric silylones **Si-14** to **Si-16** bifocal perspectives emerge. Thus, a Si=Si or Si=Ge double bond with one lone pair of electrons at the central Si atom or alternatively, a Si(0) atom coordinated by a silylene (germylene) and a NHC (phosphine) ligand with two lone pairs of electrons are under consideration.

3.2.1 SiL_2 compounds with a C-Si-C core

The synthesis of the silylone **Si-1** starts with the cyclic bis-NHC ligand and NHC-$SiCl_2$ to give the precursor [**Si—1**Cl]Cl; dehalogenation with $NaC_{10}H_8$ in THF at low temperature generates **Si-1** as a red powder. A high field signal at −83.8 ppm was recorded in the ^{29}Si NMR spectrum. The X-ray analysis revealed Si—C distances of 1.864(1) and 1.874(1) Å and a

C-Si-C angle of 89.1(1)°. Calculation on a model compound (Dipp replaced by H) gave first and second PAs of 274 and 164 kcal/mol, respectively; *DFT* calculations are presented.[43] With $GaCl_3$ the monoaddition compound Cl_3Ga-**Si-1**[44] and with two equiv. of $ZnCl_2$ the bis adduct $(Cl_2Zn)_2$-**Si-1** are formed.[24] The action of $GeCl_2$(diox) and $SnCl_2$(NHC) generate the cationic [Cl-**Si-1**]Cl.[24] With chalkogenes an oxidation reaction occurs. Thus, S_8 gives S_2-**Si-1**,[44] Stepwise addition of 4 equiv. of CO_2 produced the addition compounds O-**Si-1**, O_2-**Si-1**, and $(CO_3)_2$-**Si-1**; the latter forms upon addition of 2 molecules of CO_2 to the Si=O bond.[45]

Si(0) stabilized by cAAC (a better π-acceptor than NHC) was prepared from the dichloride Cl_2-**Si-2** and C_8K as reducing agent; blue-black rods of **Si-2** were obtained. In the ^{29}Si NMR spectrum a singlet at $\delta = 66.7$ ppm was observed. The X-ray analysis revealed Si—C distances of 1.8411(18) and 1.8417(17) Å; the C-Si-C angle is 117.70(8)°, Theoretical calculations show the shapes of the HOMO and HOMO-1 as π-type and σ-type lone pair orbitals at Si, respectively.[21]

With a cAAC and a NHC ligand **Si-3** was obtained from dehalogenation of the iodide [I-**Si-3**]I with C_8K in toluene as bright red crystals. In the ^{29}Si NMR a signal at $\delta = 2.04$ ppm was recorded. The X-ray analysis revealed Si—C distances of 1.792(4) (to cAAC) and 1.957(5) Å (to carbene) and a C-Si-C angle of 102.8(2)°. The Si—C distances reflect the different donor-acceptor properties of the cAAC and the carbene ligand.[46]

In a similar way to the preparation of **Si-2** the silylone **Si-4** was obtained from the related dichloride Cl_2-**Si-4** and C_8K. The ^{29}Si NMR spectrum shows a signal at $\delta = 71.15$ ppm; selected X-ray parameters are Si—C bond lengths of 1.8407(13) and 1.8531(14) Å and a C-Si-C angle of 118.16(6)°.[47] **Si-4** was investigated by cyclic voltammetry (CV) showing a quasi reversible reduction at $E_{1/2} = -1.55$ V indicating the formation of a radical anion.[48] Silylones with C-Si-C core are shown in Fig. 4.

3.2.2 SiL_2 compounds with a Si-Si-Si core

Several silylones with this core are reported and shown in Fig. 5. **Si-5** was first reported by Kira, prepared from the cyclic silylene [Si(II)] *via* its $SiCl_4$ addition compound [Si(IV)] followed by reduction with C_8K. The ^{29}Si NMR resonances found at 157 and 197 ppm in benzene-d_6 are assigned to the central Si atom and the adjacent Si nuclei, respectively. The X-ray single crystal analysis showed that the central Si atom is found at four positions and the population is temperature dependent; the energy

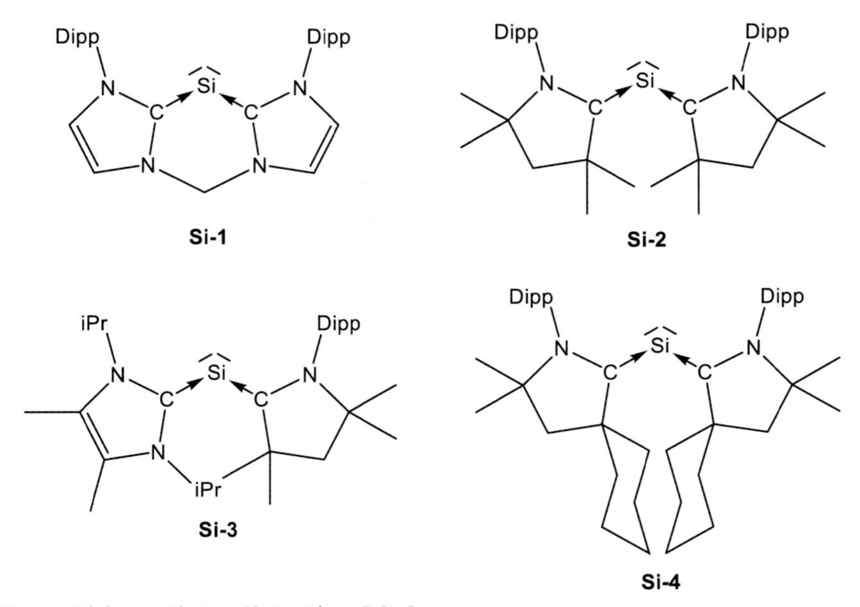

Fig. 4 Silylones **Si-1** to **Si-4** with a C-Si-C core.

difference between the four positions is within 1 kcal/mol, suggesting a dynamic disorder. The two Si—Si bond length are at 2.177(1) and 2.188 (1) Å and the Si-Si-Si angle amounts to 136.5(1)°.[49] Reactions with a series of alcohols and water end up with protonation at the central Si atom and addition of OR at the neighboring one;[50] see also a comparative discussion of $R_2E{=}E'{=}ER_2$ compounds (E=Si, Ge), using *DFT* calculations and PMO theory.[51,52]

The recently synthesized silylone **Si-6** is isoelectronic to the nitreone **N-20** and was synthesized from 1,2-$(LSi)_2$-nido-1,2-$C_2B_{10}H_{10}K_2$ and NHC-$SiCl_2$. The ^{29}Si NMR signal assignable to the central Si(0) atom was observed at $\delta = -263.8$ ppm. The Si—Si distances amount to 2.2272 (6) and 2.2225(6) Å and the Si-Si-Si angle is 82.75°. With potassium naphthalenide, coupling of two Si(0) units to Si(I)-Si(I) occurred.[53]

The silylone **Si-7** is based on a four membered ring of Si atoms and where the Si(0) is flanked by two NHC stabilized Si(II) atoms and one Si(IV) opposite to Si(0). The source is the bicyclic silicon ring compound $Si_4\{N(SiMe_3)Dipp\}_4$ with two three-coordinate Si atoms and zwitterionic character from reacting with 5 equiv. of NHC as red crystals, Si(0) resonates at 55.6 ppm in the ^{29}Si NMR spectrum. An X-ray analysis revealed Si—Si bond lengths of 2.318(5) and 2.301(5) Å and a Si-Si-Si angle of 93.43(2)°.

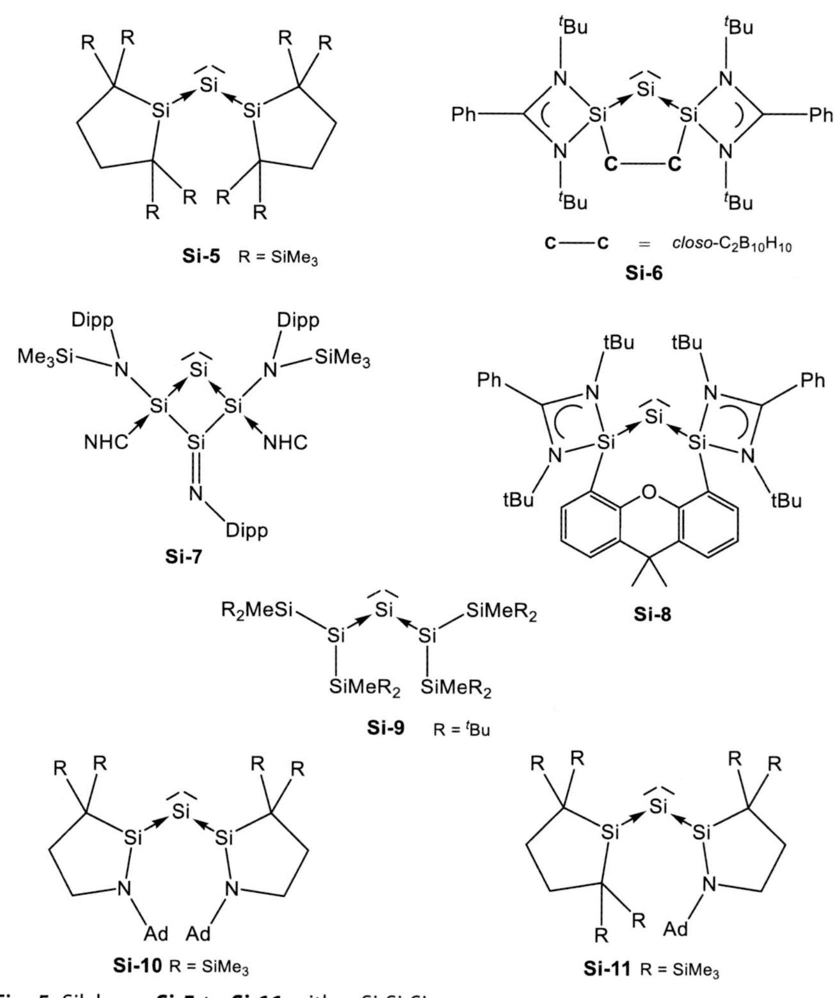

Fig. 5 Silylones **Si-5** to **Si-11** with a Si-Si-Si core.

The first PA of **Si-7** is 278.2 kcal/mol and the second PA amounts to 213.1 kcal/mol; these calculations demonstrates its silylone character.[54]

Si-8 was obtained upon dehalogenation of [Cl**Si-8**]Cl with C_8K as dark purple crystals. In the ^{29}Si NMR spectrum the signal at $\delta = -71.7$ ppm was assigned to the central Si(0) atom. The results of the X-ray analysis revealed Si—Si bond lengths of 2.526(7) and 2.586(7) Å and a Si-Si-Si angle of 104.38(3)°; the compound crystallizes with two, but nearly identical independent molecules. Reactions with N_2O lead to various oxidation products and in the presence of BPh_3 **Si-8** occurs with heterolytic H_2 cleavage and ethylene addition to form [H**Si-8**]$HBPh_3$ and [$Ph_3BC_2H_4$**Si-8**], respectively.[55]

The synthesis of **Si-9** was performed by the reaction of $Li_2Si(SiMe^tBu_2)_2$ with NHC-$SiCl_2$. In the ^{29}Si NMR spectrum a signal at $\delta=418.5$ ppm was assigned at the central Si(0) atom. An optimized structure at the B3LYP\6-31G(d) level gave Si—Si bond lengths of 2.1792 and 2.1742 Å, a Si-Si-Si angle of 164.3°. **Si-9** adds two molecules of methanol to produce $(SiMe^tBu_2)_2HSi\text{-}Si(OMe)_2\text{-}SiH(SiMe^tBu_2)_2$ with only Si(IV) atoms.[56]

Treatment of 2-bromo-2-(tribromosilyl)-1-aza-2-silacyclopentane with a sodium dispersion in the presence of cyclic (alkyl)(amino)-silylene (CAASi) at low temperatures yielded greenish black **Si-10** (a dark purple color is found in solution). The chemical shift of the central Si(0) atom appears at $\delta=157$ ppm. The X-ray analysis revealed Si—Si bond lengths of 2.2329(16) and 2.1886(16) Å and a Si-Si-Si angle of 99.78(6)°. UV–vis spectra of Si-10 were studied in KBr matrix and in hexane solution. The structure was calculated at the B3PW91-D3/B1 level of theory. Dark green **Si-11** was similarly obtained with the related cyclic silylene. In the ^{29}Si NMR spectrum the signal of the central Si(0) atom was found at $\delta=56.6$ ppm. The structural parameters are: Si—Si bond lengths, 2.2577 (5) and 2.1789(5) Å (to the Si atom of the cyclic silylene) and a Si-Si-Si angle of 119.62(2)°.[57]

3.2.3 SiL_2 compounds with a Ge-Si-Ge core

The preparation of **Si-12** follows the C_8K route starting from 1,1,2,5-tetrachloro-2,5-digerma-1-silacyclopentane (Bbt$=2,6\text{-}[CH(SiMe_3)_2]_2\text{–}4\text{-}[C(SiMe_3)_3]\text{-}C_6H_2$); yellow crystals are formed. The Si(0) atom resonates at $\delta=-16.46$ ppm. Selected structural parameters are Si—Ge bond lengths of 2.2681(18) 2.2900(18) Å and a Ge-Si-Ge angle of 80.08(4)°. *DFT* calculations were carried out.[58]

In a similar way to the preparation of **Si-5**, the germanium analog **Si-13** was obtained from the cyclic germylene [Ge(II)] *via* its $SiCl_4$ addition compound [Ge(IV)] followed by reduction with C_8K as dark greenish blue crystals. In the ^{29}Si NMR spectrum a signal at $\delta=236.6$ ppm was recorded. According to the X-ray analysis it has a crystallographic C_2 axis through the central Si atom with Ge-Si$=$2.2694(8) Å and a Ge-Si-Ge* angle of 125.71(7)°.[20] The compounds are show in Fig. 6.

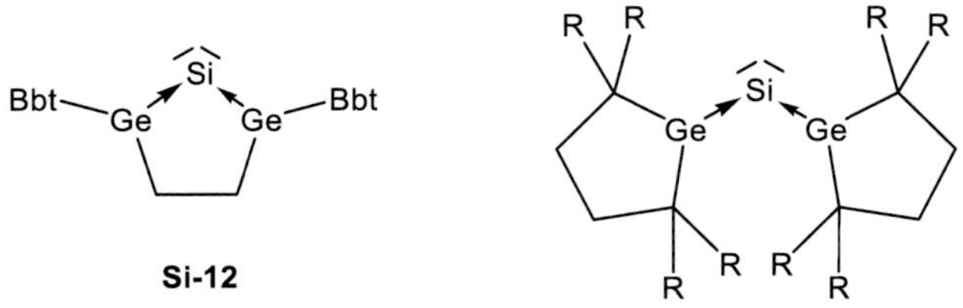

Fig. 6 Silylones **Si-12** and **Si-13** with a Ge-Si-Ge core.

3.2.4 SiL_2 compounds with a P-Si-Ge core

To synthesize **Si-14,** the starting product 1-PPh_2,2-$SiCl_3$-GeClAr*C_6H_4 was dehalogenated with {(^{mes}Nacnac)Mg}$_2$ to give red crystals after recrystallization from *n*-hexane. The signal in the ^{29}Si NMR spectrum of **Si-14** was found as a doublet at −48 ppm ($^1J_{SiP}$=217 Hz). The molecular structure exhibits Ge—Si and P—Si bond lengths of 2.2437(4) and 2.2901(6) Å, respectively, and the P-Si-Ge angle amounts to 81.5(1)°. An *NBO* analysis indicates the HOMO orbital to be the π-bond between Si and Ge atoms and the HOMO-1 orbital as a lone pair at the Si atom; the π-bond is filled with 1.69 electrons, which are almost equally shared between the germanium (47%) and silicon (53%) atoms. This is represented by formula (A) in Fig. 7; however as minor resonance contribution it might be regarded as phosphine-germylene stabilizing a formal Si(0) atom as shown by formulation (B) and thus denoted as silylone.[59]

3.2.5 SiL_2 compounds with a Si-Si-C core

The starting material for the synthesis of **Si-15** was the related tetrabromodisilane $CAr_2CH_2CH_2CAr_2SiBr$-$SiBr_3$; dehalogenation with sodium in the presence of the NHC gave yellow **Si-15**. The formal Si(0) atom resonates at 65.0 ppm in the ^{29}Si NMR spectrum. From X-ray analysis Si—Si and Si—C bond distances of 2.1787(8) and 1.962(2) Å, respectively were found, and a Si-Si-C angle of 105.40(7)° was recorded. The description of **Si-15** with a Si=Si double bond and one lone pair of electrons at the central Si atom or alternatively, a Si(0) atom coordinated by a silylene and a NHC ligand with two lone pairs of electrons is under consideration. N_2O transfers two O atom as bridging and terminal ones at the Si atoms.[60]

Si-16 was obtained from dehalogenation of the related "NHC"SiBr-$SiBr_2$(Tbb) with C_8K as a bright red microcrystalline solid. In the ^{29}Si NMR spectrum the signal of the two-coordinate Si atoms appears at 34.6 ppm. The selected Si—S— and Si—C bond lengths amount to

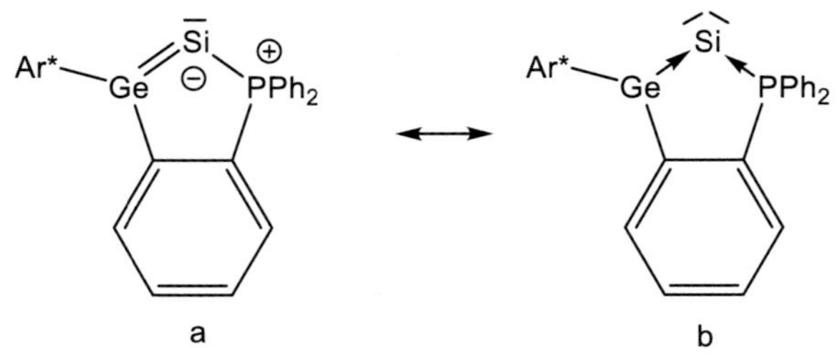

a b

Si-14 Ar* = 2,6-$Tip_2C_6H_3$
(Tip = 2,4,6-Triisopropylphenyl)

Fig. 7 Silylone **Si-14** with a Ge-Si-P core formulated with a Si=Ge double bond (A) or as silylene-phosphine stabilized Si(0) atom (B).

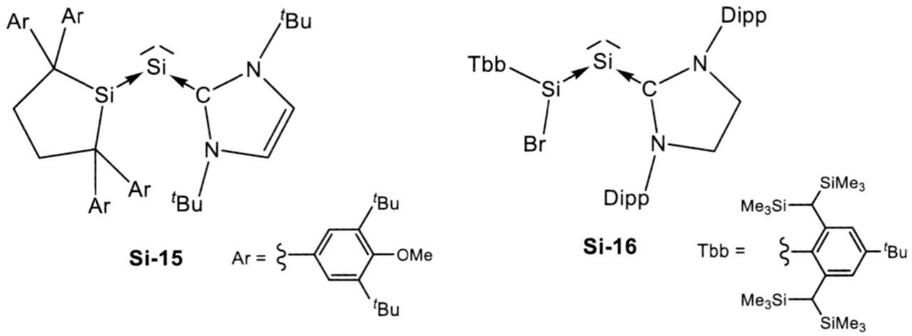

Si-15 **Si-16**

Fig. 8 Silylones **Si-15** and **Si-16** with a Si-Si-C core.

2.167(2) and 1.937(4) Å, respectively, and the Si-Si-C angle is 97.7(1)°. Results of the natural bond orbital (*NBO*) and natural resonance theory (NRT) were presented;[61] see Fig. 8.

3.3 Germylones GeL_2

3.3.1 GeL_2 compounds with a C-Ge-C core

Germylones with this core are collected in Fig. 9. Both germylones, **Ge-1** and **Ge-2** were prepared utilizing a one pot synthesis of the corresponding cAAC, $GeCl_2$(dioxane) and C_8K; dark green crystals were formed. Structural parameters for **Ge-1** are: Ge—C bond lengths = 1.9386(16) and 1.9417(15) Å and the C-Ge-C bond angle amounts to 114.71(6)–115.27(6)°. For **Ge-2** the related values are 1.945(2) and 1.9386(18) Å and 117.24(8)°. First and second proton affinity values (PAs) were theoretically calculated to be of 265.8 (**Ge-1**)/267.1 (**Ge-2**) and 180.4 (**Ge-1**)/183.8 (**Ge-2**) kcal/mol, respectively.[62]

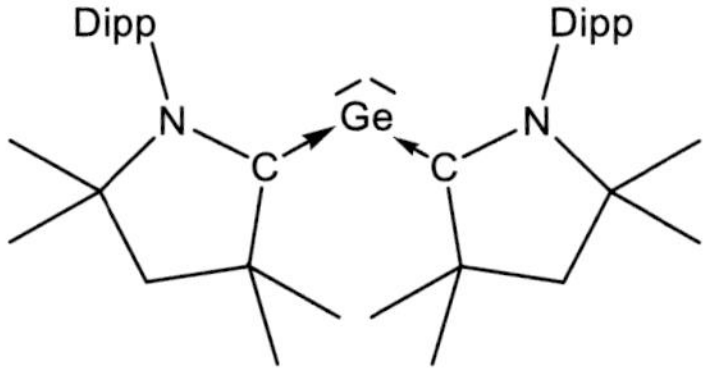

Ge-1 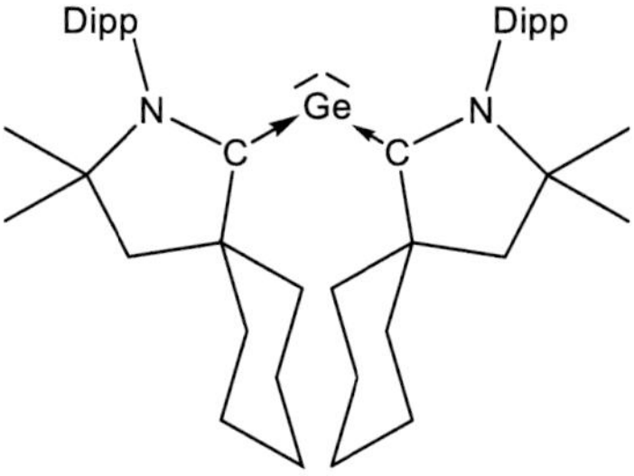 **Ge-2**

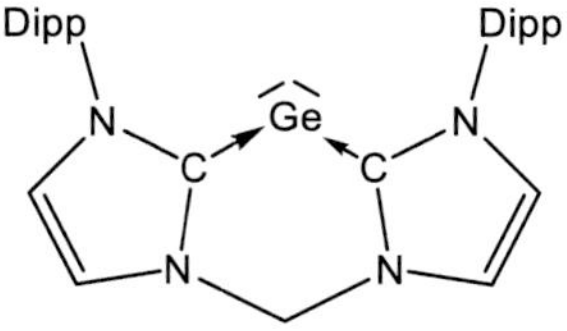

Ge-3

Fig. 9 Germylones **Ge-1** to **Ge-3** with a C-Ge-C core.

The synthesis of **Ge-3** starts from the related chelating bis-carbene and $GeCl_2$(dioxane) *via* the salt [Cl-**Ge-3**]Cl followed by dehalogenation with sodium naphthalenide. Selected interatomic distances and angles: Ge–C = 1.965(3), and 1.961(3) Å and C-Ge-C = 86.5(1)°.[22] **Ge-3** reacts with $GaCl_3$ to produce Cl_3Ga-**Ge-3**.[63]

3.3.2 GeL_2 compounds with a C-Ge-N core

Reduction of the ionic [Cl-**Ge-4**]$GeCl_3$ with C_8K yielded yellow **Ge-4**. The X-ray analysis revealed bond lengths of 1.8870(15) Å (Ge—C) and 1.9680(13) Å (Ge—N) and a C-Ge-N angle of 80.59(6)°; see Fig. 10. The nucleophilic reactivity of **Ge-4** is demonstrated by the reaction with two equivalents of MeOTf to give the corresponding dication [Me_2-**Ge-4**][OTf)$_2$, where the two lone pairs of electrons form two Ge-Me bonds.[64] Furthermore, $M(CO)_5$ addition compounds $(CO)_5M$-**Ge-4** were obtained with M = Cr, Mo, W, and [M(*cod*)Cl]$_2$, to produce the related [(cod) ClM]$_2$-**Ge-4** with M = Ir and Rh. The rhodium complex reacts with **Ge-4** in the presence of CO to give a complex with two units of **Ge-4** and a Rh-Ge-Rh-Ge-Rh chain,[65] see also synthesis of [Me-**Ge-4**]OTf in Ref. [66].

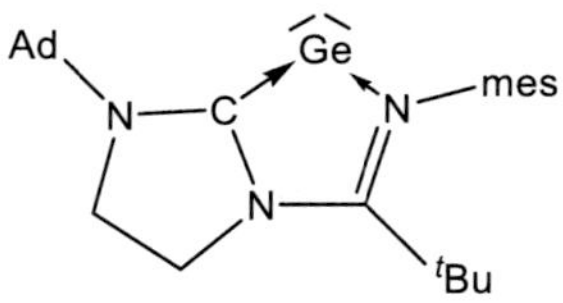

Fig. 10 Germylone **Ge-4** with a C-Ge-N core.

3.3.3 GeL_2 compounds with a N-Ge-N core

Starting with the related diiminopyridine which reacts with the germylene $GeCl_2$(dioxane) or the germylene in the presence of BCl_3 or $BF_3{\cdot}OEt_3$; this yielded the cation [Cl-**Ge-5**]$^+$ with the anions $GeCl_3^-$, BCl_4^-, or $BClF_3^-$, respectively. Reduction with C_8K gives green colored **Ge-5**. A graphical representation is found in Fig. 11. The results of the X-ray analysis revealed N—Ge bond lengths of 2.0947(7) and 2.306(7) Å to the imine nitrogen atoms and a N-Ge-N angle of 152.76(8)°; a shorter bond length of 1.8988(18) Å is found to the pyridine N atom. The Ge atom is disordered. *DFT* studies revealed partial delocalization of one of the Ge(0) lone pairs over the π^*(C=N) orbitals of the imines.[67]

3.3.4 GeL_2 compounds with a Si-Ge-Si core

Starting with a mixture of the basic silylene and $GeCl_2$(dioxane), the reaction with C_8K generates **Ge-6** as dark green crystals. The coordinating Si(II) atoms resonate at 219.4 ppm according to the ^{29}Si NMR spectrum. As depicted by an X-ray analysis similar to the central Si atom in the related trisilaallene **Si-5**, the Ge atom in **Ge-6** shows dynamic disorder in four positions indicating that four isomers exist in the solid state. Site occupancy factors (SOF) at variable temperatures were estimated. The energy difference between the most populated and the least-populated isomers in **Ge-6** is 1.55 kcal/mol. Si—Ge bond lengths are 2.2366(7) and 2.2373(7) Å and the Si-Ge-Si angle amounts to 132.38(2)° in the most populated isomer;[68] see also reference [20].

Starting from the bis silylene ligand reaction with $GeCl_2$(dioxane) the salt [Cl-**Ge-7**]Cl was obtained, which could be dechlorinated by means of C_8K to give **Ge-7** as dark blue rhombic crystals. In the ^{29}Si NMR spectrum the flanking Si atoms resonate at 51.1 ppm. The X-ray analysis revealed Si–Ge bond lengths of 2.3147(9) and 2.3190(9) Å and a Si–Ge–Si bond angle of 102.87(3)°. Reaction of **Ge-7** with $AlBr_3$ yielded the one and two addition products Br_3Al-**Ge-7** and $(Br_3Al)_2$-**Ge7**. In addition 9-borabicyclo[3.3.1]

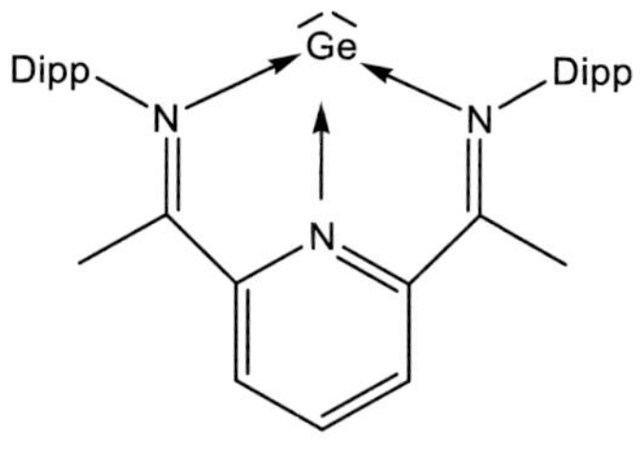

Ge-5

Fig. 11 Germylone **Ge-5** with a N-Ge-N core.

nonane (9-BBN), the adduct (9-BBN)-**Ge-7** was obtained, and H_2/BPh_3 produces the protonated product [H-**Ge-7**]$HBPh_3$. The action of $Ni(cod)_2$ gives a dimeric product with a Ge_2Ni three membered ring. **Ge-7** was studied by *DFT* calculations and *NBO* analysis, showing two lone pairs of electrons at the Ge(0) atom.[69]

Ge-8 is the germylone analog of the recently synthesized silylone **Si-6** and is isoelectronic to the nitreone **N-20**. It was synthesized similarly from 1,2-$(LSi)_2$-nido-1,2-$C_2B_{10}H_{10}K_2$ and $GeCl_2$·dioxane.The Si—Ge distances amount to 2.2896(5) and 2.2846(5) Å and the Si-Ge-Si angle is 80.59(2)°. With potassium naphthalenide, coupling of two Ge(0) units to Ge(I)-Ge(I) occurred[70]; see graphical representation in Fig. 12.

3.3.5 GeL_2 compounds with a Ge-Ge-Ge core

The germylone (trigermaallene) **Ge-9** was reported by Kira[20] and shown in Fig. 13. It was obtained in a similar manner to the procedure outlined for the Si analogs **Ge-6** or **Si-11**. According to the X-ray analysis Ge—Ge bond lengths of 2.321(2) and 2.330(2) Å were recorded; the Ge-Ge-Ge bond angle amounts to 122.61(6)°; see also [68].

3.3.6 GeL_2 compounds with a C-Ge-Si core

A mixture of Tip_2SiCl_2 and the related addition compound NHC-$GeCl_2$ was stirred with $Li/C_{10}H_8$ at low temperature to give **Ge-10**. ^{29}Si NMR: $\delta = 158.92$ ppm. An X-ray analysis revealed a Si—Ge bond length of 2.2521(5) Å and a Ge—C bond length of 2.0474(18) Å; the C-Ge-Si angle is 98.90(5)°. For the description of **Ge-10** either with a Ge=Si double bond (a) or a Ge(0) atom stabilized by silylene and NHC ligands (b), see also **Si-15**. **Ge-10** adds PhC≡CH at the Si=Ge bond to give a four membered C,C,Ge,Si ring.[71]

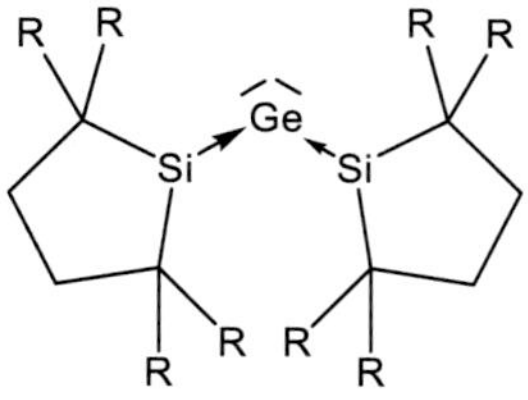

Fig. 12 Germylones **Ge-6** to **Ge-8** with a Si-Ge-Si core.

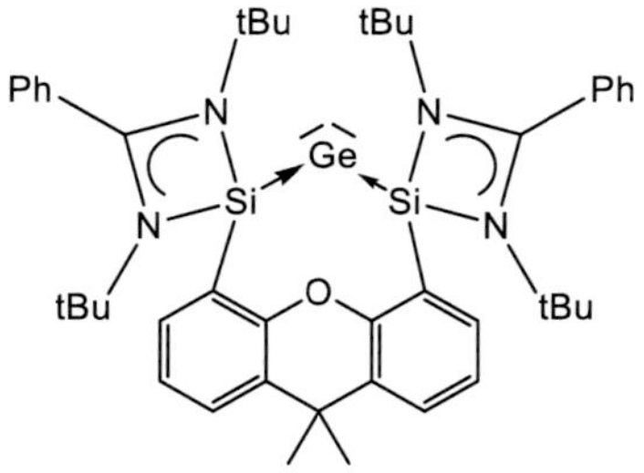

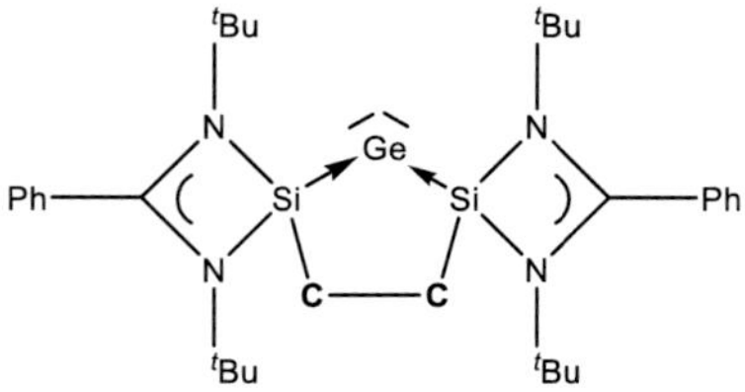

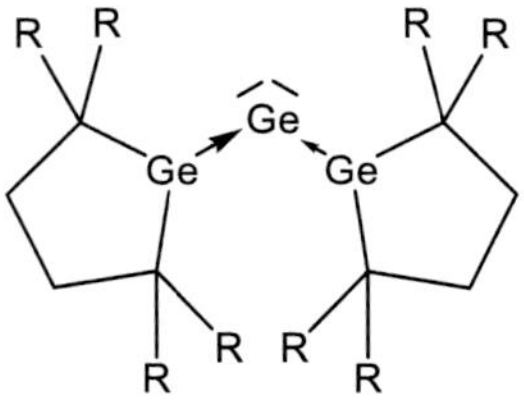

Fig. 13 Germylone **Ge-9** with a Ge-Ge-Ge core.

The red colored germylone **Ge-11** was obtained upon reacting the related NHC-$GeCl_2$ addition compound with $(Tip)_2Si{=}Si(\mathit{Tip})Li$. An E/Z equilibrium in solution exist with an E/Z ratio of 0.85:0.15. The ^{29}Si NMR spectrum of the E isomer showed a signal at $\delta = 162.50$ ppm. Relevant bond lengths are Si-Ge = 2.2757(10) Å and Ge-C = 2.061(4) Å with a Si-Ge-C angle of 101.90(10)°. Upon reaction with $Fe_2(CO)_9$ an $Fe(CO)_4$ addition compound (at Ge(0)) is obtained.[72] With $Ni(cod)_2$ a nickel complex was obtained with a four-membered Si,Si,Ge,Ni ring (disilagermirene-nickel).[73] Further nickel complexes from the reaction of **Ge-11** with $Ni(cod)_2$ were described in Ref [74]; additionally, the synthesis

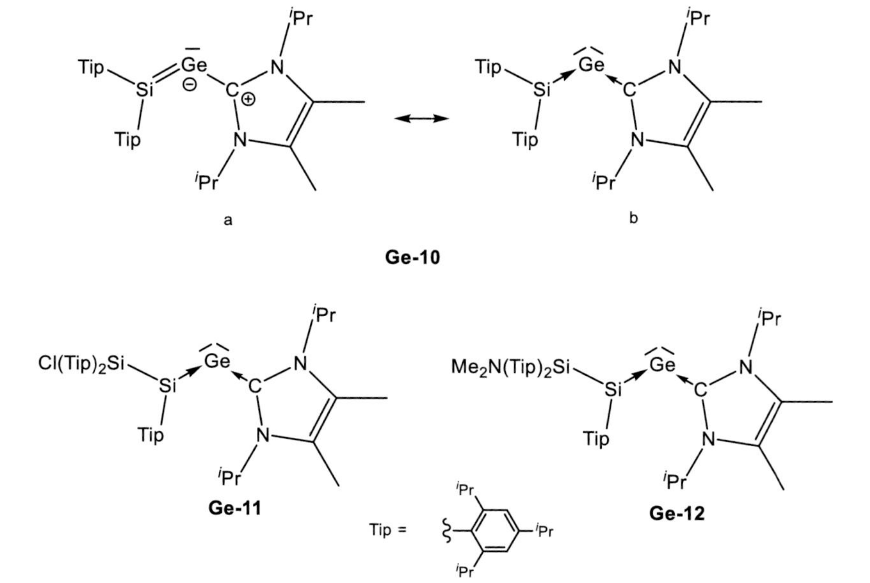

Fig. 14 Germylones **Ge-10** to **Ge-12** with a Si=Ge double bond or a silylene/NHC stabilized Ge(0) atom.

of **Ge-12** (as a mixture of the E and Z isomer) is reported from the reaction of **Ge-11** with $LiNMe_2$ and with release of LiCl. The compounds are shown in Fig. 14.

3.3.7 GeL_2 compounds with a P-Ge-Ge core

In a similar way to **Si-14** the germylone **Ge-13** can be interpreted in terms of a Ge=Ge double bond (a) or as Ge(0) atom stabilized by a chelating ligand with germylene and phosphine as pincer (b). It was prepared by dehalogenation of the related dichloro compound with {(^{mes}Nacnac) Mg}$_2$ or Na in diethyl ether. An X-ray analysis revealed Ge−Ge and Ge—P bond distances of 2.30597(19) and 2.3894(4) Å, respectively, and a Ge-Ge-P angle of 80.210(10)°. It adds P≡C-Ar across the Ge—Ge double bond with formation of a four-membered Ge,Ge,P,C ring.[75] The different views are given in Fig. 15.

3.4 Stannylones SnL_2

3.4.1 Stannylones with a N-Sn-N core

The Sn analog **Sn-1** in Fig. 16 of **Ge-5** was prepared by the reaction of 2,6-[ArN-CH_2]$_2$(NC_5H_3) (Ar = C_6H_3-2,6-$^i Pr_2$) with $Sn[N(SiMe_3)_2]_2$.

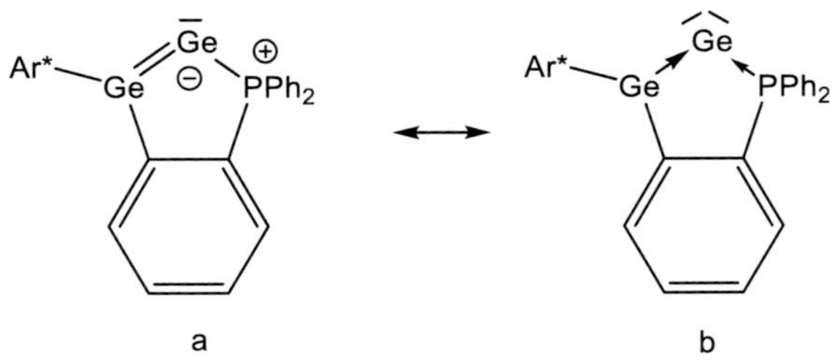

a b

Ge-13 Ar* = 2,6-$Tip_2C_6H_3$
(Tip = 2,4,6-Triisopropylphenyl)

Fig. 15 **Ge-13** with a Ge=Ge double bond (A) or the description as germylone (B).

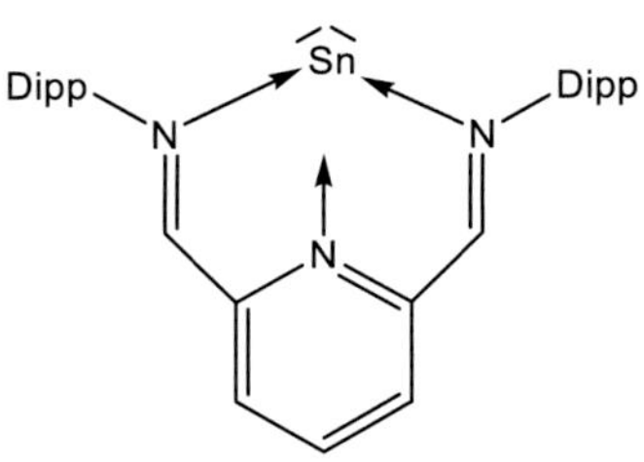

Sn-1

Fig. 16 Stannylone **Sn-1** with a N-Sn-N core.

The solution-state ^{119}Sn NMR spectrum of **Sn-1** shows one signal at $\delta = 64$ ppm with a line width of 236 Hz. The X-ray analysis showed N—Sn bond lengths of 2.397(2) and 2.315(2) Å and an N-Sn-N angle of 142.43(7)°. A shorter Sn—N bond distance to the pyridine N atom of 2.122(2) Å was found. Theoretical studies at the density functional level of theory and ^{119}Sn Mössbauer spectroscopy were performed.[76]

3.4.2 Stannylones with a Sn-Sn-Sn core

The stannylone **Sn-2** in Fig. 17 denoted earlier as tristannaallene, was prepared upon reacting the stannylene $Sn[N(SiMe_3)_2]_2$ with 6 equivalents of $NaSi^tBu_3$ as dark blue crystals. In the ^{119}Sn NMR spectrum a signal at $\delta = 2233$ ppm is due to the central Sn atom; the adjacent Sn atoms resonate at 503 ppm. X-ray analysis revealed Sn—Sn bond lengths of 2.684(1) and 2.675(1) Å and a Sn-Sn-Sn angle of 156.01(3)°. The compound crystallizes with two isomers in the unit cell; similar values were obtained for the second isomer.[77]

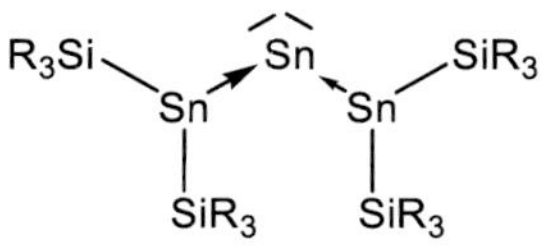

Sn-2 R = ^{t}Bu

Fig. 17 Stannylone **Sn-2** with a Sn-Sn-Sn core.

Note added in proof

After this manuscript was finished, a joint experimental and theoretical study was published, which report about the synthesis and x-ray structural characterization of a stannylone with a Si-Sn-Si core: J. Xu, C. Dai, S. Yao, J. Zhu, M. Driess, Angew. Chem. Int. Ed. 60 (2021) e202114073.

3.5 Plumbylones PbL_2

To date, no PbL_2 compound could be prepared but theoretical studies were performed on compounds with a P-Pb-P and C-Pb-C core[16,78]; proposed structures are shown in Figs. 18 and 19.

3.5.1 PbL_2 compounds with a P-Pb-P core

See Fig. 18.

3.5.2 PbL_2 compounds with a C-Pb-C core

See Fig. 19.

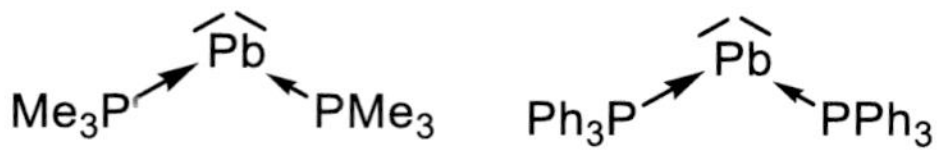

Pb-1 ***Pb-2***

Fig. 18 Plumbylone with a P-Pb-P core studied by computational methods.

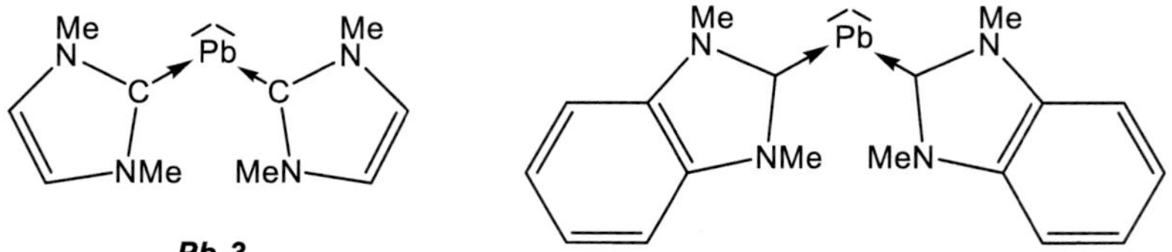

Pb-3 ***Pb-4***

Fig. 19 Plumbylones with a C-Pb-C core studied by computational methods.

4. Group 15 compounds $[EL_2]^+$, E = N, P, As, Sb, Bi

According to the positive charge, $[EL_2]^+$ compounds are less nucleophilic than the related neutral group 14 ylidones. First and second PAs have lower values and transition or main group addition compounds are less common.

4.1 Nitreone $[NL_2]^+$

In a similar manner to the bare carbon atom in carbones the neighboring and isoelectronic N(I) atom has two pairs of electrons of σ and π symmetry and can accept two additional pairs of electrons from coordination of neutral donor molecules, mimicking the properties of a metal. Like carbones, $[NL_2]^+$ compounds are referred to as nitreones; L represents Lewis bases such as phosphines, carbenes, cAACs and even silylenes. In 2018 two reviews,[32] and[79], summarized syntheses, chemistry and catalytic application of nitreones. The review[32] deals with nitreones with a C-N-C core.

4.1.1 Nitreones with a P-N-P core

The reaction of Me_3PCl with Me_3PNLi produced **N-1**;[80] it also forms upon reaction of $Me_3PNPMe_2{=}CH_2$ with HBr.[81] X-ray analysis revealed P—N distances of 1.582 Å and a P-N-P angle of 137°.[82] The structure of the cation **N-2** has been determined many times, because it is widely used as a large cation in organometallic chemistry; the PNP angle usually varies between 134° and 142°[83] but a linear arrangement (allene arrangement) was also measured showing that the PNP framewcrk is very flexible.[84] A series of cationic phosphazenes has been prepared (**N-3** to **N-6** and further examples), described in Ref[85]; they can also be interpreted as nitreones. A bent structure with both P-N-P angles of 142° in the cation **N-6** supports this view; see also Ref [86]. The known compounds with this core are collected in Fig. 20.

4.1.2 Nitreones with a P-N-As core

Mixed nitreones **N-7** and **N-8** with phosphines and arsines,[80] shown in Fig. 21 were obtained from the reaction of Me_2AsCl with $R_3P{=}NLi$, followed by methylation at the As atom with MeI.

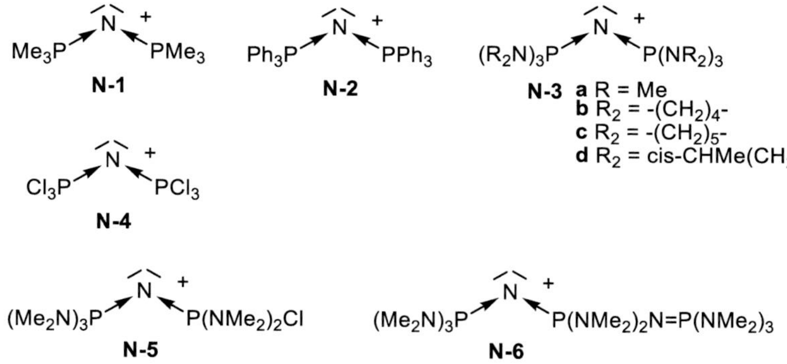

$Me_3P \rightarrow N^+ \leftarrow PMe_3$ **N-1**

$Ph_3P \rightarrow N^+ \leftarrow PPh_3$ **N-2**

$(R_2N)_3P \rightarrow N^+ \leftarrow P(NR_2)_3$ **N-3** **a** R = Me; **b** R_2 = -$(CH_2)_4$-; **c** R_2 = -$(CH_2)_5$-; **d** R_2 = cis-CHMe$(CH_2)_3$CHMe-

$Cl_3P \rightarrow N^+ \leftarrow PCl_3$ **N-4**

$(Me_2N)_3P \rightarrow N^+ \leftarrow P(NMe_2)_2Cl$ **N-5**

$(Me_2N)_3P \rightarrow N^+ \leftarrow P(NMe_2)_2N{=}P(NMe_2)_3$ **N-6**

Fig. 20 Nitreones with a P-N-P core.

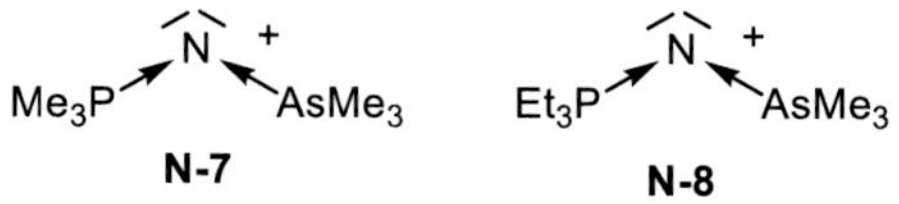

$Me_3P \rightarrow N^+ \leftarrow AsMe_3$ **N-7**

$Et_3P \rightarrow N^+ \leftarrow AsMe_3$ **N-8**

Fig. 21 Examples of nitreones with a P-N-As core.

4.1.3 Nitreones with a C-N-P core

The mixed carbene and phosphine stabilized nitreone in Fig. 22, **N-9**, was derived from reacting PCl_5 with Me_2NH in $MeNO_2$ solution, followed by hydrolysis and addition of $NaBF_4$.[86] **N-10** was prepared (the counterion is Cl^-) from the related guanidine and $[(Et_2N)_3PCl]Cl$[87]. **N-11** was prepared from the reaction of the related 2-chloro-imidazolium tetrafluoroborate and $HN{=}P(NMe_2)_3$.[88]

4.1.4 Nitreones with a C-N-C core

Nitreone with a C-N-C core were studied by *ab initio*, *MO* and *DFT* calculations and analyzed in terms of molecular orbitals, atomic charges, proton energies and complexation energies with Lewis acids;[89] see also previous studies by the same authors.[90] A quantum chemical study of drug molecules based on carbene stabilized N(I) compounds is found in Ref. [91].

The cationic N-analog of carbon suboxide, **N-12**, was obtained by condensation of FCONCO, SbF_5, and $CF_3CH_2CF_3$ in a glass vial and sealed. Crystals of the $Sb_3F_{16}^-$ salt were formed. The X-ray analysis revealed a bent structure with a C-N-C angle of 130.7° and C—N distances of 1.250(5) Å.[92] Reaction of the related chlorocyclopropenium BF_4^- salt with KHMDS in THF gave **N-13**. In the ^{15}N NMR spectrum, the central N atom resonates at −291.6 ppm. After X-ray analysis a bent structure with

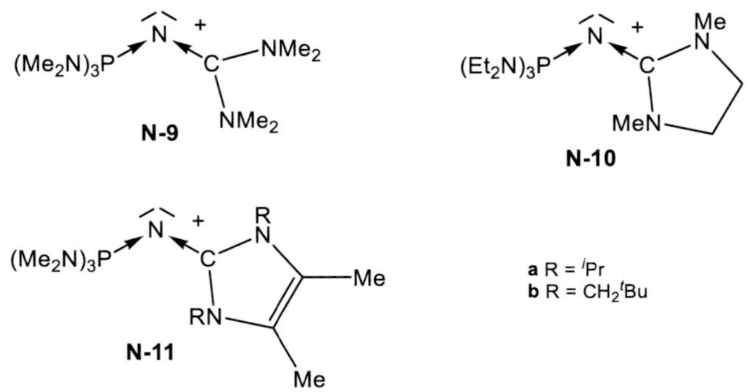

Fig. 22 Nitreones **N-9** to **N-11** with a C-N-P core.

a C—N distance of 1.332 Å was recorded. Protonation, alkylation or arylation at the central N atom produced the related dications.[93]

The chiral compounds **N-14** (Cl^- as counteranion) with different X at the benzyl group are active as phase transfer catalysts for the asymmetric alkylation of sulfonate ions to various sulfoxides with high enantioselectivities;[94] further chiral derivatives similar to **N-14** were studied as phase transfer catalysts.[95]

The asymmetric nitreone **N-15** (*N*,*N*-dimethylimidazolidino)-tetramethylguanidinium chloride) can be synthesized from dimethylimidazolinone and tetramethylguanidine; it serves as an active phase-transfer catalyst.[87] **N-16** results from reaction of the related 2-chloro-imidazolium tetrafluoroborate with $HN{=}C(NMe_2)_2$. Similarly, **N-17** stems from KF mediated coupling of the related imidazolium chloride and the corresponding amide. Results of the X-ray analysis of **N-17a** (the counteranion is OTs^-): N—C 1.332(2) Å, C-N-C 125°; similar values in **N-17b** as the iodide. Protonation to the dication proceeds with **N-17a**.[88]

The nitreone cycloguanil hydrochloride **N-18** results from protonation of cycloguanil; the nitrogen atom is stabilized by a chelating carbene ligand. According to the X-ray analysis C—N distances of 1.333 an 1.348 Å are recorded.[96]

Besides one NHC ligand, **N-19** contains the unusual ligand cyclohexa-2,5-diene-4-(diaminomethynyl)-1-ylidene as a second neutral carbene ligand for stabilizing the N(I) atom. The synthesis follows a multi-step procedure starting with methylated 2-chloro-imidazol and 4-aminobenzonitril. The first proton affinity was calculated to be 159.7 (**N-19a**) and 154.6 kcal/mol (**N-19b**). An X-ray analysis of **N-19a** (counterion is acetate) shows C—N distances of 1.308(3) Å to the NHC and 1.379(2) Å to the

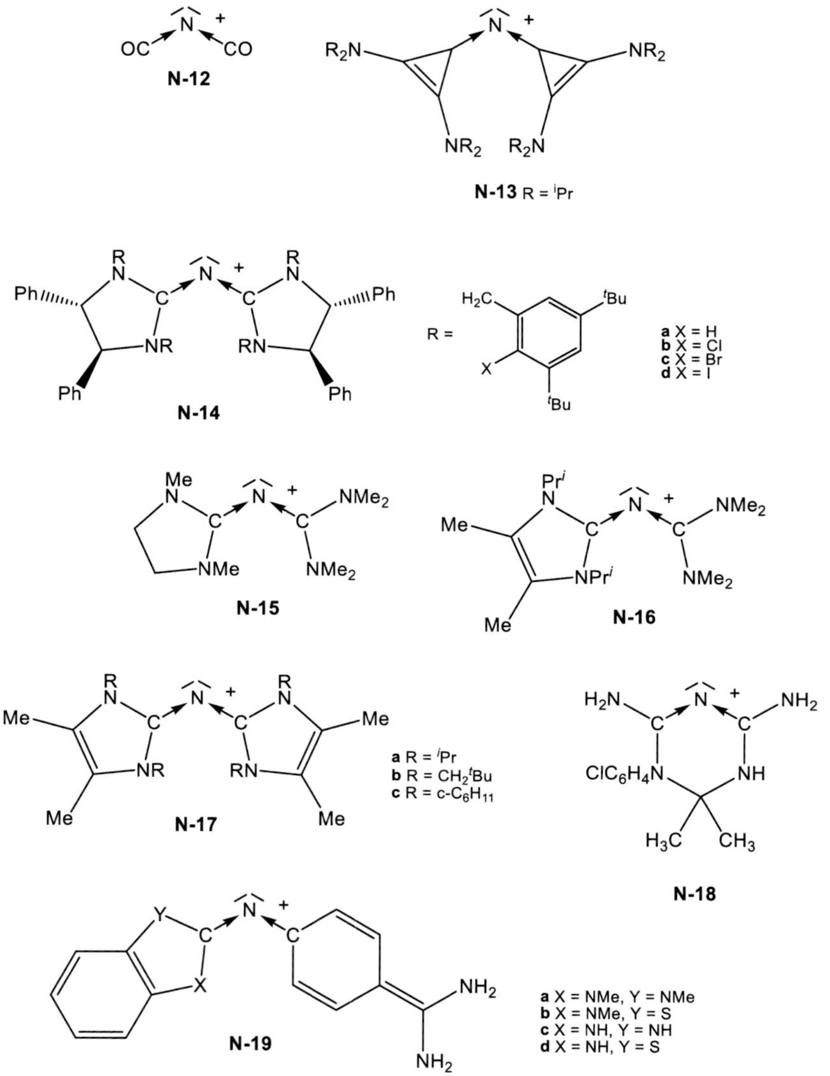

Fig. 23 Collection of nitreones with a C-N-C core.

carbene ligand, and a C-N-C angle of 120.8(2)°; the corresponding values of **N-19b** (counterion OTs^-) are 1.286(3) and 1.399(3) Å by a C-N-C angle of 124.4(2)°.[97] A graphical representation of the nitreones **N-12** to **N-19** is collected in Fig. 23.

4.1.5 Nitreone with a Si-N-Si core

The single compound **N-20** (to be prepared) in which the N(I) atom is stabilized by a chelating bis(silylene) ligand bridged by a closo-C_2B_{10} cluster is shown in Scheme 1. Pale-yellow crystals were obtained upon one-electron oxidation with silver(I). An X-ray analysis offered Si—N distances of 1.639 (3) and 1.645(3) Å and a Si-N-Si angle of 115.8(2)°.[98]

4.2 Phosphorone $[PL_2]^+$

4.2.1 Phosphorones with a P-P-P core

A review on phosphine and carbon stabilized P(I) cations has appeared in 2007.[99] In 2020 a further review from the same group updates this subject.[100] A series of cationic phosphorones with P^+ (or P(I)) stabilized by various chelating ligands or PR_3 groups had been prepared. Main preparation routes are shown in Eq. (1) (shown for the dppe stabilized cation **P-2a**) based on the reduction of PCl_3 under oxidation of $SnCl_2$. A further access to phosphorones is given in Eq. (2).

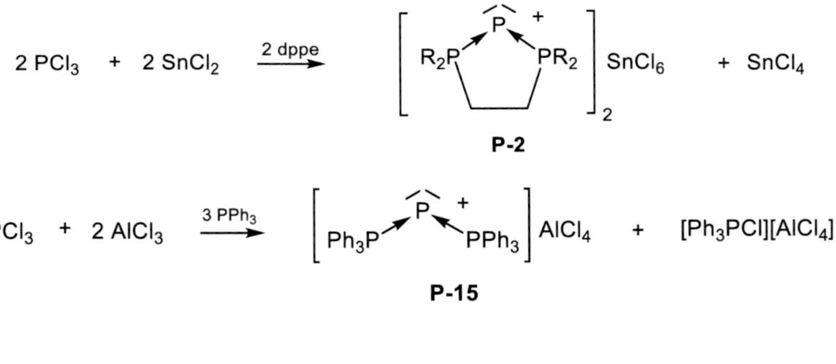

$$2\,PCl_3 + 2\,SnCl_2 \xrightarrow{2\,dppe} [(R_2P)_2P^+]_2\,SnCl_6 + SnCl_4 \quad (1)$$

P-2

$$PCl_3 + 2\,AlCl_3 \xrightarrow{3\,PPh_3} [Ph_3P \rightarrow P^+ \leftarrow PPh_3]\,AlCl_4 + [Ph_3PCl][AlCl_4] \quad (2)$$

P-15

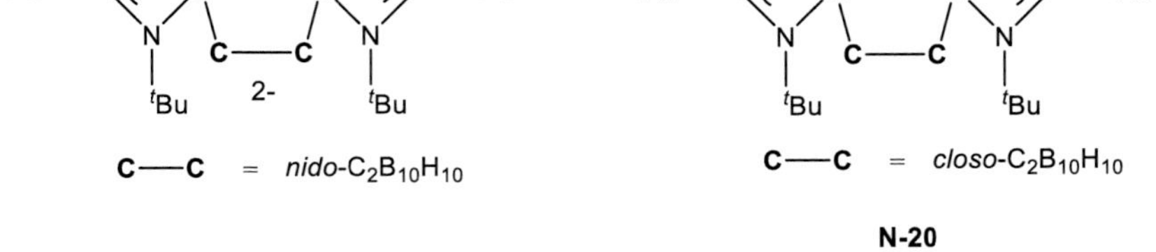

Ad, N, N, ${}^tBu$, Ph, Si, C—C, 2-, a, **N-20**

C—C = *nido*-$C_2B_{10}H_{10}$ C—C = *closo*-$C_2B_{10}H_{10}$

Scheme 1 Nitreone with a Si-N-Si core; a) AgOTf. Counterion of **N-20** is OTf^-.

$$3\,PCl_3 + 2\,R_2PCH_2PR_2 \longrightarrow 2\,[R_2P(\rightarrow P \leftarrow)PR_2(CH_2)]^+ + [R_2PClCH_2PClR_2]Cl_2 \quad (3)$$

P-1

P-1a and **P-1b** with a cyclic four-membered ring were obtained as the $SnCl_6$ salts using the route as described in Eq. (1); **P-1c** and **P-1d** were obtained with chloride as counterions from PCl_3 and the related bis-dialkylphosphino methane according to the procedure shown in Eq. (3). All compounds were identified by NMR spectroscopy. **P-1b** was stable enough to be methylated by methyl triflate to give the dicationic compound [Me**P-1b**]$_2$SbCl$_6$(SO$_3$CF$_3$)$_2$;[101] see also.[102]

The first phosphorone with a five-membered ring, [P(dppe)]$_2$SnCl$_6$ (**P-2a**, R=Ph), was prepared according to the procedure outlined in Eq. (1) and studied by ^{31}P NMR spectroscopy. The colorless compound crystallizes with two molecules of CH_2Cl_2. X-ray analysis: P-P=2.122 (1), 2.228(2) Å, P-P-P=88.9(1)°,[103] see also Ref. [104]. X-ray analysis of the bromide: P-P=2.1231(9), 2.1308(9), P-P-P=88.37(3).[105] **P-2a** as $AlCl_4$ salt could also be prepared directly from [P(PPh$_3$)$_2$]AlCl$_4$ (**P-15**) and dppe; action of HCl gave [H**P-2a**](AlCl$_4$)$_2$.[106] In the reaction of PI_3 with dppe, the iodide [P(dppe)]I formed, and with the release of I_2; anion exchange with $NaBPh_4$ gave [P(dppe)]BPh$_4$. Both compounds were studied by X-ray analyses; I/BPh$_4$: P-P=2.131(2), 2.133(2)/2.135(2), 2.122(2) Å, P-P-P=89.35(6)°/86.44(7);[107] see also [108]. Further derivatives **P-2b, P-2c, P-2d** with various anions are characterized by ^{31}P NMR spectroscopy.[109] **P-2d** with different substituents Ph and Et at the P atoms of the chelating ligand shows an ABC system. **P-2a** is the starting material for a series of phosphorones with a C-P-C and C-P-P core.[110]

Introduction of the chelating ligand 1,2-bisdiphenylphosphinobenzene (dppben) according to Eq. (1) gave the complex [P(dppben)]$_2$SnCl$_6$ (**P-3**) with the structural parameters P-P=2.124(1), 2.122(1) Å, P-P-P=90.67 (4)°.[109] Reaction with the unsaturated diphosphane Ph$_2$PCH=CHPPh$_2$ (dppE) following Eq. (1) produced **P-4**; the anions are $SnCl_6^-$ and halides.[111] Methylation with methyl triflate at the central P atom produces the related dication.[112]

With bis(2-diphenylphosphinoethyl)phenylphosphan and **P-15** both phosphines were replaced to produce a yellow powder identified as **P-5**; the counterion was $AlCl_4^-$. Variable temperature ^{31}P NMR spectra indicated a rapid equilibrium with a bicyclic form as intermediate ($\Delta H°=$ 9.6 kcal/mol, $\Delta S°=-9.9$ cal/mol K), see Fig. 24.[106] The basic cyclic

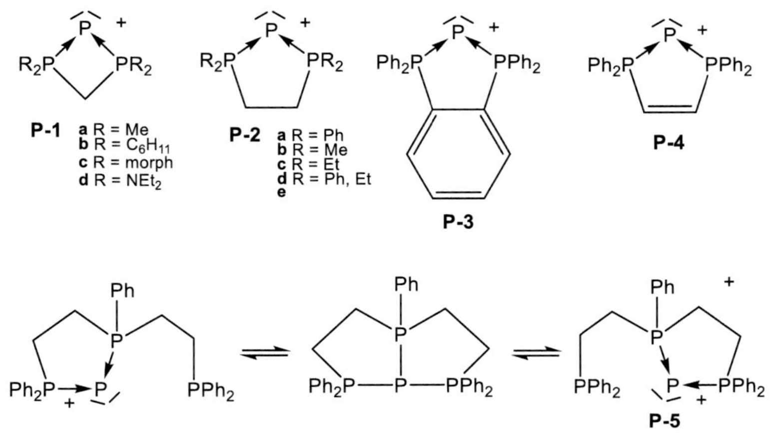

Fig. 24 Tetra and Pentacyclic cationic $[PL_2]^+$ compounds.

six-membered ring cation **P-6a** was formed according to the procedure outlined in Eq. (1) as the $SnCl_6$ salt, using dppp (*bis*-1,3-diphenylphosphinopropane). The X-ray analysis revealed the following parameter: P-P = 2.132(1), 2.132(1) Å, P-P-P = 96.44(6)°;[111] similar values were reported for the $AlCl_4^-$ salt,[113] the I^-, BPh_4^-, PF_6^-, $GaCl_4^-$, and OTf^- salts[108] and the Br-salt.[105] **P-6b** was mentioned[114] as well as its protonated dication from reacting with triflic acid.

Bearing an exocyclic methylene group, **P-7** (as the $SnCl_6$ salt) was similarly obtained following the Sn route of Eq. (1)[115] **P-8** were prepared as the iodide upon reacting of 1,8-bis(diphenylphosphino)naphthalene with P_2I_4 (1/2 equiv) in DCM. Yellow crystals with the structural parameters P-P = 2.1172(7), 2.1178(7) Å, P-P-P = 91.84(3)° were obtained.[116] Using the chelate effect, with tetrakis(diphenylphosphinomethyl)methane and **P-15** (as the $AlCl_4^-$ salt) the dication **P-9** was prepared; **P-9** could be protonated with CH_2Cl_2 or alkylated with ^{t}BuCl to the related tetracation; see also [117]. In a similar way, by using 1.1.1-tris(diphenylphosphinomethyl) ethane, **P-10** was obtained.[106] The phosphorones with a six-membered ring are collected in Fig. 25. Compounds with the seven-membered ring **P-11a**[111] and **P-11b** formed (as the $SnCl_6^-$ salt) as well as was the o-xylol derivative **P-12** according to Eq. (1); similarly, the ferrocenium derivative **P-13** is mentioned without further preparative details[114]; see Table 2. In the phosphazene derivative **P-14** the P(I) atom is part of an eight-membered ring and was prepared upon methylation of the neutral cyclophosphazene with MeI, Me_2SO_4, or $Me_3O(BF_4)$;[118] see Fig. 26.

Following Eq. (3) the triphenylphosphine derivative **P-15** was prepared in CH_2Cl_2 through reductive break-up of PCl_3 as the $AlCl_4^-$ salt. **P-15**

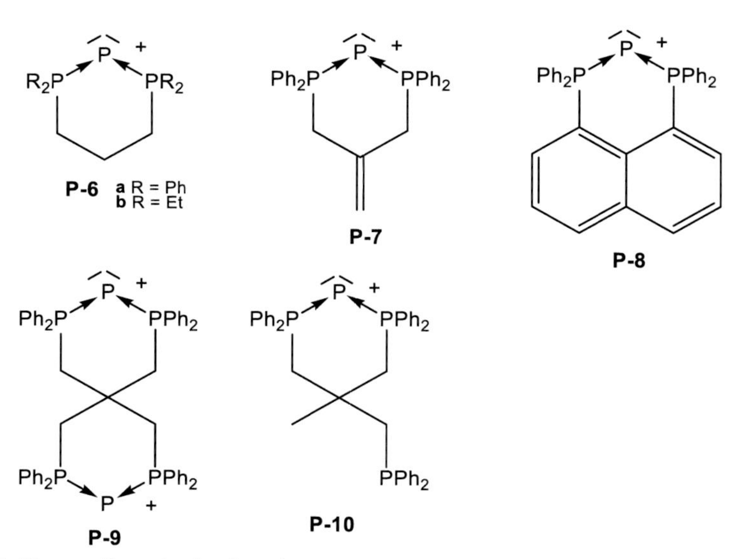

Fig. 25 Hexacyclic cationic phosphorones.

Table 2 Cyclic phosphorones.

Nr	L	d_{Px} (t)	d_{PL} (d)	$J_{Px,L}$	Ref
P-1a	dmpm	−213	60	437	[101]
P-1b	dcxpm	−290	88	458	[101]
P-1c	d(*mo*)pm	−126	56	358	[101,102]
P-1d	d(NMe_2)pm	−90	56	357	[101,102]
P-2a	dppe	−232	64	502	[103]
P-2b	dmpe	−213	60	437	[109]
P-2c	bdpe	−264	81	444	[109]
P-2d	depe	−251	79, 67	435, 461	[109]
P-3	dppben	−213	58	453	[109]
P-4	dppE	−248	72	473	[111]
P-5	bisphos[c]	−251	65	456	[106]
P-6a	dppp	−210	23	422	[111]
P-6b	dppp	−253	31	417	[114]
P-7	dpmE	−213	23	431	[115]
P-8	dpna	−216	26	527	[116]

Table 2 Cyclic phosphorones.—cont'd

Nr	L	d_{Px} (t)	d_{PL} (d)	$J_{Px,L}$	Ref
P-9	tetraphos	−250	18	441	[106]
P-10	triphos	−243	20	444	[106]
P-11a	dppb	−211	34	455	[111,114]
P-11b	dhpb	−263	48	472	[114]
P-12	dppxy	−215	25	438	[114]
P-13	dppFc	−153	33	497	[114]
P-14		−117	42	436	[118]

^{31}P MNR shifts of the cations **P-1** to **P-14** in ppm, J in Hz. dmpm = bis-dimethylphosphinomethane; dcxpm = bis-dicyclohexylphosphinomethane; d(mo) pm = bis-dimorpholinophosphinomethane; dnpm = bis-diaminophosphinomethane; depe = 1-diethylphosphino-2-diphenylphophinoethane; bdpe = 1,2-bisdiethylphosphinoethane; dmpe = 1,2-bisdimethylphosphinoethane; dppben = 1, 2-bisdiphenylphosphinobenzene; dpmE = 1,*l*-bis[(diphenylphosphanyl)methyl]-ethene; dpna = 1,8-Bis(diphenylphosphino) naphthalene; tetraphos = tetrakis (diphenylphosphinomethyl)methane; triphos = 1.1.1-tris(diphenylphosphinomethyl) ethane; dppb = 1,4-diphenylphosphinobutane; dhpb = 1,4-dicyclohexylphosphinobutane; dppxy = bis-diphenylphosphinoxylol; dppfc = diphenylphosphino ferrocenium.

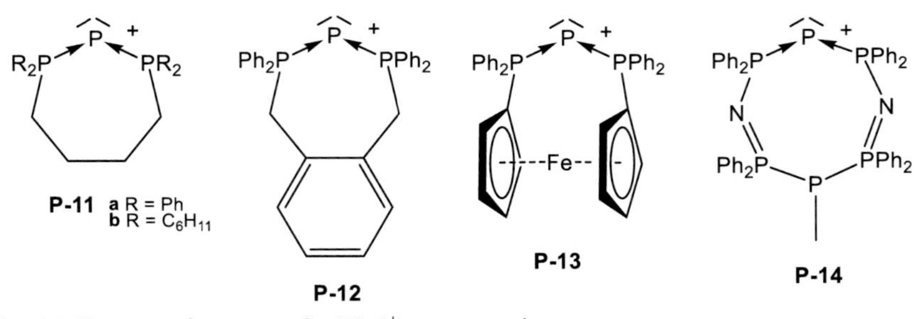

Fig. 26 Hepta and more cyclic $[PL_2]^+$ compounds.

crystallizes with one half of a molecule of CH_2Cl_2; X-ray analysis data are P-P = 2.137(6), 2.128(6)(1) Å, P-P-P = 102.2(2)°,[119] see also Ref. [104]. Treatment of **P-15** with series of more basic phosphines (each one equivalent) generated the asymmetric salts **P-26** to **P-35** depicted in Table 4. A second equivalent replaces the remaining PPh_3 group to give the symmetric compounds **P-16** to **P-25,** listed in Table 3, according to Eq. (4) The synthesis of the BPh_4^- salt of **P-22** is also described.[104] The salts of the unsymmetrical triphosphenium ions are also formed upon reaction of the salts with the symmetrical ions **P-15** and **P-16** to **P-25 (**119); for alkylation or protonation, see Ref. [120].

Table 3 Symmetric phosphorones, ^{31}P NMR (AB_2 type) shifts of the cations **P-15** to **P-22** in ppm, *J* in Hz.

Nr	L	dP_A	dP_B	$^{1}JP,P$	Refs.
P-15	PPh_3	−174	30	449	[119]
P-16	$MePh_2P$	−176	23	464	[119]
P-17	Me_2PhP	−159	12	451	[119]
P-18	Bu_3P	−229	33	473	[119]
P-19	Et_2NPh_2P	−164	79	508	[119]
P-20	$(Et_2N)_2PhP$	−163	64	510	[119]
P-21	$PhHNPh_2P$	−182	47	498	[119]
P-22	$(Me_2N)_3P$	−194	85	518	[116]
P-23	$[(CH_2)_5N]_3P$	−193	79	560	[119]
P-24	$[O(CH_2)_4N]_3P$	−207	78	566	[119]
P-25	$(EtO)_3P$	−218	82	508	[119]

Table 4 ^{31}P NMR shifts of asymmetrical $[L\text{-}P_x\text{-}PPh_3]^+$ **P-23** to **P-32**) cations in ppm, *J* in Hz.

Nr	L	P_x	PPh_3	L	$J_{PX,PPh3}$	$J_{Px,L}$	$J_{PPh3,L}$	Refs.
P-26	$MePh_2P$	−180	30	24	481	480	25	[119]
P-27	Me_2PhP	−173	31	16	482	463	26	[119]
P-28	Bu_3P	−199	32	32	502	458	41	[119]
P-29	Et_2NPh_2P	−167	30	64	510	501	30	[119]
P-30	$(Et_2N)_2PhP$	−170	30	77	524	479	27	[119]
P-31	$PhHNPh_2P$	−178	29	49	479	524	27	[119]
P-32	$(Me_2N)_3P$	−180	29	84	523	493	30	[119]
P-33	$[(CH_2)_5N]_3P$	−173	29	77	542	497	32	[119]
P-34	$[O(CH_2)_4N]_3P$	−181	28	76	526	528	32	[119]
P-35	$(EtO)_3P$	−196	32	80	473	562	15	[119]

$$\textbf{P-15} \xrightarrow[-\,PPh_3]{+\,L} \underset{\textbf{P-26 to P-35}}{[L\text{—}P\text{—}PPh_3][AlCl_4]} \xrightarrow[-\,PPh_3]{+\,L} \underset{\textbf{P-16 to P-25}}{[L\text{—}P\text{—}L][AlCl_4]} \quad (4)$$

(4)

Table 5 P(1) stabilized by NHCs; $\delta = {}^{31}$P NMR in ppm; distances in Å, angles in °.

Nr.	R	anion	P-C	P-C	C-P-C	^{31}P NMR	Refs.
P-38a	Me	OTf	1.797(1)	1.796(1)	98.2(1)	−119	124
P-38b	Dipp	OTf^-	1.773(3)	1.818(3)	109.2(1)	−66	125
P-38c	mes	Cl^-	–	–	–	−124	126
P-39a	Me	I^-	1.790(2)	1.801(2)	96.96(7)	−114	124,127
P-39b	Et	Cl^-	–	–	–	−129	126
P-39c	^{i}Pr	Cl^-	1.824(2)	1.823(2)	97.35(9)	−126	126
P-40a	Me	OTf^-, Br^-	1.783(3)	1.787(3)	99.36(12)	−101	128
P-40b	Et	Br^-	1.798(2)	1.799(2)	98.78(10)	−108	128
P-40c	^{i}Pr	Br^-	1.8002(13)	1.8032(13)	98.63(6)	−109	128
P-41a	Me	OTf^-	1.7810(14)	1.7813(15)	92.07(6)	−81 to −83	129
P-41b	Bn	BPh_4^-	1.7797(16)	1.7811(16)	92.51(7)	−81 to −83	129
P-41c	^{n}Bu	Br^-	–	–	–	−81 to −83	129
P-42		BPh_4^-	1.799(3)	1.799(3)	98.76(14)	−92.0	110
P-43		BPh_4^-	–	–	–	−118.9	110

Table 6 Arseones with a P-As-P core.

	P-As in Å	P-As in Å	P-As-P in °	^{31}P NMR in ppm	Refs.
As-1	2.266(2)	2.236(2)	103.26(2)	89.0	145
As-2	–	–	–	60.8	107,146
As-3	2.254(2)	2.252(1)	84.57(5)	54.6	109
As-4	2.2517(8)	2.2468(9)	87.37(2)	68.4	109
As-5	2.250(1)	2.244(1)	93.0(1)	18.3	115
As-6	–	–	–	23.6	109
As-7	–	–	–	24.0	109
As-8	2.253(2)	2.260(2)	97.06(7)	33.9	147

4.2.2 Posphorone with a C-P-C core

Few compounds have been prepared in which P(I) is stabilized by two molecules of carbenes, such as dimethylaminocarben, NHCs, CDCs, or others. Two pairs of free electrons and a bent structure similar to the related isoelectronic carbone denotes its bonding system as belonging to the member of phosphorones. Some examples of imidazolidinium substituted P(I) atoms are mentioned.[121] Compounds with a C-P-C core are collected in Figs. 27–31.

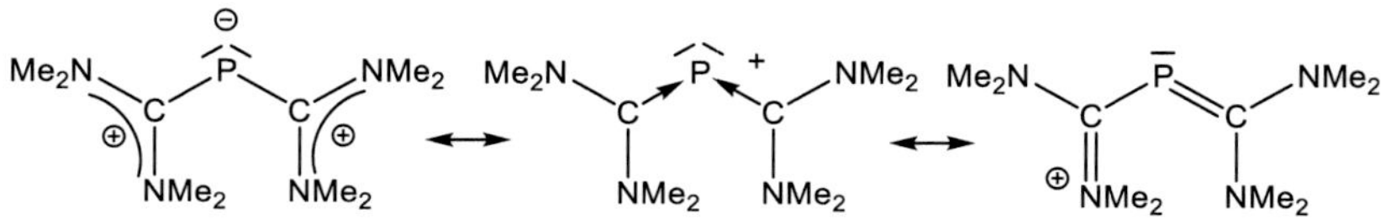

P-36

Fig. 27 Mesomeric forms of **P-36**.

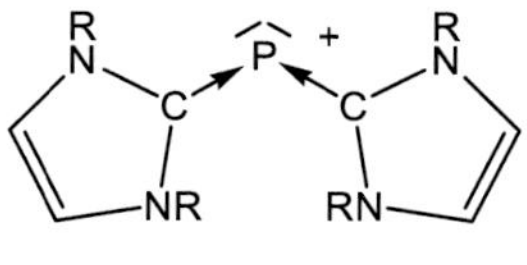

P-38 a R = Me b R = Dipp

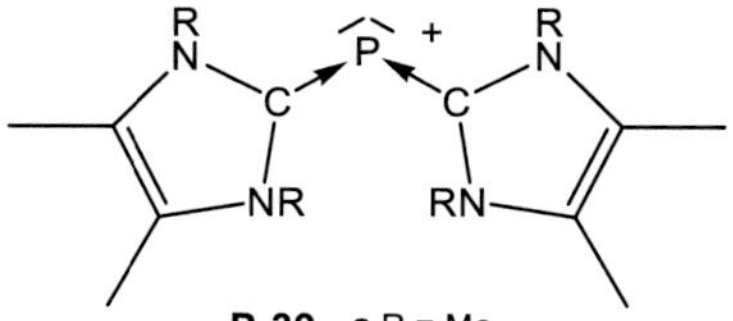

P-39 a R = Me b R = Et c R = iPr

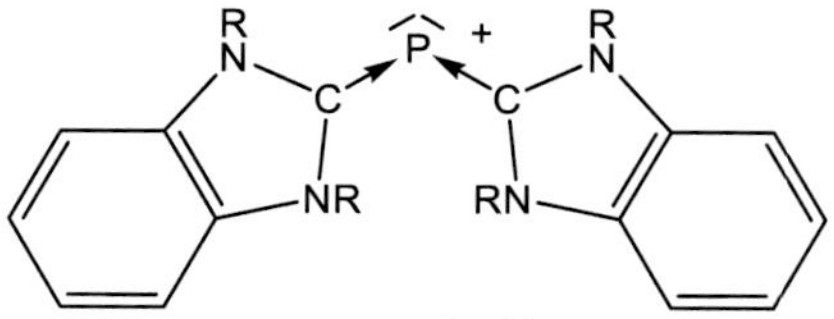

P-40 a R = Me b R = Et c R = iPr

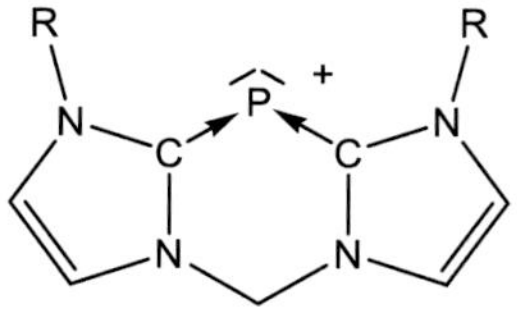

P-41 a R = Me b R = Bn c R = nBu

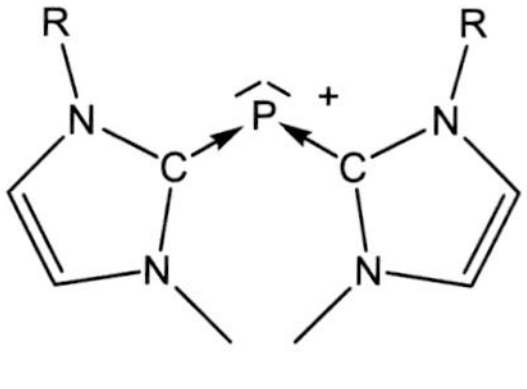

P-42 R = P^tBu_2

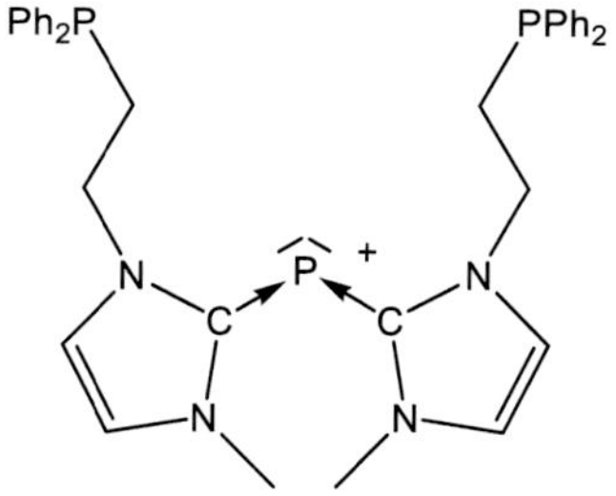

P-43

Fig. 28 P(I) atoms stabilized by NHCs.

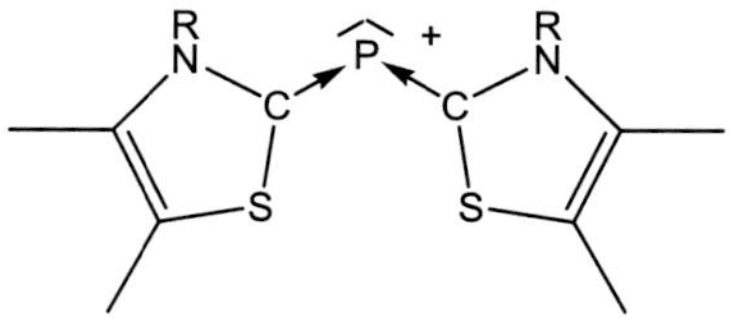

P-44 a R = Me b R = Et c R = ^iPr d R = ^nBu

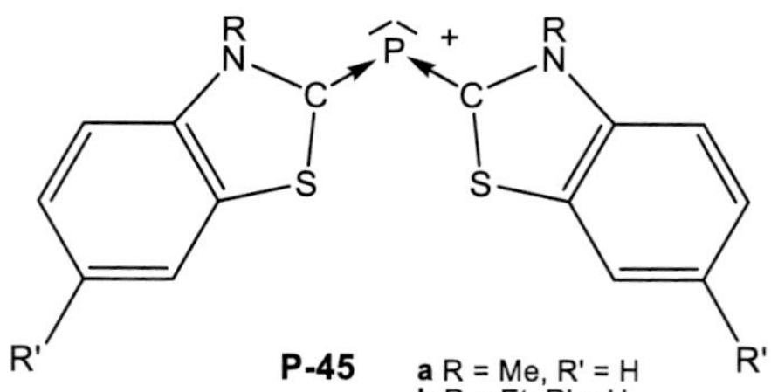

P-45 a R = Me, R' = H b R = Et, R' = H c R = Me, R' = OMe d R = Et, R' = OMe e R = Et, R' = Br

Fig. 29 P(I) stabilized by various N,S heterocyclic carbenes.

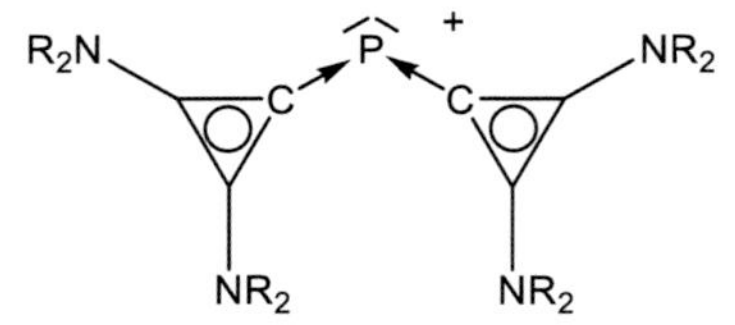

P-46 R = ^{i}Pr

Fig. 30 P(I) cation **P-46** with cyclopropylidene ligands.

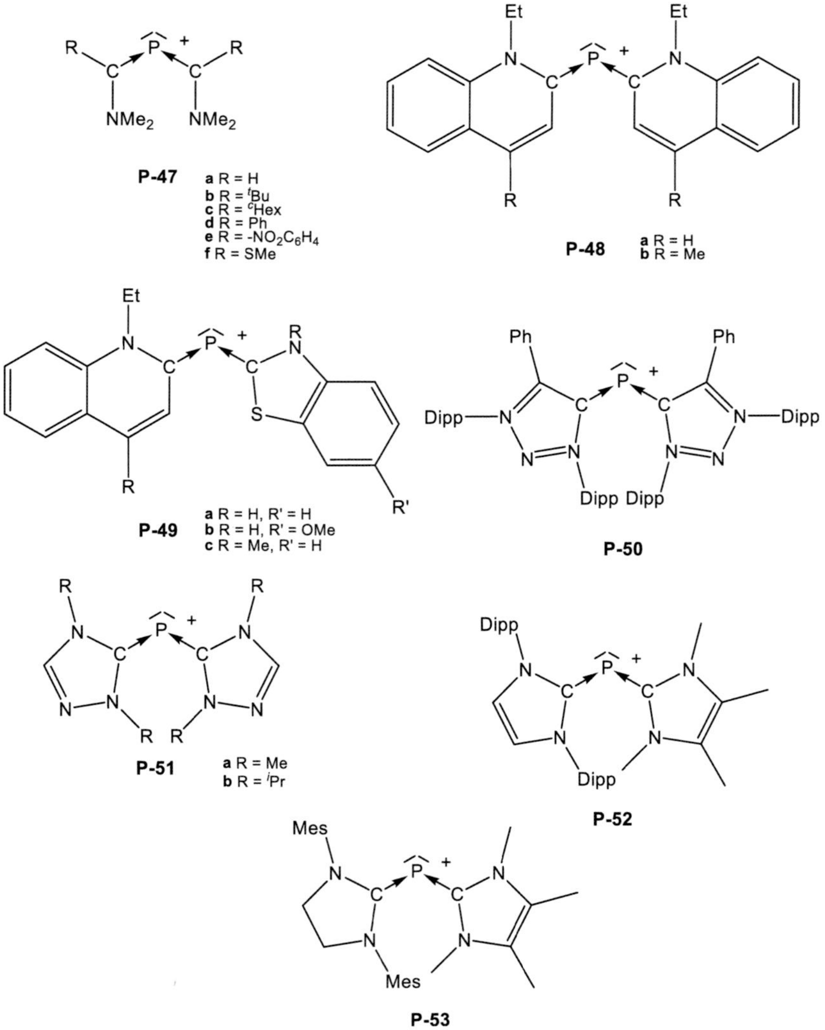

P-47 **a** R = H, **b** R = ^{t}Bu, **c** R = ^{c}Hex, **d** R = Ph, **e** R = $-NO_2C_6H_4$, **f** R = SMe

P-48 **a** R = H, **b** R = Me

P-49 **a** R = H, R' = H, **b** R = H, R' = OMe, **c** R = Me, R' = H

P-50

P-51 **a** R = Me, **b** R = ^{i}Pr

P-52

P-53

Fig. 31 P(I) cations stabilized by NHC, cAAC and similar carbene ligands.

If two equivalents of tetramethylchloroformamidinium chloride are allowed to react with $P(SiMe_3)_3$ the chloride of **P-36** is obtained with release of 3 eq. of Me_3SiCl, named as tetrakis(dimethylamino)phosphaallyl chloride. In the ^{31}P NMR spectrum, a singlet at −20 ppm was recorded.[122] **P-36** can be methylated by MeI at the P atom to give the related dication [Me-**P-36**]I_2. An X-ray analysis of the ClO_4^- salt confirms the bent structure of **P-36** with a C-P-C angle of 103.6(2)°; the C—P distance amounts to 1.796(4) Å.[123]

The cyclic carbene stabilized compounds **P-37** (**a** = Me, **b** = Et) were obtained as the BPh_4^- salt upon ligand exchange from reacting [**P-22**]BPh_4^- with the related electron rich olefin biimidazolidinylidene; a low field shift at −93 ppm for the phosphorus atom was found (Scheme 2).[122]

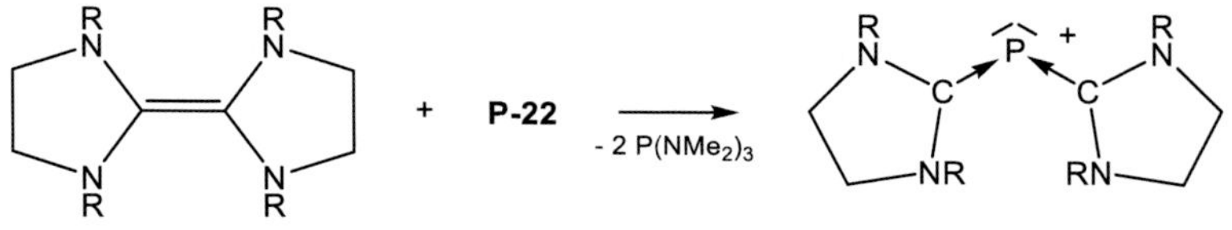

Scheme 2 Synthesis of **P-37**.

Several compounds with P(I) stabilized by NHCs were reported; important data were collected in Table 5. Thus, the iodide of **P-38a** was obtained as bright yellow crystals upon reacting the related imidazolidinium chloride with P_4/KO^{t}Bu.[124] Reduction of the related dichloro derivative by KC_8 produced **P-38b** as the OTf^- or $GaCl_4^-$ salt; **P-38b** gives a series of mono addition products M**P-38b** (M = CuBr, AuCl, and AgOTf) and with AuCl the bis adduct $(AuCl)_2$**P-38b**.[125] **P-38c** is mentioned in Ref. [126] **P-39a** was similarly obtained as **P-38a**[124] and also prepared with the anion $[(TerN)_2P]^-$; similar crystallographic parameters were recorded.[127] The compounds **P-40** could be synthesized by ligand exchange of **P-2a** with the corresponding benzimidazole-2-ylidenes and were fully characterized.[128]

Chelating bis-N-heterocycliccarbene ligated P(I) salts have been described. The salts **P-41** (**a**, R = Me; **b**, R = benzyl; **c**, R = ^{n}Bu) form upon reacting the corresponding chelating ligand at −78° with the bromide of **P-22**. According to the ^{31}P NMR spectra the P(I) atoms are less shielded than the acyclic variants.[129] The BPh_4^- salt of **P-2**a was allowed to react with two equivalents of the related NHC to give yellow **P-42**. A similar procedure gave **P-43**.[110]

Compounds with a P(I) atom stabilized by S,N heterocyclic carbenes have been known since 1964 and are collected in Fig. 29. Thus, **P-44b** was obtained as a deep red salt (I^- as the counterion) upon reacting N-ethyl-4,5-dimethylthiazolium iodide/KH at −78° in THF and addition of **P-2a** as the bromide. The ^{31}P NMR spectrum features a singlet at 17 ppm. A similar procedure with related dimethylthiazolium iodide gave **P-44a**, **P-44c**, and **P-44d**, respectively. Crystallographic parameters of **P-42b** are: P-C = 1.7866(15) Å, C-P-C = 102.23(10)°; the P atom sits on a C_2 axis. All compounds react with elementary sulfur to give [S_2-**P-44**]I salts; the two sulfur atoms occupy the free pairs of electrons at P(I).[130]

A series of **P-45** cations was prepared by Dimroth through reaction of N-ethyl-2-chlorobenzothiazolium fluoroborate (or N-Methyl) and $P(CH_2OH)_3$ to give **P-45a** and **P45b**, respectively. The related reaction produced **P-45c** and **P-45d**. The ^{31}P NMR signal was located at −26 ppm;[131] see also[132,133]. The crystal structure analysis was performed for the perchlorate of **P-45b** (R = Et, R′ = H); P-C = 1.76 Å and C-P-C = 105°. The planes of both benzthiazole rings form an angle of 3°.[134]

The chloride of **P-46** shown in Fig. 30 was obtained from the reaction of P_4 with the related carbene as a dark orange powder; a high field singlet in the ^{31}P NMR spectrum at −93 ppm was recorded. The X-ray analysis revealed P—C distances of 1.787(2) and 1.788(2) Å and a C-P-C angle of 104.68(9)°.[135] **P-46** was also obtained [O-**P-44**]BF_4 and [O_2-**P-44**]BF_4.[136]

Carbenes stabilized by only one N-electron pair, similar to cAACs, also coordinate to a P(I) atom. A series of P(I) compounds, **P-47a** to **P-47f**, were prepared upon reacting the corresponding imidoyl chloride with $P(SiMe_3)_3$. The yellow to orange red colored compounds show singlets in the ^{31}P NMR spectra in the range of 16–117 ppm.[137] The phosphamethincyanine **P-48a** and **P-48b** were reported in Ref. [132] and the crystal structure of **P-48a** (as the perchlorate) gave P—C distances of 1.81 and 1.76 Å and a C-P-C angle of 104°.[138] The asymmetrical compounds **P-49** were reported in [132].

Starting from 4-phenyl-1,2,3-triazol-5-ylidene (a mesoionic carbene) which was allowed to react with PBr_3/KPF_6 gave **P-50** as a red PF_6^- salt. The ^{31}P NMR spectrum shows a singlet at $\delta = -89.8$ ppm. An X-ray analysis revealed C–P bond lengths of 1.796(3) and 1.797(3) Å and a C-P-C angle of 104.58(12)°. **P-50** reacts with O_2 and S_8 to give [O_2-**P-50**]PF_6 and [S-**P-50**]PF_6, respectively.[139]

Two moles of 1,4-dimethyl-1,2,4-triazolium iodide [^{Me}Taz][I] and K^tBuO were allowed to react with **P-2a** iodide to give **P-51a** with release of dppe; products with the anions Br^- and BPh_4^- were obtained similarly. The X-ray analysis of **P-51a** showed P—C bond lengths of 1.795(4) Å and a C-P-C angle of 95.1(2)°. With [^{iPr}Taz][I] the related violet salts **P-51b** was obtained; the crystallographic parameters of the BPh_4^- salt are P—C bond lengths of 1.799(2) Å and a C-P-C angle of 99.72(12)°. ^{31}P NMR shifts are $\delta = -125.9$ ppm. Comparison with the related salts **P-38** revealed that TAZs are better π-acceptors than IMIDs.[140]

The asymmetric phosphorone **P-53** (BPh_4^- salt) was prepared upon reaction of a solution of **P-54** with ^{Me}NHC^{Me}. The compound resonates at $\delta = 94.3$ ppm in the ^{31}P NMR spectrum. Important structural parameters are 1.772(3) Å to SIMes and 1.828(3) Å to ^{Me}NHC^{Me} and a C-P-C bond angle of 98.21(15)°. A 1:1 mixture of the related NHCs was allowed to react with the OTf^- salt of **P-2a** to give yellow **P-52**; it resonates at $\delta = -103.1$ ppm. The bond lengths from an X-ray analysis are P–C (NHC^{Dipp}) = 1.7944(18) Å and P–C (^{Me}NHC^{Me}) = 1.821(2) Å; the C–P–C angle is 101.99(8)°.[110] The compounds are collected in Fig. 31.

4.2.3 *Phosphorone with a C-P-P core*

The phosphorone **P-54**, also denoted as [(^{Cl}Im^{Dipp})P-P(Dipp) = N(SiMe$_3$)]$^+$ was prepared from [(^{Cl}Im^{Dipp})P = P(Dipp)]OTf upon reacting with Me_3SiN_3. The ^{31}P NMR spectrum reveals an AX spin system with a large $^1J_{PP}$ coupling constant of −665 Hz; P(I) resonates at $\delta = -51.5$ ppm, while the adjacent P atom resonates at $\delta = 179.4$ ppm. The molecular structure displays C—P and P—P bond lengths of 1.829(4) and 2.0367(16) Å, respectively.[141]

The phosphorone **P-55** could only be obtained in solution as the BPh_4^- salt upon reacting of the related carbene with **P-2a**. In the ^{31}P NMR spectrum a signal at about −125 ppm was assigned to the P(I) atom. Large yellow crystals of **P-56** as the BPh_4^- salt were obtained according to the procedure outlined in Fig. 32. A signal at −161.1 ppm in the ^{31}P NMR spectrum was assigned to the P(I) atom. C—P and P—P bond lengths of 1.7955(11) and 2.1294(4) Å, respectively, and a C-P-P angle of 96.37(4)° result from X-ray analysis.[110] Compounds with a C-P-P core are collected in Fig. 32.

4.2.4 *Phosphorone with a N-P-N core*

Reaction of the related 1,2-bis(arylimino)acenaphthene with a mixture of $PCl_3/SnCl_2$ generated the dark green phosphorone **P-57** as the $SnCl_5^-$ salt; see Fig. 33. It can be seen as a phosphorone in which the P(I) atom with two lone pairs of electrons is coordinated by two immino functions (a), or as a

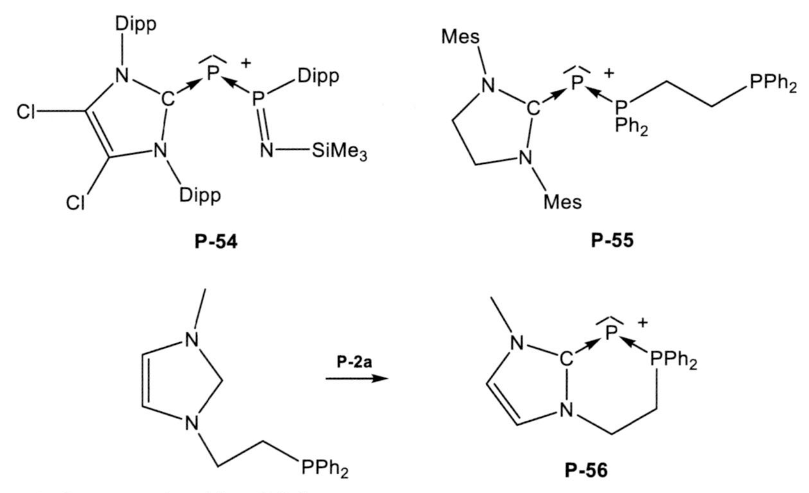

Fig. 32 Compounds with a C-P-P core.

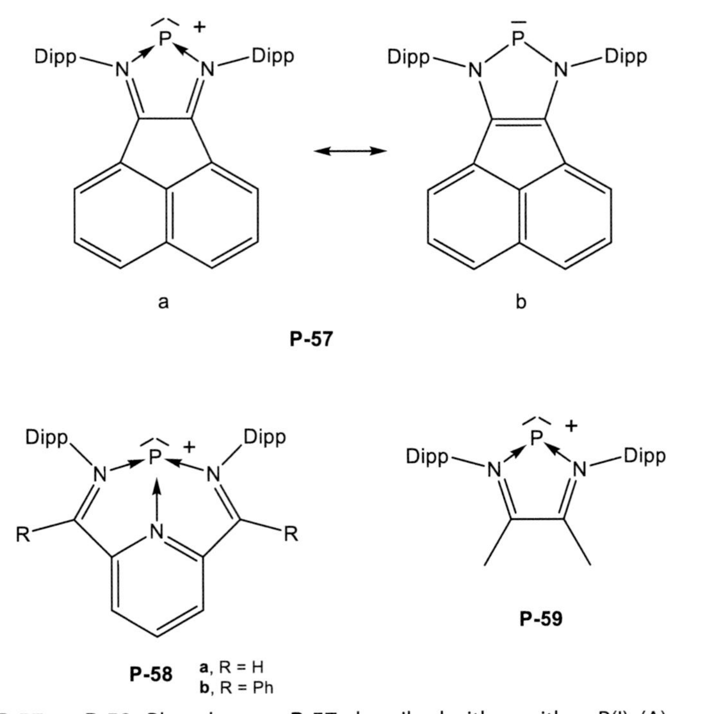

Fig. 33 **P-57** to **P-59**; Phosphorone **P-57** described either with a P(I) (A) or a P(III) atom (B).

P(III) atom with one lone pair of electrons and two P—N bonds. An X-ray analysis revealed P—N bond lengths of 1.694(4) and 1.689(4) Å; the N-P-N angle amounts to 90.23(17)°. Similar values are found for the I_3^- salt. For the related arseone **As-11** from the same working group, see Fig. 36.[142]

A red powder of the phosphorone **P-58a** was prepared upon addition of a DCM solution of PI_3 to a solution of the related ligand in the same solvent. **P-58b** was obtained analogously. The X-ray analysis of **P-58a** (as the I_3^- salt) revealed P—N bond distances of 1.877(7) and 1.975(8) Å; a shorter bond distance to the pyridine N atom of 1.722(6) Å was found. The related values of **P-58b** are 1.936(6) Å for the flanking P—N bonds and 1.722(8) Å to the pyridine N-atom. In the Br^- and $B_{12}Cl_{12}^-$ salts of **P-58a** the P—N distances to the imine nitrogen atoms are more asymmetrical (Br^- salt: 1.755(3) and 2.318(3) Å; $B_{12}Cl_{12}^-$ salts: 1.808(2) and 2.177(2) Å). In the ^{31}P NMR spectrum **P-58a** resonate at $\delta = 169$ ppm and **P-58b** at $\delta = 154$ ppm.[143]

Reaction of 1,4-diaryl-2,3-dimethyl-1,4-diazabutadiene with PCl_3/$SnCl_2$ in DCM yielded pink colored **P-59** as the $SnCl_5^-$ salt; the analogous reaction with PI_3 gave the I_3^- salt. In the ^{31}P NMR spectra the compounds resonate at $\delta = 201$ ppm. The X-ray analyses of both salts revealed C—P bond lengths of 1.667(3) and 1.668(3) Å with N-P-N angles of 88.88 (17)° and 88.90(19)°. Calculations were performed using the *Gausian-98* suite of programs.[144] The compounds with the N-P-N core are summarized in Fig. 33.

4.3 Arseone $[AsL_2]^+$

A series of arseones with P-As-P and C-As-C cores is reported, and reviewed in Ref. [99].

4.3.1 Arseone with a P-As-P core

Compounds with a P-As-P core show singlets in the ^{31}P NMR spectra ranging between ca. 20 and 90 ppm; the P—As distances range from 2.236 to 2.266 Å. They are collected in Fig. 34. The BPh_4^- salt of **As-1**[145] was obtained by reduction of $AsCl_3$ with $P(NMe_2)_3$ (3 equiv.) in the presence of $NaBPh_4$. Data of the X-ray analysis of colorless crystals are given in Table 6. [As(dppe)]I (**As-2**) was obtained by room temperature reaction of dppe with AsI_3 in dichloromethane to produce an orange solution. After 2 h of stirring the volatile products were removed in vacuum and

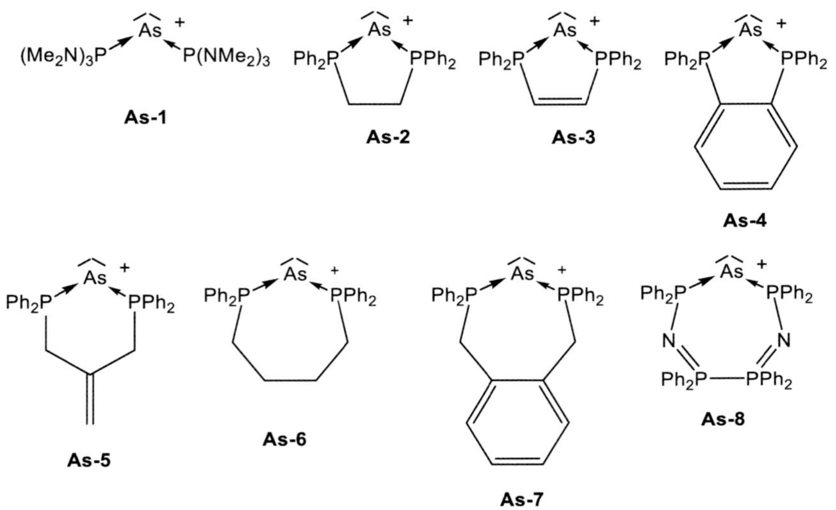

Fig. 34 Collection of the arseones **As-1** to **As-7** with a P-As-P core.

the solid material was washed with ether to remove I_2 and then with pentane to remove residual dppe. Concentration of the solution gave yellow solid of the arseone [As(dppe)]I.[107] The crystal structures of **As-2** with various anions are reported in Ref. [146].

Reaction of *cis*-1,2-Bisdiphenylphosphinoethene (dppE) and $SnCl_2$ in DCM followed by addition of $AsCl_3$ afforded **As-3**. For the crystal structure of the $C_2H_2(Ph_2POSnCl_5)^{2-}$ salt, see Table 6. **As-4** as the $SnCl_6^{2-}$ salt was similarly obtained from 1,2-Bisdiphenylphosphinobenzene, $SnCl_2$ and $AsCl_3$ to produce fine yellow crystals.[109] In **As-5** the As atom is part of a six-membered ring; the yellow powder formed as the $SnCl_6^{2-}$ salt upon reacting the related chelating phosphine with $AsCl_3$ and $SnCl_2$.[115] **A-6** and **A-7** were similarly obtained from the chelating phosphine, $AsCl_3$ and $SnCl_2$ and characterized by their ^{31}P NMR spectra.[109] AsI_3 and 2 equivalents of $LiN(PPh_2)_2$ gave **As-8** as the I^- salt. The crystal structure of the iodide with 4 THF molecules was determined.[147]

4.3.2 Arseones with a C-As-C core

The BF_4^- salts of **As-9** (arsa-methin-cyanine) and **As-10** stem from the reaction of $As(SiMe_3)_3$ with 2-chloro-N-alkyl-benzthiazolium fluoroborate or 2-chloro-N-alkyl- chinolinium fluoroborate, respectively.[133] No structural details are reported. Arseones with this core are depicted in Fig. 35.

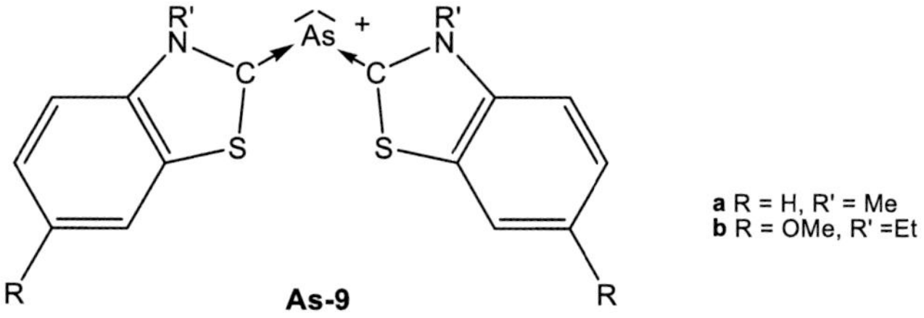

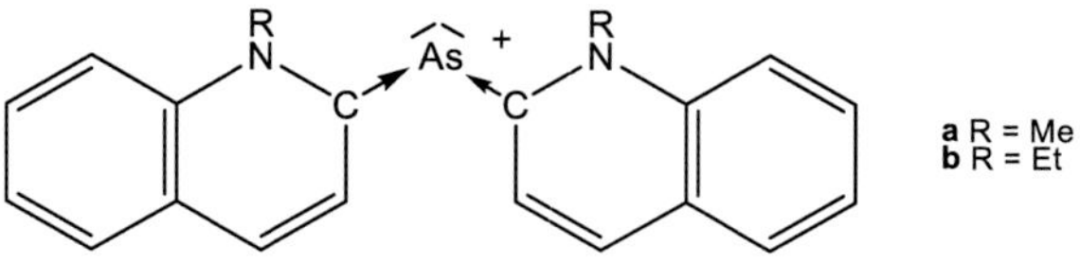

a R = H, R' = Me
b R = OMe, R' =Et

As-9

a R = Me
b R = Et

As-10

Fig. 35 Arseones **As-9** and **As-10** with a C-As-C core.

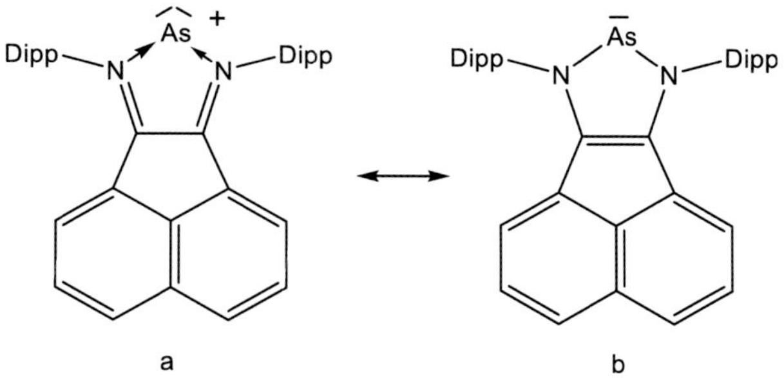

a b

As-11

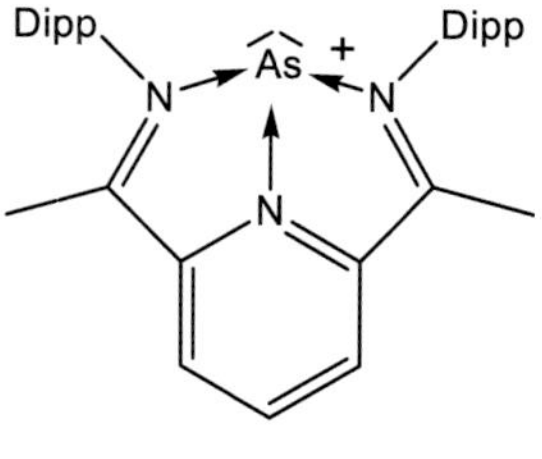

As-12

Fig. 36 Arseones with an N-As-N core. **As-11** described either with an As(I) (A) or an As(III) atom (B); see also **P-57** in Fig. 33.

4.3.3 Arseone with a N-As-N core

Arseone with this core are mentioned in a review[99] and are collected in Fig. 36. **As-11** with the $SnCl_5^-$ anion (prepared from the basic diimino compound and a mixture of $AsCl_3/SnCl_2$) can be seen as an arseone in which the As(I) atom with two lone pairs of electrons is coordinated by two immino functions (a), or as a As(III) atom with one lone pair of electrons and two amino bonds, (b) An X-ray analysis of green crystals revealed As—N bond lengths of 1.839(3) and 1.857(4) Å; the N-As-N angle amounts to 84.89 (15)°; see the related phosphorone **P-57**.[142]

Synthesis of **As-12** follows the typical procedure above mentioned for **As-11** (diimin/$AsCl_3$/$SnCl_2$).[148]

4.4 Stibione $[SbL_2]^+$

The stibione **Sb-1** could be stabilized by a strong σ-donor and π-acceptor ligand such as the cAAC, shown in Fig. 37.

4.4.1 Stibione with a C-Sb-C core

Purple crystals of the OTf^- salt of **Sb-1** were obtained upon reduction of SbF_3 with C_8K in the presence of two equivalents of the related cAAC. The molecular structure reveals Sb—C bond lengths of 2.145(2) and 2.1498(18) Å and a C-Sb-C angle of 111.87(7)°. The salt was characterized by multi NMR spectroscopy. Calculation of the charge distribution with the *NBO* method gives a positive partial charge at the Sb(I) atom of +0.72 e.[149]

The reaction of $[(\eta^5\text{-Cp}^*)\text{Sb(tol)}][\text{B}(\text{C}_6\text{F}_5)_4]_2$ ($\text{Cp}^* = \text{C}_5\text{Me}_5^-$) with bis(diisopropylamino)cyclopropenylidenes (BAC) gave yellowish crystals of **Sb-2**. As the $[\text{B}(\text{C}_6\text{F}_5)_4]^-$ salt. The X-ray analysis revealed a symmetric structure with a C—Sb bond lengths of 2.141(3) Å and a C-Sb-C angle of 89.8(1)°.[150] Compounds with the C-Sb-C core are summarized in Fig. 37.

4.4.2 Stibione with a P-Sb-P core

The stibione **Sb-3** was obtained as the OTf^- salt upon reaction of the chelating ligand 5,6-bis(diisopropylphosphino)acenaphthene with $SbCl_3$ and 3 equivalents of Me_3SiOTf as orange crystals. The molecular structure shows P—Sb bond lengths of 2.4595(9) and 2.4598(9) Å and a P-Sb-P angle of 89.53(3)°. In the ^{31}P NMR spectrum a peak at $\delta = 16$ ppm was attributed to **Sb-3**. **Sb-3** exhibits nucleophilic behavior towards the coinage metals

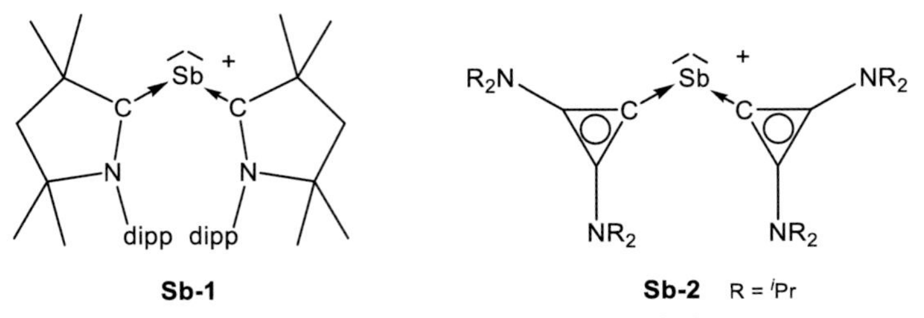

Fig. 37 Graphical representation of the stibiones **Sb-1** and **Sb-2**.

Cu^{I}, Ag^{I}, and Au^{I}. Thus, the trications [M-(**Sb-3**)$_2$]$^{3+}$ (M = Au, Ag) with a linear Sb-M-Sb array result from **Sb-3** and AuCl or $AgSbF_6$, respectively. MOTf (M = Cu, Ag) produced the compounds [(OTf)$_2$M-**Sb-3**]$^{2+}$. CuCl gave complexes with Cu_2Cl_2 and Cu_2ClOTf centers coordinated by two **Sb-3** units. *DFT* calculations have been performed on **Sb-3**; *NBO* analysis revealed two lone pairs of elecrons of σ- an π-type on Sb. First and second PAs of 168.4 and 47.4 kcal mol^{-1}, respectively, were calculated.[151] The graphical representation is outlined in Fig. 38.

4.5 Bismutone [BiL$_2$]$^+$

4.5.1 Bimutones with a C-Bi-C core

The isostructural bismutone **Bi-1** was similarly obtained as the stibione **Sb-1** from $BiCl_3$, C_8K and the related cAAC at −100°. Deep blue rod-shaped crystals of the OTf^- salt, containing two molecules of LiOTf were isolated. The molecular structure (see graphical representation in Fig. 39) reveals Bi—C bond lengths of 2.270(3) and 2.314(3) Å and a C-Bi-C angle of 111.89(10)°. The product was characterized by multi NMR and UV–vis spectroscopies. Calculation of the charge distribution with the *NBO* method gives a positive partial charge at the Bi(I) atom of +0.69 e.[149]

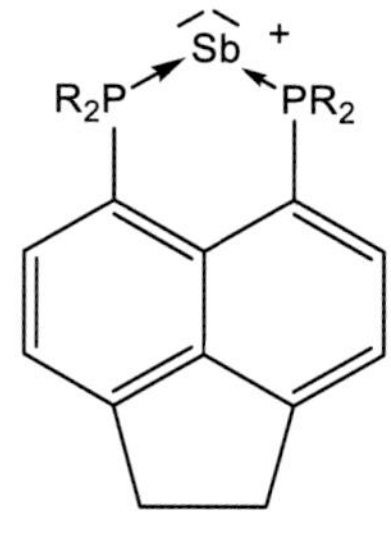

Sb-3 R = ^{i}Pr

Fig. 38 Graphical representation of the stibione **Sb-3**.

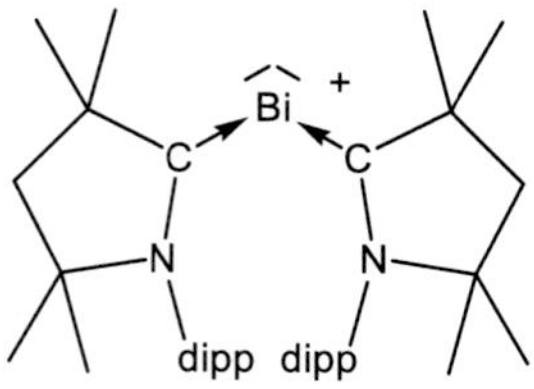

Bi-1

Fig. 39 Graphical representation of the bismutone **Bi-1**.

5. Group 16 compounds $[EL_2]^{2+}$, E = O, S, se, Te

According to the twofold positive charge $[EL_2]^{2+}$ compounds are less nucleophilic and addition compounds at E are less common. Part of the review in [79] is dedicated to group 16 $[EL_2]^{2+}$ compounds.

5.1 Oxygeone $[OL_2]^{2+}$

5.1.1 Oxygeone wth a P-O-P core

A colorless solid of **O-1** was obtained as the SO_3F^- salt upon reaction of triphenylphosphine oxide with excess fluorosulfonic anhydride.[152] The crystal structure of the $SO_3CF_3^-$ salt of **O-1** is reported;[153] a slightly bent structure with a P-O-P angle of 164.5(2)°, and P—O distances of 1.597(2), and 1.598 (2) Å were found. A similar reaction between $(Me_2N)_3PO$ and trifluoromethanesulfonic anhydride afforded $[O\{P(NMe_2)_3\}_2][SO_3CF_3]_2$; see **O-1** and **O-2** in Fig. 40.[154]

5.1.2 Oxygeone with a C-O-C core

The oxygen atom O(II) stabilized by aminocarbene ligands **O-3** was prepared from two equivalents of tetramethyl urea and Tf_2O; the counterion is OTf^- A signal at 156.7 ppm for the carbene C atom in the ^{13}C NMR spectrum was recorded. **O-4** with the counterion OTf^- contains six-membered cAAC ligands and was obtained similarly upon reacting with *N*-methylpyridone. The ^{13}C signal for the donating C atom, $\delta = 155.2$ ppm. **O-5a** was analogously obtained from the related urea and the reloaded ^{13}C NMR signal appears at $\delta = 154.7$ ppm. The X-ray analysis of **O-5a** as the OTf- salt revealed C—O bond distances of about 1.349(10) and 1.345(9) Å and a C-O-C angle of 120.3(7)°.[155] The results for **O-5b** [156] and **O-5c** were very similar.[157] The X-ray analysis of **O-6a** revealed O—C bond distances of 1.346(11) and 1.348(11) Å and a C-O-C angle of 116.0(7)°. The compound **O-6a** was obtained from reacting 2,3-bis(dimethylamino) propanone with $(CF_3SO_2)_2O$. A analogous procedure gave **O-6b;**[155] see also **O-6a** as starting material for mixed sulfurones and selenones;[155] see Fig. 41.

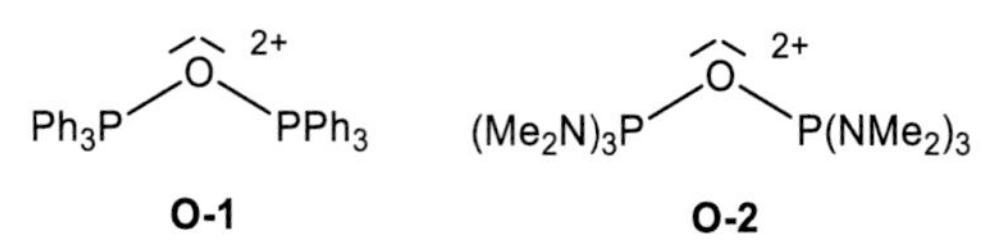

Fig. 40 Graphical representation of $[O(PPh_3)_2][SO_3F]_2$ (**O-1**) and **O-2**.

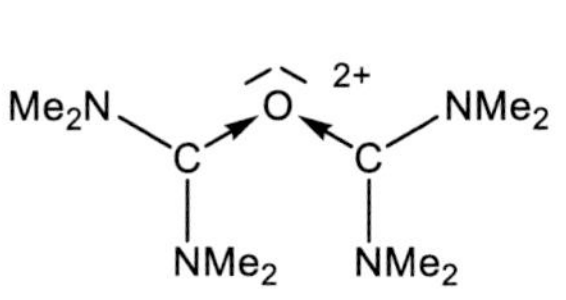

Fig. 41 Oxygeone **O-3** to **O-6** with a C-O-C core.

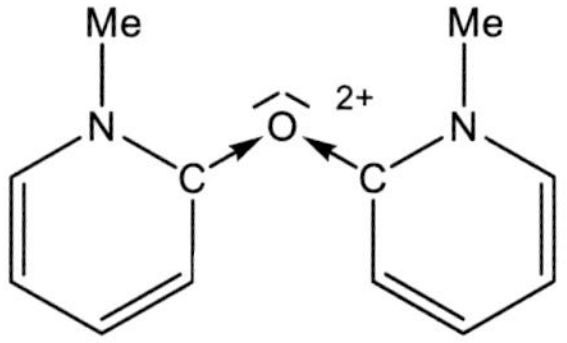

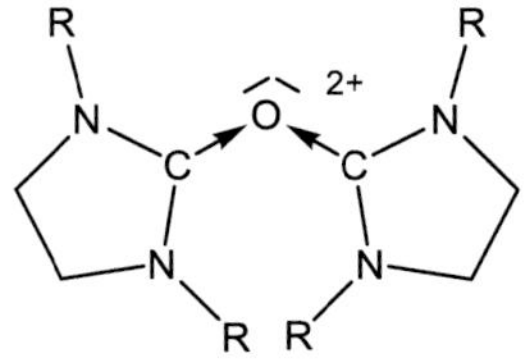

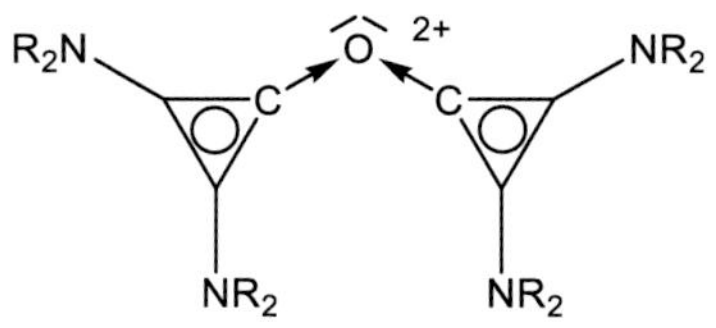

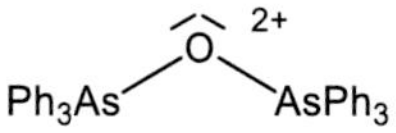

Fig. 42 Oxygeone **O-7** with a As-O-As core.

5.1.3 Oxygeone with a As-O-As core

Following the procedure outlined for **O-1** with triphenylarsine oxide colorless **O-7** was obtained as the SO_3F^- salt;[152] see Fig. 42.

5.2 Sulfurones $[SL_2]^{2+}$

5.2.1 Sulfurone with a C-S-C core

The sulfurone **S-1a** was prepared upon reacting **O-5c** with the related thione as the OTf^- salt. The asymmetrical **S-2** stems from **O-5c** with the corresponding S,S substituted thione. The reaction of **O-3** with the related thione generates **S-3**; the compounds were characterized by IR spectroscopy and elemental analyses.[157] **S-4** was described in Ref. [158].

The crystal structure analysis of the accidentally obtained sulfurone **S-5** reveals a S—C bond length of 1.761 Å; the anion is $Sb_2Cl_{10}^-$.[159] The X-ray analysis of **S-6** shows C—S bond distances of 1.70(1) Å and a C-S-C angle of 103.3(5)°,[160] for the preparation see Ref. [161]; see Fig. 43.

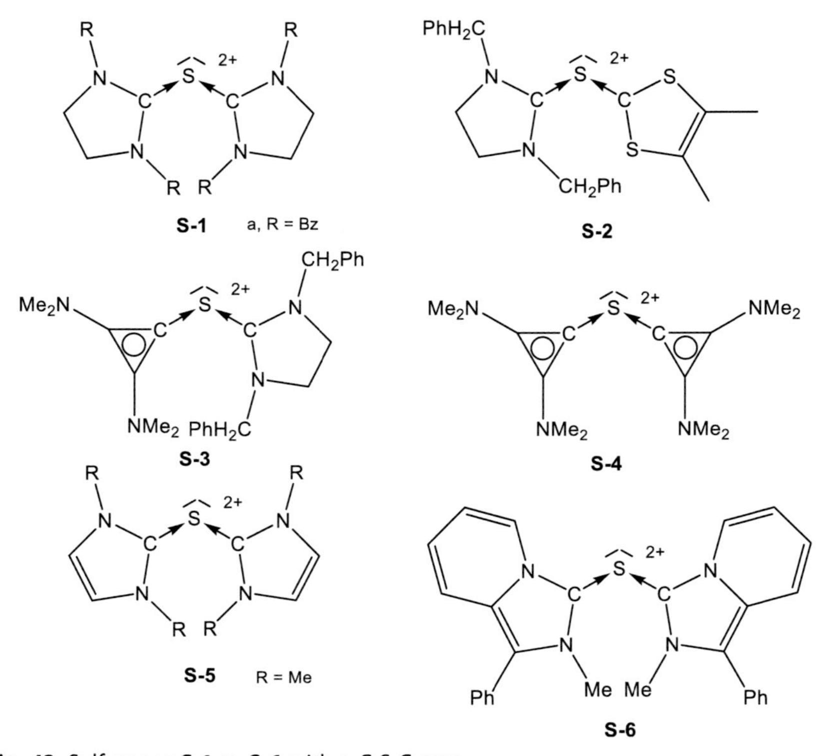

Fig. 43 Sulfurones **S-1** to **S-6** with a C-S-C core.

5.3 Selenone $[SeL_2]^{2+}$

5.3.1 Selenones with a C-Se-C core

Light yellow **Se-1** was obtained by the reaction of the related NHC with **Se-4** (see N-Se-N core). An X-ray analysis of **Se-1** as the OTf^- salt revealed C—Se bond distances of 1.915(3) and 1.920(3) Å and a C-Se-C angle of 96.3(1)°.[162]

The starting material for the OTf^- salt of **Se-2** was the oxygeone **O-5c** which was allowed to react with 1,3-dibenzyl-2-imidazolidineselenone. **Se-3** was similarly obtained from **O-6a** as the starting material. The compounds were characterized by ^1H, ^{13}C NMR and IR spectroscopies.[157] Selenones with a C-Se-C core are collected in Fig. 44.

5.3.2 Selenone with an N-Se-N core

The diamine stabilized Se(II) in **Se-4a** (as the OTf^- salt) was obtained as a colorless powder upon dehalogenation of the related dichloro compound

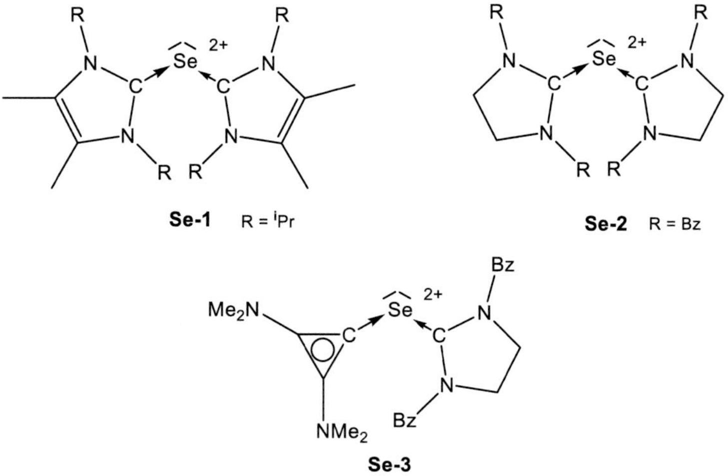

Se-1 R = ^{i}Pr

Se-2 R = Bz

Se-3

Fig. 44 Selenones **Se-1** to **Se-3** with a C-Se-C core.

with two equivalents of TMS-OTf. In the ^{77}Se NMR spectrum it resonates at $\delta = 1335$ ppm. An X-ray analysis revealed Se—N bond lengths of 1.843 (3) and 1.845(3) Å and a N-Se-N angle of 84.1(2)°.[162] The related **Se-4b** (as the $SnCl_6^{2-}$ salt) was prepared by reacting a mixture of 1,4-(2,6-diisopropyl) phenyl-1,4-diaza-1,3-butadiene) and $SnCl_2$ with $SeCl_4$. In the ^{77}Se NMR spectrum a signal at $\delta = 1268$ ppm was recorded. Selected structural data are Se—N bond lengths of 1.890(4) Å and a N-Se-N angle of 81.8(3)°.[163] **Se-5** was proposed as an intermediate upon dehalogenation of the related dibromo compound by AgOTf and was not isolated.[162] The compounds are summarized in Fig. 45.

5.4 Telurone $[TeL_2]^{2+}$

5.4.1 Telurone with a C-Te-C core

Te-1 in Fig. 46 was prepared from **Te-3** by ligand exchange, the counter-anions are OTf^-. To a solution of **Te-3** was added the related NHC to give a colorless solid of **Te-1**. An X-ray analysis of **Te-1** revealed Te—C distances of 2.163(4) and 2.368(3) Å and a C-Te-C angle of 91.5(1)°.[164]

5.4.2 Telurone with a N-Te-N core

Te-2 was mentioned in a review.[79] **Te-3** in Fig. 47 was prepared from the related TeI_2 complex upon reaction with AgOTf and obtained as a dark red powder. An X-ray analysis revealed N—Te bond lengths of 2.151(4) and

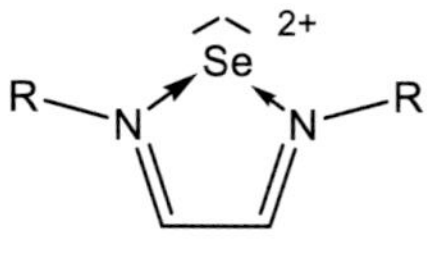

Se-4

a, R = Cy
b, R = Dipp

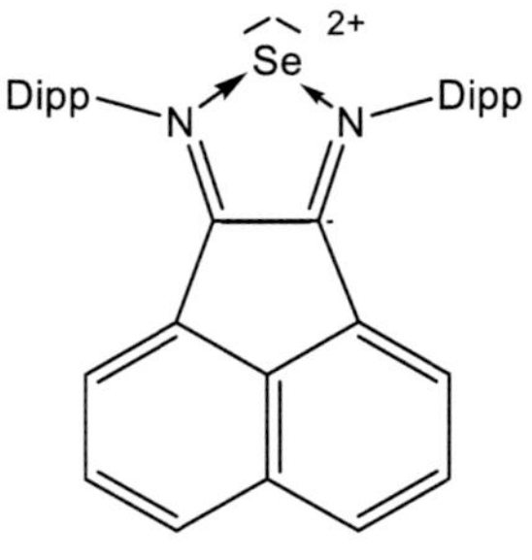

Se-5

Fig. 45 Selenones with a N-Se-N core.

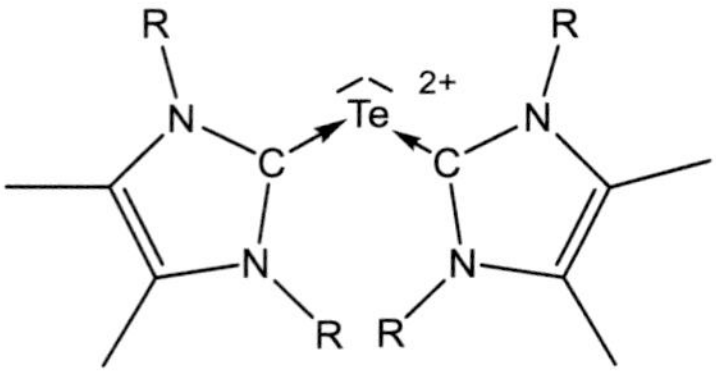

R = ^{i}Pr

Te-1

Fig. 46 Telurone **Te-1** with a C-Te-C core.

Te-2

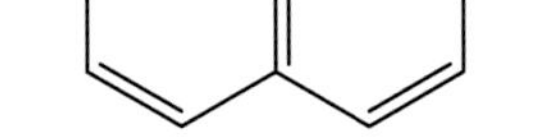

Te-3

Fig. 47 Telurones **Te-2** and **Te-3** with a N-Te-N core.

2.182(4) Å and a N-Te-N angle of 75.9(2)°; strong ion-pairing gave additionally Te—O contacts of 2.329(4) and 2.471(4) Å to the anion OTf^{-}.[164]

6. Conclusion and future perspectives

The present summary of theoretical and experimental work shows that great progress has been made in the field of the title compounds, which is a dynamic area of inorganic chemistry. The account may be considered as a momentary summary of the rapidly developing field of low-valent main-group chemistry. The synthesis and isolation particularly of heavy analogs of compounds with the general formula EL_2 or ELL' is a challenge for the experimental skills of the inventive chemist. The characterization of the bonding type between the atom or ion E and the ligands L poses an equally challenging problem for the theoretical analysis of the interatomic interactions and the traditional heuristic bonding models. It may be expected that the present review will need to be updated already within a few years.

References

1. (a) Dewar, M. J. S. *Bull. Soc. Chim. Fr.* **1951**, *18*, C79; (b) Chatt, J.; Duncanson, L. A. *J. Chem. Soc.* **1953**, 2929.
2. Frenking, G.; Fröhlich, N. *Chem. Rev.* **2000**, *100*, 717–774.
3. Tonner, R.; Öxler, F.; Neumüller, B.; Petz, W.; Frenking, G. *Angew. Chem. Int. Ed.* **2006**, *45*, 8038–8042.
4. Ramirez, F.; Desai, N. B.; Hansen, B.; McKelvie, N. *J. Am. Chem. Soc.* **1961**, *83*, 3539–3540.
5. Varshavskii, Y. S. *Russ. J. Gen. Chem.* **1980**, *50*, 406, (*Zhurnal Obshchei Khimii*, **1980**, *50*, 514–528 English version https://www.academia.edu/12112367/Complexes_of_Carbon_in_Oxidation_States_0_II_and_IV.
6. Hardy, G. E.; Zink, J. I.; Kaska, W. C.; Baldwin, J. C. *J. Am. Chem. Soc.* **1978**, *100*, 8001–8002.
7. Dewar, M. J. S. *J. Am. Chem. Soc.* **1984**, *106*, 669–682.
8. Mohapatra, C.; Kundu, S.; Paesch, A. N.; Herbst-Irmer, R.; Stalke, D.; Andrada, D. M.; Frenking, G.; Roesky, H. W. *J. Am. Chem. Soc.* **2016**, *138*, 10429–10432.
9. Carbon suboxide C3O2 has a bent equilibrium geometry in the gas phase, but it becomes linear in the solid state due to intermolecular interactions (a) Gas phase value Jensen, P; Johns, J.W.C J. *Mol. Spectrosc* **1986**, *118*, 248–266; (b) Solid state structure Ellern, A; Drews, T; Seppelt, K Z. *Anorg. Allg. Chem* **2001**, *627,* 73–76; (c) CCSD (T) calculations using large basis sets; Koput, J. *Chem. Phys. Lett.* **2000**, *320*, 237–244.
10. Havenith, R. W. A.; Cunha, A. V.; Klein, J. E. M. N.; Perolari, F.; Feng, X. *PhysChemChemPhys* **2021**, *23*, 3327–3334.
11. (a) Schmidbaur, H. *Angew. Chem. Int. Ed.* **2007**, *46*, 2984–2985; (b) Frenking, G.; Neumüller, B.; Petz, W.; Tonner, R.; Öxler, F. *Angew. Chem. Int. Ed.* **2007**, *46*, 2986–2987.
12. (a) Himmel, D.; Krossing, I.; Schnepf, A. *Angew. Chem. Int. Ed.* **2014**, *53*, 370–374; (b) Frenking, G. *Angew. Chem. Int. Ed.* **2014**, *53*, 6040–6064; (c) Himmel, D.; Krossing, I.; Schnepf, A. *Angew. Chem. Int. Ed.* **2014**, *53*, 6047–6048.
13. (a) Tonner, R.; Frenking, G. *Chem. Eur. J.* **2008**, *14*, 3260–3272; (b) Tonner, R.; Frenking, G. *Chem. Eur. J.* **2008**, *14*, 3273–3289; (c) Tonner, R.; Frenking, G. *Pure Appl. Chem.* **2009**, *81*, 597–614; (d) Klein, S.; Tonner, R.; Frenking, G. *Chem. Eur. J.* **2010**, *16*, 10160–10170; (e) Esterhuysen, C.; Frenking, G. *Chem. Eur. J.* **2011**, *17*, 9944–9956.
14. Tonner, R.; Frenking, G. *Angew. Chem. Int. Ed.* **2007**, *46*, 8695–8698.
15. Dyker, C. A.; Lavallo, V.; Donnadieu, B.; Bertrand, G. *Angew. Chem. Int. Ed.* **2008**, *47*, 3206–3209.

16. (a) Fürstner, A.; Alcarazo, M.; Goddard, R.; Lehmann, C. W. *Angew. Chem. Int. Ed.* **2008**, *47*, 3210–3214; (b) Alcarazo, M.; Lehmann, W.; Anoop, A.; Thiel, W.; Fürstner, A. *Nat. Chem.* **2009**, *1*, 295–301.
17. Frenking, G.; Tonner, R. *WIREs Comput Mol Sci* **2011**, *1*, 869–878.
18. (a) Chen, W.-C.; Shen, J.-S.; Jurca, T.; Peng, C.-J.; Lin, Y.-H.; Wang, Y.-P.; Shih, W.-C.; Yap, G. P. A.; Ong, T.-G. *Angew. Chem. Int. Ed.* **2015**, *54*, 15207–15227; (b) Hsu, Y.-C.; Shen, J.-S.; Lin, B.-C.; Chen, W.-C.; Chan, Y.-T.; Ching, W.-M.; Yap, G. P. A.; Hsu, C.-P.; Ong, T.-G. *Angew. Chem. Int. Ed.* **2015**, *54*, 2420–2424; (c) Chen, W.-C.; Lee, C.-Y.; Lin, B.-C.; Hsu, Y.-C.; Shen, J.-S.; Hsu, C.-P.; Yap, G. P. A.; Ong, T.-G. *J. Am. Chem. Soc.* **2014**, *136*, 914–917; (d) Chen, W.-C.; Hsu, Y.-C.; Lee, C.-Y.; Yap, G. P. A.; Ong, T.-G. *Organometallics* **2013**, *32*, 2435–2442; (e) Chen, W.-C.; Shih, W.-C.; Jurca, T.; Zhao, L.; Andrada, D. M.; Peng, C.-J.; Chang, C.-C.; Liu, S.-K.; Wang, Y.-P.; Wen, Y.-S.; Yap, G. P. A.; Hsu, C.-P.; Frenking, G.; Ong, T.-G. *J. Am. Chem. Soc.* **2017**, *139*, 12830–12836.
19. (a) Takagi, N.; Shimizu, T.; Frenking, G. *Chem. Eur. J.* **2009**, *15*, 3448–3456; (b) Takagi, N.; Shimizu, T.; Frenking, G. *Chem. Eur. J.* **2009**, *15*, 8593–8604; (c) Takagi, N.; Frenking, G. *Theoret. Chem. Acc.* **2011**, *129*, 615–623.
20. (a) Ishida, S.; Iwamoto, T.; Kabuto, C.; Kira, M. *Nature* **2003**, *421*, 725–727; (b) Iwamoto, T.; Masuda, H.; Kabuto, C.; Kira, M. *Organometallics* **2005**, *24*, 197–199.
21. (a) Mondal, K. C.; Roesky, H. W.; Klinke, F.; Schwarzer, M. C.; Frenking, G.; Niepötter, B.; Wolf, H.; Herbst-Irmer, R.; Stalke, D. *Angew. Chem. Int. Ed.* **2013**, *52*, 2963–2967.
22. Xiong, Y.; Yao, S.; Tan, G.; Inoue, S.; Driess, M. *J. Am. Chem. Soc.* **2013**, *135*, 5004–5007.
23. (a) Kuwabara, T.; Nakada, M.; Hamada, J.; Guo, J. D.; Nagase, S.; Saito, M. *J. Am. Chem. Soc.* **2016**, *138*, 11378–11382; (b) Arrowsmith, M.; Braunschweig, H.; Celik, M. A.; Dellermann, T.; Dewhurst, R. D.; Ewing, W. C.; Hammond, K.; Kramer, T.; Krummenacher, I.; Mies, J.; Radacki, K.; Schuster, J. K. *Nat. Chem.* **2016**, *8*, 890–894.
24. Yao, S.; Xiong, Y.; Driess, M. *Acc. Chem. Rev.* **2017**, *50*, 2026–2037.
25. Celik, M. A.; Sure, R.; Klein, S.; Kinjo, R.; Bertrand, G.; Frenking, G. *Chem. Eur. J.* **2012**, *18*, 5676–5692.
26. Kinjo, R.; Donnadieu, B.; Celik, M. A.; Frenking, G.; Bertrand, G. *Science* **2011**, *333*, 610–613.
27. (a) Lavallo, V.; Canac, Y.; Prasang, C.; Donnadieu, B.; Bertrand, G. *Angew. Chem. Int. Ed.* **2005**, *44*, 5705–5709; (b) Soleilhavoup, M.; Bertrand, G. *Acc. Chem. Res.* **2015**, *48*, 256–266.
28. Braunschweig, H.; Dewhurst, R. D.; Hupp, F.; Nutz, M.; Radacki, K.; Tate, C. W.; Vargas, A.; Ye, Q. *Nature* **2015**, *522*, 327–330.
29. Zhang, Q.; Li, W.-L.; Xu, C.; Chen, M.; Zhou, M.; Li, J.; Andrada, D. M.; Frenking, G. *Angew. Chem. Int. Ed.* **2015**, *54*, 11078–11083.
30. Sidgwick, N. V. *Chem. Rev.* **1931**, *9*, 77–88.
31. (a) Frenking, G.; Tonner, R.; Klein, S.; Takagi, N.; Shimizu, T.; Krapp, A.; Pandey, K. K.; Parameswaran, P. *Chem. Soc. Rev.* **2014**, *43*, 5106–5139; (b) Frenking, G.; Hermann, M.; Andrada, D. M.; Holzmann, N. *Chem. Soc. Rev.* **2016**, *45*, 1129–1144.
32. Patel, N.; Sood, R.; Bharatam, P. V. *Chem. Rev.* **2018**, *118*, 8770–8785.
33. Zhang, L.; Dong, J.; Zhou, M.; Qin, Q. *J. Chem. Phys.* **2000**, *113*, 10169–10173.
34. Takagi, N.; Tonner, R.; Frenking, G. *Chem. Eur. J.* **2012**, *18*, 1772–1780.
35. Andrada, D. M.; Holzmann, N.; Frenking, G. *Can. J. Chem.* **2016**, *94*, 1006–1014.
36. Purushothaman, I.; De, S.; Parameswaran, P. *Chem. Eur. J.* **2018**, *24*, 3816–3824.

37. Majhi, P. K.; Sasamori, T. *Chem. Eur. J.* **2018**, *24*, 9441–9455.
38. Fustier-Boutignon, M.; Nebra, N.; Mézailles, N. *Chem. Rev.* **2019**, *119*, 8555–8570.
39. Petz, W.; Frenking, G. *Top. Organomet. Chem.* **2010**, *30*, 49–92.
40. Petz, W. *Coord. Chem. Rev.* **2015**, *291*, 1–27.
41. Alcarazo, M. *Struct. Bond.* **2018**, *177*, 25–50.
42. Liu, S.; Chen, W. C.; Ong, T. G. *Struct. Bond.* **2018**, *177*, 51–71.
43. Xiong, Y.; Yao, S.; Inoue, S.; Epping, J. D.; Driess, M. *Angew. Chem.* **2013**, *125*, 7287–7291.
44. Xiong, Y.; Yao, S.; Mîller, R.; Kaupp, M.; Driess, M. *Angew. Chem. Int. Ed.* **2015**, *54*, 10254–10257.
45. Burchert, A.; Yao, S.; Miller, R.; Schattenberg, C.; Xiong, Y.; Kaupp, M.; Driess, M. *Angew. Chem. Int. Ed.* **2017**, *56*, 1894–1897.
46. Li, Y.; Chan, Y.-C.; Li, Y.; Purushothaman, I.; De, S.; Parameswaran, P.; So, C.-W. *Inorg. Chem.* **2016**, *55*, 9091–9098.
47. Mondal, K. C.; Samuel, P. P.; Tretiakov, M.; Singh, A. P.; Roesky, H. W.; Stückl, A. C.; Niepötter, B.; Carl, E.; Wolf, H.; Herbst-Irmer, R.; Stalke, D. *Inorg. Chem.* **2013**, *52*, 4736–4743.
48. Roy, S.; Mondal, K. C.; Krause, L.; Stollberg, P.; Herbst-Irmer, R.; Stalke, D.; Meyer, J.; Stückl, A. C.; Maity, B.; Koley, D.; Vasa, S. K.; Xiang, S. Q.; Linser, R.; Roesky, H. W. *J. Am. Chem. Soc.* **2014**, *136*, 16776–16779.
49. Ishida, S.; Iwamoto, T.; Kabuto, C.; Kira, M. *Nature* **2003**, *421*, 725–727.
50. Iwamoto, T.; Abe, T.; Ishida, S.; Kabuto, C.; Kira, M. *J. Organomet. Chem.* **2007**, *692*, 263–270.
51. Kira, M.; Iwamoto, T.; Ishida, S.; Masuda, H.; Abe, T.; Kabuto, C. *J. Am. Chem. Soc.* **2009**, *131*, 17135–17144.
52. Kira, M. *Chem. Commun.* **2010**, *46*, 2893–2903.
53. Yao, S.; Kostenko, A.; Xiong, Y.; Ruzicka, A.; Driess, M. *J. Am. Chem. Soc.* **2020**, *142*, 12608–12612.
54. Keuter, J.; Hepp, A.; Mück-Lichtenfeld, C.; Lips, F. *Angew. Chem. Int. Ed.* **2019**, *58*, 4395–4399.
55. Wang, Y.; Karni, M.; Yao, S.; Kaushansky, A.; Apeloig, Y.; Driess, M. *J. Am. Chem. Soc.* **2019**, *141*, 12916–12927.
56. Tanaka, H.; Inoue, S.; Ichinohe, M.; Driess, M.; Sekiguchi, A. *Organometallics* **2011**, *30*, 3475–3478.
57. Koike, T.; Nukazawa, T.; Iwamoto, T. *J. Am. Chem. Soc.* **2021**, *143*, 14332–14341.
58. Sugahara, T.; Sasamori, T.; Tokitoh, N. *Angew. Chem. Int. Ed.* **2017**, *56*, 9920–9923.
59. Wilhelm, C.; Raiser, D.; Schubert, H.; Sindlinger, C. P.; Wesemann, L. *Inorg. Chem.* **2021**, *60*, 9268–9272.
60. Kobayashi, R.; Ishida, S.; Iwamoto, T. *Organometallics* **2021**, *40*, 843–847.
61. Ghana, P.; Arz, M. I.; Das, U.; Schnakenburg, G.; Filippou, A. C. *Angew. Chem. Int. Ed.* **2015**, *54*, 9980.
62. Li, Y.; Mondal, K. C.; Roesky, H. W.; Zhu, H.; Stollberg, P.; Herbst-Irmer, R.; Stalke, D.; Andrada, D. M. *J. Am. Chem. Soc.* **2013**, *135*, 12422–12428.
63. Xiong, Y.; Yao, S.; Karni, M.; Kostenko, A.; Burchert, A.; Apeloig, Y.; Driess, M. *Chem. Sci.* **2016**, *7*, 5462–5469.
64. Su, B.; Ganguly, R.; Li, Y.; Kinjo, R. *Angew. Chem. Int. Ed.* **2014**, *53*, 13106–13109.
65. Su, B.; Ota, K.; Li, Y.; Kinjo, R. *Dalton Trans.* **2019**, *48*, 3555–3559.
66. Su, B.; Ganguly, R.; Lib, Y.; Kinjo, R. *Chem. Commun.* **2016**, *52*, 613–616.
67. Chu, T.; Belding, L.; van der Est, A.; Dudding, T.; Korobkov, I.; Nikonov, G. I. *Angew. Chem. Int. Ed.* **2014**, *53*, 2711–2715.
68. Iwamoto, T.; Abe, T.; Kabuto, C.; Kira, M. *Chem. Commun.* **2005**, 5190–5192.
69. Wang, Y.; Karni, M.; Yao, S.; Apeloig, Y.; Driess, M. *J. Am. Chem. Soc.* **2019**, *141*, 1655–1664.

70. Yao, S.; Kostenko, A.; Xiong, Y.; Lorent, C.; Ruzicka, A.; Driess, M. *Angew. Chem. Int. Ed.* **2021**, *60*, 14864–14868.
71. Jana, A.; Huch, V.; Scheschkewitz, D. *Angew. Chem. Int. Ed.* **2013**, *52*, 12179–12182.
72. Jana, A.; Majumdar, M.; Huch, V.; Zimmer, M.; Scheschkewitz, D. *Dalton Trans.* **2014**, *43*, 5175–5181.
73. Majhi, P. K.; Zimmer, M.; Morgenstern, B.; Scheschkewitz, D. *J. Am. Chem. Soc.* **2021**, *143*, 8981–8986.
74. Majhi, P. K.; Zimmer, M.; Morgenstern, B.; Huch, V.; Scheschkewitz, D. *J. Am. Chem. Soc.* **2021**, *143*, 13350–13357.
75. Krebs, K. M.; Hanselmann, D.; Schubert, H.; Wurst, K.; Scheele, M.; Wesemann, L. *J. Am. Chem. Soc.* **2019**, *141*, 3424–3429.
76. Flock, J.; Suljanovic, A.; Torvisco, A.; Schoefberger, W.; Gerke, B.; Pöttgen, R.; Fischer, R. C.; Flock, M. *Chem. Eur. J.* **2013**, *19*, 15504–15517.
77. Wiberg, N.; Lerner, H.-W.; Vasisht, S.-K.; Wagner, S.; Karaghiosoff, K.; Nöth, H.; Ponikwar, W. *Eur. J. Inorg. Chem.* **1999**, 1211–1218.
78. Cabrera-Trujillo, J. J.; Fernández, I. *Inorg. Chem.* **2019**, *58*, 7828–7836.
79. Nesterov, V.; Reiter, D.; Bag, P.; Frisch, P.; Holzner, R.; Porzelt, A.; Inoue, S. *Chem. Rev.* **2018**, *118*, 9678–9842.
80. Schmidbaur, H.; Jonas, G. *Chem Ber* **1968**, *101*, 1271–1285.
81. Schmidbaur, H.; Füller, H.-J. *Chem Ber* **1977**, *110*, 3528–3535.
82. Rankin, D. W. H.; Walkinshaw, M. D.; Schmidbaur, H. *J. Chem. Soc. Dalton Trans.* **1982**, 2320–2371.
83. Chin, H. B.; Smith, M. B.; Wilson, R. D.; Bau, R. *J. Am. Chem. Soc.* **1974**, *96*, 5285–5287.
84. Wilson, R. D.; Bau, R. *J. Am. Chem. Soc.* **1974**, *96*, 7601–7602.
85. Schwesinger, R.; Schlemper, H.; Hasenfratz, C.; Willaredt, J.; Dambacher, T.; Breuer, T.; Ottaway, C.; Fletschinger, M.; Boele, J.; Fritz, H.; Putzas, D.; Rotter, H. W.; Bordwell, F. G.; Satish, A. V.; Ji, G.-Z.; Peters, E.-M.; Peters, K.; von Schnering, H. G.; Walz, L. *Liebigs Ann.* **1996**, 1055–1081.
86. Schwesinger, R.; Link, R.; Wenzl, P.; Kossek, S.; Keller, M. *Chem. Eur. J.* **2006**, *12*, 429–437.
87. Pleschke, A.; Marhold, A.; Schneider, M.; Kolomeitsev, A.; Röschenthaler, G.-V. *J. Fluor. Chem.* **2004**, *125*, 1031–1038.
88. Kunetskiy, R. A.; Císařová, I.; Šaman, D.; Lyapkalo, I. M. *Chem. Eur. J.* **2009**, *15*, 9477–9485.
89. Patel, D. S.; Bharatam, P. V. *J. Phys. Chem. A* **2011**, *115*, 7645–7655.
90. Patel, D. S.; Bharatam, P. V. *Chem. Commun.* **2009**, 1064–1066.
91. Kathuria, D.; Arfeen, M.; Bankkar, A. A.; Bharatam, P. V. *J. Chem. Sci.* **2016**, *128*, 1607–1614.
92. Bernhardi, I.; Drews, T.; Seppelt, K. *Angew. Chem. Int. Ed.* **1999**, *38*, 2232–2233.
93. Kozma, Á.; Gopakumar, G.; Farès, C.; Thiel, W.; Alcarazo, M. *Chem. Eur. J.* **2013**, *19*, 3542–3546.
94. Zong, L.; Ban, X.; Kee, C. W.; Tan, C.-H. *Angew. Chem. Int. Ed.* **2014**, *53*, 11849–11853.
95. Ma, T.; Fu, X.; Kee, C. W.; Zong, L.; Pan, Y.; Huang, K.-W.; Tan, C.-H. *J. Am. Chem. Soc.* **2011**, *133*, 2828–2831.
96. Schwalbe, C. H.; Hunt, W. E. *J. Chem. Soc. Chem. Commun.* **1978**, 188–190.
97. Bharatam, P. V.; Arfeen, M.; Patel, N.; Jain, P.; Bhatia, S.; Chakraborti, A. K.; Khullar, S.; Gupta, V.; Mandal, S. K. *Chem. Eur. J.* **2016**, *22*, 1088–1096.
98. Yao, S.; Szilvasi, T.; Xiong, Y.; Lorent, C.; Ruzicka, A.; Driess, M. *Angew. Chem. Int. Ed.* **2020**, *59*, 22043–22047.
99. Ellis, B. D.; Macdonald, C. L. B. *Coord. Chem. Rev.* **2007**, *251*, 936–973.

100. Dionisi, E. M.; Binder, J. F.; LaFortune, J. H. W.; Macdonald, C. L. B. *Dalton Trans.* **2020**, *49*, 12115–12127.
101. Dillon, K. B.; Monks, P. K.; Olivey, R. J.; Karsch, H. H. *Heteroat. Chem.* **2004**, *15*, 464–467.
102. Schmidpeter, A.; Lochschmidt, S.; Burget, G. *Phosphorus und Sulfur* **1983**, *29*, 23–26.
103. Schmidpeter, A.; Lochschmidt, S.; Sheldrick, W. S. *Angew. Chem. Int. Ed.* **1982**, *21*, 63–64.
104. Schmidpeter, A.; Lochschmidt, S.; Cowley, A. H.; Pakulski, M. *Inorg. Synth.* **1990**, *27*, 253–258.
105. Norton, E. L.; Szekely, K. L. S.; Dube, J. W.; Bomben, P. G.; Macdonald, C. L. B. *Inorg. Chem.* **2008**, *47*, 1196–1203.
106. Lochschmidt, S.; Schmidpeter, A. *Z. Naturforsch.* **1985**, *40b*, 765–773.
107. Ellis, B. D.; Carlesimo, M.; Macdonald, C. L. B. *Chem. Commun.* **2003**, 1946–1947.
108. Ellis, B. D.; Macdonald, C. L. B. *Inorg. Chem.* **2006**, *45*, 6864–6874.
109. Barnham, R. J.; Deng, R. M. K.; Dillon, K. B.; Goeta, A. E.; Howard, J. A. K.; Puschmann, H. *Heteroat. Chem.* **2001**, *12*, 501–510.
110. Binder, J. F.; Swidan, A.; Macdonald, C. L. B. *Inorg. Chem.* **2018**, *57*, 11717–11725.
111. Boon, J. A.; Byers, H. L.; Dillon, K. B.; Goeta, A. E.; Longbottom, D. A. *Heteroat. Chem.* **2000**, *11*, 226–231.
112. Dillon, K. B.; Olivey, R. J. *Heteroat. Chem.* **2004**, *15*, 150–154.
113. Deng, R. M. K.; Dillon, K. B.; Goeta, A. E.; Thompson, A. L. *Acta Crystallogr.* **2005**, *E61*, m296–m298.
114. Burton, J. D.; Deng, R. M. K.; Dillon, K. B.; Monks, P. K.; Olivey, R. J. *Heteroat. Chem.* **2005**, *16*, 447.
115. Gamper, S. F.; Schmidbaur, H. *Chem Ber* **1993**, *126*, 601–604.
116. Kilian, P.; Slawin, A. M. Z.; Woollins, J. D. *Dalton Trans.* **2006**, 2175–2183.
117. Lochschmidt, S.; Schmidpeter, A. *Z. Naturforsch.* **1985**, *40b*, 765–773.
118. Schmidpeter, A.; Steinmüller, F.; Sheldrick, W. S. *Z. Anorg. Allg. Chem.* **1989**, *579*, 158–172.
119. Schmidpeter, A.; Lochschmidt, S.; Sheldrick, W. S. *Angew. Chem. Int. Ed.* **1985**, *24*, 226–227.
120. Schmidpeter, A.; Lochschmidt, S.; Karaghiosoff, K.; Sheldrick, W. S. *J. Chem. Soc. Chem. Commun.* **1985**, 1447–1448.
121. Schwedtmann, K.; Zanoni, G.; Weigand, J. J. *Chem. Asian J.* **2018**, *13*, 1388–1405.
122. Schmidpeter, A.; Lochschmidt, S.; Willhalm, A. *Angew. Chem. Int. Ed.* **1983**, *22*, 545–546.
123. Day, R. O.; Willhalm, A.; Holmes, J. M.; Holmes, R. R.; Schmidpeter, A. *Angew. Chem. Int. Ed.* **1985**, *24*, 764–765.
124. Cicač-Hudi, M.; Bender, J.; Schlindwein, S. H.; Bispinghoff, M.; Nieger, M.; Grützmacher, H.; Gudat, D. *Eur. J. Inorg. Chem.* **2016**, 649–658.
125. Schwedtmann, K.; Holthausen, M. H.; Feldmann, K.-O.; Weigand, J. J. *Angew. Chem. Int. Ed.* **2013**, *52*, 14204–14208.
126. Ellis, B. D.; Dyker, C. A.; Deckenc, A.; Macdonald, C. L. B. *Chem. Commun.* **2005**, 1965–1967.
127. Hinz, A.; Schulz, A.; Villinger, A. *Chem. Commun.* **2016**, 6328–6331.
128. Macdonald, C. L. B.; Binder, J. F.; Swidan, J. A.; Nguyen, H.; Kosnik, S. C.; Ellis, B. D. *Inorg. Chem.* **2016**, *55*, 7152–7166.
129. Binder, J. F.; Swidan, A.; Tang, M.; Nguyen, J. H.; Macdonald, C. L. B. *Chem. Commun.* **2015**, *51*, 7741–7744.
130. Binder, J. F.; Corrente, A. M.; Macdonald, C. L. B. *Dalton Trans.* **2016**, *45*, 2138–2147.
131. Dimroth, K.; Hoffmann, P. *Angew. Chem. Int. Ed.* **1964**, *3*, 384.

132. Dimroth, K.; Hoffmann, P. *Chem. Ber.* **1966**, *99*, 1325–1331.
133. Märkl, G.; Lieb, F. *Tetrahedron Lett.* **1967**, *36*, 3489–3493.
134. Allmann, R. *Angew. Chem. Int. Ed.* **1965**, *4*, 150–151.
135. Back, O.; Kuchenbeiser, G.; Donnadieu, B.; Bertrand, G. *Angew. Chem. Int. Ed.* **2009**, *48*, 5530–5533.
136. Zhou, J.; Liu, L. L.; Cao, L. L.; Stephan, D. W. *Angew. Chem. Int. Ed.* **2019**, *58*, 18276–18280.
137. Schmidpeter, A.; Willhalm, A. *Angew. Chem. Int. Ed.* **1984**, *23*, 903–904.
138. Kawada, I.; Allmann, R. *Angew. Chem. Int. Ed.* **1968**, 7, 69.
139. Huang, L.; Huang, S.; Zhang, Z.; Cao, L.; Xu, X.; Yan, X. *Organometallics* **2021**, *40*, 1190–1194.
140. Elnajjar, F. O.; Binder, J. F.; Kosnik, S. C.; Macdonald, C. L. B. *Z. Anorg. Allg. Chem.* **2016**, *642*, 1251–1258.
141. Schwedtmann, K.; Hennersdorf, F.; Bauza, A.; Frontera, A.; Fischer, R.; Weigand, J. J. *Angew. Chem. Int. Ed.* **2017**, *56*, 6218–6222.
142. Reeske, G.; Hoberg, C. R.; Hill, N. J.; Cowley, A. H. *J. Am. Chem. Soc.* **2006**, *128*, 2800–2801.
143. Martin, C. D.; Ragogna, P. J. *Dalton Trans.* **2011**, *40*, 11976–11980.
144. Ellis, B. D.; Macdonald, C. L. B. *Inorg. Chim. Acta* **2007**, *360*, 329–344.
145. Driess, M.; Ackermann, H.; Aust, J.; Merz, K.; von Wüllen, C. *Angew. Chem. Int. Ed.* **2002**, *41*, 450–453.
146. Ellis, B. D.; Macdonald, C. L. B. *Inorg. Chem.* **2004**, *43*, 5981–5986.
147. Dotzler, M.; Schmidt, A.; Ellermann, J.; Knoch, F. A.; Moll, M.; Bauer, W. *Polyhedron* **1996**, *15*, 4425–4433.
148. Reeske, G.; Cowley, A. H. *Chem. Commun.* **2006**, 1784–1786.
149. Siddiqui, M. M.; Sarkar, S. K.; Nazish, M.; Morganti, M.; Köhler, C.; Cai, J.; Zhao, L.; Herbst-Irmer, R.; Stalke, D.; Frenking, G.; Roesky, H. W. *J. Am. Chem. Soc.* **2021**, *143*, 1301–1306.
150. Zhou, J.; Kim, H.; Liu, L. L.; Cao, L. L.; Stephan, D. W. *Chem. Commun.* **2020**, 12935–12936.
151. Kumar, V.; Gonnade, R. G.; Yildiz, C. B.; Majumdar, M. *Angew. Chem. Int. Ed.* **2021**. https://doi.org/10.1002/anie.202111339, in press.
152. Niyogi, D. G.; Singh, S.; Verma, R. D. *Can. J. Chem.* **1989**, *67*, 1895–1897.
153. You, S.-L.; Razavi, H.; Kelly, J. W. *Angew. Chem. Int. Ed.* **2003**, *42*, 83–85.
154. Aaberg, A.; Gramstad, T.; Husebye, S. *Tetrahedron Lett.* **1979**, *24*, 2263–2264.
155. Maas, G.; Stang, P. J. *J. Organomet. Chem.* **1983**, *48*, 3038–3043.
156. Stang, P. J.; Maas, G.; Smith, D. L.; McCloskey, J. A. *J. Am. Chem. Soc.* **1981**, *103*, 4837–4845.
157. Maas, G.; Singer, B. *Chem Ber* **1983**, *116*, 3659–3674.
158. Yoshida, Z.-I.; Konishi, H.; Ogoshi, H. *Israel J. Chem.* **1981**, *21*, 139–144.
159. Williams, D. J.; Poor, P. H.; Ramirez, G.; Vanderveer, D. *Inorg. Chim. Acta* **1989**, *165*, 167–172.
160. Pointer, D. J.; Wilford, J. B. *J. Chem. Soc. Chem. Commun.* **1987**, 816–817.
161. Glover, E. E.; Vaughan, K. D.; Bishop, D. C. *J. Chem. Soc. Perkin. I* **1973**, 2959–2999.
162. Dutton, J. L.; Battista, T. L.; Sgro, M. J.; Ragogna, P. J. *Chem. Commun.* **2010**, *46*, 1041–1043.
163. Dutton, J. L.; Tuononen, H. M.; Jennings, M. C.; Ragogna, P. J. *J. Am. Chem. Soc.* **2006**, *128*, 12624–12625.
164. Dutton, J. L.; Tuononen, H. M.; Ragogna, P. J. *Angew. Chem. Int. Ed.* **2009**, *48*, 4409–4413.

CHAPTER EIGHT

Recent advances in electrocatalytic CO_2 reduction with molecular complexes

Sergio Fernández[a,b], Geyla C. Dubed Bandomo[a,b], and Julio Lloret-Fillol[a,c,*]

[a]Institute of Chemical Research of Catalonia (ICIQ), The Barcelona Institute of Science and Technology, Tarragona, Spain
[b]Departament de Química Física i Inorgànica, Universitat Rovira i Virgili, Tarragona, Spain
[c]Catalan Institution for Research and Advanced Studies (ICREA), Barcelona, Spain
*Corresponding author: e-mail address: jlloret@iciq.es

Contents

Abstract

Solutions are urgently needed to mitigate our dependence on fossil fuels and climate change. In this sense, reducing CO_2 electrocatalytically to fuels and fine chemicals is a promising strategy when powered by renewable energy. However, there are multiple challenges to overcome to have a practical application, such as improving the catalysts in terms of activity (turnover numbers, turnover frequencies) while displaying high energy efficiency (low overpotential), excellent selectivity for a preferred product (Faraday efficiencies), and long-term stability. To meet these demanding goals, a better understanding of the reaction mechanism is needed, to make more favorable the pathways that yield desired products and avoid the undesired ones. Indeed, different strategies to increase the activity, without compromising the overpotential or selectivity, have emerged from a rational design during the last years. In particular, this review focusses on molecular catalysts as the mainstay in this logical evolution of catalysts. Since mechanistic information relates to the CO_2 electrocatalytic reduced product, the chapter is divided into two main sections, a first main body of information of two-electron reduced products, such as CO and formic acid, and a second section with

Advances in Inorganic Chemistry, Volume 79
ISSN 0898-8838
https://doi.org/10.1016/bs.adioch.2022.01.001

more than two-electron reduction products (CH_2O, CH_3OH and CH_4). Selected aspects of interest for catalyst development of this quickly evolving research area are discussed.

1. Introduction

The current environmental emergency and the increasing energy demand have led to a rising interest in developing new and more effective methods for renewable energy storage.[1] The conversion of renewable energy into chemical energy through an electrical potential is one of the most promising approaches for that purpose. In this circular economy context, CO_2 has become a target molecule to utilize as a C_1 building block in the synthesis of fuels and fine chemicals.[2] However, the CO_2 molecule is linear with stable delocalized π-HOMO orbitals and a highly energetic LUMO orbital. The electronic and geometric structure of CO_2 makes it stable against electrophiles and one-electron reduction (Eq. 1).[3] Nevertheless, CO_2 typically reacts with nucleophiles like amines or hydroxide. Consequently, to produce effectively the multielectron multiproton CO_2 reduction to fuels needs to be catalyzed.[4]

Heterogeneous catalysts are envisioned as the most suitable ones for large-scale CO_2 reduction in water.[5] They can also provide access to multiple products, from CO to CH_4 (Eqs. 2–6) and products with more than one carbon (C_{2+}), but product selectivity is still problematic. Moreover, since the CO_2 reduction reaction (CO_2RR) is preferably performed in water, the hydrogen evolution reaction (HER) is an undesired competitive process to avoid (Eq. 7). In contrast, most homogeneous catalysts are limited to forming two-electron reduction products from the CO_2RR, i.e., CO and formic acid (FA).[6] Although recent reports illustrate that well-defined molecular complexes can also reduce CO_2 beyond the two-electron reduction products. Homogeneous catalysts also showed promising results in terms of durability and catalytic performance when heterogenized.[7–11] However, the main advantage of homogeneous catalysts is that their general well-defined nature and equivalence of all catalytic sites and solubility, facilitate their study by spectroscopic techniques. In this regard, the CO_2 reduction mechanisms have been studied extensively during the last years, detecting fundamental steps and identifying thermodynamic and kinetic factors governing the CO_2 reduction reaction. Those studies are demanding but crucial to establish the fundamentals to develop faster and more selective catalysts.[12] In general, a combination of electrochemistry, spectroscopy, and computational modeling is required to extract meaningful information.

$$CO_2 + e^- \rightarrow CO_2^{\cdot -} \quad E^{o\prime} = -1.90 \text{ V vs NHE} \tag{1}$$

$$CO_2 + 2H^+ + 2e^- \rightarrow CO + H_2O \quad E^{o\prime} = -0.53 \text{ V vs NHE} \tag{2}$$

$$CO_2 + 2H^+ + 2e^- \rightarrow HCO_2H \quad E^{o\prime} = -0.61 \text{ V vs NHE} \tag{3}$$

$$CO_2 + 4H^+ + 4e^- \rightarrow H_2CO + H_2O \quad E^{o\prime} = -0.48 \text{ V vs NHE} \tag{4}$$

$$CO_2 + 6H^+ + 6e^- \rightarrow CH_3OH + H_2O \quad E^{o\prime} = -0.38 \text{ V vs NHE} \tag{5}$$

$$CO_2 + 8H^+ + 8e^- \rightarrow CH_4 + 2H_2O \quad E^{o\prime} = -0.24 \text{ V vs NHE} \tag{6}$$

$$2H^+ + 2e^- \rightarrow H_2 \quad E^{o\prime} = -0.42 \text{ V vs NHE} \tag{7}$$

Techniques to highlight are cyclic voltammetry (CV) analysis and spectroelectrochemistry (SEC), which are very convenient to shed light on the electrochemical reaction mechanisms. CV, together with controlled potential electrolysis (CPE), facilitates the evaluation and catalyst benchmarking in terms of overpotential (η), maximum turnover frequency (TOF_{max}), and Faradaic yield (FY%). For a given electrochemical reduction (A is reduced to B), the formal parameters can be defined as,

$$\eta = E^{o\prime}{}_{A/B} - E_{app} \tag{8}$$

$$TOF_{max} = k_{obs} = k_{cat}\left[A^0\right] \tag{9}$$

$$FY\% = \frac{Q_B}{Q_{tot}} \cdot 100 \tag{10}$$

where $E^{o\prime}{}_{A/B}$ and E_{app} correspond to the thermodynamic reduction potential from A to B at the working conditions and the applied potential, respectively.[13] k_{obs} is the observed first-order kinetic rate constant for the catalytic reaction in s^{-1} units, which presents a linear dependence on the concentration of the reactant, giving the slope of the linear function as ($M^{-1}\ s^{-1}$). In the calculation of the FY%, Q_B and Q_{tot} are the charge associated with the formation of product B and the total charge passed during the experiment, respectively.

In addition, *in situ* spectroelectrochemistry (SEC) has emerged as a powerful technique to investigate the electrocatalytic CO_2 reduction mechanism on the electrode surface under working conditions. In spectroelectrochemical experiments, information upon the reaction is obtained simultaneously from time-resolved, and *in situ* electrochemical and spectroscopic measurements, Infrared (IR) spectroscopy, coupled with electrochemistry (IR-SEC), provide access to the detection of potential CO_2 reduction intermediates such as metal carboxylates and carbonyls. Moreover, the IR vibration is sensitive to the chemical environment and the metal oxidation state.[14]

In this regard, this review article addresses a selection of advances within the last 3 years, focusing on the rational catalyst development driven by the mechanistic understanding of the CO_2RR, focusing on well-defined homogeneous catalysts. Therefore, we have divided this review article into two main sections considering the products produced, two-electron (CO and formic acid) and more than two-electron reduction products (CH_2O, CH_3OH and CH_4). This classification helps to introduce the common critical aspects that share different catalysts for obtaining a specific and selective product. The second section presents more than two electrons CO_2 reduction reactions; however, the number of examples is still limited because it is an emerging research area.

2. Catalysts for the two-electron reduction of CO_2

2.1 CO_2 reduction to CO

CO is a relevant chemical platform widely used in industrial processes such as the Fischer-Tropsch synthesis of hydrocarbons or carbonylation reactions.[15] Molecular catalysts found effective for CO_2 reduction to CO hold the promise to be developed further toward application.[16] A driving force for the development of those molecular catalysts was the understanding of the reaction mechanism. The current insightful understanding of the mechanism originated from the large variety of catalytic systems studied, including homogeneous and heterogenized molecular catalysts based on first-, second- and third-row transition metals with a large variety of ligands, i.e., macrocyclic, pyridine-based, organometallic (CO, phosphine, carbene), among others.[17] Despite the diversity of catalysts, most of them share common intermediates such as low valent metal species $[Mn]^n$, metal-carboxylate $[M\text{-}CO_2]^n$, and metal-carbonyl $[M\text{-}CO]^{n/n+1}$ intermediates (Fig. 1).[3]

The general catalytic cycle for the CO_2-to-CO is shown in Fig. 1. The first step is commonly considered the CO_2 binding to an $[M]^n$ low-valent intermediate, forming a metal-carboxylate adduct $[M\text{-}CO_2]^n$. In the presence of a Brønsted acid $[M\text{-}CO_2]^n$ can be protonated to the corresponding $[M\text{-}CO_2H]^{n+1}$ intermediate (Fig. 1A, Eq. 2). A second protonation step can trigger the C—O bond cleavage forming a metal-carbonyl species $[M\text{-}CO]^{n+2}$. This is called the "protonation first" mechanism. Instead, if $[M\text{-}CO_2H]^{n+1}$ is reduced first, the mechanism follows the "reduction first" pathway. In this case, the $[M\text{-}CO_2H]^n$ protonation triggers the C—O bond cleavage, forming $[M\text{-}CO]^{n+1}$, a common intermediate in both mechanisms.

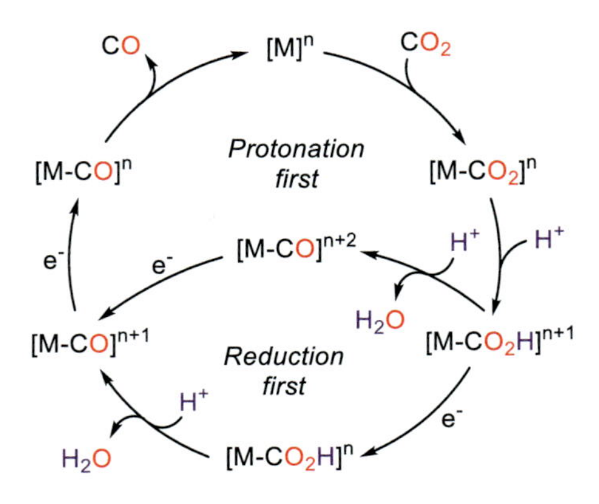

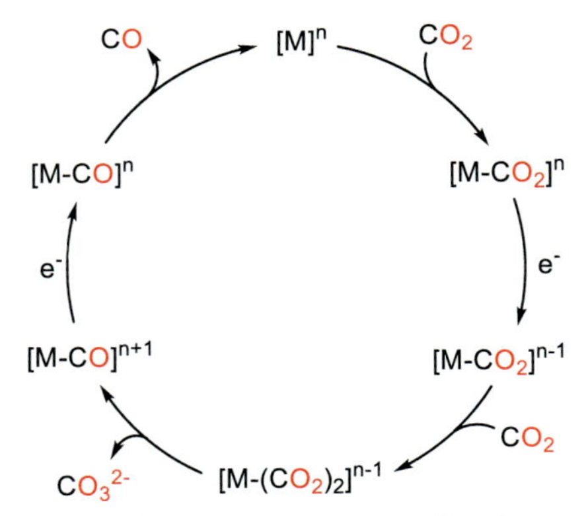

Fig. 1 General catalytic cycle for the "proton-assisted" reduction of CO_2 to CO (A) and the "reductive disproportionation" mechanism (B).

Usually, a subsequent reduction promotes the CO release from the reduced metal-carbonyl intermediate ($[M\text{-}CO]^n$). The cleavage of the carbon–oxygen bond is critical and usually identified as the rate-determining step (rds) of the electrocatalytic CO_2-to-CO reduction mechanism.[18–26] However, the CO release can also play a significant role in the kinetic barrier of the catalytic cycle since metal carbonyls could be thermodynamic sinks.[27,28] Depending on the electronic structure of the molecular complex, the CO_2RR can proceed through the "protonation first" or the "reduction first" mechanism. In general, electron-rich metals bound to electron-donating ligands favor the "protonation first" mechanism due to the increased basicity of the metal-carboxylate adduct. Besides, the "reduction first" mechanism is favored by electron-poor metals bound to less donating ligands.[26]

Besides the general catalytic CO_2 to CO reduction mechanisms where protonation is crucial, the CO_2 activation can also occur without protons. In this case, the postulated pathway for the aprotic reaction described as a reductive disproportionation mechanism involves a second CO_2 molecule acting as a Lewis acid to promote the O abstraction breaking the C—O bond forming CO_3^{2-} (Fig. 1B, Eq. 11).[29]

$$2CO_2 + 2e^- \rightarrow CO + CO_3^{2-} \qquad (11)$$

Nevertheless, according to the nature of the ligand within the metal complexes and ligand structure, the mechanisms present differences. Therefore, in the following sections selected representative families of CO_2-to-CO reduction catalysts are described.

2.1.1 Porphyrin and phthalocyanine complexes

The versatility of the structures of porphyrins and phthalocyanines, together with their capacity to accommodate different metal centers, yielded access to fundamental catalyst design principles.[3,7,12,30–35] Another notable aspect of metal complexes based on these ligands is that they present a large variety of accessible oxidation states due to the redox noninnocent nature of the ligands.

Metalloporphyrins: We will mainly cover Fe and Co representative examples.[4,5,18,36–38]

Iron porphyrins: The influential work of Savéant and coworkers on the electrochemical investigation of iron tetra-phenylporphyrin (**FeP1**) and derivates (Fig. 2) is significant.[4,18,36–38,40–44] Their work contributed to understanding ligand effects (through-structure) and the secondary coordination shell (through-space) on the overpotential and turnover frequency (TOF), including "catalytic Tafel plots" of catalytic processes.[18,19,39–41,45,46]

FeP1 illustrated the characteristic CO_2 catalytic reduction cycle. The starting $[Fe^{III}(TPP)]^+$ is first reduced by three electrons to $[Fe^0(TPP)]^{2-}$ which then can react with CO_2 yielding $[Fe^I(TPP)(^{\bullet}CO_2)]^{2-}$ (Fig. 2). Recent studies have proposed that the electronic structure of $[Fe^0(TPP)]^{2-}$ is described better as a double reduced ligand ($[Fe^{II}(TPP^{\bullet\bullet})]^{2-}$), which has implications for its reactivity.[47] A crucial aspect of the mechanism is the CO_2 reduction enhancement by adding an acid (Fig. 2B),[42] due to an ion-pair (Lewis) or H-bonding (Brønsted) interaction that stabilizes the

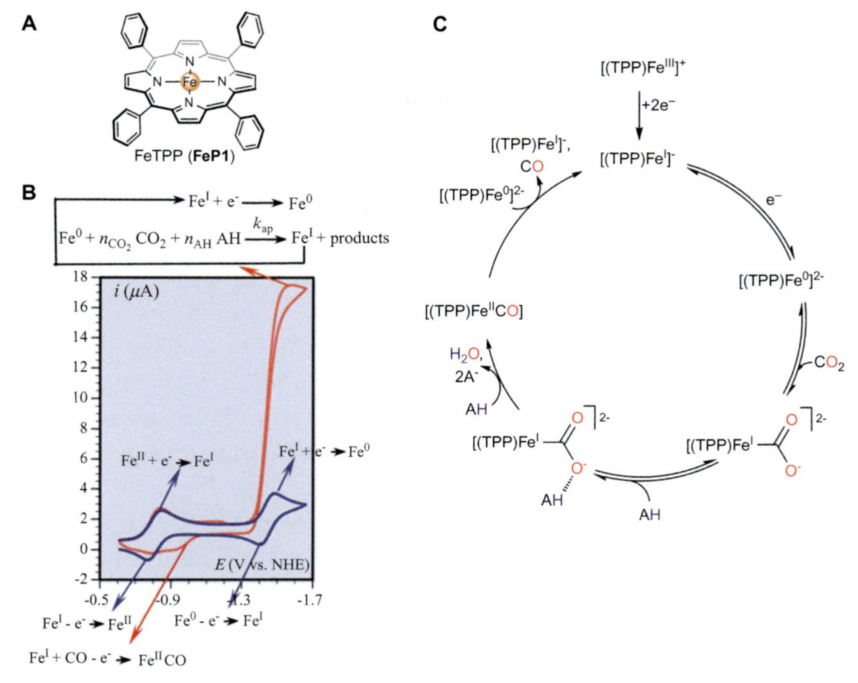

Fig. 2 (A) Line-drawing of [Fe(TPP)]Cl (**FeP1**). (B) CV of **FeP1** under Ar (blue) and CO_2 and an acid (red). (C) The proposed catalytic mechanism for CO_2 reduction to CO mediated by **FeP1** in the presence of Brønsted acids.[39] *Panel (B) reproduced from Costentin, C.; Savéant, J.-M., Nat. Rev. Chem. 2017, 1, 0087, with permission from the American Chemical Society.*

$[Fe^I(TPP)(^{\bullet}CO_2)]^{2-}$ intermediate,[36–38,48] facilitating the C—O cleavage (rds).[18,19,49] Then, the resulting $[Fe^{II}(TPP)(CO)]$ undergoes comproportionation with $[Fe^{II}(TPP^{\bullet\bullet})]^{2-}$ to release CO and regenerate $[Fe^I(TPP)]^-$, concluding the catalytic cycle.[50]

The influence of the through-structure effect on CO_2 reduction was studied by systematically evaluating electron-withdrawing and donating substituents in the iron tetraphenyl porphyrin.[40,51] These studies correlate the catalytic rate (TOF_{max}) and the standard catalytic redox potentials (E^0_{cat}).[52–54] Electron-withdrawing substituents stabilize the formal Fe^0, reducing the overpotential needed for catalysis to occur (E^0_{cat}). However, at the same time, the catalytic rate (TOF_{max}) decreases due to the lower basicity of the formal Fe^0, and thus the formation of iron-carboxylate adduct is disfavored.[43,44] In turn, electron-donating substituents, o,o′-methoxy mesoaryl groups tend to enhance the TOF_{max} at the expense of a less favorable catalytic potential (E^0_{cat}). These electronic through-structure effects showed characteristic linear free-energy relationships, such that a lowering of the driving force of the homogeneous reaction is typically associated with a lowering of TOF_{max} and vice versa.[24]

Secondary coordination shell (through-space) effects are related to those observed with intermolecular Brønsted acids.[51,55,56] In this regard, incorporating local proton sources, or more generally, H-bond donors on the porphyrin, induces an improvement of the catalytic performance (TOF_{max}) without harming the catalytic potential (E^0_{cat}).[43] For instance, replacing the *ortho*-methoxy mesoaryl groups with phenolic ones at the porphyrin (**FeP2**) led to a ca. 300-fold increase in TOF_{max} at 1.1 V overpotential.[44] The role of the -OH groups was ascribed to intramolecular proton relays, facilitating the successive protonation and C—O bond cleavage steps and by stabilizing {Fe–CO_2} intermediates by H-bonding (Fig. 3A).[44]

Alternatively, the stabilization of {Fe–CO_2} intermediates has been proposed via through-space Coulombic interaction with positively charged pendant groups spatially oriented in the proximity of the active site.[39] An informative example is the TPP iron porphyrin with four trimethylanilinium groups in the *para* position of the phenyl rings (**FeP3**, Fig. 3B). The electroanalytical study revealed a break in the linear free-energy relationships between TOF_{max} and E^0_{cat}.[51] This effect increased when quaternary ammonium groups were located in ortho positions, closer to the metal center.

Chang and coworkers also investigated a series of four positional isomers with amide functionalities in the *ortho*- or *para*-positions of the mesophenyl ring.[57] The catalytic experiments revealed that the presence of the amide group in the *ortho* position significantly increases the activity, with the distal isomer being the most active across the series, as a result of the more favorable spatial location of the amide group for the H-bonding stabilization of the {Fe–CO_2} intermediate. The presence of local urea functional groups was reported to have an even more substantial effect on catalysis than the amide groups due to multipoint H-bonding.[58] Analogously, pendant methylimidazolium moieties positively affected CO_2 reduction in a DMF/H_2O mixture due to the electrostatic stabilization of the reaction intermediate.[59]

A recent report suggests that the presence of tertiary amines in solution in the presence of a weak proton source is an effective strategy to drive the selectivity of the formal $[Fe^0(TPP)]^{2-}$ toward HCOOH.[60] The tertiary amine appears to increase the basicity of the central C-atom of the substrate in the $[Fe(CO_2)(TPP)]^{2-}$ adduct, allowing for its protonation and

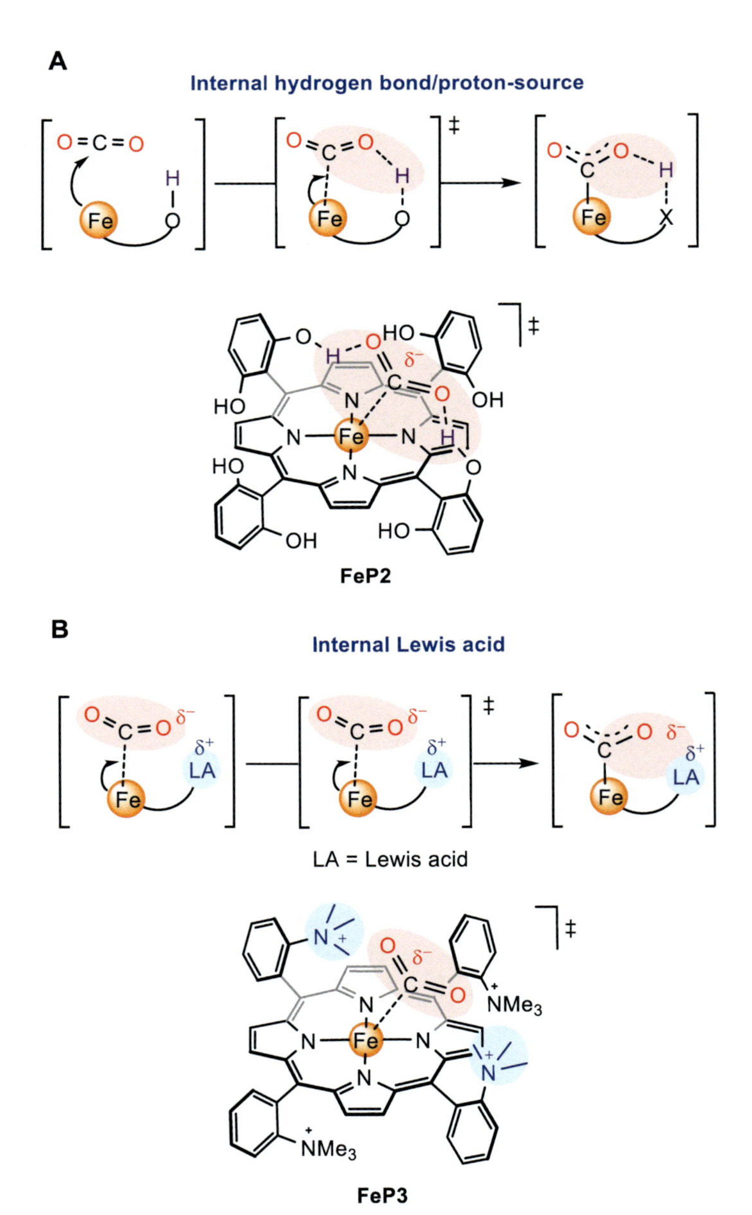

Fig. 3 Secondary coordination shell (through-space) effect. (A) By introducing phenolic functionalities (**FeP2**) or (B) by Coulombic interaction with positively charged pendant groups (**FeP3**).

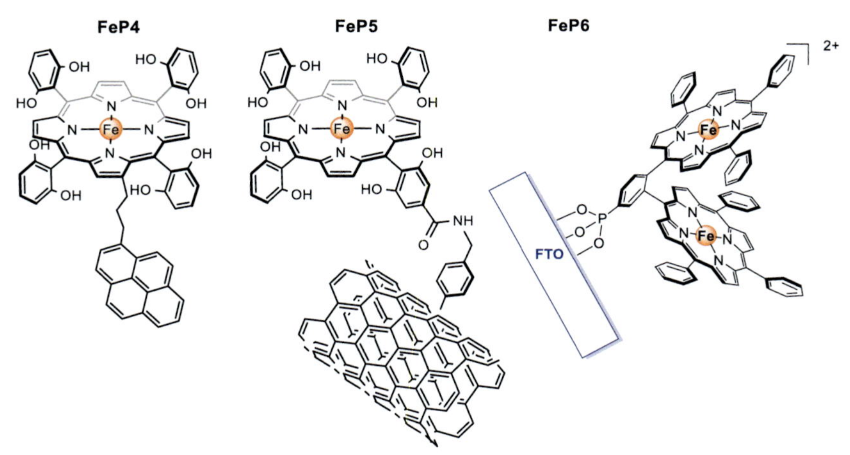

Fig. 4 Examples of heterogenized Iron porphyrins FeP4, FeP5, FeP6.[61–63]

facilitating dissociation of format, consistent with the strong *trans*-coordination of tertiary amines to the Fe center.

Heterogenization of iron porphyrin catalysts: Use of *ortho* and *ortho'* -OH phenyl substituted iron porphyrins when heterogenized on multiwalled carbon nanotubes (CNTs) (i) through π-stacking **FeP4** (Fig. 4)[61,62] and (ii) covalently **FeP5** (Fig. 4)[62,64] yielded excellent catalytic results with selective CO formation (90% and 95%) in neutral pH unbuffered water at low overpotential (330 and 510 mV). Another interesting heterogenization example was the immobilization of a monolayer of the cofacial Fe porphyrin dimer via a phosphonic acid to a fluorine-doped tin oxide electrode, FTO/**FeP6** (Fig. 4).[63] The catalyst showed activity in both organic (DMF/5% H_2O/0.1 M $TBAPF_6$) and aqueous solutions (0.1 M borate solution, pH 7.0), of (0.25 and 1.5 mA cm^{-2}) for CO_2 to CO conversion (93% and 90%) at low overpotential (510 and 420 mV).

FeP3 and reduced graphene oxide can self-assemble to produce a highly porous 3D hierarchical composite with good catalytic activity and selectivity for CO, and Faradaic efficiency (97%) at the typical overpotential (480 mV).[65] Interestingly, iron porphyrin-based graphene hydrogel operates at low overpotential (280 mV) with excellent Faradaic efficiency (96%).[66] Another approach reported is the incorporation of Fe porphyrin into reticular materials.[10,67–71] For instance, thin films of Fe porphyrin-based MOF-525 could be attached on a conductive electrode with an effective catalyst surface concentration.[72] The electrochemical characterization showed a catalytic wave consistent with the redox potential of the $Fe^{I/0}$ couple. Bulk electrolysis showed high current densities (4 mA cm^{-2} at 650 mV) for CO and H_2 mixtures (1.2:1).

Cobalt porphyrins: In a comparable manner to the iron tetraphenylporphyrin, the cobalt analog (**CoP1**) catalytically reduces the CO_2 to CO and formic acid.[73] However, as a homogeneous catalyst in organic media, **CoP1** requires a larger overpotential (>1 V) due to the requirement of forming the doubly reduced $[Co^0(TPP)]^{2-}$ species (Fig. 5). Therefore, electron-withdrawing groups, such as F and CF_3, are usually employed to improve the overpotential.[73] However, **CoP1** analogs are very efficient catalysts once immobilized on conductive carbon-based electrodes, such as gas diffusion electrodes (GDE),[75] pyrolytic graphite (PG)[76] or reduced graphene oxide (rGO).[77] For instance, Daasbjerg and coauthors compared the electrocatalytic CO_2 reduction of **CoP1** under homogeneous and heterogeneous conditions.[74] Under homogeneous conditions employing DMF as the solvent, **CoP1** yielded poor catalytic activity (0.17 mA cm^{-2}) with low product selectivity for CO (50%) and formation of H_2 (2%), formate (4%), acetate (2%), and oxalate (0.4%) at a high overpotential (1.02 V). Electrolysis experiments showed that conversion through $[Co^0(TPP)]^{2-}$ causes a severe catalyst deactivation. However, the immobilization of **CoP1** on CNTs permits an aqueous medium, and as a consequence, CO_2 to CO occurs at a lower overpotential (550 mV) and interestingly with high Faradaic yield (>90%). The difference in reactivity and selectivity between aqueous and nonaqueous conditions was attributed to the role of water in stabilizing the anionic $[Co(TPP)(CO_2)]^{2-}$ adduct through solvation, increasing the reaction rate of $[Co^I(TPP)]^-$ reacting with CO_2 (Fig. 5). These results were consistent with DFT mechanistic studies.[78,79]

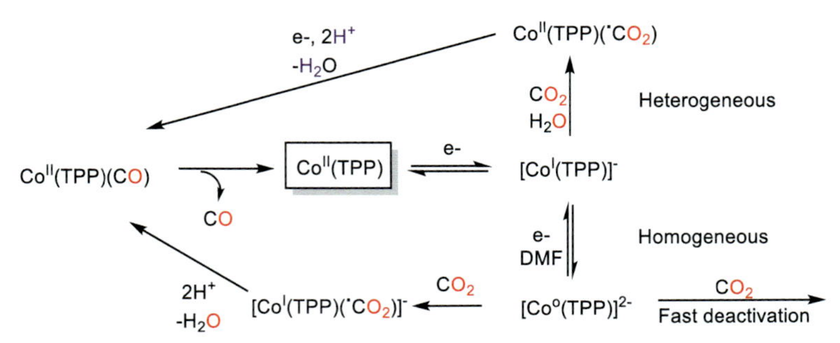

Fig. 5 Proposed mechanism for CO_2 electroreduction catalyzed by **CoP1** under homogeneous and heterogeneous conditions.[74]

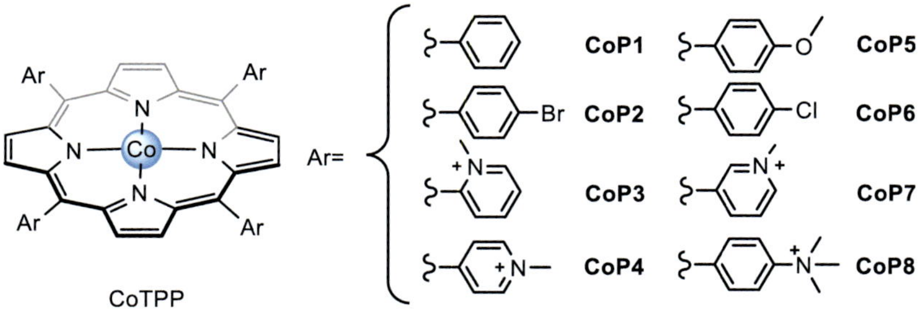

Fig. 6 Inductive and electrostatic effects on cobalt porphyrins for heterogeneous electrocatalytic CO_2 reduction.[81]

The fine-tuning of the ligand in cobalt porphyrins provides a rational design similar to that in iron-based porphyrins. Recently, Manthiram and coworkers[80] illustrated the impact of the peripheral functionalization of the aromatics in cobalt porphyrins on CO_2 reduction (Fig. 6). The introduction of cationic peripheral substituents in cobalt porphyrins (**CoP3**, **CoP4**, **CoP7**, **CoP8**) increased the CO_2 reduction rates. Thus, the increase in the catalytic activity of immobilized cobalt porphyrins was related to their capacity to stabilize the {Co-(CO_2)} adduct. Additionally, the heterogenization of the cobalt [5,10,15,20-(tetra-*N*-methyl-4-pyridyl)porphyrin] tetrachloride (**CoP4**) in reduced graphene oxide (rGO) was reported as a highly stable film for electrocatalytic CO_2 reduction to CO (45%) and HCO_2H (45%) in an aqueous electrolyte at 590 mV of overpotential.[77] The proposed CO_2RR mechanism proceeds as in the previous case.[76,79]

Additional strategies have been developed to study cobalt-based porphyrins in aqueous media, such as covalent attachment of Co porphyrins to carbon-based electrodes,[82–87] and on reticular materials such as MOFs[10,68–71] and COFs.[10,88–90] In particular, the 2D MOF nanosheets (or metal–organic layers; MOLs) are interesting because they facilitate the exposure of the Co porphyrin active sites.[69,81] Then, the post-modification of the nodes of the 2D MOF (unsaturated Zr_6 clusters) allowed to tune the microenvironment around the Co porphyrin favoring CO_2 reduction over H_2 evolution. A remarkable improvement in the catalytic performance was obtained by incorporating a cobalt porphyrin (**CoP9**) as a COF building block (Fig. 7).[10] In an aqueous electrolyte, the resulting catalyst presented high catalytic activity and selectivity toward CO (90%). In a further study, the electronic character of the Co–COFs was tuned by replacing the linkers and improving the catalytic activity (Fig. 7).[70] The activity and selectivity were correlated with the inductive effects of the appended functionality and the electronic character of the reticulated

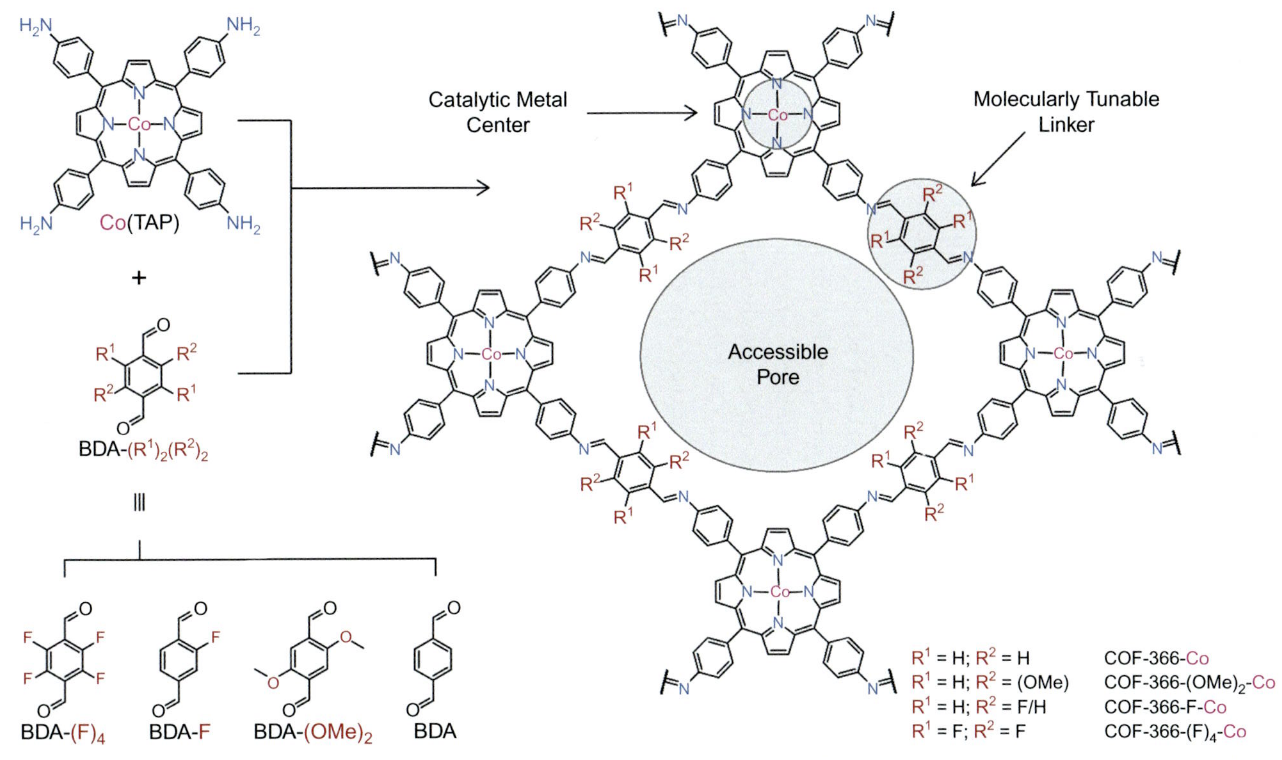

Fig. 7 Covalent organic frameworks (COF) based on cobalt-porphyrin catalytic sites (**CoP9**) with electronic modulation. *Reproduced from Dierecks, C. S.; Lin, S.; Kornienko, N.; Kapustin, E. A.; Nichols, E. M.; Zhu, C.; Zhao, Y.; Chang, C. J.; Yaghi, O. M., J. Am. Chem. Soc. 2018, 140, 1116–1122, with permission from the American Chemical Society.*

molecular active sites. The extent of this effect was proportional to the electronegativity, and the number of functional groups installed.

Recently, metal porphyrin-based covalent organic frameworks (Co, Ni, Fe) covalently linked on CNTs were reported as efficient catalysts for CO_2 reduction.[91] Among all those considered, the Co-COF exhibited superior activity and CO selectivity (99%) with high current density (6.5 mA cm^{-2}) and good durability (50 h). Alternatively, a crown ether and cobalt-porphyrin-based COF (TAPP(Co)-B18C6-COF) was evaluated.[88] The crown ether units in the backbone of the COF played a crucial role by enhancing the hydrophilicity of the frameworks and promoting electron transfer from the crown ether to the Co-porphyrin. The COF enhances the CO_2 binding ability and showed good Faradaic efficiencies for CO (93%) at −0.90 V vs RHE.

On metal–organic layers (MOLs) (**CoP10**, Fig. 8), between cobalt protoporphyrin catalytic sites and pyridine/pyridinium (py/pyH^+) a synergistic effect has been reported that enhances the catalysis affording a high CO selectivity (92%).[92] The process resembles the second or outer coordination sphere effects by forming a [pyH^+-CO_2^--CoPP] species. The addition of divalent cations (Ca^{2+} or Zn^{2+}) decreased the CO selectivity, attributed to the blockage of the terpyridyl units. *In situ* electrochemical diffuse reflectance infrared Fourier transform spectroscopy (DRIFTS) measurements during reduction of **CoP10** under CO_2, provided evidence of the pyridine protonation to form pyridinium units.[92] In the proposed mechanism, the

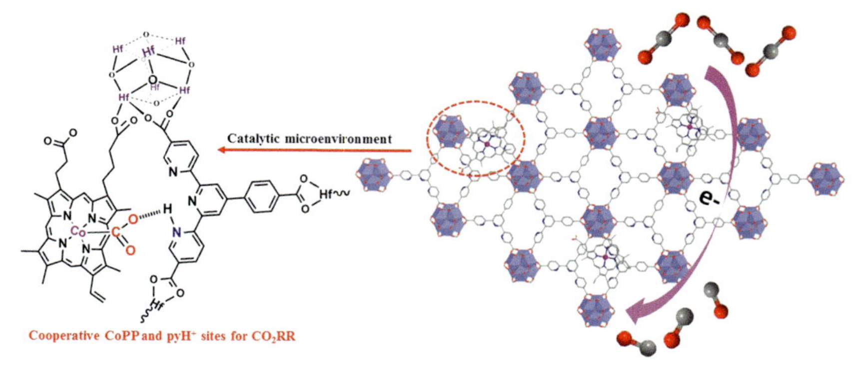

Fig. 8 Structure of TPY-MOL-CoPP (**CoP10**) and the cooperative activation of CO_2 by CoPP and pyH^+. *Reproduced from Guo, Y.; Shi, W.; Yang, H.; He, Q.; Zeng, Z.; Ye, J. Y.; He, X.; Huang, R.; Wang, C.; Lin, W., J. Am. Chem. Soc. 2019, 141, 17875–17883, with permission from the American Chemical Society.*

one-electron reduction of CoPP to $[CoPP]^-$ is followed by the CO_2 binding and protonation, leading to the formation of the key $[pyH–Co(PP)-CO_2^-]^0$ adduct, which is stabilized by a pre-positioned pyridine moiety to favor CO_2 reduction over H_2 evolution. Then, a second electron reduction follows the C—O bond cleavage, forming $[Co(PP)(CO)]^0$, which undergoes CO release, closing the catalytic cycle.[92]

Phthalocyanine complexes: Applications of metal phthalocyanines (M(Pc), M = Co,[12,20,93–95] Fe,[96,97] Ni,[96,98] Cu[99–101]; Pc = phthalocyanine) in electrocatalytic CO_2 reduction have attracted considerable attention due to the advantages of low cost, facile synthesis, and excellent chemical stability. However, due to the low solubility, their study is mainly restricted as immobilized on conductive supports. Simple π–π interaction can be used for the immobilization of π-conjugated materials. As in the studies with porphyrin complexes, the effect of substituent on the ligand plays an important role and has been investigated.[93,102,103]

Co(Pc) and their derivatives have presented great potential for application in electrochemical devices (Fig. 9). An earlier report showed that **CoPc1** is a poor electrocatalyst for CO_2 reduction when used in a homogeneous phase.[32] However, attached to carbon materials via non-covalent interactions, covalent bonding or periodic immobilization significantly improved the electrocatalytic activity and allowed the use of an aqueous medium.[8,12,99,103–107] Taking advantage of the strong π-stacking interaction

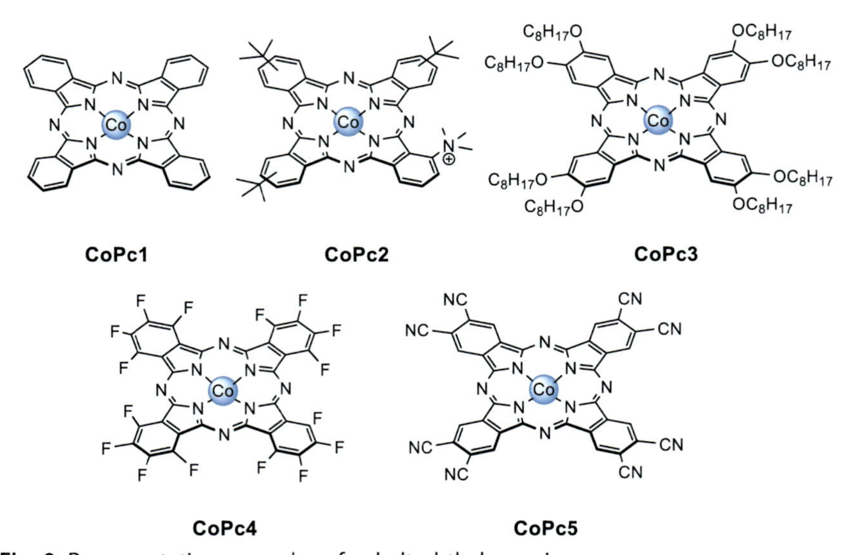

Fig. 9 Representative examples of cobalt phthalocyanines.

between **CoPc1** and carbonaceous materials, dip-coating or drop-casting methods are the most popular deposition techniques to attach the catalyst on a carbon support.[32,98,102,104,106,108,109] Nevertheless, the catalyst loading and dispersion of cobalt phthalocyanine catalysts are vital for having excellent catalytic performance.[109] As a result of their planar structure, cobalt phthalocyanine generally tends to aggregate; these aggregates (J and/or H) reduce the number of accessible catalytic sites and could have different products selectivity.[110] Zhu et al. have demonstrated that poor electron transport through phthalocyanine aggregates also hampers electron accessibility to catalytically active CO_2 reduction centers.[109] For instance, however, the hybridization of **CoPc1** with carbon nanotubes minimizes the aggregation.[111] The **CoPc2** species bearing one positively charged trimethyl ammonium moiety and three tert-butyl groups appended on the phthalocyanine macrocycle, selectively reduced the CO_2 to CO in water (Fig. 9).[8] The electrochemical characterization of **CoPc2** presented three reversible redox waves assigned to the Co^{II}/Co^{I} redox couple and two subsequent ligand reduction peaks. The catalyst produces a high activity over a broad pH range (4–14), with a selective CO production (91%) and good stability (10 h) and current density (20 mA cm^{-2}).[8]

A strategy to increase the dispersibility of cobalt phthalocyanines was to increase steric hindrance by introducing extended alkoxy groups at the aromatic moieties (**CoPc3**). Its immobilization was successful on graphene. Indeed, compared to an analogous cobalt phthalocyanine/graphene catalyst, the steric hindrance of the long alkoxy chains helped suppress the phthalocyanine aggregation, resulting in significantly greater number of accessible catalytic sites and improved catalytic activity. Stable production of CO (77%) was achieved for more than 30 h of electrolysis at -0.59 V vs RHE ($\eta = 480$ mV) with high current density (20 mA cm^{-2}).[112] A limited formation of aggregates was also obtained for **CoPc1** by the *in situ* polymerization of a thin conformal coating layer around CNTs.[95] The hybrid electrocatalyst selectively reduces CO_2-to-CO (90%) at 500 mV of overpotential with excellent long-term durability (24 h).

The catalytic performance and selectivity were also improved by varying the nature of the complex **CoPc1**. The functionalization of the CoPc structure with electron-withdrawing substituents gave rise to beneficial effects on catalysis due to facilitating the $Co^{II/I}$ redox couple and CO release step. For example, a perfluorinated CoPc complex (**CoPc4**) adsorbed on carbon cloth served as a robust catalyst for simultaneous CO_2/CO conversion and H_2O/O_2 splitting.[113] A similar enhancing catalytic effect was observed by introducing –CN groups to the CoPc molecule (**CoPc5**) (Fig. 9).

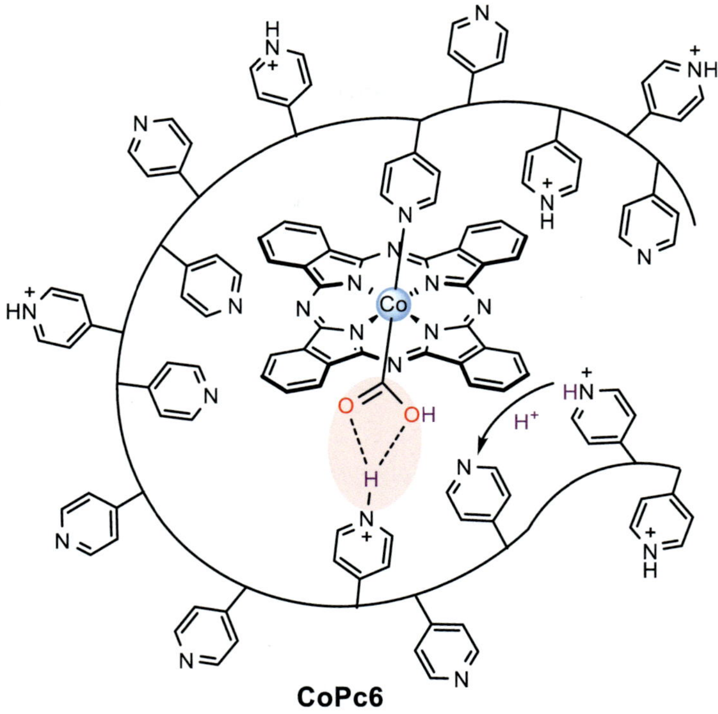

Fig. 10 CoPc immobilized in P_4VP (**CoPc6**).[114,115]

As an alternative, a **CoPc1** was encapsulated using the poly-4-vinylpyridine (P_4VP) polymer as a ligand (**CoPc6**) and taking advantage of the axial coordination sites of the **CoPc1** (Fig. 10).[114,115] The resulting material was better in terms of activity and selectivity compared with **CoPc1** alone. The authors proposed that the presence of P_4VP contributes to enhancing the **CoPc1** catalytic activity due to: (i) the axial coordination of the pyridine increases the nucleophilicity of the Co^{I} center facilitating the CO_2 binding step and (ii) the interaction with partially protonated peripheral pyridyl residues of the polymeric film that may undergo secondary coordination sphere effects (e.g., stabilizing H-bonding interactions, proton relays). These studies evidenced a synergistic relationship between the primary and outer coordination sphere effects.

In a recent report, the introduction of Co phthalocyanine catechol building blocks into a novel metal-catecholate framework (MOF-1992) was described.[116] The system displayed more accessible CoPc sites and improved charge transfer properties, leading to a significantly higher electroactive surface area than that of the previously reported reticular Co/Fe catalysts. At neutral pH, MOF-1992 yields CO_2 to CO with 80% Faradaic efficiency at η of 520 mV with current densities of 20 mA cm^{-2}. In a similar way, the decoration of the external surface of the zeolite ZIF-90 with active Co tetraminonaphthalocyanine units also produced a selective CO_2-to-CO catalyst.[117]

It is very noteworthy that the real potential of cobalt phthalocyanine was achieved when operating under flow conditions using an aqueous electrolyte.[8,9,118] For example, the evaluation of **CoPc1** in a tandem flow cell with nickel foam in the anode led to excellent CO selectivity at industrial relevant current densities (200 mA cm^{-2}) at an overall cell voltage of ca. 2.5 V, out-performing Ag nanoparticles as a CO_2-to-CO reduction catalyst.[9,119]

2.1.2 Non-heme macrocyclic, polypyridyl and aminopyridyl complexes

In this section, we will describe three different families of catalysts: non-heme macrocyclic, polypyridyl, and aminopyridine complexes of catalysts. In all these catalysts, the metal center is coordinated by nitrogen-donor ligands such as amines or pyridines. This structural feature leads to common reactivity patterns.

The Ni-cyclam catalyst: Among this group of complexes, $[Ni(cyclam)]^{2+}$ (**Ni1**) is the flagship electrocatalyst for the CO_2 reduction to CO. In the 1980s, Sauvage and coworkers first reported this molecular complex as a selective CO_2 reduction catalyst in water.[120] Nonetheless, Eisenberg and coworkers previously showed that other cyclam-like Co (**Co1**) and Ni (**Ni9**) (shown in Fig. 12) complexes demonstrate catalytic activity toward the same reaction.[121] Since these two pioneering studies, many groups have reported mechanistic studies and developed strategies to improve the **Ni1** catalytic performance (Fig. 11). The catalytic activity of the latter complex is susceptible to the nature of the solvent and the working electrode material. For instance, **Ni1** shows outstanding performances in water (pH 4) using a Hg pool electrode with excellent selectivity for CO production over H_2 (FY_{CO} = 94%). Nonetheless, there is strong evidence for the adsorption of the $[Ni]^+$ active species over the Hg electrode surface, which leads to a fast catalytic rate and a low overpotential.[122] The same system and working electrode but in DMF produce formic acid in a 75% FY.[123] Instead, it has demonstrated high selectivity and a homogeneous behavior using glassy carbon as the working electrode either in aqueous KCl solution, aqueous acetonitrile (4:1 MeCN:H_2O), and ionic liquids as both solvent and electrolyte.[124,125]

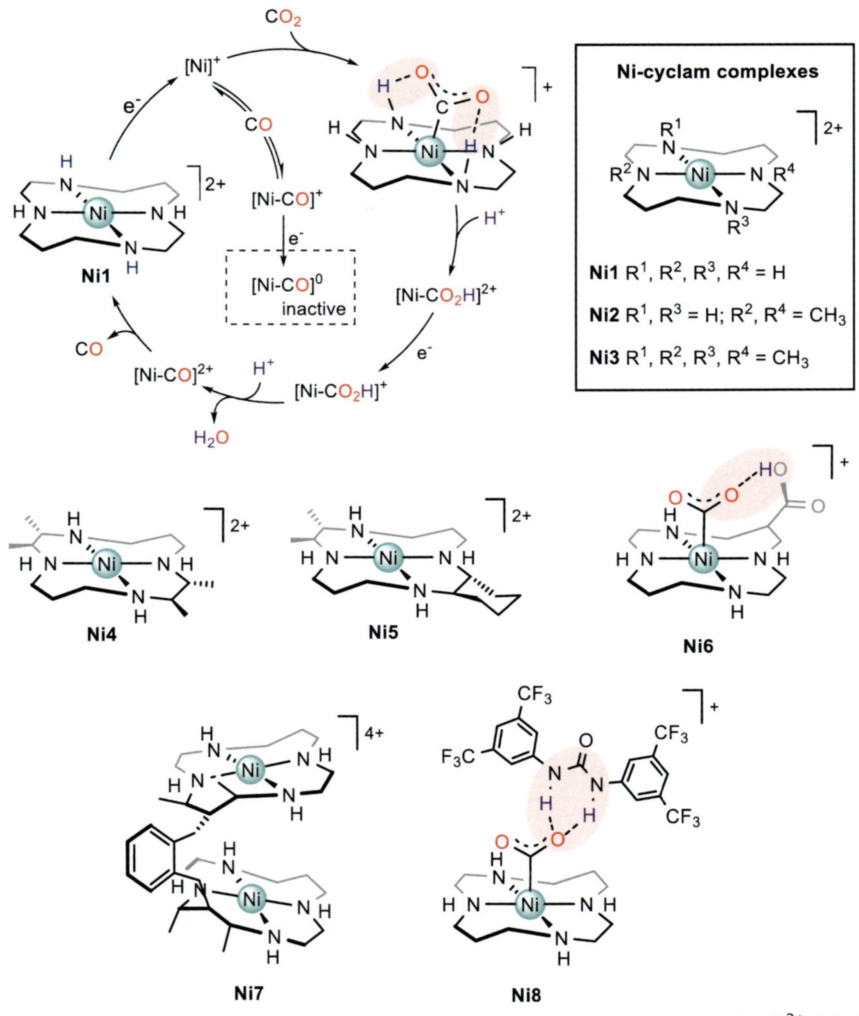

Fig. 11 Catalytic cycle for the CO_2 reduction to CO catalyzed by $[Ni(cyclam)]^{2+}$ (**Ni1**) over a Hg electrode and selection of Ni cyclam-like complexes and intermediates (**Ni2–Ni8**) for the reduction of CO_2 to CO.

Both experimental and computational mechanistic studies propose that the CO_2 binding step is thermodynamically and kinetically feasible at the $[Ni^I]^+$ intermediate stage ($K_{CO2} \approx 6\ M^{-1}$ in acetonitrile). In the resulting Ni-carboxylate adduct, the η^1-CO_2 ligand is stabilized through hydrogen bonding interactions with the -NH groups of the cyclam macrocycle. Indeed, substituting the –NH with methyl groups (**Ni2**, **Ni3**) is detrimental for the catalytic activity in terms of overpotential, current and selectivity.[124,126] After the reduction and protonation of $[Ni\text{-}CO_2]^+$, a second protonation event triggers the C—O bond cleavage giving $[Ni\text{-}CO]^{2+}$, which undergoes CO release and regenerates the initial $[Ni]^{2+}$ compound. After several ligand modifications, only a few of the cyclam-like catalysts improved the catalytic activity of $[Ni(cyclam)]^{2+}$. The introduction of chiral centers in the ligand (**Ni4**, **Ni5**) favors the formation of the flat isomer of Ni-cyclam, which presents an optimal geometry for the adsorption to Hg and minimizes the steric hindrance for the CO_2 binding.[127] The location of a carboxylic acid functionality in the cyclam ligand (**Ni6**) outperforms the unmodified catalyst under very acidic conditions (pH <2) in water with a remarkable selectivity for CO ($FY_{CO} = 66\%$) production over H_2 ($FY_{H2} = 15\%$).[128] More recently, the design of a dinuclear Ni-cyclam analog (**Ni7**) shows the cooperation between the two Ni centers nearby, as previously shown for a Fe porphyrin dimer catalyst.[129,130] Chang and cocorkers proposed to use an organic urea cocatalyst to enhance the catalytic activity of **N1** in aqueous acetonitrile media keeping an excellent selectivity for CO production. The organic urea can stabilize the metal-carboxylate intermediate through selective two-point hydrogen bonding interactions (**Ni8**).[131]

FTIR-SEC experiments allowed the identification of $[Ni\text{-}CO]^+$ as a crucial intermediate in the **Ni1** deactivation pathway. At the catalytic applied potentials, the two Ni^{II} intermediates, $[Ni]^{2+}$ and $[Ni\text{-}CO]^{2+}$ can be reduced to $[Ni]^+$ and $[Ni\text{-}CO]^+$, respectively. Additionally, the CO produced after successive catalytic cycles can favor the formation of $[Ni\text{-}CO]^+$ from $[Ni]^+$. Furthermore, the strong π-backbonding from Ni^I to the CO ligand prevents the CO liberation from $[Ni\text{-}CO]^+$. Finally, $[Ni\text{-}CO]^+$ is reduced *in situ* to the unstable $[Ni\text{-}CO]^0$ intermediate, which decomposes to $Ni(CO)_4$ as observed by FTIR-SEC.[125,132] Kubiak and coworkers proposed to use **Ni1** as a catalyst together with increasing amounts of **Ni3** (up to 20 eq.) as a sacrificial CO scavenger to boost the catalytic activity for CO_2 reduction.

More advanced strategies have been employed to increase the performance of Ni cyclam-like catalysts, such as the covalent binding of the catalyst in carbon electrodes or metal oxide semiconductors.[133,134] An effective strategy consisted of the incorporation of Ni-cyclam units in a polymeric allylamine matrix.[135] The **Ni1** binding through a pendant histidine residue of azurin formed an artificial enzyme active in photocatalytic CO_2 reduction to CO.[136] Recently, Machan and coworkers successfully implemented **Ni1** as a homogeneous electrocatalyst in a flow-cell setup using organic solvents (acetonitrile or DMF) and NH_4PF_6 as both an electrolyte and a proton donor.[137] It is worth noting that only traces of formate were found in DMF, in contrast to the previously reported finding under conventional static conditions. Under flow conditions, **Ni1** shows improved durability and selectivity than in the conventional H-type configuration. However, flow FTIR analysis of the solution also reveals the same $[Ni\text{-}CO]^+$ and $Ni(CO)_4$ decomposition products. Catalyst **Ni10** (shown in Fig. 12), based on a nonmacrocyclic six-coordinate aminopyridine ligand, also undergoes demetallation after a few turnovers, giving $Ni(CO)_4$ as the final decomposition product.

Catalysts based on imino/aminopyridine and related ligands: As previously mentioned, $[Co(HMD)]^{2+}$ (**Co1**) was one of the first macrocyclic complexes reported for the electrocatalytic CO_2 reduction by Eisenberg and coworkers back in 1980.[121] **Co1** shows lower selectivity than **Ni1** for the production of CO. However, the CO_2 binding constant to the corresponding $[Co^I]^+$ intermediate ($K_{CO2} \approx 10^4\ M^{-1}$) is within four orders

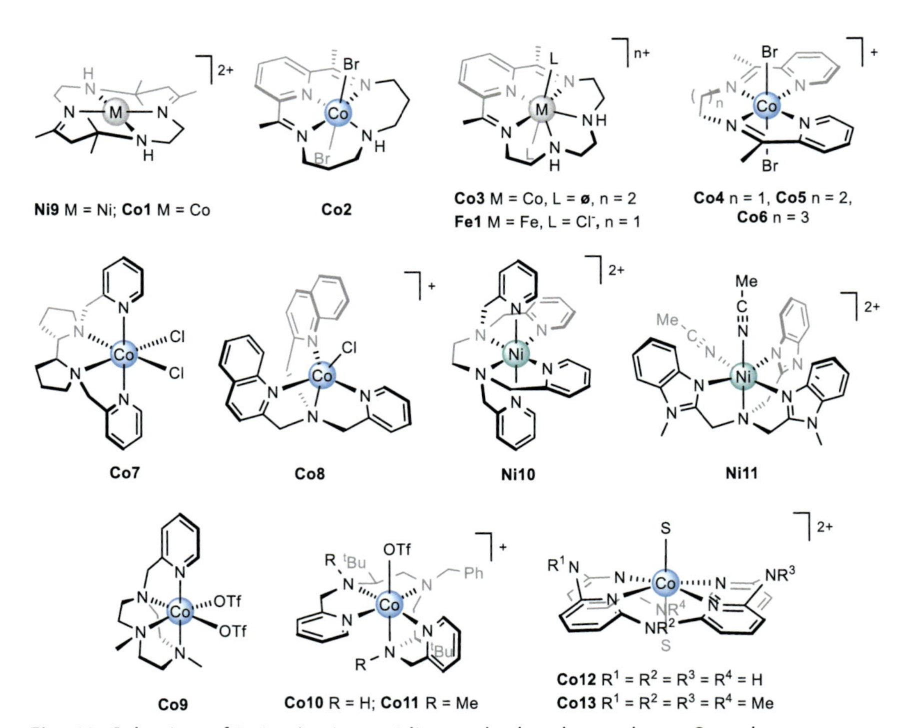

Fig. 12 Selection of imino/aminopyridine and related complexes. S = solvent.

of magnitude higher than in the case of Ni-cyclam.[138] As in the cyclam ligand, polar –NH groups are crucial for stabilizing the η^1-CO_2 ligand.[139] Fujita and coworkers made remarkable contributions to CO_2 reduction by using **Co1** as a model catalyst.[3] The isolation and characterization of the $[Co–CO_2]^+$ adduct by XANES allowed determining that the carboxylate adduct electronic structure is described better as $[Co^{II}(CO_2^{\bullet-})]^+$ in contrast to the $[Ni^{III}$-$CO_2]^+$ adduct proposed in the case of **Ni1**.[140] Moreover, an intriguing bimetallic $[Co$-$C(\mathit{OH})O$-$Co]^{3+}$ adduct was formed in the solid-state after sparging a $[Co^I]^+$ solution with CO_2.[141] After several days, the same reaction mixture evolves, giving $[Co^I$-$CO]^+$.[142]

Later, many molecular catalysts based on non-heme ligand platforms have been reported to reduce CO_2 to CO (Fig. 12). Peters and coworkers electrochemically reported a 45% FY_{CO} and 30% FY_{H2} in aqueous acetonitrile using a bis(imino)pyridine Co complex (**Co2**) as a catalyst.[143] In this case, a doubly reduced intermediate is responsible for the CO_2 binding in agreement with the non-innocent nature of the ligand. Further investigations of the same system suggest that the bound –NH group plays a significant role in stabilizing the Co-CO_2 adduct through hydrogen bonding interactions.[144] Robert, Lau and coworkers report a dramatic change in the product selectivity in neat DMF by replacing Co by Fe in a bis(imino)pyridine complex (**Co3**, **Fe1**). While **Co3** gives CO as the primary product (82% FY_{CO}), **Fe1** produces a 75–80% FY of formate.[145] More recently, McCrory and coworkers reported a family of non-macrocyclic bis(imino)pyridine complexes (**Co4**-**Co6**) in which the ligand is redox noninnocent and also plays a key role in the catalytic activity and stability. In the absence of a Brønsted acid (or with a low concentration), **Co4** shows poor catalytic activity for the CO_2 reduction to CO, and formate is generated in a non-negligible amount. The authors propose that intramolecular ligand deprotonation in the Co-CO_2 intermediate is responsible for the catalyst deactivation giving formate as a byproduct (Fig. 13).[146] Instead, in the presence of acid (11 M H_2O or 5.5 M TFE), **Co4** is a

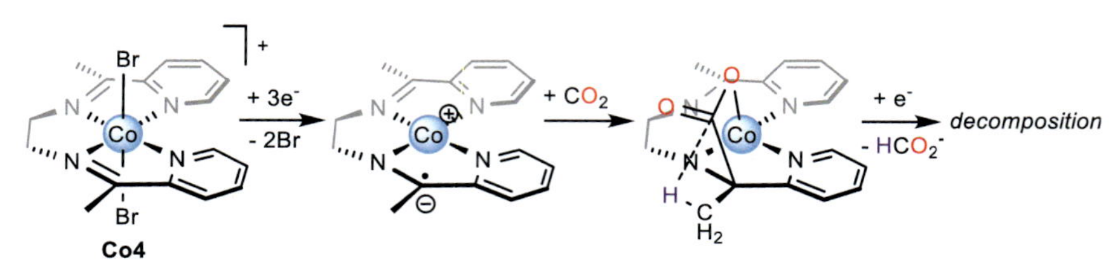

Fig. 13 Proposed decomposition pathway of **Co4**.

selective electrocatalyst for the CO_2 reduction to CO. They showed that a rigid ligand backbone favors the planar coordination of the metal, which seems to be crucial for a higher current and FY_{CO} (**Co4** > **Co5** > **Co6**). They concluded that flexible ligands induce a strong CO affinity to the reduced Co complex, leading to catalyst poisoning.[147]

Tetracoordinate aminopyridine Co (**Co7**–**Co11**) and Ni (**Ni11**) have shown catalytic activity for CO_2 reduction to CO. These ligands are considered highly basic and redox innocent platforms which generate highly nucleophilic Co^I and Ni^I intermediates that easily bind CO_2. The mechanistic proposals for **Co7**, **Co8** and **Ni11** are very similar to Ni-cyclam, starting with the CO_2 binding to the corresponding M^I intermediate to form the metal-carboxylate adduct. After two protonations and one reduction event, the C—O bond cleavage leads to formation of a M^{II}-CO intermediate that releases CO.[148–150] In our group, we have investigated the catalytic behavior of **Co9** in anhydrous acetonitrile, and with small amounts of added water (0.5 M).[27] IR-SEC, together with DFT calculations and labeling studies, allowed the identification of the formation of a $[Co^I\text{-}CO]^+$ species even at the non-catalytic $Co^{II/I}$ reduction wave even in the absence of protons (Fig. 14, black mechanism). This phenomenon has significant mechanistic implications: (1) CO_2 binds to $[Co^I]^+$; (2) the C—O bond cleavage takes place at low overpotentials; (3) the resulting $[Co^{II}\text{-}CO]^{2+}$ is reduced in a one electron step to form a thermodynamic sink intermediate $[Co^I\text{-}CO]^+$. DFT calculations illustrate that the CO release from the former intermediate is endergonic and contributes to the kinetic barrier of the mechanism. Therefore, we propose an alternative catalytic cycle starting with the one-electron reduction of $[Co^I\text{-}CO]^+$ followed by the CO_2 binding. Subsequent protonation and reduction events trigger the C—O bond cleavage giving a Co biscarbonyl species $[Co^I(CO)_2]^+$ that can undergo CO dissociation (Fig. 14, blue mechanism). Complex **Co9** presents an excellent catalytic performance under photocatalytic conditions, suggesting that $[Co^{II}\text{-}CO]^{2+}$ can release CO before forming $[Co^I\text{-}CO]^+$. Investigations on blue light irradiation under controlled potential electrolysis conditions reveal a beneficial effect of blue light in the catalytic activity, selectivity, and durability of **Co9**. We propose that light activates the Co—CO bond favoring the CO dissociation, leading to the differences in catalytic activity between photo- and electrocatalytic conditions. The effect of light irradiation was even more evident in the case of complex **Co10** in which the catalytic production of CO was observed at the $Co^{II/I}$ redox potential.[28]

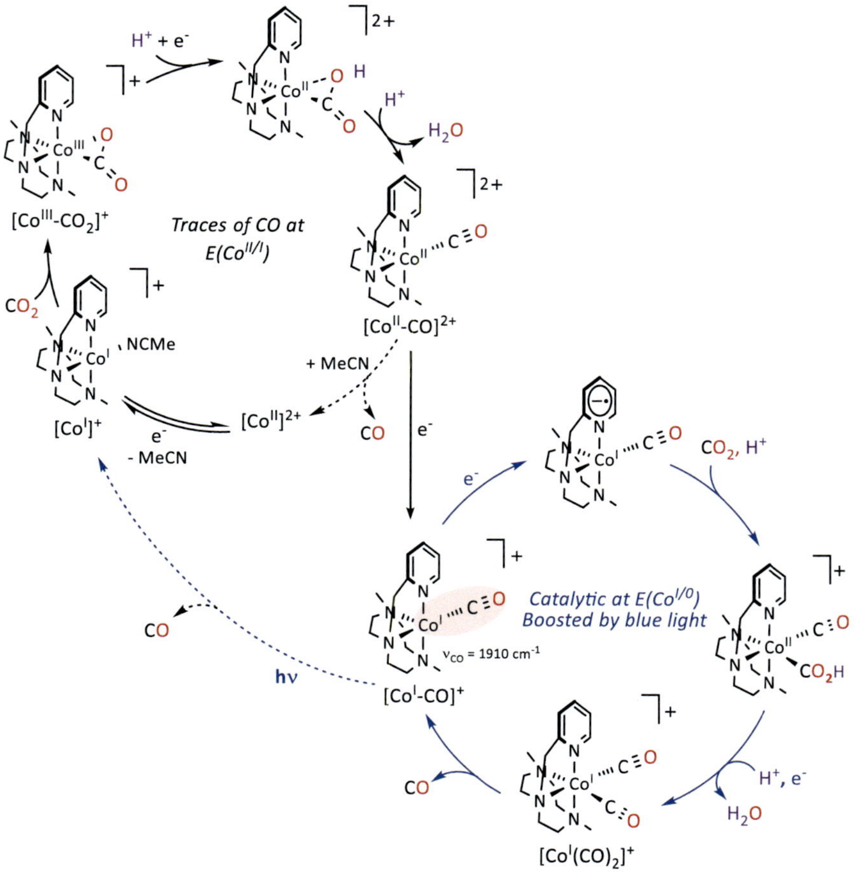

Fig. 14 Mechanistic proposal for the electrochemical reduction of CO_2 to CO catalyzed by **Co9**.

The intramolecular effect of –NH groups has also been studied with **Co10** and **Co11**.[28] Complex **Co10** shows a better catalytic performance than its methylated counterpart **Co11**. However, the hydrogen bonding effect could not be discriminated from the strong inductive effect of the coordinated –NH groups. In this context, Marinescu and coworkers investigated the second coordination sphere effect of pendant –NH groups in a macrocyclic tetrapyridine-based Co catalyst (**Co12**).[56] The systematic methylation of the different secondary amine groups reveals a monotonic decrease in catalytic activity.[151] Despite its high overpotential ($E_{cat} = -2.8$ V vs $Fc^{+/0}$), **Co12** shows excellent selectivity for CO production in DMF solution with added TFE as a proton source. Instead, the fully

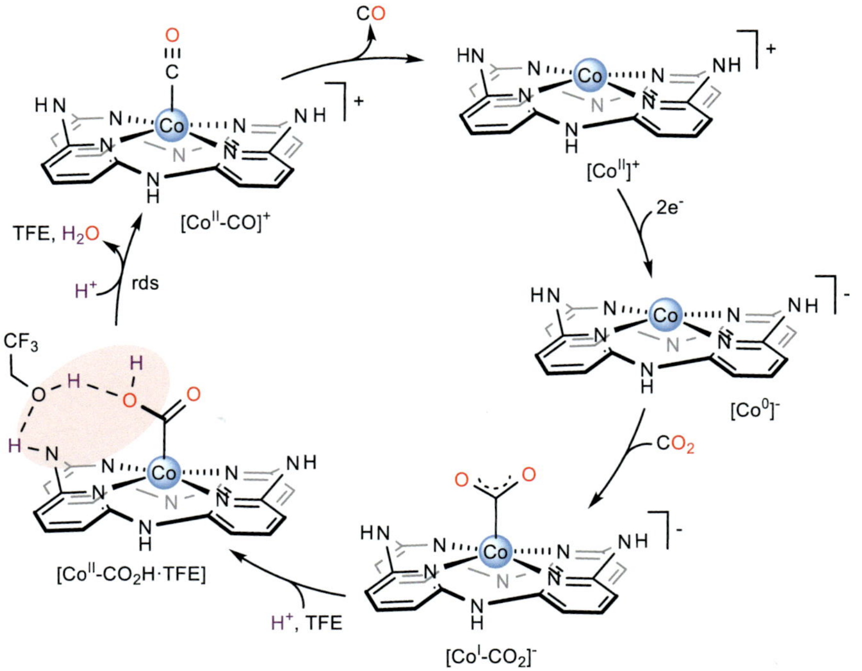

Fig. 15 Mechanistic proposal for the CO_2 reduction to CO catalyzed by **Co12**.

methylated counterpart (**Co13**) does not generate a catalytic response. DFT calculations on the catalytic intermediates suggest that the origin of such intramolecular effect is hydrogen bonding interactions with the Brønsted acid forming a [Co^{II}-CO_2H·TFE] adduct (Fig. 15), alternatively to the expected stabilization of the η^1-CO_2 ligand in the Co-CO_2 adduct proposed for cyclam-like complexes. Note that in this particular family of catalysts, the CO_2 binding step occurs at the Co^0 intermediate.

Catalysts based on polypyridyl ligands: This family of complexes is based on ligands such as *2,2′*-bipyridine (bpy), *2,2′:6′,2″*-terpyridine (tpy), *2,2′:6′,2″:6″,2‴*-quaterpyridine (qpy) and their derivatives (Fig. 16). The non-innocent character of the polypyridyl moiety makes them less basic than aminopyridine ligands providing an extra-stabilization to the reduced species.

Fontecave and coworkers studied the catalytic behavior of a series of homoleptic tpy first-row transition metal complexes in aqueous DMF (5% H_2O). Both $[Ni(tpy)_2]^{2+}$ and $[Co(tpy)_2]^{2+}$ showed CO_2 reduction catalytic activity with ca. 20% FY_{CO}.[152] In the case of **Ni12** the first reductions

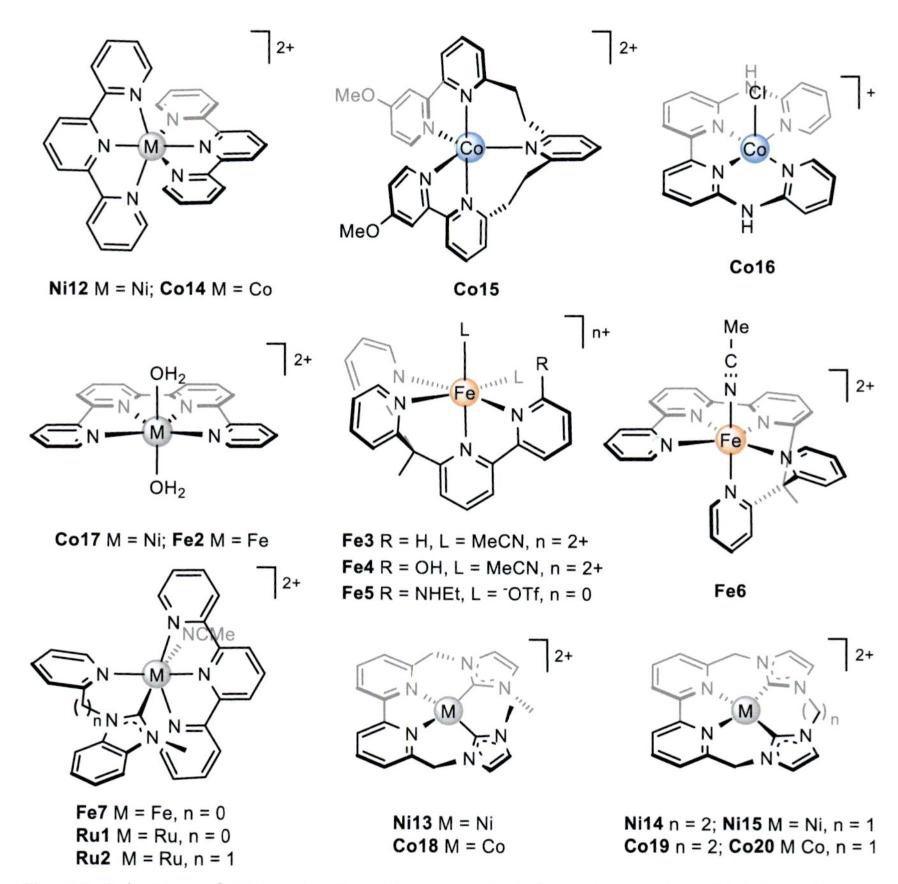

Fig. 16 Selection of CO_2 reduction electrocatalysts based on polypyridyl ligands.

are described better as ligand-centered. Whereas in the case of **Co14**, the first electrochemical process is metal-centered, the second one is a ligand-centered process. Perhaps, therefore H_2 evolution is a competitive reaction in the case of Co. The rational substitution of the tpy ligand with electron-donating ^{t}Bu groups allowed to minimize the H_2 evolution pathway while increasing the selectivity for CO.[153] The proposed electrocatalytic cycle for CO_2 and H^+ with **Co14** is described in Fig. 17. Note that the active species are formed after losing a neutral tpy ligand. Fujita and coworkers reported the trigonal bipyramidal cobalt complex **Co15** in which the CO_2 binding process occurs over a two-electron reduced species with radical ligand character $[Co^IL^{\bullet-}]$. This is also the case for catalyst **Co16** and it seems the general behavior of Co catalysts is based on non-innocent ligands (**Co2–6**, **Co14–20**).[154,155]

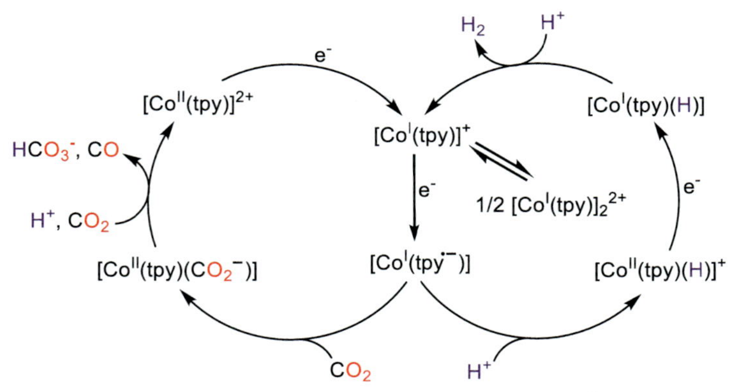

Fig. 17 Mechanistic proposal for the electrochemical CO_2 and H^+ reduction catalyzed by **Co14**.

Robert, Lau and coworkers showed that the polypyridine Co complex **Co17** was also highly active and selective for the production of CO in acetonitrile using PhOH (3 M) as a proton source. Indeed, **Co16** is one of the fastest performing electrocatalysts with a TOF_{max} of $3.3 \times 10^4\ s^{-1}$ at 300 mV overpotential.[156] The same catalyst supported over MWCNTs revealed one of the best catalytic performances for heterogeneized molecular catalysts in aqueous media (pH 7.3) with selectivities close to 100% for CO in an extended range of overpotentials (240–440 mV).[157] The iron-based version of this catalyst (**Fe2**) exhibits some differences for **Co17**. **Fe2** shows very similar behavior to **Co9** due to product inhibition which leads to the formation of an inactive $[Fe^0(CO)(qpy)]$ species and a drop in FY_{CO} (48%). The FY_{CO} was improved to 70% by shining visible light under electrolysis conditions. Based on experimental evidence, the authors proposed an EEC and ECD mechanism for **Co17** and **Fe2**, respectively (Fig. 18).[158] A recent computational mechanistic study by Head-Gordon and coworkers corroborated the two mechanistic pathways.[159] However, from the orbital analysis of one- and two-electron reduced species, they conclude that the two reductions over **Fe2** are ligand centered whereas the second reduction over **Co17** is delocalized onto the metal through metal–ligand bonding interactions. With respect to the metal-carboxylate adduct, according to the DFT analysis, the strong Lewis acidic character of Fe favors the η^2-CO_2 while it is a η^1-CO_2 in the case of Co.

In the last 2 years, other Fe catalysts have been reported as efficient electrocatalysts for the CO_2 reduction to CO. Long, Chang and coworkers achieved the control of selectivity in aqueous acetonitrile solution

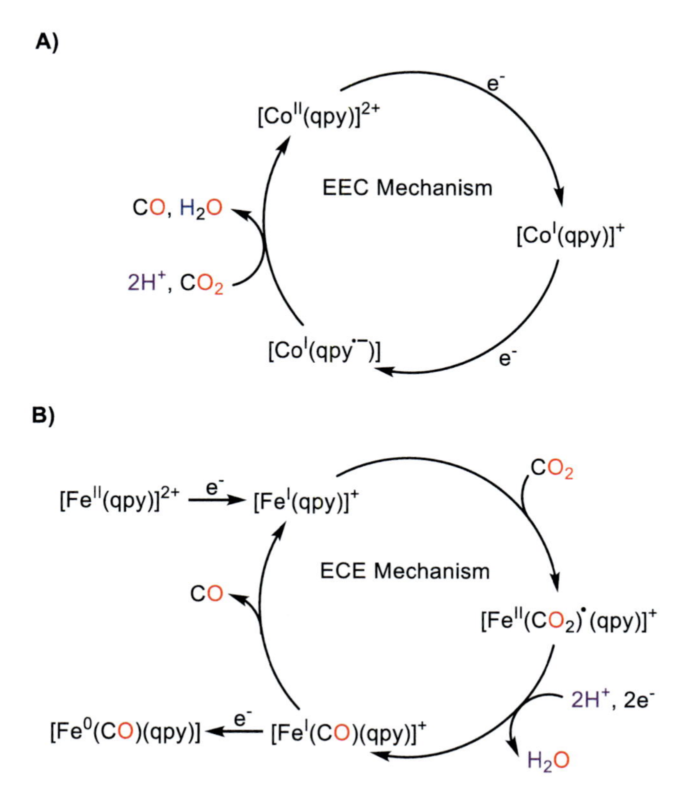

Fig. 18 An experimental-based mechanistic proposal for the electrochemical CO_2 reduction catalyzed by **Co16** (A) and **Fe2** (B).

(11 M H_2O) through the rational modification of catalyst **Fe3**.[160] The ligand modification with a pendant -OH group promotes the HER (**Fe4**, ca. 20% FY_{CO}, ca. 50% FY_{H2}) while the introduction of a milder hydrogen bond donor such as a secondary amine, favors the CO_2 reduction to CO (**Fe5**, ca. 80% FY_{CO}, ca. 10% FY_{H2}). The phenolic proton in **Fe4** is too acidic and protonates the two-electron reduced intermediate to give the corresponding hydride [Fe—H]. Instead, the -NHEt group can stabilize the CO_2 adduct [Fe-CO_2] via hydrogen bonding, and this was also corroborated by DFT calculations (Fig. 19). The Lewis acidic character of Fe favors the η^2-CO_2 coordination in the Fe-carboxylate adduct in agreement with **Fe2**.

The same group exploited the metal–ligand cooperativity by designing a five-coordinate polypyridilamino Fe complex based on the redox-active tpy scaffold (**Fe6**).[161] This catalyst resulted in being highly selective toward

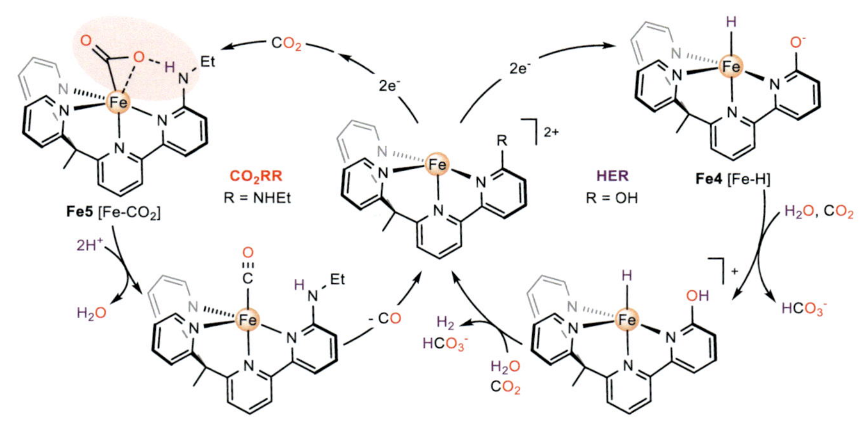

Fig. 19 Mechanistic proposal for the electrochemical HER catalyzed by **Fe4** and CO_2 reduction catalyzed by **Fe5**.

CO_2 reduction to CO (94% FY_{CO}) at only 190 mV overpotential in an organic solvent, but it was also highly selective in aqueous media. The two-electron reduced active species has been isolated and characterized as an open-shell singlet Fe^{II} complex with an antiferromagnetically coupled biradical tpy ligand. The strong exchange metal–ligand coupling was proposed to be crucial for the CO_2 reduction at low overpotential and the suppression of the HER.

In the group of catalysts based on polypyridyl ligands, we have also included metal–carbene complexes such as **Fe7**, **Ru1**–2, **Ni13**–**15**, **Co18**–**20**. Although carbene complexes are considered organometallic, these are examples that contain redox-active tpy or bpy ligands. Therefore, their reactivity is more similar to the previously presented catalysts than the classical organometallic catalysts. For Fe and Ru pyridyl-carbene tpy complexes, the geometric isomerization of the pyridyl-carbene ligand is crucial in the CO release step. For catalyst **Fe7**, the catalytic CO_2 reduction to CO occurs selectively at only 150 mV with excellent selectivity (96% FY_{CO2}) in aqueous acetonitrile under constant CO_2 flow. Under static CPE conditions, the FY_{CO} drops to 33%, indicating a specific product inhibition that can be minimized by simply sparging the solution with CO_2. The mechanistic proposal is shown in Fig. 20.[162] Structurally similar Ru complexes **Ru1** and **Ru2** allowed for the systematic investigation of the *trans* effect in the electrocatalytic CO_2 reduction.[163,164] Introducing the highly donating carbene ligand in *trans* position to the coordination vacancy favors the CO_2 binding step and the CO release. The stronger *trans* effect induced by the *trans*-carbene is critical in CO

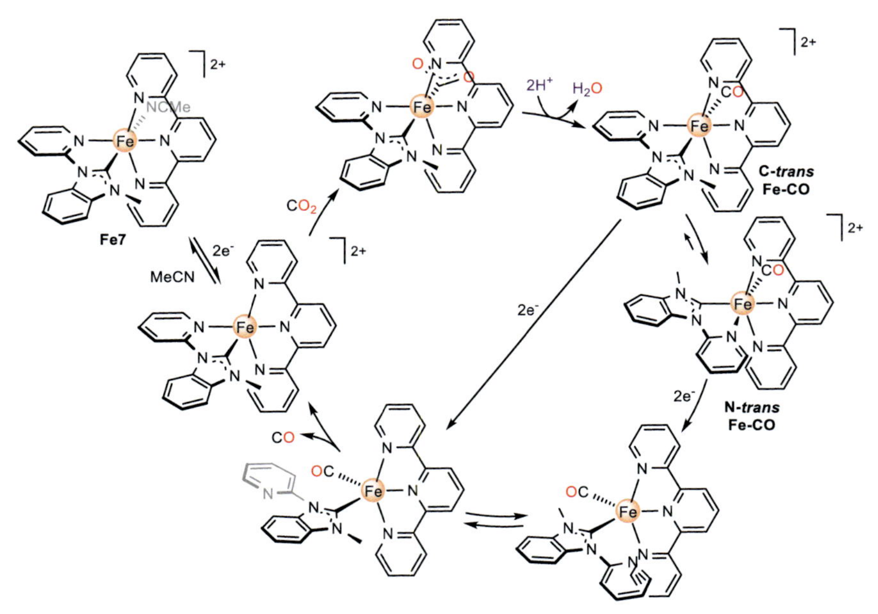

Fig. 20 Mechanistic proposal for the electrochemical CO_2 reduction to CO catalyzed by **Fe7**.

dissociation being much faster in the case of C-*trans* than in the N-*trans* isomer. These results suggest that combining a redox-active ligand (e.g., tpy), and a highly donating ligand (e.g., NHC) could maintain low overpotentials, suppressing the undesired ligand dissociation, while maintaining the basicity to activate the CO_2. Previous work from Meyer and coworkers demonstrated that a phosphate derivatized tpy Ru pyridyl–carbene complex anchored onto FTO treated with TiO_2 and carbon nanotubes can perform in water producing CO in more than 300 TON.[165] However, the system's durability was probably limited due to the formation of an inactive N-*trans* carbonyl complex.

Jurss, Panetier and coworkers investigated the effect of the ligand rigidity on a family of *bis*-NHC Ni (**Ni13–15**) and Co (**Co18–20**) bpy electrocatalysts. Their results revealed that the less rigid non-macrocyclic Ni complex produces H_2 as the major product in aqueous acetonitrile. Catalysis becomes more selective for the CO_2 reduction to CO by increasing the rigidity of the macrocyclic ligand (**Ni13** < **Ni14** < **Ni15**).[166] Electronic structure calculations suggest that the one-electron reduced species is more metal-centered in the more flexible complex **Ni13**, favoring a reactive hydride intermediate. The analogous Co complexes outperformed their

Ni counterparts and showed the same trend regarding the selectivity for CO production (**Co18** < **Co19** < **Co20**).[23] The most rigid Co complex, **Co20**, shows excellent selectivity (>90% FY_{CO}) and stability in an aqueous electrolyte for 50 h CPE. DFT calculations on this series reveal that the first one-electron reduction occurs on the ligand in the three cases. However, the second reduction is more metal-centered for **Co20**, resulting in a more reactive intermediate toward CO_2.

A striking example is the Cr bpy bis-phenolate catalyst (**Cr1**, Fig. 21) reported by Machan and coworkers in 2020.[167] This is the first Cr complex to reduce CO_2 electrochemically to CO with a quantitative Faradaic efficiency and a very low overpotential (15 TON, $FY_{CO} = 96\%$, $\eta = 110$ mV). Interestingly, the CV of **Co1** under CO_2 in the presence of PhOH shows a textbook *S*-shaped voltammogram which is symptomatic of a steady-state catalytic process. Further CV experiments and simulations led to the proposal of an ECEC mechanism in which the C—O bond cleavage is the rds of the catalytic cycle. In a more recent study, they explored the addition of dibenzothiophene-5,5-dioxide (DBTD) as a co-catalyst to induce inner-sphere electron transfer, which results in the activation of the "reductive disproportionation" pathway described previously (Fig. 1B).[168] This remarkable example demonstrates that Group 6 elements can work efficiently in CO_2 reduction, although they are required to achieve negative valencies to become catalytically active. Indeed, the low electron count of the reduced intermediates can be crucial to prevent catalyst inhibition due to CO binding processes.

2.1.3 Organometallic complexes

This section will be devoted to the most recent advances in CO_2 reduction to CO using organometallic catalysts, i.e., metal complexes based on classical organometallic ligands such as CO, phosphine, NHC, among others.[17] As we will show, some of them are also based on pyridyl ligands, although they show particular features due to their organometallic nature.

Remarkable mechanistic investigations have been performed with a cyclopentadienone tricarbonyl Fe^0 catalyst (**Fe8**), which produces CO selectively (96% FY) in a dry acetonitrile solution with high rates.[169] The metal–ligand cooperation between the protonated reduced cyclopentadienone ligand and the Fe center is crucial in the C—O bond cleavage step of the CO_2 reduction mechanism (Fig. 22). In this case, the formation of a stable [Fe—H] species has been identified as the main deactivation pathway.[170]

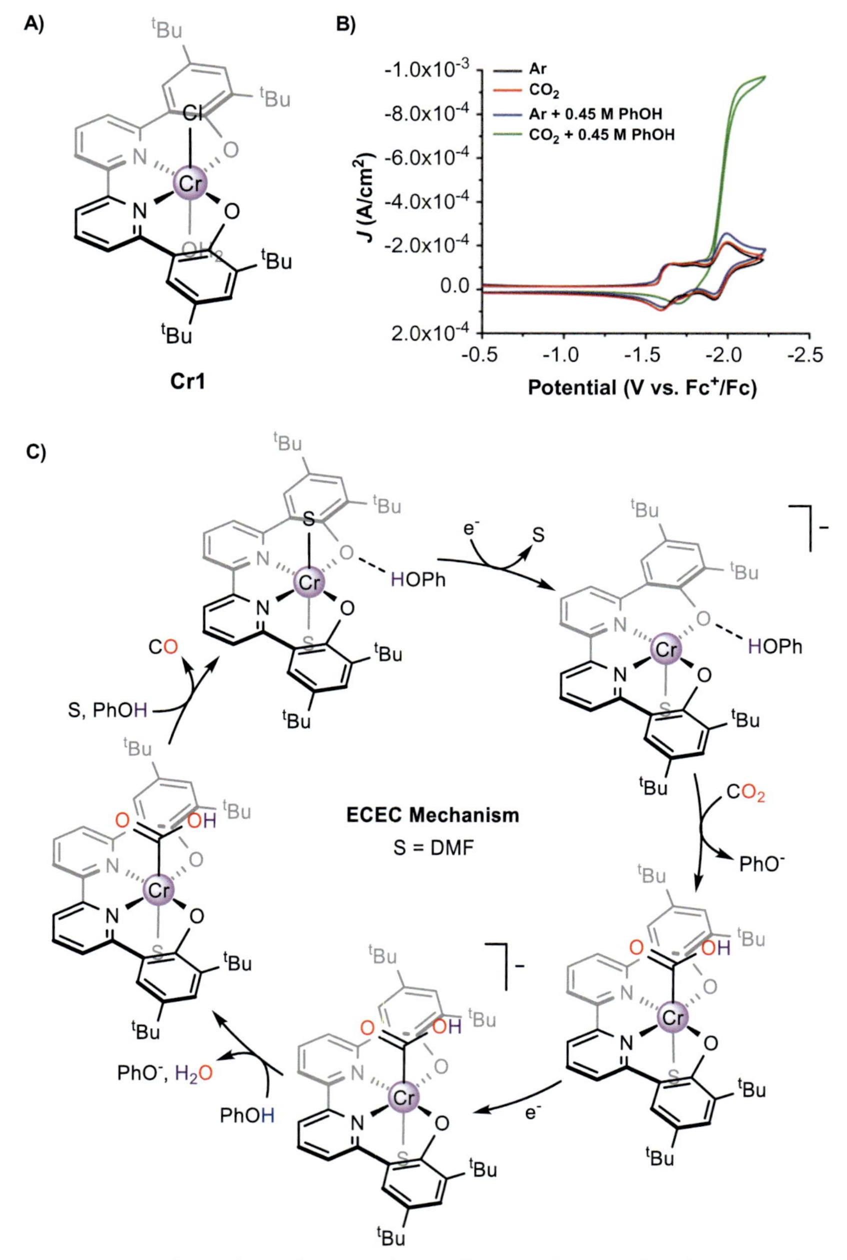

Fig. 21 Structure (A) and CV of **Cr1** catalyst under Ar and CO_2 in the absence and presence of PhOH in DMF/TBAPF_6 solution (B). Proposed catalytic cycle (C). *Adapted with permission from Hooe, S. L.; Dressel, J. M.; Dickie, D. A.; Machan, C. W., ACS Cat. 2020, 10, 1146–1151. Copyright 2020 American Chemical Society.*

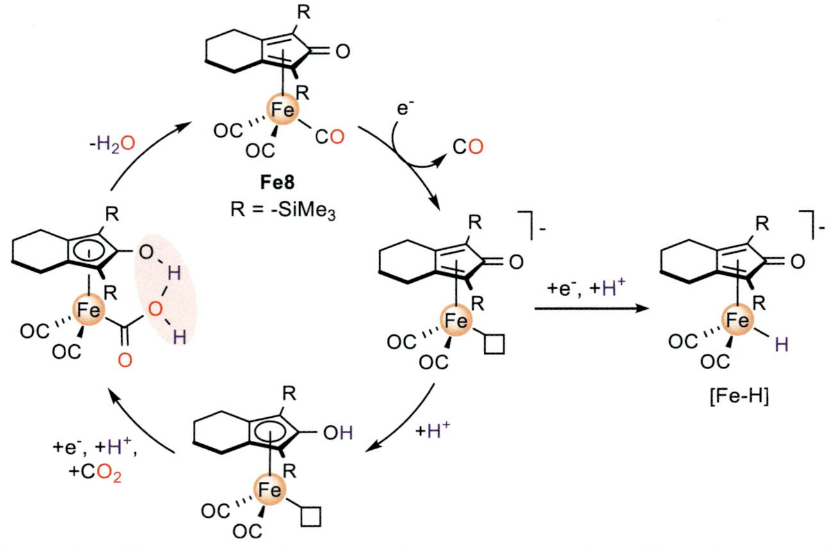

Fig. 22 Simplified mechanistic proposal for the electrocatalytic CO_2 reduction to CO with the **Fe8** catalyst.

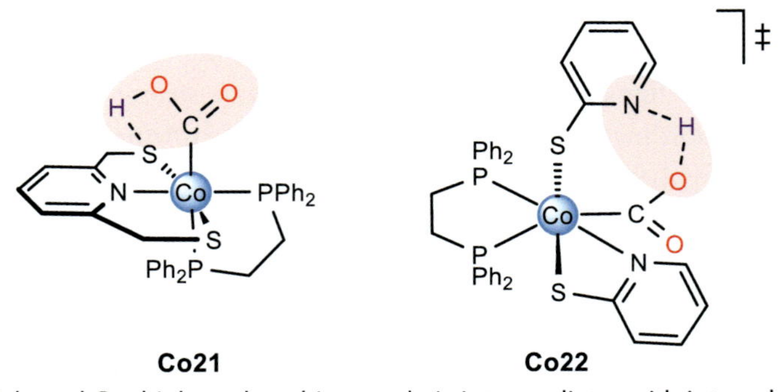

Fig. 23 Selected Co-thiolate phosphine catalytic intermediates with internal hydrogen bond/proton source second coordination sphere interactions.

Following the same idea of metal–ligand cooperation and second sphere coordination effects, the group of Dey and coworkers reported two examples of Co-thiolate catalysts based on di-phosphine ligands that are selective for the CO_2 reduction to CO ($FY_{CO} > 90\%$, Fig. 23). In the case of catalyst **Co21**, the protonated thiolate ligand in the catalyst promotes the intramolecular protonation assisting the C—O bond cleavage step.[24] This catalyst shows a $TOF_{max} = 1559\ s^{-1}$ at a remarkably low overpotential of 70 mV in acetonitrile/H_2O mixtures. In the case of catalyst **Co22**, the protonation of the ligand takes place over the pyridyne fragment, which assists both the CO_2 binding and C—O bond cleavage steps.[171] Interestingly, both complexes show a high $KIE_{H/D} > 5$, which confirms that a protonation event is involved in the rds of the catalytic cycle.

Mn and Re fac-tricarbonyl complexes: The *fac*-$[M(bpy)(CO)_3(L)]$ (M=Re, Mn; L=Br^-, Cl^-) is one of the most widely studied families of catalysts for the selective reduction of CO_2-to-CO. The pioneering work by Lehn and coworkers on the electrochemical CO_2 reduction catalyzed by *fac*-$[Re(bpy)(CO)_3Cl]$ dates from 1984.[172] It was not until 2011 that Deronzier, Chardon-Noblat and coworkers discovered that *fac*-$[Mn(bpy)(CO)_3Br]$ is also an excellent electrocatalyst for the selective CO_2-to-CO reduction working at 400 mV less overpotential than its Re counterpart.[173]

Since 2011, many groups have worked on the mechanistic understanding of the CO_2 reduction catalytic cycle (Fig. 24).[26] The different intermediates are easy to identify by the characteristic IR absorption pattern of the *fac*-CO ligands. Thus, IR-SEC has been widely used to identify and characterize catalytic intermediates in Re and Mn *fac*-tricarbonyl complexes since they share common intermediates.[174] However, there are remarkable differences, such as the fast dimerization of the one-electron reduced Mn radical species $[Mn^{\bullet}]$ to give [Mn—Mn]. Electronic structure calculations suggest that the first one-electron reduction has a strong ligand character in the pentacoordinate $[Re^{\bullet}]$ intermediate while it is more metal-centered in the case of $[Mn^{\bullet}]$ which are represented better as *fac*-$[Re^{I}(bpy)^{\bullet -}(CO)_3]$ and *fac*-$[Mn^{0}(bpy)(CO)_3]$, respectively.[175] A common five-coordinate anion $[M]^-$ is formed after a second one-electron reduction. In this case, both Re and Mn complexes are described better as *fac*-$[M^{0}(bpy)^{\bullet -}(CO)_3]$ species which are responsible for the binding of CO_2. In this regard, another relevant difference between Re and Mn is that while the CO_2 binding is exergonic for Re, it is usually endergonic for *fac*-$[Mn(bpy\text{-}R)(CO)_3Br]$ catalysts, which then require the addition of an acid to function as a catalyst.

As a favorable consequence of the extensive mechanistic studies on this family of catalysts, some strategies based on modifying the catalyst environment provided promising results in catalytic performance and even a change in the product selectivity. In the particular case of Re, the introduction of second coordination sphere effects by introducing hydrogen bonding groups such as a tethered thiourea (**Re2**)[176] or a positively charged imidazolium group (**Re3**)[177] led to an improvement in the catalytic

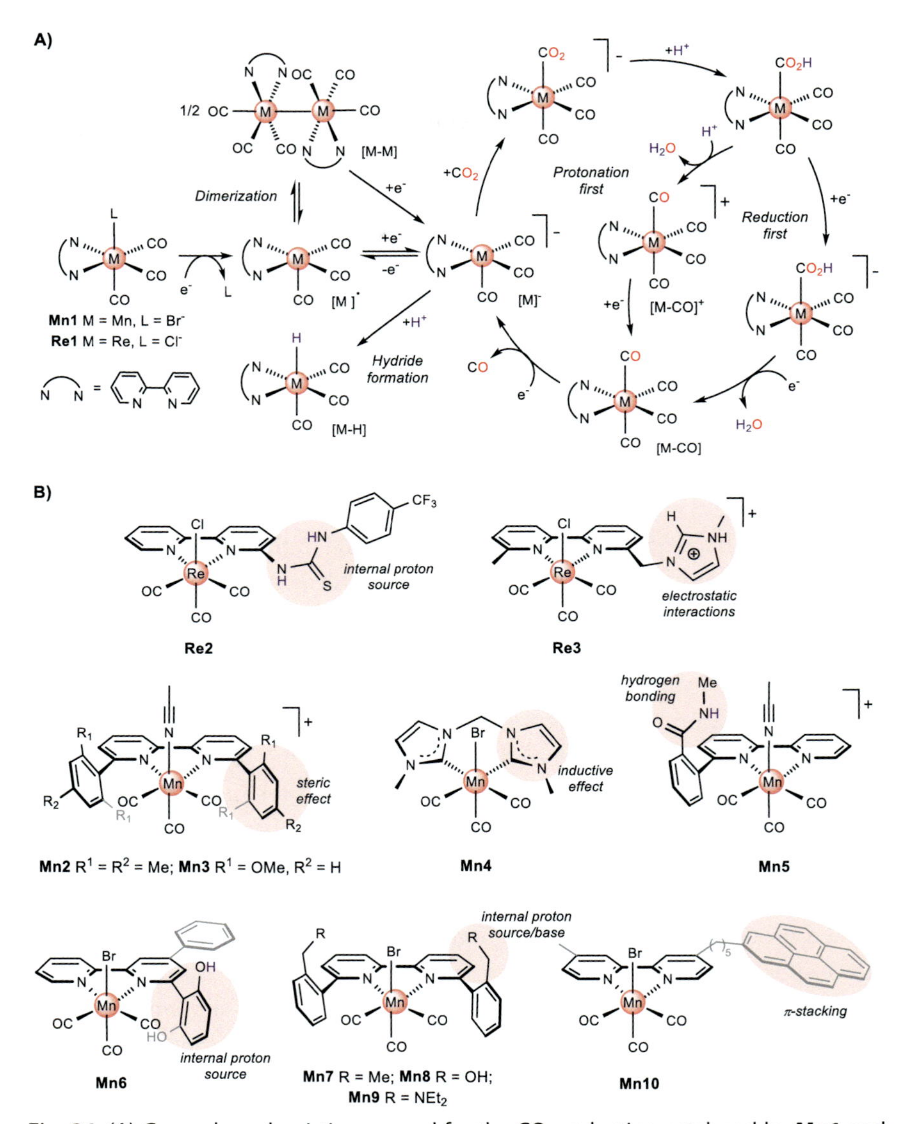

Fig. 24 (A) General mechanistic proposal for the CO_2 reduction catalyzed by **Mn1** and **Re1** catalysts and derivates. (B) Selected Re and Mn bpy *fac*-tricarbonyl catalysts.

performance. The introduction of highly conjugated groups in the bpy ligand such as nanographenes,[178] naphthalimide[179] or perylene diimide[180] moieties, resulted in an effective decrease of the catalyst onset potential concerning the unmodified Lehn's catalyst. Marinescu and coworkers incorporated catalytically active *fac*-[$Re(CO)_3Cl$] units in a COF based on

5,5′-diamine-2,2′-bpy linkers, although the catalytic activity dropped after 30 min CPE in acetonitrile solution.[181] Recently, Kubiak and coworkers studied the CO_2 reduction catalytic activity of *fac*-[Re(bpy-^{t}Bu)$(CO)_3$Cl] supported on MWCNT in water (pH 7.3) which constitutes a robust, scalable and selective electrode for the catalytic production of CO.[182]

One of the first effective modifications of the *fac*-[Mn(bpy)$(CO)_3$Br] species consisted of the introduction of bulky substituents such as mesityl groups at the 6 and 6′ position of the bpy ligand (**Mn2**).[183] The steric effect induced by the bulky groups prevents the formation of the [Mn—Mn] dimeric species, which results in a direct reduction of the [Mn$^•$] in one more electron to give $[Mn]^-$. Thus, the CO_2 binding occurs at very low overpotentials, and the protonation-first mechanism can be activated by adding Mg^{2+} as a Lewis acid. On the same theme, Grills, Ertem, Rochford and coworkers reported catalyst **Mn3**, which can perform selectively through the protonation-first pathway with a 0.55 V decrease in overpotential to the reduction-first pathway.[25] The authors propose an allosteric hydrogen bonding interaction between the methoxy groups and the added Brønsted acid, facilitating the C-OH bond cleavage step. The protonation-first pathway is also the operative mechanism with the highly active Mn(I) *bis*-NHC **Mn4** catalyst reported by Lloret-Fillol, Royo and coworkers (Fig. 25).[184] In this case, IR-SEC allowed the detection of the $[Mn\text{-}CO]^+$ intermediate, which was proposed to be the protonation-first pathway's resting state. The highly donating *bis*-NHC ligand does not favor the formation of the [Mn—Mn] dimeric species; simultaneously, it increases the nucleophilic character of the metal-carboxylate, which becomes catalytic even in the absence of added protons. **Mn4** realizes the highest TOF_{max} (320,000 s^{-1} in the presence of 0.55 M H_2O) reported so far for Mn-based catalysts.

As was mentioned previously for some polypyridyl and aminopyridyl-based catalysts, the substitution of the bpy ligand in *fac*-[Mn(bpy-R)$(CO)_3$Br] complexes with hydrogen bonding groups, is an effective strategy for the improvement of the catalytic activity in terms of catalytic current and overpotential. This is the case for catalyst **Mn5** in which an amide group has been introduced in the bpy ligand.[185] Hydrogen bonding interactions between the amide group and a weak acid such as water (see Fig. 26), decrease the C—O bond cleavage barrier, thus favoring the protonation-first mechanism at low overpotentials. Further attempts have been devoted to changing selectivity from CO to HCO_2^- as the main CO_2 reduction products (**Mn6–9**). These results will be described in more detail in the following section focusing on the CO_2 to formate reduction mechanism.

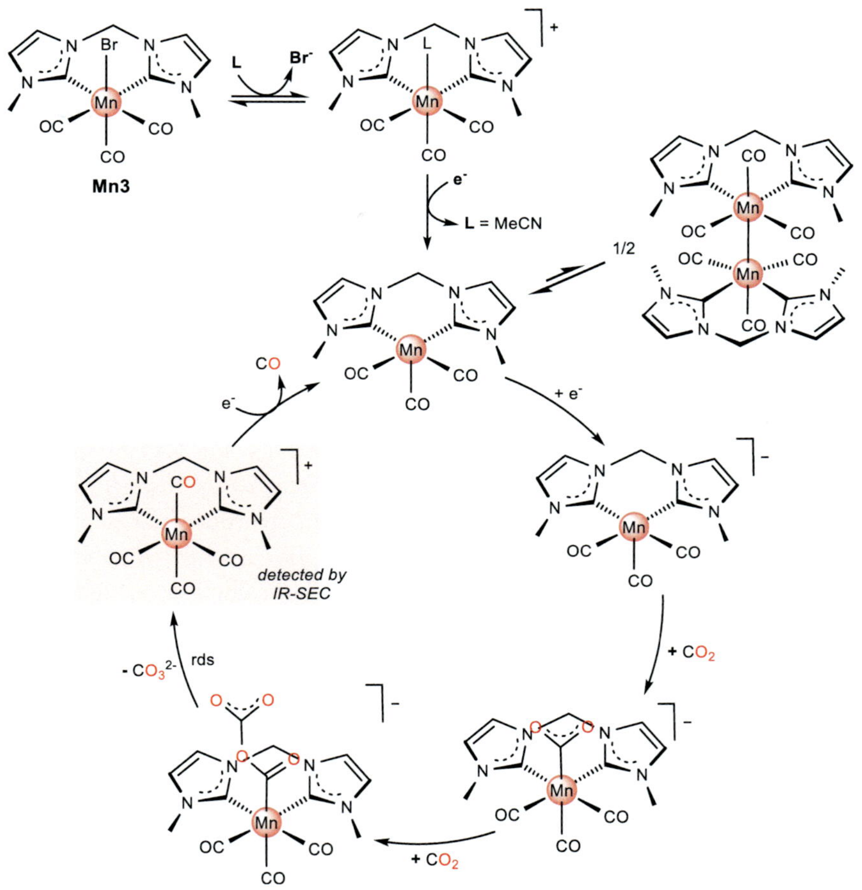

Fig. 25 Mechanistic proposal for the CO_2 reduction to CO catalyzed by **Mn4** in anhydrous acetonitrile.

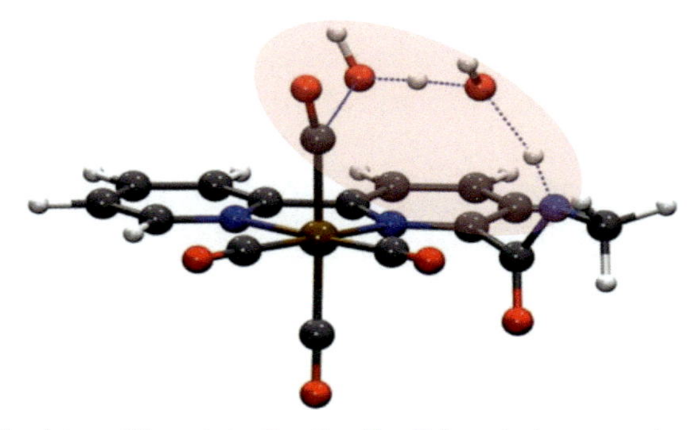

Fig. 26 Optimized transition state for the C—O bond cleavage step from the [Mn-$CO_2H{\cdot}H_2O$] adduct of **Mn5**. *Reproduced from Yang, Y.; Ertem, M. Z.; Duan, L., Chem. Sci. 2021, 12, 4779–4788, with permission from the Royal Society of Chemistry.*

Mn complexes to be applied for CO_2 reduction have been immobilized following different strategies, including the simple drop-casting of *fac*-[Mn(bpy-R)$(CO)_3$Br] derivatives mixed with Nafion® or Nafion®/MWCNTs on top of glassy carbon.[186,187] Reisner and coworkers anchored catalyst **Mn10** into MWCNTs *via* π-stacking interactions taking advantage of the pyrene fragment tethered to the bpy ligand.[188] More sophisticated technologies have been employed, such as the introduction of the catalytic units into metal–organic conjugated microporous polymers,[189,190] covalently attached to carbon paper,[191] MOFs,[94,192] or on top of silicon nanowires (SiNWs) photocathodes,[193] allowing to function with supported catalysts in water.[194] This year our group has reported the first example of COF based on *fac*-[Mn(bpy-R)$(CO)_3$Br] linkers (**Mn11**).[11] This catalyst shows catalytic activity in water, a 72% selectivity for CO_2 reduction, including CO (ca. 50% FY_{CO}) and formic acid (ca. 20% FY_{HCO2-}), as well as extended durability and recyclability with an overpotential of 190 mV (140 mV earlier than the molecular catalyst). IR-SEC in the solid-state demonstrated that the catalytic units are mechanically constrained inside the reticular material, and thus, the formation of the [Mn—Mn] species is disfavored (Fig. 27).

2.2 CO_2 reduction to FA

Another possible pathway for CO_2 electro-reduction involves the formation of a metal-hydride intermediate. The reduced species reacts with a proton source instead of CO_2 to afford a metal hydride species (Fig. 28). Formate can be produced by hydride transfer from the M-H to solvated CO_2 or via a CO_2 insertion into the M-H bond.[195] This M-H is also a common intermediate in HER; therefore, CO_2 reduction to formate competes with H_2 evolution and with the classical CO_2-to-CO reduction pathway.[4,12,60] This triple competition lies at the basis of the selectivity issues, which can be addressed by tuning the hydricity (ΔG_{H-}) of the hydride intermediate and the proton activity (pK_a) of the acid.[196] The Ir (**Ir1**), Fe (**Fe9**), and Pt (**Pt1**) catalysts reported by the groups of Meyer and Brookhart, Berben, and Yang, respectively, are pivotal examples of highly selective electrocatalysts for CO_2 reduction to formate (>90% Faradaic efficiency, Fig. 28).[197–200]

Inspired by the active site of natural formate dehydrogenase (FDH) and CO dehydrogenase (CODH) enzymes, Fontecave, Li and coworkers

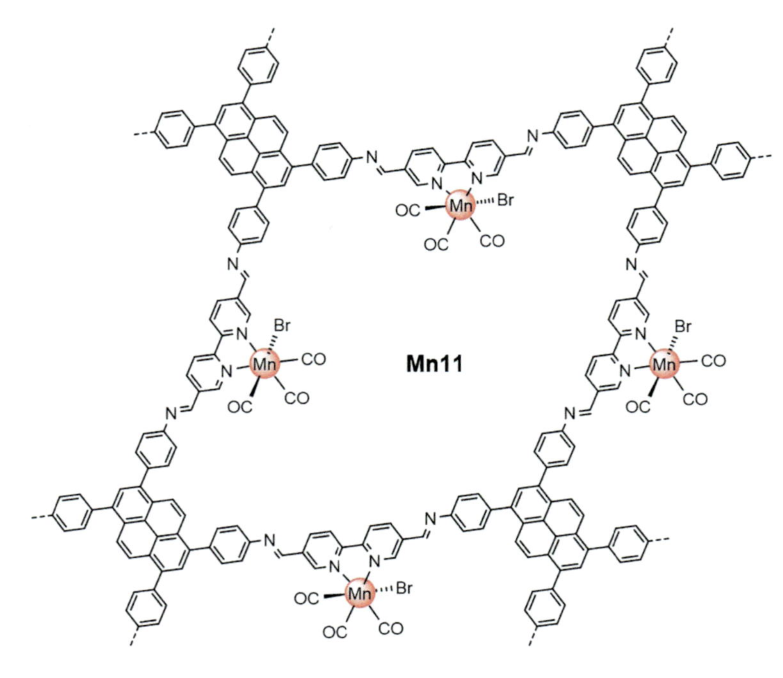

- Well-defined Mn coordination center
- Mechanical constraint
- CO_2 electroreduction in water at low η
- Accessible catalytic centers

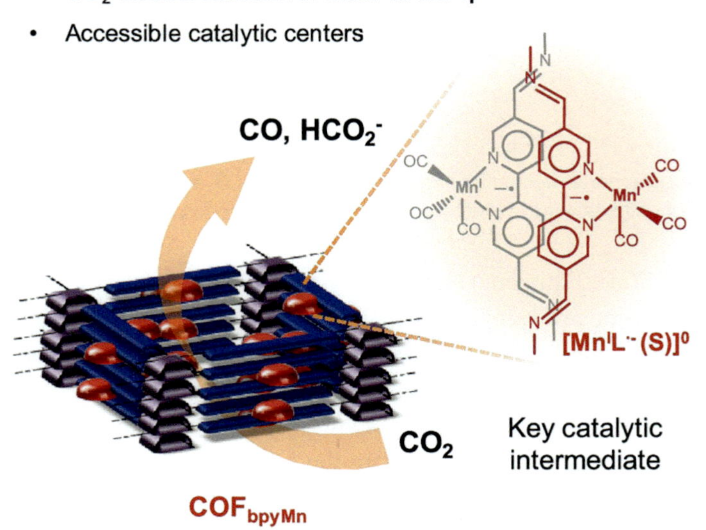

Fig. 27 Structure of the **Mn11** catalyst based on a COF with *fac*-[Mn(bpy-R)$(CO)_3$Br] linkers (top). Schematic representation of the catalyst in 3D and close-up of the reduced species (bottom). *Adapted with permission from Dubed Bandomo, G. C.; Mondal, S. S.; Franco, F.; Bucci, A.; Martin-Diaconescu, V.; Ortuño, M. A.; van Langevelde, P. H.; Shafir, A.; López, N.; Lloret-Fillol, J., ACS Cat. 2021, 11, 7210–7222. Copyright 2021 American Chemical Society.*

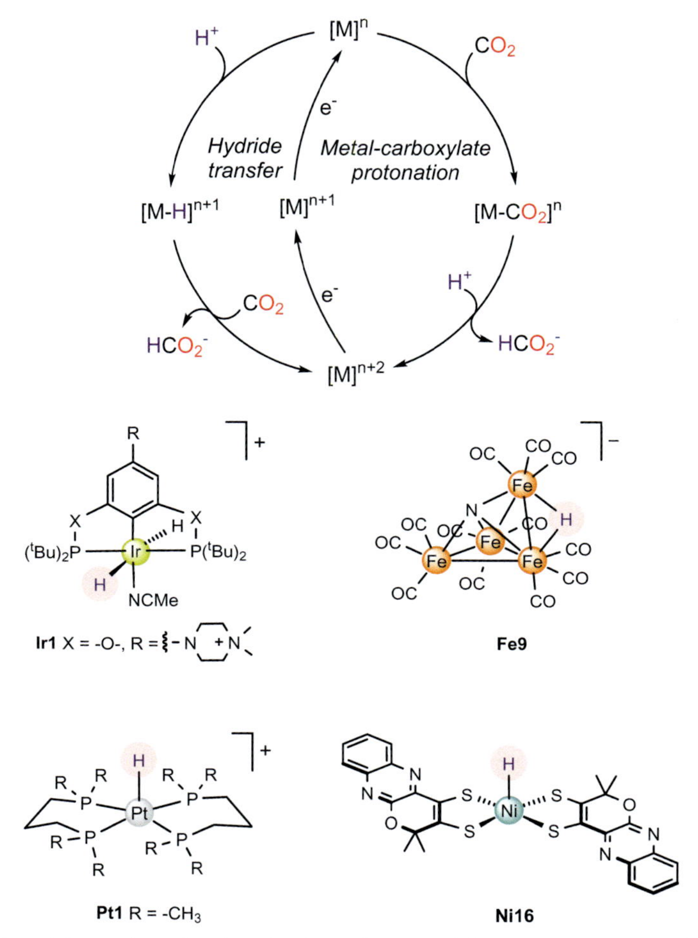

Fig. 28 (Top) General mechanism for the reduction of CO_2 to HCO_2^-. (Bottom) Selected examples of catalytic hydride intermediates for the reduction of CO_2 to HCO_2H/HCO_2^-.

reported a Ni(bis-dithiolene) electrocatalyst (**Ni16**) that reduces CO_2 to formic acid as the major product (60% FY in acetonitrile with 2 M TFE), together with CO (19% FY_{CO}) and H_2 (9% FE_{H2}).[201] The proposed mechanism for these electrocatalysts involves protonation of the metal center in water or by an external proton source with a modest proton activity in organic solvents and a subsequent hydride transfer from the M-H to the CO_2 molecule.

An alternative strategy consists of tailoring the second coordination sphere to drive changes in product selectivity. In 2017, Artero and coworkers reported a new family of cyclopentadienyl cobalt complexes based on diphosphine ligands functionalized by pendant amine groups (**Co23**).[202] The internal pendant base stabilizes the hydride transfer TS (Fig. 30) through hydrogen bond interactions with H_2O molecules, rather than being directly involved as proton relays, enabling high selectivity (98% FY_{HCO2-}). Likewise, pendant bases or proton relays on the supporting bpy ligand of the classic *fac*-$Mn(CO)_3$ CO_2-to-CO reduction electrocatalysts (**Mn5–7**) provide changes in selectivity of the reaction towards the generation of formate. For example, the substitution of the bpy with acidic phenol groups (**Mn6**) facilitates the intramolecular protonation of the anion intermediate to give the corresponding [Mn—H] species, leading to formate formation.[203,204] More recently, Daasbjerg, Skrydstrup, Baik and coworkers reported a series of manganese catalysts with various pendant groups tethered to the bpy ligand (**Mn7–9**).[205] In this series of catalysts, there is a strong effect of the second sphere effects in the product distribution in acetonitrile solution with added TFE (2.0 M). While bpy substitution with an alkyl group (**Mn7**) or OH (**Mn8**) promotes the CO production, the substitution with an amine base (**Mn9**) favors the formate production with Faradaic yields above 70%. Experimental mechanistic studies showed the formation of an Mn—H species and, according to DFT, the amine groups are involved as proton shuttles in the protonation of the metal. Fig. 29 summarizes the proposed catalytic cycles for the formate production with **Mn9** catalyst. The formate production can proceed at low rates and low overpotential through the insertion into [Mn-H] (Fig. 29, blue mechanism) or at higher overpotentials through the fast insertion into $[Mn\text{-}H]^-$ (Fig. 29, red mechanism).

Robert, Lau and coworkers proposed an elegant strategy based on metal–metal cooperation to force selectivity towards formic acid under photocatalytic conditions. The authors designed a dimeric version of the highly active CO_2-to-CO reduction $[Co(qpy)]^{2+}$ electrocatalyst by using a xanthene group as a bridging scaffold (**Co24**).[206] Under basic conditions, HCO_2^- is selectively produced, while in the presence of phenol, the selectivity shifts toward CO. *In situ* IR-SEC revealed the formation of a $4e^-$-reduced CO_2 adduct stabilized by the second Co center, which is proposed as a common intermediate in the formation of both CO and HCO_2^- (Fig. 30).

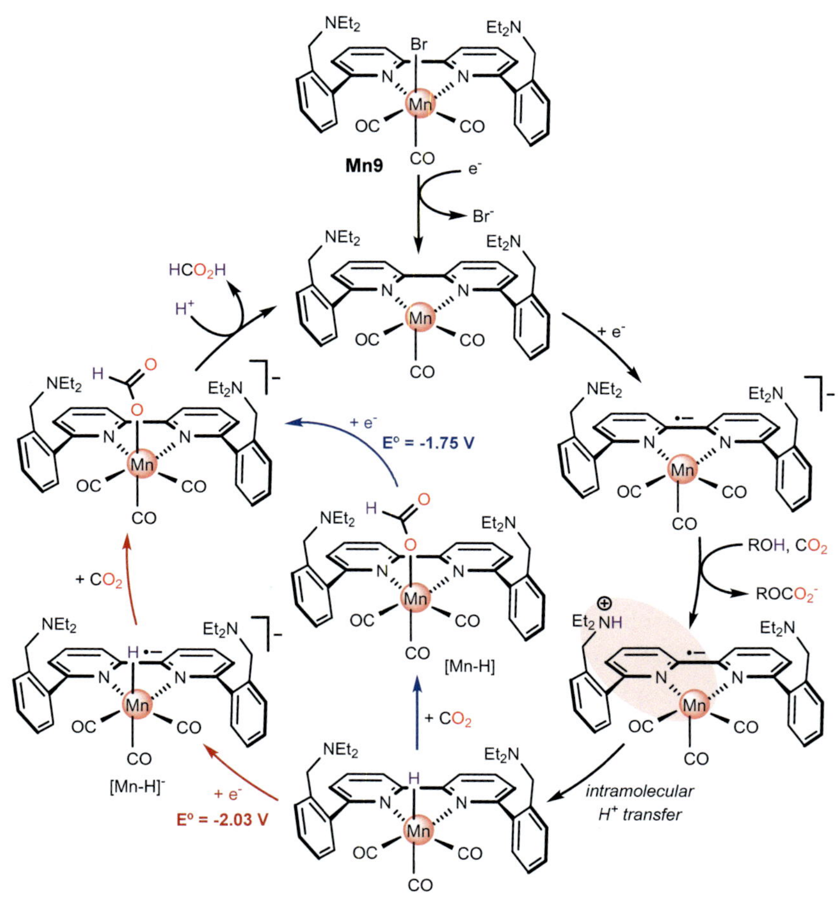

Fig. 29 Mechanistic proposal for the CO_2 reduction to HCO_2^- catalyzed by **Mn9**. Redox potentials *vs* $Fc^{+/0}$.

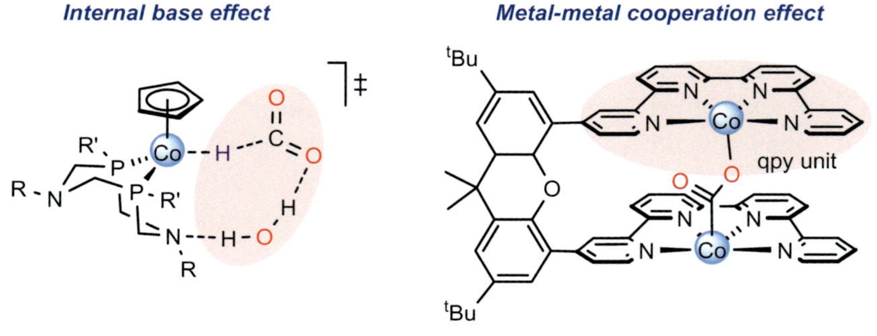

Internal base effect

Metal-metal cooperation effect

qpy unit

Co23 R = $-CH_2Ph$; R' = -Cy

Co24 Co-Co_2-Co adduct

Fig. 30 Second coordination sphere effects in CO_2 reduction to formate. Relevant TS for **Co23** or intermediate for **Co24**.

3. Beyond the two-electron reduction of CO_2

The multielectron reduction of CO_2 beyond CO and H_2CO_2/HCO_2^- using well-defined molecular catalysts is still rare. The monometallic nature of most molecular CO_2 reduction electrocatalysts also limits the formation of C_2 products (e.g., acetic acid), which is even less ubiquitous.[6,207] However, there are recent examples of homogeneous and heterogenized molecular complexes that can produce CH_2O ($4e^-$), CH_3OH ($6e^-$), CH_4 ($8e^-$) (Fig. 31).[5] For example, Koper and coworkers reported a cobalt protoporphyrin immobilized (**CoPX**) on a pyrolytic graphite (PG) electrode, which produces CO (major product) and CH_4 (ca. 2.5% FY at pH 1, $P_{CO2} = 10$ atm) together with small amounts of HCO_2H and CH_3OH in water (pH 1–3) at moderate redox potentials (−0.8 V vs RHE).[76] As shown in the general mechanism of Fig. 32, CO is proposed to be an intermediate in reducing CO_2 to CH_3OH and CH_4. DFT studies on Co porphyrins suggest that the weak binding of CO to the catalyst hinders the reduction to CH_3OH and CH_4. Computational and experimental results confirm that lower pH values favor the protonation of the M-CO intermediate favoring the CO_2 reduction beyond $2e^-$.[79]

In the last few years, cobalt phthalocyanine (CoPc) derivatives heterogenized over carbon materials have emerged as the most promising catalysts for the electrochemical reduction of CO_2-to-CO (ca. 95% selectivity, 165 mA cm^{-2} at −0.92 V vs RHE).[8,9,94] Furthermore, the **CoPc1** complex

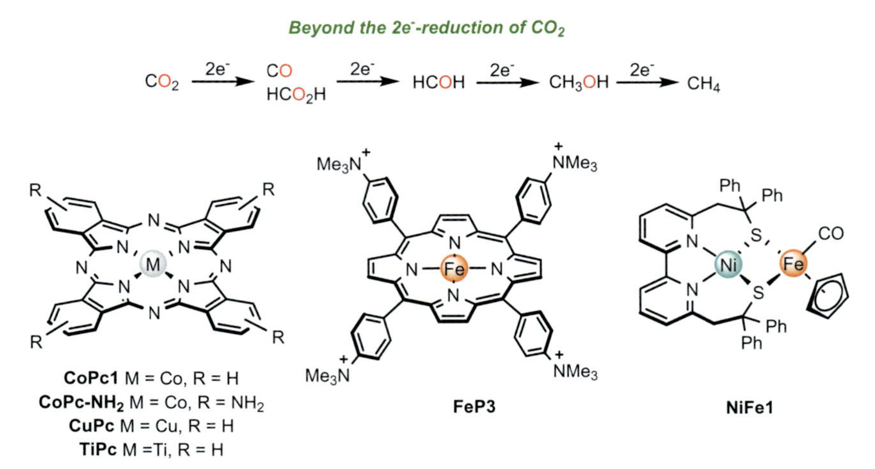

Fig. 31 Selected molecular catalysts active for the reduction of CO_2 involving more than two electrons.

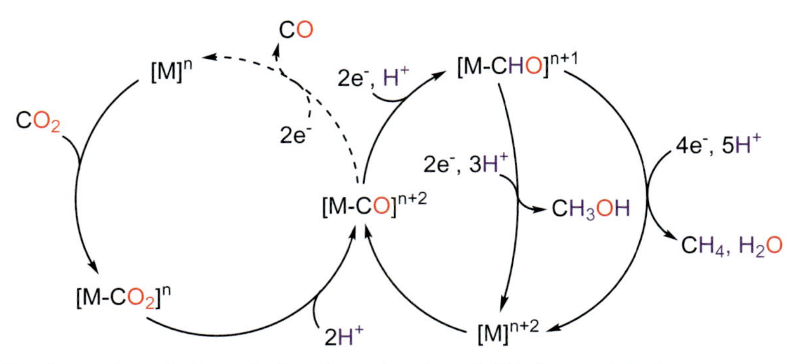

Fig. 32 General catalytic cycle for the reduction of CO_2 beyond $2e^-$.

deposited onto multiwall carbon nanotubes produces MeOH in H_2O, confirming that both CO and CH_2O are intermediate products in the reduction.[208] Wang and coworkers reported a 44% FY for MeOH production after 1 h of CPE at −0.94 V vs RHE.[209] Other metal phthalocyanines such as **CuPc** or **TiPc** produce CH_4 in moderate yields (28% FY_{CH4}) under electrochemical conditions when deposited in gas diffusion electrodes.[98] However, a recent study showed that **CuPc** could evolve to metallic copper particles under reducing electrochemical conditions.[101] Although catalysts producing CH_4 are of broad interest, well-defined systems are more suitable for understanding reactivity. Analysis of the nature of the catalysts after performing catalysis, or detection of catalytic intermediates are the desired targets. A potential pitfall when investigating molecular catalysts for electroreduction, specially beyond $2e^-$-reduction, is the ill-defined origin of the products formed. Herein, the comparison of the products formed using $^{12}CO_2$ and $^{13}CO_2$ is mandatory.

The Fe porphyrin complex functionalized with trimethylanilinium groups (**FeP3**),[51] produces a remarkable amount of CH_4 (up to 159 TON_{CH4}) under mild photocatalytic conditions using photosensitizers ($Ir(ppy)_3$ (ppy = 2-phenylpyridine) or an organic dye derived from phenoxazine).[210,211]

Similar behavior was observed in a recent study using a series of Ni bipyridyl-NHC complexes as catalysts (**Ni13–15**) under photocatalytic conditions.[212] The best results were obtained with the macrocyclic complex **Ni14**. In this case, CH_4 evolution was also observed after 1 h of bulk electrolysis (100% FY_{CH4}, −2.67 V vs $Fc^{+/0}$) under a 1:1 CO/H_2 atmosphere. More recently, the bioinspired Ni-CODH cofactor model **NiFe1** has shown catalytic activity for the CO_2 reduction to CH_4 (12% FY_{CH4}) in water at pH 4.[213] Labeling studies using $^{13}CO_2$ corroborate that CH_4 originates from CO_2 reduction. CPE in the presence of CO delivers only

H_2 suggesting that CO is not an intermediate in the CO_2-to-CH_4 reduction process. Based on this result, the authors proposed that the two sites are involved in methane formation: one binds CO_2, and the other delivers hydrides to the bound CO_2.

The lack of a mechanistic understanding of those catalysts should prompt more profound studies to disclose the structural basis, the thermodynamic and kinetic factors that drive the more than $2e^-$ CO_2 reduction, as the key features for designing highly efficient catalysts for the synthesis of sustainable fuels.

4. Conclusions and perspective

The multiproduct nature of the CO_2RR hinders a generalization of the reaction mechanisms. However, the same design principles as in other pivotal redox reactions, such as the oxygen evolution reaction (OER) and the hydrogen evolution reaction (HER) have been proposed to enhance the catalytic activity and selectivity of the CO_2RR. In the CO_2 reduction to CO, the classical tuning of the electronic effects resulted in the linear modification of the catalytic activity with the overpotential, i.e., a catalyst with lower overpotentials works at slower rates. The most popular strategies to break these linear scaling relationships are based on the introduction of hydrogen bonds or electrostatic interactions to facilitate the CO_2 binding by stabilizing the M-CO_2 type of intermediates and activating the C—O bond cleavage steps. Indeed, the through-space stabilization of rate-determining intermediates or transition states is key to enhancing the reaction kinetics even at lower overpotentials. Still the number of available examples based on a logical tuning of catalysts to influence the key steps of the catalytic cycle is scarce. The same principles also apply to the CO_2 reduction to formate, and more examples or rational tuning the selectivity are envisioned. Instead, the CO_2 reduction by more than $2e^-$ is an emerging field, and the reaction mechanisms and strategies are not yet defined. Nevertheless, advances in mechanistic understanding have been proved as powerful tools to improve CO_2 reduction catalysis.

Acknowledgments

The authors acknowledge the financial support of the ICIQ Foundation, the CERCA Program/Generalitat de Catalunya, MICINN through Severo Ochoa Excellence Accreditation 2020-2023 (CEX2019-000925-S, MIC/AEI), AGAUR 2017-SGR-1647 (J.L.-F.), MICINN (PID2019-110050RB-I00, J.L-F.) and the Spanish Ministry of Universities for an FPU fellowship FPU16/04234 (S.F.).

References

1. Luna, P. D.; Hahn, C.; Higgins, D.; Jaffer, S. A.; Jaramillo, T. F.; Sargent, E. H. *Science* **2019**, *364*, eaav3506.
2. Bushuyev, O. S.; De Luna, P.; Dinh, C. T.; Tao, L.; Saur, G.; van de Lagemaat, J.; Kelley, S. O.; Sargent, E. H. *Joule* **2018**, *2*, 825–832.
3. Schneider, J.; Jia, H.; Muckerman, J. T.; Fujita, E. *Chem. Soc. Rev.* **2012**, *41*, 2036–2051.
4. Franco, F.; Fernández, S.; Lloret-Fillol, J. *Curr. Opin. Electrochem.* **2019**, *15*, 109–117.
5. Franco, F.; Rettenmaier, C.; Jeon, H. S.; Roldan Cuenya, B. *Chem. Soc. Rev.* **2020**, *49*, 6884–6946.
6. Boutin, E.; Robert, M. *Trends Chem.* **2021**, *3*, 359–372.
7. Dalle, K. E.; Warnan, J.; Leung, J. J.; Reuillard, B.; Karmel, I. S.; Reisner, E. *Chem. Rev.* **2019**, *119*, 2752–2875.
8. Wang, M.; Torbensen, K.; Salvatore, D.; Ren, S.; Joulié, D.; Dumoulin, F.; Mendoza, D.; Lassalle-Kaiser, B.; Işci, U.; Berlinguette, C. P.; Robert, M. *Nat. Commun.* **2019**, *10*, 3602.
9. Ren, S.; Joulié, D.; Salvatore, D.; Torbensen, K.; Wang, M.; Robert, M.; Berlinguette Curtis, P. *Science* **2019**, *365*, 367–369.
10. Lin, S.; Diercks Christian, S.; Zhang, Y.-B.; Kornienko, N.; Nichols Eva, M.; Zhao, Y.; Paris Aubrey, R.; Kim, D.; Yang, P.; Yaghi Omar, M.; Chang Christopher, J. *Science* **2015**, *349*, 1208–1213.
11. Dubed Bandomo, G. C.; Mondal, S. S.; Franco, F.; Bucci, A.; Martin-Diaconescu, V.; Ortuño, M. A.; van Langevelde, P. H.; Shafir, A.; López, N.; Lloret-Fillol, J. *ACS Cat.* **2021**, *11*, 7210–7222.
12. Boutin, E.; Merakeb, L.; Ma, B.; Boudy, B.; Wang, M.; Bonin, J.; Anxolabéhère-Mallart, E.; Robert, M. *Chem. Soc. Rev.* **2020**, *49*, 5772–5809.
13. Stratakes, B. M.; Dempsey, J. L.; Miller, A. J. M. *ChemElectroChem* **2021**, *8*, 4161–4180.
14. Machan, C. W. *Curr. Opin. Electrochem.* **2019**, *15*, 42–49.
15. Jensen, M. T.; Rønne, M. H.; Ravn, A. K.; Juhl, R. W.; Nielsen, D. U.; Hu, X.-M.; Pedersen, S. U.; Daasbjerg, K.; Skrydstrup, T. *Nat. Commun.* **2017**, *8*, 489.
16. Elgrishi, N.; Chambers, M. B.; Wang, X.; Fontecave, M. *Chem. Soc. Rev.* **2017**, *46*, 761–796.
17. Kinzel, N. W.; Werlé, C.; Leitner, W. *Angew. Chem. Int. Ed.* **2021**, *60*, 11628–11686.
18. Costentin, C.; Drouet, S.; Passard, G.; Robert, M.; Savéant, J.-M. *J. Am. Chem. Soc.* **2013**, *135*, 9023–9031.
19. Costentin, C.; Robert, M.; Savéant, J.-M. *Acc. Chem. Res.* **2015**, *48*, 2996–3006.
20. Bonin, J.; Maurin, A.; Robert, M. *Coord. Chem. Rev.* **2017**, *334*, 184–198.
21. Savéant, J.-M. *Chem. Rev.* **2008**, *108*, 2348–2378.
22. Keith, J. A.; Grice, K. A.; Kubiak, C. P.; Carter, E. A. *J. Am. Chem. Soc.* **2013**, *135*, 15823–15829.
23. Su, X.; McCardle, K. M.; Chen, L.; Panetier, J. A.; Jurss, J. W. *ACS Cat.* **2019**, *9*, 7398–7408.
24. Dey, S.; Ahmed, M. E.; Dey, A. *Inorg. Chem.* **2018**, *57*, 5939–5947.
25. Ngo, K. T.; McKinnon, M.; Mahanti, B.; Narayanan, R.; Grills, D. C.; Ertem, M. Z.; Rochford, J. *J. Am. Chem. Soc.* **2017**, *139*, 2604–2618.
26. Grills, D. C.; Ertem, M. Z.; McKinnon, M.; Ngo, K. T.; Rochford, J. *Coord. Chem. Rev.* **2018**, *374*, 173–217.
27. Fernández, S.; Franco, F.; Casadevall, C.; Martin-Diaconescu, V.; Luis, J. M.; Lloret-Fillol, J. *J. Am. Chem. Soc.* **2020**, *142*, 120–133.
28. Fernández, S.; Cañellas, S.; Franco, F.; Luis, J. M.; Pericàs, M.À.; Lloret-Fillol, J. *ChemElectroChem* **2021**, *8*, 1–11.
29. Lee, G. R.; Maher, J. M.; Cooper, N. J. *J. Am. Chem. Soc.* **1987**, *109*, 2956–2962.

30. Windle, C. D.; Perutz, R. N. *Coord. Chem. Rev.* **2012**, *256*, 2562–2570.
31. Takeda, H.; Cometto, C.; Ishitani, O.; Robert, M. *ACS Cat.* **2017**, 7, 70–88.
32. Meshitsuka, S.; Ichikawa, M.; Tamaru, K. *J. Chem. Soc. Chem. Commun.* **1974**, 158–159.
33. Benson, E. E.; Kubiak, C. P.; Sathrum, A. J.; Smieja, J. M. *Chem. Soc. Rev.* **2009**, *38*, 89–99.
34. Costentin, C.; Robert, M.; Savéant, J.-M. *Chem. Soc. Rev.* **2013**, *42*, 2423–2436.
35. Francke, R.; Schille, B.; Roemelt, M. *Chem. Rev.* **2018**, *118*, 4631–4701.
36. Bhugun, I.; Lexa, D.; Saveant, J.-M. *J. Am. Chem. Soc.* **1994**, *116*, 5015–5016.
37. Bhugun, I.; Lexa, D.; Savéant, J.-M. *J. Phys. Chem.* **1996**, *100*, 19981–19985.
38. Hammouche, M.; Lexa, D.; Momenteau, M.; Saveant, J. M. *J. Am. Chem. Soc.* **1991**, *113*, 8455–8466.
39. Costentin, C.; Savéant, J.-M. *Nat. Rev. Chem.* **2017**, *1*, 0087.
40. Azcarate, I.; Costentin, C.; Robert, M.; Savéant, J.-M. *J. Phys. Chem. C* **2016**, *120*, 28951–28960.
41. Costentin, C.; Savéant, J.-M. *J. Am. Chem. Soc.* **2017**, *139*, 8245–8250.
42. Hammouche, M.; Lexa, D.; Savéant, J. M.; Momenteau, M. *J. Electroanal. Chem. Interfacial Electrochem.* **1988**, *249*, 347–351.
43. Costentin, C.; Drouet, S.; Robert, M.; Savéant, J.-M. *Science* **2012**, *338*, 90–94.
44. Costentin, C.; Passard, G.; Robert, M.; Savéant, J. M. *Proc. Natl. Acad. Sci. U. S. A.* **2014**, *111*, 14990–14994.
45. Nichols, A. W.; Machan, C. W. *Front. Chem.* **2019**, 7.
46. Koenig, J. D. B.; Willkomm, J.; Roesler, R.; Piers, W. E.; Welch, G. C. *ACS Appl. Energy Mater.* **2019**, *2*, 4022–4026.
47. Römelt, C.; Song, J.; Tarrago, M.; Rees, J. A.; van Gastel, M.; Weyhermüller, T.; DeBeer, S.; Bill, E.; Neese, F.; Ye, S. *Inorg. Chem.* **2017**, *56*, 4746–4751.
48. Bhugun, I.; Lexa, D.; Savéant, J.-M. *J. Am. Chem. Soc.* **1996**, *118*, 1769–1776.
49. Ramos Sende, J. A.; Arana, C. R.; Hernandez, L.; Potts, K. T.; Keshevarz-K, M.; Abruna, H. D. *Inorg. Chem.* **1995**, *34*, 3339–3348.
50. Costentin, C.; Passard, G.; Robert, M.; Savéant, J.-M. *J. Am. Chem. Soc.* **2014**, *136*, 11821–11829.
51. Azcarate, I.; Costentin, C.; Robert, M.; Savéant, J.-M. *J. Am. Chem. Soc.* **2016**, *138*, 16639–16644.
52. Wang, Y.-H.; Pegis, M. L.; Mayer, J. M.; Stahl, S. S. *J. Am. Chem. Soc.* **2017**, *139*, 16458–16461.
53. DuBois, D. L. *Inorg. Chem.* **2014**, *53*, 3935–3960.
54. Costentin, C.; Savéant, J.-M. *J. Am. Chem. Soc.* **2018**, *140*, 16669–16675.
55. Shaw, W. J.; Helm, M. L.; DuBois, D. L. *Biochim. Biophys. Acta - Bioenerg.* **2013**, *1827*, 1123–1139.
56. Chapovetsky, A.; Do, T. H.; Haiges, R.; Takase, M. K.; Marinescu, S. C. *J. Am. Chem. Soc.* **2016**, *138*, 5765–5768.
57. Nichols, E. M.; Derrick, J. S.; Nistanaki, S. K.; Smith, P. T.; Chang, C. J. *Chem. Sci.* **2018**, *9*, 2952–2960.
58. Gotico, P.; Boitrel, B.; Guillot, R.; Sircoglou, M.; Quaranta, A.; Halime, Z.; Leibl, W.; Aukauloo, A. *Angew. Chem. Int. Ed.* **2019**, *58*, 4504–4509.
59. Khadhraoui, A.; Gotico, P.; Boitrel, B.; Leibl, W.; Halime, Z.; Aukauloo, A. *Chem. Commun.* **2018**, *54*, 11630–11633.
60. Margarit, C. G.; Asimow, N. G.; Costentin, C.; Nocera, D. G. *ACS Energy Lett.* **2020**, *5*, 72–78.
61. Tatin, A.; Comminges, C.; Kokoh, B.; Costentin, C.; Robert, M.; Savéant, J.-M. *Proc. Natl. Acad. Sci. U. S. A.* **2016**, *113*, 5526.
62. Maurin, A.; Robert, M. *J. Am. Chem. Soc.* **2016**, *138*, 2492–2495.
63. Mohamed, E. A.; Zahran, Z. N.; Naruta, Y. *Chem. Mater.* **2017**, *29*, 7140–7150.

64. Maurin, A.; Robert, M. *Chem. Commun.* **2016**, *52*, 12084–12087.
65. Choi, J.; Wagner, P.; Jalili, R.; Kim, J.; MacFarlane, D. R.; Wallace, G. G.; Officer, D. L. *Adv. Energy Mater.* **2018**, *8* (1801280).
66. Choi, J.; Kim, J.; Wagner, P.; Gambhir, S.; Jalili, R.; Byun, S.; Sayyar, S.; Lee, Y. M.; MacFarlane, D. R.; Wallace, G. G.; Officer, D. L. *Energ. Environ. Sci.* **2019**, *12*, 747–755.
67. Hu, X.-M.; Salmi, Z.; Lillethorup, M.; Pedersen, E. B.; Robert, M.; Pedersen, S. U.; Skrydstrup, T.; Daasbjerg, K. *Chem. Commun.* **2016**, *52*, 5864–5867.
68. Cao, L.; Wang, C. *ACS Cent. Sci.* **2020**, *6*, 2149–2158.
69. Kornienko, N.; Zhao, Y.; Kley, C. S.; Zhu, C.; Kim, D.; Lin, S.; Chang, C. J.; Yaghi, O. M.; Yang, P. *J. Am. Chem. Soc.* **2015**, *137*, 14129–14135.
70. Diercks, C. S.; Lin, S.; Kornienko, N.; Kapustin, E. A.; Nichols, E. M.; Zhu, C.; Zhao, Y.; Chang, C. J.; Yaghi, O. M. *J. Am. Chem. Soc.* **2018**, *140*, 1116–1122.
71. Wang, Y.-R.; Huang, Q.; He, C.-T.; Chen, Y.; Liu, J.; Shen, F.-C.; Lan, Y.-Q. *Nat. Commun.* **2018**, *9*, 4466.
72. Hod, I.; Sampson, M. D.; Deria, P.; Kubiak, C. P.; Farha, O. K.; Hupp, J. T. *ACS Cat.* **2015**, *5*, 6302–6309.
73. Behar, D.; Dhanasekaran, T.; Neta, P.; Hosten, C. M.; Ejeh, D.; Hambright, P.; Fujita, E. *Chem. A Eur. J.* **1998**, *102*, 2870–2877.
74. Hu, X.-M.; Rønne, M. H.; Pedersen, S. U.; Skrydstrup, T.; Daasbjerg, K. *Angew. Chem. Int. Ed.* **2017**, *56*, 6468–6472.
75. Sonoyama, N.; Kirii, M.; Sakata, T. *Electrochem. Commun.* **1999**, *1*, 213–216.
76. Shen, J.; Kortlever, R.; Kas, R.; Birdja, Y. Y.; Diaz-Morales, O.; Kwon, Y.; Ledezma-Yanez, I.; Schouten, K. J. P.; Mul, G.; Koper, M. T. M. *Nat. Commun.* **2015**, *6*, 8177.
77. Bochlin, Y.; Korin, E.; Bettelheim, A. *ACS Appl. Energy Mater.* **2019**, *2*, 8434–8440.
78. Leung, K.; Nielsen, I. M. B.; Sai, N.; Medforth, C.; Shelnutt, J. A. *Chem. A Eur. J.* **2010**, *114*, 10174–10184.
79. Shen, J.; Kolb, M. J.; Göttle, A. J.; Koper, M. T. M. *J. Phys. Chem. C* **2016**, *120*, 15714–15721.
80. Zhu, M.; Yang, D.-T.; Ye, R.; Zeng, J.; Corbin, N.; Manthiram, K. *Cat. Sci. Technol.* **2019**, *9*, 974–980.
81. Hansch, C.; Leo, A.; Taft, R. W. *Chem. Rev.* **1991**, *91*, 165–195.
82. Marianov, A. N.; Jiang, Y. *Appl. Catal. B-Environ.* **2019**, *244*, 881–888.
83. Atoguchi, T.; Aramata, A.; Kazusaka, A.; Enyo, M. *J. Chem. Soc. Chem. Commun.* **1991**, 156–157.
84. Yao, S. A.; Ruther, R. E.; Zhang, L.; Franking, R. A.; Hamers, R. J.; Berry, J. F. *J. Am. Chem. Soc.* **2012**, *134*, 15632–15635.
85. Tanaka, H.; Aramata, A. *J. Electroanal. Chem.* **1997**, *437*, 29–35.
86. Quezada, D.; Honores, J.; García, M.; Armijo, F.; Isaacs, M. *New J. Chem.* **2014**, *38*, 3606–3612.
87. Soto, M.; Gotor-Fernández, V.; Rodríguez-Solla, H.; Baratta, W. *ChemCatChem* **2021**, *13*, 2152–2157.
88. An, S.; Lu, C.; Xu, Q.; Lian, C.; Peng, C.; Hu, J.; Zhuang, X.; Liu, H. *ACS Energy Lett.* **2021**, *6*, 3496–3502.
89. Johnson, E. M.; Haiges, R.; Marinescu, S. C. *ACS Appl. Mater. Interfaces* **2018**, *10*, 37919–37927.
90. Liang, Z.; Qu, C.; Xia, D.; Zou, R.; Xu, Q. *Angew. Chem. Int. Ed.* **2018**, *57*, 9604–9633.
91. Dong, H.; Lu, M.; Wang, Y.; Tang, H.-L.; Wu, D.; Sun, X.; Zhang, F.-M. *Appl. Catal. B-Environ.* **2022**, *303*, 120897.

92. Guo, Y.; Shi, W.; Yang, H.; He, Q.; Zeng, Z.; Ye, J. Y.; He, X.; Huang, R.; Wang, C.; Lin, W. *J. Am. Chem. Soc.* **2019**, *141*, 17875–17883.
93. Yang, S.; Yu, Y.; Gao, X.; Zhang, Z.; Wang, F. *Chem. Soc. Rev.* **2021**.
94. Zhang, X.; Wu, Z.; Zhang, X.; Li, L.; Li, Y.; Xu, H.; Li, X.; Yu, X.; Zhang, Z.; Liang, Y.; Wang, H. *Nat. Commun.* **2017**, *8*, 14675.
95. Han, N.; Wang, Y.; Ma, L.; Wen, J.; Li, J.; Zheng, H.; Nie, K.; Wang, X.; Zhao, F.; Li, Y.; Fan, J.; Zhong, J.; Wu, T.; Miller, D. J.; Lu, J.; Lee, S.-T.; Li, Y. *Chem* **2017**, *3*, 652–664.
96. Bockris, J. O. M.; Wass, J. C. *Mater. Chem. Phys.* **1989**, *22*, 249–280.
97. Torbensen, K.; Han, C.; Boudy, B.; von Wolff, N.; Bertail, C.; Braun, W.; Robert, M. *Chem. A Eur. J.* **2020**, *26*, 3034–3038.
98. Furuya, N.; Matsui, K. *J. Electroanal. Chem. Interfacial Electrochem.* **1989**, *271*, 181–191.
99. Latiff, N. M.; Fu, X.; Mohamed, D. K.; Veksha, A.; Handayani, M.; Lisak, G. *Carbon* **2020**, *168*, 245–253.
100. Kusama, S.; Saito, T.; Hashiba, H.; Sakai, A.; Yotsuhashi, S. *ACS Cat.* **2017**, *7*, 8382–8385.
101. Weng, Z.; Wu, Y.; Wang, M.; Jiang, J.; Yang, K.; Huo, S.; Wang, X.-F.; Ma, Q.; Brudvig, G. W.; Batista, V. S.; Liang, Y.; Feng, Z.; Wang, H. *Nat. Commun.* **2018**, *9*, 415.
102. Sun, L.; Reddu, V.; Fisher, A. C.; Wang, X. *Energ. Environ. Sci.* **2020**, *13*, 374–403.
103. Yue, Z.; Ou, C.; Ding, N.; Tao, L.; Zhao, J.; Chen, J. *ChemCatChem* **2020**, *12*, 6103–6130.
104. Mahmood, M. N.; Masheder, D.; Harty, C. J. *J. Appl. Electrochem.* **1987**, *17*, 1223–1227.
105. Karapinar, D.; Zitolo, A.; Huan, T. N.; Zanna, S.; Taverna, D.; Galvão Tizei, L. H.; Giaume, D.; Marcus, P.; Mougel, V.; Fontecave, M. *ChemSusChem* **2020**, *13*, 173–179.
106. Lieber, C. M.; Lewis, N. S. *J. Am. Chem. Soc.* **1984**, *106*, 5033–5034.
107. Zhu, M.; Chen, J.; Guo, R.; Xu, J.; Fang, X.; Han, Y.-F. *Appl. Catal. B-Environ.* **2019**, *251*, 112–118.
108. Abe, T.; Taguchi, F.; Yoshida, T.; Tokita, S.; Schnurpfeil, G.; Wöhrle, D.; Kaneko, M. *J. Mol. Catal. A: Chem.* **1996**, *112*, 55–61.
109. Zhu, M.; Ye, R.; Jin, K.; Lazouski, N.; Manthiram, K. *ACS Energy Lett.* **2018**, *3*, 1381–1386.
110. Hunter, C. A.; Sanders, J. K. M. *J. Am. Chem. Soc.* **1990**, *112*, 5525–5534.
111. Jiang, Z.; Wang, Y.; Zhang, X.; Zheng, H.; Wang, X.; Liang, Y. *Nano Res.* **2019**, *12*, 2330–2334.
112. Choi, J.; Wagner, P.; Gambhir, S.; Jalili, R.; MacFarlane, D. R.; Wallace, G. G.; Officer, D. L. *ACS Energy Lett.* **2019**, *4*, 666–672.
113. Morlanés, N.; Takanabe, K.; Rodionov, V. *ACS Cat.* **2016**, *6*, 3092–3095.
114. Kramer, W. W.; McCrory, C. C. L. *Chem. Sci.* **2016**, *7*, 2506–2515.
115. Liu, Y.; McCrory, C. C. L. *Nat. Commun.* **2019**, *10*, 1683.
116. Matheu, R.; Gutierrez-Puebla, E.; Monge, M.Á.; Diercks, C. S.; Kang, J.; Prévot, M. S.; Pei, X.; Hanikel, N.; Zhang, B.; Yang, P.; Yaghi, O. M. *J. Am. Chem. Soc.* **2019**, *141*, 17081–17085.
117. Yang, Z.; Zhang, X.; Long, C.; Yan, S.; Shi, Y.; Han, J.; Zhang, J.; An, P.; Chang, L.; Tang, Z. *CrstEngComm* **2020**, *22*, 1619–1624.
118. Lu, X.; Wu, Y.; Yuan, X.; Huang, L.; Wu, Z.; Xuan, J.; Wang, Y.; Wang, H. *ACS Energy Lett.* **2018**, *3*, 2527–2532.
119. Kutz, R. B.; Chen, Q.; Yang, H.; Sajjad, S. D.; Liu, Z.; Masel, I. R. *Energ. Technol.* **2017**, *5*, 929–936.
120. Beley, M.; Collin, J.-P.; Ruppert, R.; Sauvage, J.-P. *J. Chem. Soc. Chem. Commun.* **1984**, 1315–1316.

121. Fisher, B. J.; Eisenberg, R. *J. Am. Chem. Soc.* **1980**, *102*, 7361–7363.
122. Balazs, G. B.; Anson, F. C. *J. Electroanal. Chem.* **1992**, *322*, 325–345.
123. Collin, J. P.; Jouaiti, A.; Sauvage, J. P. *Inorg. Chem.* **1988**, *27*, 1986–1990.
124. Froehlich, J. D.; Kubiak, C. P. *Inorg. Chem.* **2012**, *51*, 3932–3934.
125. Honores, J.; Quezada, D.; García, M.; Calfumán, K.; Muena, J. P.; Aguirre, M. J.; Arévalo, M. C.; Isaacs, M. *Green Chem.* **2017**, *19*, 1155–1162.
126. Song, J.; Klein, E. L.; Neese, F.; Ye, S. *Inorg. Chem.* **2014**, *53*, 7500–7507.
127. Schneider, J.; Jia, H.; Kobiro, K.; Cabelli, D. E.; Muckerman, J. T.; Fujita, E. *Energ. Environ. Sci.* **2012**, *5*, 9502–9510.
128. Neri, G.; Aldous, I. M.; Walsh, J. J.; Hardwick, L. J.; Cowan, A. J. *Chem. Sci.* **2016**, *7*, 1521–1526.
129. Cao, L.-M.; Huang, H.-H.; Wang, J.-W.; Zhong, D.-C.; Lu, T.-B. *Green Chem.* **2018**, *20*, 798–803.
130. Mohamed, E. A.; Zahran, Z. N.; Naruta, Y. *Chem. Commun.* **2015**, *51*, 16900–16903.
131. Nichols, E. M.; Chang, C. J. *Organometallics* **2019**, *38*, 1213–1218.
132. Froehlich, J. D.; Kubiak, C. P. *J. Am. Chem. Soc.* **2015**, *137*, 3565–3573.
133. Zhanaidarova, A.; Moore, C. E.; Gembicky, M.; Kubiak, C. P. *Chem. Commun.* **2018**, *54*, 4116–4119.
134. Neri, G.; Walsh, J. J.; Wilson, C.; Reynal, A.; Lim, J. Y. C.; Li, X.; White, A. J. P.; Long, N. J.; Durrant, J. R.; Cowan, A. J. *Phys. Chem. Chem. Phys.* **2015**, *17*, 1562–1566.
135. Saravanakumar, D.; Song, J.; Jung, N.; Jirimali, H.; Shin, W. *ChemSusChem* **2012**, *5*, 634–636.
136. Schneider, C. R.; Shafaat, H. S. *Chem. Commun.* **2016**, *52*, 9889–9892.
137. Jiang, C.; Nichols, A. W.; Walzer, J. F.; Machan, C. W. *Inorg. Chem.* **2020**, *59*, 1883–1892.
138. Gangi, D. A.; Durand, R. R. *J. Chem. Soc. Chem. Commun.* **1986**, 697–699.
139. Creutz, C.; Schwarz, H. A.; Wishart, J. F.; Fujita, E.; Sutin, N. *J. Am. Chem. Soc.* **1991**, *113*, 3361–3371.
140. Fujita, E.; Furenlid, L. R.; Renner, M. W. *J. Am. Chem. Soc.* **1997**, *119*, 4549–4550.
141. Fujita, E.; Szalda, D. J.; Creutz, C.; Sutin, N. *J. Am. Chem. Soc.* **1988**, *110*, 4870–4871.
142. Szalda, D. J.; Fujita, E.; Creutz, C. *Inorg. Chem.* **1989**, *28*, 1446–1450.
143. Lacy, D. C.; McCrory, C. C. L.; Peters, J. C. *Inorg. Chem.* **2014**, *53*, 4980–4988.
144. Sheng, H.; Frei, H. *J. Am. Chem. Soc.* **2016**, *138*, 9959–9967.
145. Chen, L.; Guo, Z.; Wei, X.-G.; Gallenkamp, C.; Bonin, J.; Anxolabéhère-Mallart, E.; Lau, K.-C.; Lau, T.-C.; Robert, M. *J. Am. Chem. Soc.* **2015**, *137*, 10918–10921.
146. Nie, W.; McCrory, C. C. L. *Chem. Commun.* **2018**, *54*, 1579–1582.
147. Nie, W.; Wang, Y.; Zheng, T.; Ibrahim, A.; Xu, Z.; McCrory, C. C. L. *ACS Cat.* **2020**, *10*, 4942–4959.
148. Wang, F.; Cao, B.; To, W.-P.; Tse, C.-W.; Li, K.; Chang, X.-Y.; Zang, C.; Chan, S. L.-F.; Che, C.-M. *Cat. Sci. Technol.* **2016**, *6*, 7408–7420.
149. Wang, J.-W.; Huang, H.-H.; Sun, J.-K.; Ouyang, T.; Zhong, D.-C.; Lu, T.-B. *ChemSusChem* **2018**, *11*, 994.
150. Wang, J.-W.; Huang, H.-H.; Sun, J.-K.; Zhong, D.-C.; Lu, T.-B. *ACS Cat.* **2018**, *8*, 7612–7620.
151. Chapovetsky, A.; Welborn, M.; Luna, J. M.; Haiges, R.; Miller, T. F.; Marinescu, S. C. *ACS Cent. Sci.* **2018**, *4*, 397–404.
152. Elgrishi, N.; Chambers, M. B.; Artero, V.; Fontecave, M. *Phys. Chem. Chem. Phys.* **2014**, *16*, 13635–13644.
153. Elgrishi, N.; Chambers, M. B.; Fontecave, M. *Chem. Sci.* **2015**, *6*, 2522–2531.
154. Queyriaux, N.; Abel, K.; Fize, J.; Pécaut, J.; Orio, M.; Hammarström, L. Sustain. *Energy Fuel* **2020**, *4*, 3668–3676.

155. Queyriaux, N. *ACS Cat.* **2021**, *11*, 4024–4035.
156. Cometto, C.; Chen, L.; Lo, P.-K.; Guo, Z.; Lau, K.-C.; Anxolabéhère-Mallart, E.; Fave, C.; Lau, T.-C.; Robert, M. *ACS Cat.* **2018**, *8*, 3411–3417.
157. Wang, M.; Chen, L.; Lau, T.-C.; Robert, M. *Angew. Chem. Int. Ed.* **2018**, *57*, 7769–7773.
158. Cometto, C.; Chen, L.; Anxolabéhère-Mallart, E.; Fave, C.; Lau, T.-C.; Robert, M. *Organometallics* **2019**, *38*, 1280–1285.
159. Loipersberger, M.; Cabral, D. G. A.; Chu, D. B. K.; Head-Gordon, M. *J. Am. Chem. Soc.* **2021**, *143*, 744–763.
160. Zee, D. Z.; Nippe, M.; King, A. E.; Chang, C. J.; Long, J. R. *Inorg. Chem.* **2020**, *59*, 5206–5217.
161. Derrick, J. S.; Loipersberger, M.; Chatterjee, R.; Iovan, D. A.; Smith, P. T.; Chakarawet, K.; Yano, J.; Long, J. R.; Head-Gordon, M.; Chang, C. J. *J. Am. Chem. Soc.* **2020**, *142*, 20489–20501.
162. Gonell, S.; Lloret-Fillol, J.; Miller, A. J. M. *ACS Cat.* **2021**, *11*, 615–626.
163. Gonell, S.; Massey, M. D.; Moseley, I. P.; Schauer, C. K.; Muckerman, J. T.; Miller, A. J. M. *J. Am. Chem. Soc.* **2019**, *141*, 6658–6671.
164. Gonell, S.; Assaf, E. A.; Duffee, K. D.; Schauer, C. K.; Miller, A. J. M. *J. Am. Chem. Soc.* **2020**, *142*, 8980–8999.
165. Wang, Y.; Marquard, S. L.; Wang, D.; Dares, C.; Meyer, T. J. *ACS Energy Lett.* **2017**, *2*, 1395–1399.
166. Su, X.; McCardle, K. M.; Panetier, J. A.; Jurss, J. W. *Chem. Commun.* **2018**, *54*, 3351–3354.
167. Hooe, S. L.; Dressel, J. M.; Dickie, D. A.; Machan, C. W. *ACS Cat.* **2020**, *10*, 1146–1151.
168. Hooe, S. L.; Moreno, J. J.; Reid, A. G.; Cook, E. N.; Machan, C. W. *Angew. Chem. Int. Ed.* **2022**, *61*, e202109645.
169. Rosas-Hernández, A.; Junge, H.; Beller, M.; Roemelt, M.; Francke, R. Catal. *Sci. Technol.* **2017**, 7, 459–465.
170. Oberem, E.; Roesel, A. F.; Rosas-Hernández, A.; Kull, T.; Fischer, S.; Spannenberg, A.; Junge, H.; Beller, M.; Ludwig, R.; Roemelt, M.; Francke, R. *Organometallics* **2019**, *38*, 1236–1247.
171. Ahmed, M. E.; Rana, A.; Saha, R.; Dey, S.; Dey, A. *Inorg. Chem.* **2020**, *59*, 5292–5302.
172. Hawecker, J.; Lehn, J.-M.; Ziessel, R. *J. Chem. Soc. Chem. Commun.* **1984**, 328–330.
173. Bourrez, M.; Molton, F.; Chardon-Noblat, S.; Deronzier, A. *Angew. Chem. Int. Ed.* **2011**, *50*, 9903–9906.
174. Machan, C. W.; Sampson, M. D.; Chabolla, S. A.; Dang, T.; Kubiak, C. P. *Organometallics* **2014**, *33*, 4550–4559.
175. Riplinger, C.; Sampson, M. D.; Ritzmann, A. M.; Kubiak, C. P.; Carter, E. A. *J. Am. Chem. Soc.* **2014**, *136*, 16285–16298.
176. Haviv, E.; Azaiza-Dabbah, D.; Carmieli, R.; Avram, L.; Martin, J. M. L.; Neumann, R. *J. Am. Chem. Soc.* **2018**, *140*, 12451–12456.
177. Sung, S.; Kumar, D.; Gil-Sepulcre, M.; Nippe, M. *J. Am. Chem. Soc.* **2017**, *139*, 13993–13996.
178. Qiao, X.; Li, Q.; Schaugaard, R. N.; Noffke, B. W.; Liu, Y.; Li, D.; Liu, L.; Raghavachari, K.; Li, L.-S. *J. Am. Chem. Soc.* **2017**, *139*, 3934–3937.
179. Franco, F.; Cometto, C.; Garino, C.; Minero, C.; Sordello, F.; Nervi, C.; Gobetto, R. *Eur. J. Inorg. Chem.* **2015**, *2015*, 296–304.
180. Koenig, J. D. B.; Dubrawski, Z. S.; Rao, K. R.; Willkomm, J.; Gelfand, B. S.; Risko, C.; Piers, W. E.; Welch, G. C. *J. Am. Chem. Soc.* **2021**, *143*, 16849–16864.
181. Popov, D. A.; Luna, J. M.; Orchanian, N. M.; Haiges, R.; Downes, C. A.; Marinescu, S. C. *Dalton Trans.* **2018**, *47*, 17450–17460.

182. Zhanaidarova, A.; Jones, S. C.; Despagnet-Ayoub, E.; Pimentel, B. R.; Kubiak, C. P. *J. Am. Chem. Soc.* **2019**, *141*, 17270–17277.
183. Sampson, M. D.; Kubiak, C. P. *J. Am. Chem. Soc.* **2016**, *138*, 1386–1393.
184. Franco, F.; Pinto, M. F.; Royo, B.; Lloret-Fillol, J. *Angew. Chem. Int. Ed.* **2018**, *57*, 4603–4606.
185. Yang, Y.; Ertem, M. Z.; Duan, L. *Chem. Sci.* **2021**, *12*, 4779–4788.
186. Walsh, J. J.; Smith, C. L.; Neri, G.; Whitehead, G. F. S.; Robertson, C. M.; Cowan, A. J. *Faraday Discuss.* **2015**, *183*, 147–160.
187. Neri, G.; Donaldson, P. M.; Cowan, A. J. *Phys. Chem. Chem. Phys.* **2019**, *21*, 7389–7397.
188. Reuillard, B.; Ly, K. H.; Rosser, T. E.; Kuehnel, M. F.; Zebger, I.; Reisner, E. *J. Am. Chem. Soc.* **2017**, *139*, 14425–14435.
189. Smith, C. L.; Clowes, R.; Sprick, R. S.; Cooper, A. I.; Cowan, A. J. Sustain. *Energy Fuel* **2019**, *3*, 2990–2994.
190. Sato, S.; Saita, K.; Sekizawa, K.; Maeda, S.; Morikawa, T. *ACS Cat.* **2018**, *8*, 4452–4458.
191. Sun, C.; Rotundo, L.; Garino, C.; Nencini, L.; Yoon, S. S.; Gobetto, R.; Nervi, C. *ChemPhysChem* **2017**, *18*, 3219–3229.
192. Fei, H.; Sampson, M. D.; Lee, Y.; Kubiak, C. P.; Cohen, S. M. *Inorg. Chem.* **2015**, *54*, 6821–6828.
193. Torralba-Peñalver, E.; Luo, Y.; Compain, J.-D.; Chardon-Noblat, S.; Fabre, B. *ACS Cat.* **2015**, *5*, 6138–6147.
194. Walsh, J. J.; Neri, G.; Smith, C. L.; Cowan, A. J. *Organometallics* **2019**, *38*, 1224–1229.
195. Taheri, A.; Berben, L. A. *Chem. Commun.* **2016**, *52*, 1768–1777.
196. Waldie, K. M.; Ostericher, A. L.; Reineke, M. H.; Sasayama, A. F.; Kubiak, C. P. *ACS Cat.* **2018**, *8*, 1313–1324.
197. Ceballos, B. M.; Yang, J. Y. *Proc. Natl. Acad. Sci. U. S. A.* **2018**, *115*, 12686–12691.
198. Kang, P.; Cheng, C.; Chen, Z.; Schauer, C. K.; Meyer, T. J.; Brookhart, M. *J. Am. Chem. Soc.* **2012**, *134*, 5500–5503.
199. Kang, P.; Meyer, T. J.; Brookhart, M. *Chem. Sci.* **2013**, *4*, 3497–3502.
200. Taheri, A.; Thompson, E. J.; Fettinger, J. C.; Berben, L. A. *ACS Cat.* **2015**, *5*, 7140–7151.
201. Fogeron, T.; Todorova, T. K.; Porcher, J.-P.; Gomez-Mingot, M.; Chamoreau, L.-M.; Mellot-Draznieks, C.; Li, Y.; Fontecave, M. *ACS Cat.* **2018**, *8*, 2030–2038.
202. Roy, S.; Sharma, B.; Pécaut, J.; Simon, P.; Fontecave, M.; Tran, P. D.; Derat, E.; Artero, V. *J. Am. Chem. Soc.* **2017**, *139*, 3685–3696.
203. Franco, F.; Cometto, C.; Ferrero Vallana, F.; Sordello, F.; Priola, E.; Minero, C.; Nervi, C.; Gobetto, R. *Chem. Commun.* **2014**, *50*, 14670–14673.
204. Franco, F.; Cometto, C.; Nencini, L.; Barolo, C.; Sordello, F.; Minero, C.; Fiedler, J.; Robert, M.; Gobetto, R.; Nervi, C. *Chem. A Eur. J.* **2017**, *23*, 4782–4793.
205. Rønne, M. H.; Cho, D.; Madsen, M. R.; Jakobsen, J. B.; Eom, S.; Escoudé, É.; Hammershøj, H. C. D.; Nielsen, D. U.; Pedersen, S. U.; Baik, M.-H.; Skrydstrup, T.; Daasbjerg, K. *J. Am. Chem. Soc.* **2020**, *142*, 4265–4275.
206. Guo, Z.; Chen, G.; Cometto, C.; Ma, B.; Zhao, H.; Groizard, T.; Chen, L.; Fan, H.; Man, W.-L.; Yiu, S.-M.; Lau, K.-C.; Lau, T.-C.; Robert, M. *Nat. Catal.* **2019**, *2*, 801–808.
207. De, R.; Gonglach, S.; Paul, S.; Haas, M.; Sreejith, S. S.; Gerschel, P.; Apfel, U.-P.; Vuong, T. H.; Rabeah, J.; Roy, S.; Schöfberger, W. *Angew. Chem. Int. Ed.* **2020**, *59*, 10527–10534.
208. Boutin, E.; Wang, M.; Lin, J. C.; Mesnage, M.; Mendoza, D.; Lassalle-Kaiser, B.; Hahn, C.; Jaramillo, T. F.; Robert, M. *Angew. Chem. Int. Ed.* **2019**, *58*, 16172–16176.

209. Wu, Y.; Jiang, Z.; Lu, X.; Liang, Y.; Wang, H. *Nature* **2019**, *575*, 639–642.
210. Rao, H.; Schmidt, L. C.; Bonin, J.; Robert, M. *Nature* **2017**, *548*, 74–77.
211. Rao, H.; Lim, C.-H.; Bonin, J.; Miyake, G. M.; Robert, M. *J. Am. Chem. Soc.* **2018**, *140*, 17830–17834.
212. Shirley, H.; Su, X.; Sanjanwala, H.; Talukdar, K.; Jurss, J. W.; Delcamp, J. H. *J. Am. Chem. Soc.* **2019**, *141*, 6617–6622.
213. Ahmed, M. E.; Adam, S.; Saha, D.; Fize, J.; Artero, V.; Dey, A.; Duboc, C. *ACS Energy Lett.* **2020**, *5*, 3837–3842.

CHAPTER NINE

Polyoxometalate systems to probe catalyst environment and structure in water oxidation catalysis☆

Q. Yin[a], Yurii V. Geletii[a], Tianquan Lian[a], Djamaladdin G. Musaev[a,b], and Craig L. Hill[a,*]
[a]Department of Chemistry, Emory University, Atlanta, GA, United States
[b]Cherry L. Emerson Center for Scientific Computation, Emory University, Atlanta, GA, United States
*Corresponding author: e-mail address: chill@emory.edu

Contents

Abstract

From multiple perspectives, d-block-metal-substituted polyoxometalates (POMs) represent synthetically tunable homogeneous analogs of many transition metal oxides. Both classes of materials have a range of catalytic activities derived from the presence of redox active metals in all-oxygen ligand environments. These carbon-free environments provide for oxidative and other forms of stability. Subsequently, our group and others, demonstrated that metal-substituted POMs can function as effective water oxidation catalysts (WOCs), frequently termed oxygen evolution catalysts (OECs), and hydrogen evolution catalysts (HECs). Most of the Hill group research in this area has focused on making POM OECs of 3d metals, Ru and Ir and studying the mechanisms of their dark reactions using multiple equivalents of an oxidant or light-driven reactions using a photosensitizer

☆ Portions of the text in three sections of this article were taken from the as-yet-unpublished Ph.D. dissertation of Qiushi Yin, Emory University, 2021 with permission from this author.

Advances in Inorganic Chemistry, Volume 79
ISSN 0898-8838
https://doi.org/10.1016/bs.adioch.2021.12.009

and a terminal oxidant, frequently persulfate. Background information related to POM OEC chemistry is addressed including photoelectrochemical cells (PECs), homogeneous versus heterogeneous catalysis, energetic and mechanistic features of water oxidation/oxygen evolution, and factors that control rates and stability of OECs including POM OECs.

Abbreviations

CoPi cobalt oxide phosphate
DFT density function theory
EXAFS Extended X-ray absorption fine structure
HEC hydrogen evolution catalyst
I2M metal oxyl/oxo coupling
MOF metal organic framework
NMR nuclear magnetic resonance
OEC oxygen evolution catalyst
PEC photoelectrochemical cell
POM polyoxometalate
PT proton transfer
TOF turnover frequency
WNA water nucleophilic attack
WOC water oxidation catalyst
XANES X-ray absorption near edge structure
XPS X-ray photoelectron spectroscopy

1. General considerations

Our access to energy underlies the expansion of civilizations and access to improved technologies. The current mode of energy acquisition comes predominantly from fossil fuels.[1–4] Specifically, of the world's total energy supply, 81.3% originates from fossil fuels (26.9% from coal, 31.6% from petroleum, and 22.8% from natural gas).[1] The issue of resource scarcity presents in a few different ways as a result of our current energy economy. The most obvious problem is the limited supply of fossil fuels. As its name suggests, fossil fuels are buried carbonaceous fuel sources mainly the result of millions of years of transformation of organic matter.[5–7] While the accessibility of fossil fuels appears to be quite high, economically accessible deposits differ vastly based on geography.[8] Energy independence of nation-states are thus not guaranteed, often leading to potential geopolitical conflicts.

The far more pressing issue is the scarcity of our atmospheric carbon budget.[9–13] The utilization of fossil fuels produces enormous quantities of greenhouse gasses mainly in the form of carbon dioxide, leading to global climate change. The impacts of climate change are varied and largely negative.[9,14–16] We can already observe some of its effects in the form of sea level rise, deoxygenation of lakes,[17] extreme weather patterns, and higher temperatures.[9–17] The anticipated effect in the coming century includes, but is not limited to, food and water crises, rising human health issues, ocean acidification, and loss of biodiversity.[9,14–16] Adaptations will be required in all aspects of human society, the direct cost of which is estimated to be $8 trillion dollars per year by 2050.[1,2] The solution to such a large scale problem would likely require multifaceted approaches, with the ultimate objective being the lowering of carbon emissions to zero. The main stated goal of the International Energy Agency including its 38 member and association countries is to achieve zero emissions by 2050.[1,2] The only way to achieve this goal is to drastically alter our energy source and emancipate ourselves from fossil fuels.

If we trace the sources of energy used by modern human civilization to its roots, about 90% of it comes from our Sun. This includes all fossil fuels, hydropower, biofuels, wind, tide/wave/ocean, and direct solar energy. Of these, the most promising mode of energy acquisition is direct solar energy conversion. It is the fuel source that facilitates zero carbon emissions and could match humanity's energy demands.[1–4,9,18–20] Additional advantages of the solar energy source, include wide availability and off-the-grid compatibility.[18–20] The 2021 US annual energy outlook predicts that about 10% of all the annual energy supply could come from direct solar energy by 2050.[3] In these predictive models, the direct solar energy conversion is assumed to be solar photovoltaic technology. Photovoltaics are the most technologically mature solar conversion technology, converting light energy to electric potential energy. But one main drawback of this technology is the storage of the resulting electricity as current solutions center around constructing large arrays of batteries. The low cost of photovoltaic electricity is one driver in the extremely active area of battery technology. An alternative proposed method of direct solar energy conversion is the photoelectrochemical cell (PEC).[18–21] A photovoltaic-electrolyzer combination will readily convert sunlight to hydrogen and oxygen. However, it does suffer from a few drawbacks when compared to potential PEC systems. Mainly, photovoltaic-electrolyzers have an intrinsically lower upper limit on energy conversion efficiency than PECs, with much of this due to the high operating current density of electrolyzers. In addition, PECs are potentially more versatile in their deployment and compatibility with CO_2 reduction reactions. However, PECs present a number of tractable challenges across many different areas of chemistry and chemical engineering that require advances in fundamental science. As such, PEC systems continue to be attractive for a wide range of research scientists and engineers.[19,21] We note that water reduction to hydrogen produces a stoichiometric amount of oxygen by water oxidation. Water oxidation is a complex chemical process, which requires using a very efficient and oxidatively robust catalyst.

In the following sections, we discuss issues impacting our recent research. These are currently important in catalytic water oxidation and associated oxygen evolution and applicable not just to PEC approaches but to water oxidation catalysis in general. We first discuss PECs, then different types of OECs including the current pros and cons of each. Subsequently, we introduce polyoxometalates (POMs) and show how various POM derivatives can function as tunable molecular versions of heterogeneous metal oxide multielectron, multiproton catalysts. We then discuss fundamental energetic and mechanistic considerations of OEC systems, and finally the several aspects of OEC and system stability.

2. Photoelectrochemical cells (PECs)

Photoelectrochemical cells differ from traditional photovoltaics in a few aspects. PECs typically include a p-type photocathode and an n-type photoanode. The electric potential generated from the photoelectric effect is directly converted to chemical energy, eliminating the need for battery storage. A basic example of PEC design is shown in Fig. 1. Typically, water splitting is the primary reaction used to achieve chemical energy storage. Its two component half-reactions, water oxidation and water reduction, can be seen in Fig. 1. The US Department of Energy evaluates PEC research as more a long-term research target rather than an imminent technology for a hydrogen economy. Recent efforts in PEC research have shifted the focus onto CO_2 reduction in conjunction with water oxidation.[21] Not surprisingly, such a system would mimic natural photosynthesis, the main mode of energy acquisition underlying most life on Earth. Unfortunately, Nature's photosynthesis has a very low energy efficiency (<6%) and mere mimicry will not be sufficient. Our artificial photosynthesis must beat Nature's photosynthesis by a significant margin to be competitive with other renewables.[19–21] In any case, water oxidation is a cornerstone reaction in future of PEC research and construction.

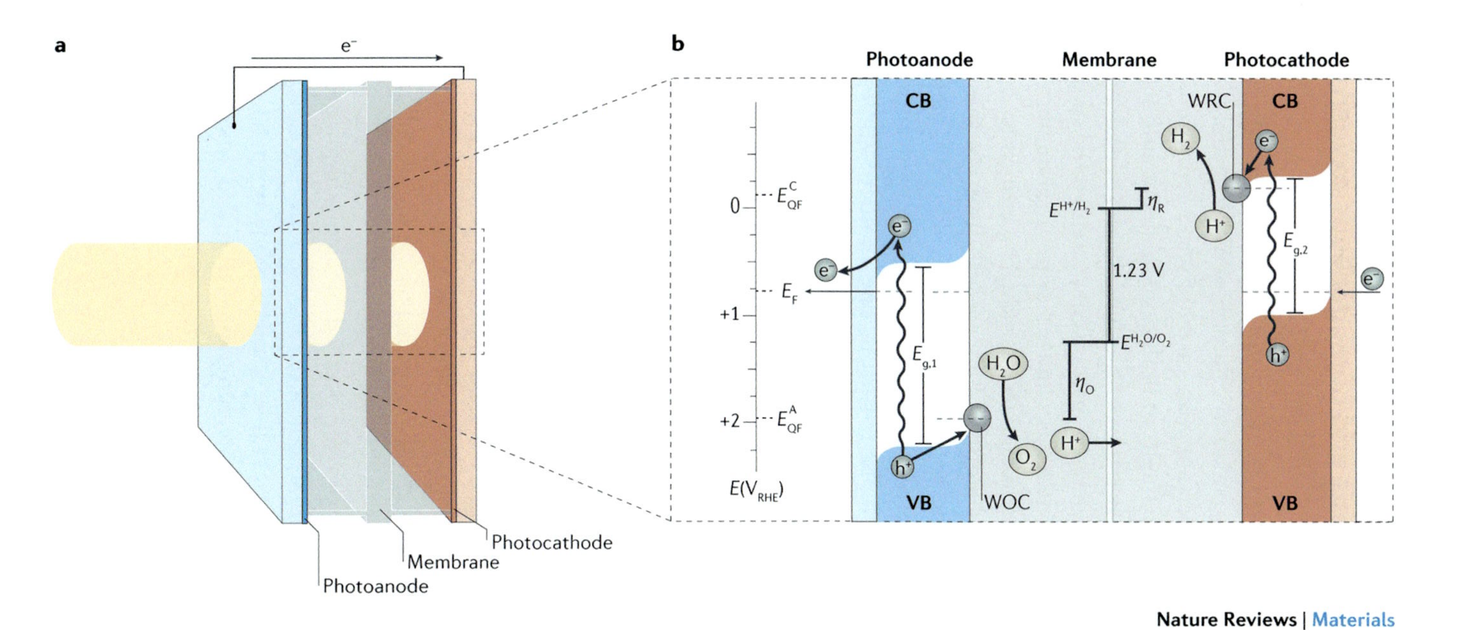

Fig. 1 A simplified scheme for a PEC containing all the basic requisite components for a functional device. The major components (A) and a more detailed cross-section (B) are shown. *This figure is taken from Sivula, K.; van de Krol, R., Nat. Rev. Mater. 2016, 1, 15010 and reproduced with permission.*

3. Heterogeneous and homogeneous oxygen evolution catalysts (OECs)

Our research developing transition-metal-substituted POMs as OECs (also referred to as water oxidation catalysts, or WOCs) and using them to address several issues associated with catalytic water oxidation and oxygen evolution: they exhibit elements of both homogeneous and heterogeneous catalysis; thus, we first review the contrasting foci and methods in these two general areas of catalysis. Heterogeneous systems tend to be studied by electrochemists, photo-electrochemists and other physical chemists, while the latter, homogeneous, systems tend to be studied by inorganic, organometallic and physical organic chemists. The technical thrusts of these two communities tend to be different in part because different techniques and skillsets are largely operative. Significantly, the two types of materials and their investigation are complementary to one another. While the heterogeneous oxides/hydroxides are more robust and thus more amenable to practical application as OECs, they are more difficult to study experimentally and computationally. There is one stability concern regarding many metal oxides, including the leading mixed-3d-metal oxide OECs: they dissolve in acidic aqueous medium. Others, due to their amphoterism, dissolve in base as well as acid. A PEC that functions in strong acid would be of much potential practical interest. Hydrogen is more easily generated in acid, and acid in general can catalyze both the reduction and oxidation of water. However, water is harder to oxidize thermodynamically as acidity increases. To date, the hydrolytic instability of 3d metal oxides in acid has conspired to prevent formulation of effective solid anodes or photoanodes for catalytic water oxidation in these conditions. Some 3d metal-oxide OEC electrodes can be protected to some extent with overlayers of more acid stable material but the current densities at pH 0 are very low.[22]

Many techniques that provide detailed molecular information, such as geometrical and electronic structures of catalytic active sites, require soluble samples and thus are not applicable to metal oxide or metal oxide-nitride OECs. X-ray spectroscopy techniques, XPS and more so, XAFS (XANES and EXAFS), can provide some structural information at the molecular level of heterogeneous catalyst active sites. As a consequence, the use of these techniques in such research has increased in recent years. However, soluble OEC complexes and their geometrical and electronic structures are far more amenable to elucidation at the molecular level by a range of techniques than their insoluble counterparts. Homogeneous OECs can be easily studied by solution 1D and 2D solution NMR methods, electronic absorption spectroscopy and single crystal X-ray diffraction, among other methods, that are not effectively applied to heterogeneous catalysts. Furthermore, the electronic structures and other attributes of soluble OECs can be calculated more precisely and more easily than those of heterogeneous OECs. In addition, homogeneous transition-metal-complex OECs can frequently be prepared in both compositionally and isomerically pure forms, which is difficult to achieve for most heterogeneous OECs. The usual synthetic approaches to metal oxides and other insoluble materials of interest in solar fuels technology (metal nitrides, mixed oxy-nitrides and others), can generate multiple phases which can be difficult to separate, and there is rarely the opportunity to purify the particular solid phase of interest by re-crystallization. The same challenge is present in metal organic framework (MOF) chemistry.

The best performing OECs historically are ruthenium oxides and iridium oxides.[23–26] They have some of the lowest overpotentials for four-electron water oxidation at all pH ranges and current densities. For these reasons, formulations of Ru and/or Ir materials are widely employed in current commercial electrolyzers. For PECs, however, Ru- and Ir-based oxygen evolution catalysts (OECs) are less attractive mainly due to the scarcity of ruthenium and iridium in the earth's crust. The terrestrial scarcity and associated cost disadvantages of Ru- and Ir-based systems has directed most research on OEC development in the last decade to 3d-metal-containing systems. There are some monomeric and polymeric organic molecules that could in principle provide the requirement of water oxidation to dioxygen, namely, the capability to take up 4 electrons and 4 protons, but there has been very little attention to the use of such matrices in this catalytic context presumably because of the high bar for stability, and particularly, oxidative stability for a viable OEC.

Mixed d- and p-block OECs, such as cobalt oxide phosphate (CoPi)[27–29] and more recently mixed 3d-metal oxides, including cobalt-nickel mixed-metal oxyhydroxides, have been the focus of many recent heterogeneous catalytic or photoelectrochemical OEC studies.[30–40] Significantly, the mixing of cobalt and nickel has a synergistic effect that allows the resulting Co-Ni oxyhydroxides to perform much better than either cobalt oxides/oxyhydroxides or nickel oxides/oxyhydroxides alone. However, the relationship between these changes in catalyst composition and catalytic activity is still unclear. There have been many studies, including ours on both heterogeneous and homogeneous OECs.[24,27,28,30–83] However, good molecular models of the most promising mixed-3d-metal-oxide heterogeneous OECs are lacking.

A more detailed understanding of how the local environment of OEC transition metal active sites affects the reaction mechanism and the reaction rates in metal oxide water oxidation will be very useful in the further development of OECs and their implementation in PECs. An in-depth understanding of the relevant water oxidation catalysis mechanism is crucial to improving the catalytic performance and the operating stability of the catalysts. And while much progress has been made clarifying the actual mechanism of water oxidation catalysis (see Section 5), the exact mechanism is always dependent on the transition metal centers, the associated oxidation states, the ligands, and the reactive environment.

4. Polyoxometalate systems for water oxidation/oxygen evolution

Transition metal-oxygen cluster polyanions, otherwise known as polyoxometalates (POMs; Fig. 2) are ideal molecular analogs for studying bulk metal oxides.[84–89] POMs most frequently comprise W(VI), Mo(VI), V(V), and Nb(V) centers bridged by oxygens (formally oxide ions or their protonated counterparts, hydroxide ions). They vary widely in composition, geometrical structure, shape, and negative charge density (Figs. 2 and 3). Most POMs are synthetically accessible, and the syntheses can frequently be easily scaled up. Significantly, this huge and rapidly growing class of inorganic compounds are essentially soluble analogs of heterogeneous metal oxides. Most importantly, many POMs allow atomic-level control over the chemical composition as well as geometrical structure. Defect forms of POMs, most commonly monovacant (lacunary) or trivacant forms of Keggin and Wells-Dawson derivatives, can be used as multidentate ligands that bind one, two or more metals (d- or f-block) in precisely defined locations and ligand environments.

Most POMs are also redox active. Their rich redox chemistry is rendered tunable by substituting d- and/or f-block metals for one or more of the d^0 metals, e.g., W(VI) centers in the POM framework. The acid–base properties of POMs are adjustable, largely dictated by the intrinsic aqueous acid–base chemistry of the framework metal: MO_6 octahedra, where

POM hydrolytic equilibria:

$$n\,[WO_4]^{2-} \underset{OH^-}{\overset{H^+}{\rightleftharpoons}} \underbrace{[W_xO_y]^{n-} + nH_2O}_{\text{Polyoxometalates (POMs)}} \underset{OH^-}{\overset{H^+}{\rightleftharpoons}} \underset{\text{metal oxide}}{WO_3} + nH_2O$$

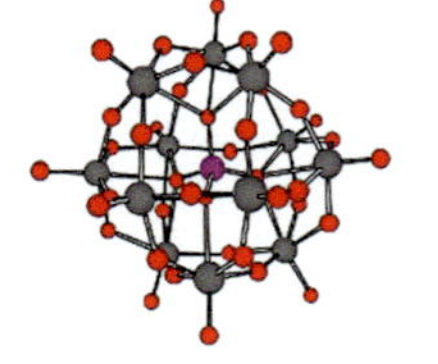

$[PW_{12}O_{40}]^{6-}$ (Keggin)

P, purple; W, gray, O, red

General POM properties:

Stable to O_2, H_2O; readily prepared in bulk; nontoxic; inexpensive.

Versatility of POMs. Properties readily altered:

Elemental compositions, potentials, acidities, solubilities, shape, size, charge, counterions, others.

Fig. 2 General properties of polyoxometalates (POMs). Top: general hydrolytic equilibria that link monomeric oxometalates (in this case WO_4^{2-}) to both POMs and metal oxides/hydroxides. Bottom, left: an exemplary POM, the Keggin structure. Bottom right: general and alterable properties of POMs.

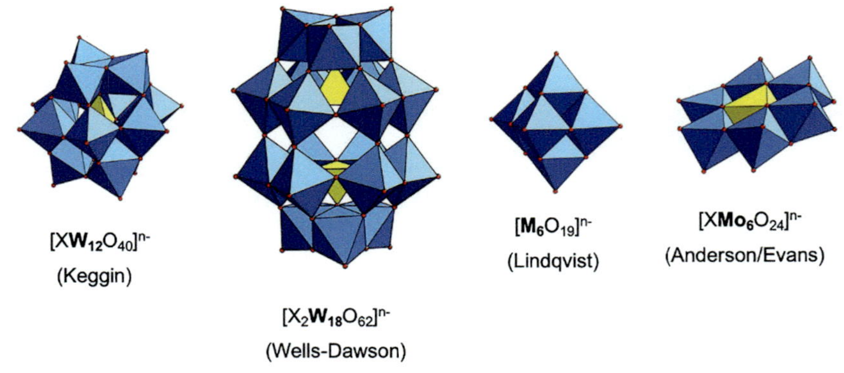

Fig. 3 Four parent (plenary) POM structural families. A large variety of derivatives of all four of these families has been made. The Keggin and Wells-Dawson structures each have several isomers from which defect forms are readily made, and from these, many d- and f-block-metal substituted derivatives are accessible.

M = W(VI), Mo(V) or V(V) are largely compatible with acid, while Nb(V), and the rarely investigated Ta(V) centers, are compatible in aqueous base. Mixtures of these metals confer neutral-pH stability to a POM and thus some medical applications.[90–94] Finally, all POMs, be they parent (plenary) or defect (lacunary) structures, bear high negative charges which are balanced by the requisite number of counterions. These collective properties have made POMs very successful as redox catalysts and several processes have been commercialized in recent decades.[95–97]

Our many years of structural, medical and catalytic studies on POMs made it evident that POMs substituted with appropriate redox-active transition-metal ions might function as OECs that would offer the advantages of both homogeneous catalysts and heterogeneous catalysts, possibly without the disadvantages of either. We reasoned that POMs are compositionally and structurally tunable well beyond what is possible with heterogeneous metal oxide systems because they are soluble. In general, solubility makes POMs amenable to extensive experimental and computational investigation at the molecular level. Our group and that of Bonchio published the first POM WOC, $[\{Ru_4O_4(OH)_2(H_2O)_4\}(\gamma\text{-}SiW_{10}O_{36})_2]^{10-}$, in 2008.[98,99] At that time, POMs were the only homogeneous OECs that were carbon-free and thus stable to oxidative degradation. Other advantages were also evident to us at the outset of POM WOC chemistry, including the presence of many proximal oxygens to assist potentially in taking up protons as the POM WOC transition-metal active-sites sequentially accumulate oxidizing equivalents. POMs, OEC mechanisms and the latter process, termed "redox leveling" are addressed in the subsequent sections. Not surprisingly, a considerable number of POM-based OECs have been reported and these complexes have been reviewed.[100–103]

5. Energetic and mechanistic issues in catalytic water oxidation

Water oxidation is a seemingly straightforward reaction, with the oxygen in water molecules consuming four holes sequentially, ultimately leading to the evolution of molecular oxygen. This process is well studied and has been known for a long time. However, its thermodynamics and kinetics make the catalysis of the oxygen evolution reaction nontrivial. Three main half-cell reactions are typically discussed when examining water oxidation:

One-electron oxidation:

$$H_2O \rightarrow \bullet OH(aq) + (H^+ + e^-) \quad E^\circ = 2.73\,V \qquad [1]$$

Two-electron oxidation:

$$2H_2O \rightarrow H_2O_2 + 2(H^+ + e^-) \quad E^\circ = 1.76\,V \qquad [2]$$

Four-electron oxidation:

$$2H_2O \rightarrow O_2 + 4(H^+ + e^-) \quad E^\circ = 1.23\,V \qquad [3]$$

Although reactions [1]–[3] all lead to stable or semi-stable products, the four-electron process in Eq. [3] is usually the desired reaction in PECs as it has the lowest reduction potential. Realistically, reactions [2] and [3] require catalysts to proceed at potentials close to the standard reduction potentials due to their multielectron mechanisms. A conspicuous question arises at this point. What determines the selectivity of the products among reactions [1]–[3]? A qualitative thermodynamic analysis addresses the underlying issue.[104–106] Consider a typical metal oxide water oxidation catalyst. In a generalized scheme, the following reaction steps can occur in water oxidation:

One-election oxidation:

$$H_2O \rightarrow \bullet OH(aq) + (H^+ + e^-) \qquad \Delta G_{\bullet OH} \qquad [4]$$

Two-electron oxidation:

$$M - OH_2 \rightarrow M - OH + (H^+ + e^-) \qquad \Delta G_{MOH} \qquad [5]$$

$$2M - OH + 2(H^+ + e^-) \rightarrow 2M + H_2O_2 + 2(H^+ + e^-) \qquad \Delta G_{H2O2} \qquad [6]$$

Four-electron oxidation:

$$M - OH_2 \rightarrow M - OH + (H^+ + e^-) \qquad \Delta G_{MOH} \qquad [5]$$

$$M - OH + (H^+ + e^-) \rightarrow M - O + 2(H^+ + e^-) \qquad \Delta G_{MO} \qquad [7]$$

$$M - O + H_2O + 2(H^+ + e^-) \rightarrow M - OOH + 3(H^+ + e^-) \qquad \Delta G_{MOOH} \qquad [8]$$

$$M - OOH + 3(H^+ + e^-) \rightarrow M + O_2 + 4(H^+ + e^-) \qquad \Delta G_{O2} \qquad [9]$$

Hydroxyl radical formation (Eq. [4]) occurs when $\Delta G_{\bullet OH} < \Delta G_{MOH}$. Otherwise, reaction [5] is preferred and either a two-electron oxidation or four-electron oxidation occurs. Similarly, the four-electron oxidation reaction will only occur when $\Delta G_{\bullet OH} < \Delta G_{MOH}$ and $\Delta G_{MO} < \Delta G_{H2O2}$. This insight demonstrates that the critical function of a four-electron water oxidation catalyst vs a one-electron or two-electron water oxidation catalyst is lowering the Gibbs free energy of the bound oxygen atom. Descriptively, we might say that a four-electron oxidation catalyst must be able to bind water molecules and deprotonate them completely. Experimental measurements on the binding energies of metal oxide/hydroxide species and density function theory (DFT) calculations confirm this line of reasoning.[104–106]

A picture of the exact pathway of the oxygen evolution reaction at transition-metal centers only emerged in the last decade. Most of the advances came as a result of studying molecular water oxidation catalysts.[69,107] Much work has been done on ruthenium-based molecular catalysts.[107] The discovery of mononuclear transition-metal water oxidation catalysts in particular demonstrated that it is not necessary to have four consecutive oxidation events on the transition-metal centers in order to effectuate catalytic oxygen evolution. Typically, a water molecule is activated through a series of electron transfers (ET) and proton transfers (PT) or proton-coupled electron transfers (PCET) to form oxidized metal oxyl/oxo species. This metal oxyl/oxo then enables oxygen–oxygen bond formation, either through a water nucleophilic attack (WNA) from the aqueous solvent or through a radical-coupling interaction with another metal oxyl/oxo (I2M). Further ET and PT or disproportionation then leads to oxygen evolution to complete the catalytic cycle.

Multinuclear transition metal centers are not obligatory for a four-electron water oxidation catalyst. Two main classifications exist for molecular water oxidation to oxygen: WNA and I2M describe two different modes of O—O bond formation. This step is critical for catalytic water oxidation in general and often determines the catalytic rate.[107]

I2M should result in the most facile O—O bond formation for the respective mechanisms under general conditions. These insights are typically in line with what we observe for both homogeneous and heterogeneous oxygen evolution processes.

In actuality, the precise mechanism of water oxidation may not always fit nicely into this paradigm. This is especially evident when we examine heterogeneous OECs.[19,21,23,108]

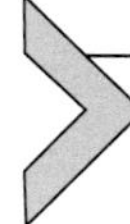

6. Rates issues involving catalysts for multielectron processes

Most of the research in the past decade on catalysts for solar fuels, i.e., OECs, HECs and CO_2 reduction catalysts, has focused on optimization of rates and product selectivity more than on catalyst stability and system durability. All this research to date has resulted in several more efficient catalysts of all three types. However, the development of CO_2 reduction catalysts has lagged behind the development of OECs, and OEC development has lagged behind that of HECs. There are few if any CO_2 reduction catalysts that perform with good selectivity at high rates and low driving force with viable levels of durability under sustained operating conditions, particularly those that produce products other than CO. Satisfactory performance for any multielectron catalyst entails high rates and selectivity for the desired product at low overpotential for electro-catalytic and photo-electro-catalytic systems and low additional driving force for homogeneous catalytic systems.

Comparing rates of different OECs, even comparisons of only heterogeneous systems or only homogeneous ones, should be discouraged because there are far too many parameters that impact observed TOFs for such comparisons to be meaningful. These include solution pH, ionic strength, applied potential, but also the type of experiment and the measurement parameters for each type. Turnover frequency, or "TOF" is an ill-defined term. Commonly TOF is measured as TON/time, which is a time- and concentration-dependent value, thus comparison of TOFs from different papers is not highly meaningful. In addition, insightful comparison of different OEC rates also depends on catalyst stability because many OECs, both heterogeneous and more so homogeneous, degrade to other catalytically competent species during turnover. For example, some Ru- and Co-based homogeneous catalysts under some conditions degrade to Ru and Co oxide/hydroxides which are good OECs. Any such transformation of the active catalyst renders rigorous evaluation of TOF values problematical at best, particularly when evidence of initial OEC degradation is evident. However, mechanistic inferences are defensible if all the water oxidation processes are demonstrated to be faster than the catalyst degradation processes, and the rate data are collected using initial rates.

Effective homogeneous or heterogeneous OECs alike must process the capability to reduce accumulating positive charge on the catalyst during the four-electron oxidation required for oxidation of water to dioxygen. This process, referred to as redox leveling, is generally accomplished by deprotonation of ligands at or near the active site after oxidation of the OEC. Other mechanisms for redox leveling are also possible including the breaking of ion pairs during the multielectron charge accumulation process. This charge neutralization makes the subsequent, successive catalyst-active-site oxidation events more thermodynamically facile and generally brings all four redox events to within a fairly narrow range of potentials that are accessible by the external oxidant. This "oxidant" can be a proximal excited state, oxidizing anode, oxidizing photoanode or an oxidant molecule. A caveat to be noted in operation of a PEC or a sustained homogeneous catalytic water oxidation process is that, if there is an accumulation of acid (proton concentration) under actual catalytic water oxidation conditions, some ligands that are needed in the deprotonation/charge-neutralization process enabling redox leveling can fail to deprotonate slowing or terminating turnover.

A key consideration for ultimate formulation of optimal catalytic multielectron processes in solar fuels chemistry is that these processes generally involve both light-dependent steps and subsequent dark steps.

The light-dependent steps entail initial photophysical and photochemical steps such as intersystem crossing of the initially-generated singlet excited states to the longer lived and reactive triplet states in the case of photosensitized homogeneous catalytic water oxidation. The longer-lived states ultimately lead to four-electron-oxidized forms for the catalyst that facilitate several dark (non-light-assisted) steps including oxidation of water and intermediate species such as peroxo derivatives and oxygen–oxygen bond formation (see Section 5). The typical relative rates are photophysical processes (fastest) > photogeneration of reactive intermediates (often multiple equivalents of an oxidant) > dark bond-cleaving and bond-forming steps (slowest). The intermediate intermolecular redox processes such as *in situ* generation of oxidizing equivalents, or, more frequently, the catalyzed dark steps, can be rate-limiting. Several factors dictate which steps are, in fact, slowest. Many papers report TOFs of some new OECs, but it is not clear the measured rates of actual O_2 evolution reflect the intrinsic TOF of the catalyst or the rate of photogeneration of the multiple oxidizing equivalents.

There is acknowledged to be general merit in matching the rates of the photo-driven processes and the dark catalytic ones, but this is practically difficult to accomplish in most cases. However, much useful information is typically obtained by studying the photophysical and photochemical processes by fast kinetic methods. To this end, our team has used transient absorption and transient reflection spectroscopies, in the absence of a catalyst to study POM OECs in solution.[109,110] By the same token, several OEC properties can be elucidated by studying the rates of water oxidation driven by multiple equivalents of oxidant in the dark over a range of experimental conditions.[100,111]

7. Multiple aspects of OEC stability

For OECs, HECs and CO_2 reduction catalysts to be viable, several aspects of stability must be considered and designed into the catalysts themselves as well as the catalyst interfaces in the operating device (PEC or other). Hydrolytic, oxidative and thermal stability must be considered along with simultaneous consideration of rates and product selectivity. With structural–activity insights on OEC rates at a reasonably advanced level now, there is growing and defensible perception that optimizing catalyst and system stability, simultaneous with attractive kinetics, may become success-limiting in development of solar fuel generation systems. Out of this perception is the importance of experiments that probe hydrolytic and other stabilities under turnover conditions and after increasingly long operation times. Thermal stability is centrally important to thermochemical water splitting, but not typically for photo-electro-catalytic water splitting.

Unlike most homogeneous OECs, transition-metal-substituted POM complexes and the corresponding metal oxides, contain the catalytic active site metal(s), e.g., Co, Mn, Cu, and Ru, in an all-oxygen ligand environments. In contrast to carbon-based ligand systems and d-block complexes with these ligands, metal-substituted POM derivatives are stable to oxidative degradation and also constitute the thermodynamically most stable form under some aqueous solution conditions (pH, ionic strength).[112] The defect POM ligands have no valence electrons and the oxygens are formally bridging oxide oxygens. This rationale of using POMs as oxidatively stable multidentate ligands for transition-metal complexes was first described by Brown and Hill in 1986 where mono-3d-metal-substituted Keggin POMs, $[(TM)PW_{11}O_{39}]^{5-}$, TM = Co(II) and Mn(II), were shown to catalyze oxo transfer oxidation of organic substrates.[113] As such they were proposed to be carbon-free and thus oxidatively stable analogs of metalloporphyrin oxo transfer catalysts such as cytochrome P-450 and its many metalloporphyrin model complexes at that time. Both metalloporphyrin complexes and $[PW_{11}O_{39}]^{5-}$ complexes bind one 3d metal by four ligands in a roughly co-planer geometry yet facilitate binding of axial ligands and the transfer of oxo groups from a range of donors to alkenes and other substrates in organic solvents.[113–115]

Hydrolytic stability of POM ligands and their 3d-metal complexes varies with the central heteroatom (P in the case of $[PW_{11}O_{39}]^{5-}$), pH, ionic strength, type of buffer, particularly the buffer anion, the buffer concentration and other parameters. Hydrolytic stability of POM OECs varies with these parameters and also with the active-site 3d metal(s) and the structural type and consequent negative charge density on the defect POM ligand. In addition, this type of stability varies with the oxidation state of the active-site 3d metal. Since our original papers on Ru-[98,99] and Co-containing POMs[55] as OECs, there have been scores of analogous studies by many other groups on other POM OECs.

Ru-based POM WOCs are stable for extended periods over a range of pH values and in immobilized forms[102,116] including anode-immobilized forms.[117–119] In contrast, most 3d-metal-substituted POM WOCs are generally unstable under electro-catalytic conditions. Finally, hydrolytic stability varies with mode of operation: POM OECs of 3d metals tend to be less stable when immobilized on anodes and used for electrocatalytic water oxidation than when used for homogeneous water oxidation with a soluble oxidant, such as $[Ru(bpy)_3]^{3+}$, or photo-generated oxidant such as the much-used $[Ru(bpy)_3]^{2+}$/light/persulfate system.

Many techniques have been brought to bear on assessing the stability of POM WOCs under both homogeneous and heterogeneous conditions. Each of these techniques must be used with some caution because many phenomena can impact the integrity of the findings. One generic concern that many investigators tend to ignore is that by adding another reagent, the conditions of the reaction are now different, but more significantly, the catalyst itself can be degraded by the presence of the new reagent making interpretation of the results very problematical. One example is the use of bipyridine (bpy) to trap freely-diffusing solvated Co(II) to eliminate water oxidation resulting from oxidation catalyzed by such species vs the starting Co-containing POM OEC. Bpy can remove many 3d metals from other complexes or materials. In the case of the POM WOC, $[Co_4(PW_9O_{34})_2]^{10-}$, the first carbon-free homogeneous catalyst,[55] now studied by many research groups, bpy does not extract Co from the complex, but in the case of isostructural WOC, $[Co_4(VW_9O_{34})_2]^{10-}$, it does extract Co making comparison of rates and other properties of these complexes moot.[120]

In the many papers that address, in part POM OEC stability, there is one underlying phenomenon which has been generally ignored: POMs are essentially equilibrium systems under some conditions. Monomeric oxometalates, the multinuclear POMs themselves and insoluble, extended metal oxides interconvert with each other at different values of pH and metal concentration in aqueous media.[112,121,122] For some 3d metals, such as vanadium(V), which is common in minerals, oxidation catalysts and other materials, different monomeric, i.e., monovanadium polyoxovanadates, equally termed "polyvanadates," dominate in both acid and base, with POMs being the thermodynamically dominant form of vanadium(V) at intermediate pH values and high vanadium concentrations. With this POM-metal oxide equilibration in mind, our multi-institution team demonstrated that POMs, or more precisely, an equilibrating ensemble of POMs was produced by simply heating a mixture of metal oxide or soluble precursors of the POM elements (V, W, P) in water at high temperature.[112] This ensemble selectively oxidized and removed the lignin from wood at high temperature under anaerobic conditions leaving cellulose for paper and paper-products manufacture. In a second step, O_2 (air) was introduced and the solubilized lignin fragments in the water were mineralized to

CO_2, H_2O and N_2 catalyzed by the equilibrated ensemble of POMs. This prototype system taught that thermodynamically stable and equilibrating POM ensembles could function as catalysts.[112,123] Under such conditions, catalyst self-repair is intrinsic to the system because any concentration deviation of the POM species present during use as a catalyst would adjust (re-equilibrate) to the functional and stable ensemble. The concept of self-repair, which mandates some equilibration in the catalyst formation process was invoked later by Nocera and his coworkers in the heterogeneous OEC, cobalt oxide-phosphate.[27,124] In a recent detailed stability study of the WOC, $[Co_4(V^VW_9O_{34})_2]^{10-}$, we noted that there is demonstrable reversibility in the formation of these POMs from other POMs and species present as the pH is reversibly adjusted.[120]

8. Looking forward in OEC development

A bottom line in water oxidation and multielectron solar fuel-relevant reactions in general is evident: formulate catalysts that are commercially viable at scale. This requires the use of readily available and inexpensive elements only, and catalysts that are fast, selective and last as long as the other components of the solar fuel generation device. There is no OEC, homogeneous or heterogeneous, that shows long-term stability under both applied oxidative bias and irradiation. Even TiO_2, a paragon of stability under photoelectrochemical conditions, degrades in time under both applied potential and irradiation. Additional progress needs to be made in several areas to address the above ongoing limitations. Perhaps formulation of thermodynamically stable, equilibrating solar fuel catalysts (OECs, HECs, CO_2 reduction catalysts) is possible, at least under some conditions. This possibility has not been fully explored to date. Current combinatorial and machine-learning-driven catalyst discovery approaches do not address the possibilities of multiple catalyst species and their possible equilibration. For conventional non-equilibrating, non-self-assembling and non-self-repairing catalysts, experiments that vary specific properties of the ligands, counterions and the microenvironment of OEC active site metals in general, to assess how these factors impact both TOF and stabilities under turnover conditions are needed.

Acknowledgments

We thank the Department of Energy, Office of Basic Sciences, Solar Photochemistry program (grant number: DE-FG02-07ER15906) for support of this work.

References

1. Agency, I. E; 2021.
2. Agency, I. E; 2020.
3. Administration, U. E. I., 2021. https://www.eia.gov/outlooks/aeo/.
4. BP; 2020.
5. Berner, R. A. *Nature* **2003**, *426*, 323–326.
6. Ourisson, G.; Albrecht, P.; Rohmer, M. *Sci. Am.* **1984**, *251*, 44–51.
7. Schobert, H. H. *Cambridge University Press*; 2013.
8. Oakleaf, J. R.; Kennedy, C. M.; Baruch-Mordo, S.; Gerber, J. S.; West, P. C.; Johnson, J. A.; Kiesecker, J. *Sci. Data* **2019**, *6*, 101.
9. IPCC; IPCC: Geneva, Switzerland, 2014.
10. Rogelj, J.; Schaeffer, M.; Friedlingstein, P.; Gillett, N. P.; van Vuuren, D. P.; Riahi, K.; Allen, M.; Knutti, R. *Nat. Climate Change* **2016**, *6*, 245–252.
11. Friedlingstein, P.; Jones, M. W.; O'Sullivan, M.; Andrew, R. M.; Hauck, J.; Peters, G. P.; Peters, W.; Pongratz, J.; Sitch, S.; Le Quéré, C.; Bakker, D. C. E.; Canadell, J. G.; Ciais, P.; Jackson, R. B.; Anthoni, P.; Barbero, L.; Bastos, A.; Bastrikov, V.; Becker, M.; Bopp, L.; Buitenhuis, E.; Chandra, N.; Chevallier, F.; Chini, L. P.; Currie, K. I.; Feely, R. A.; Gehlen, M.; Gilfillan, D.; Gkritzalis, T.; Goll, D. S.; Gruber, N.; Gutekunst, S.; Harris, I.; Haverd, V.; Houghton, R. A.; Hurtt, G.; Ilyina, T.; Jain, A. K.; Joetzjer, E.; Kaplan, J. O.; Kato, E.; Klein Goldewijk, K.; Korsbakken, J. I.; Landschützer, P.; Lauvset, S. K.; Lefèvre, N.; Lenton, A.; Lienert, S.; Lombardozzi, D.; Marland, G.; McGuire, P. C.; Melton, J. R.; Metzl, N.; Munro, D. R.; Nabel, J. E. M. S.; Nakaoka, S. I.; Neill, C.; Omar, A. M.; Ono, T.; Peregon, A.; Pierrot, D.; Poulter, B.; Rehder, G.; Resplandy, L.; Robertson, E.; Rödenbeck, C.; Séférian, R.; Schwinger, J.; Smith, N.; Tans, P. P.; Tian, H.; Tilbrook, B.; Tubiello, F. N.; van der Werf, G. R.; Wiltshire, A. J.; Zaehle, S. *Earth Syst. Sci. Data* **2019**, *11*, 1783–1838.
12. Le Quéré, C.; Andres, R. J.; Boden, T.; Conway, T.; Houghton, R. A.; House, J. I.; Marland, G.; Peters, G. P.; van der Werf, G. R.; Ahlström, A.; Andrew, R. M.; Bopp, L.; Canadell, J. G.; Ciais, P.; Doney, S. C.; Enright, C.; Friedlingstein, P.; Huntingford, C.; Jain, A. K.; Jourdain, C.; Kato, E.; Keeling, R. F.; Klein Goldewijk, K.; Levis, S.; Levy, P.; Lomas, M.; Poulter, B.; Raupach, M. R.; Schwinger, J.; Sitch, S.; Stocker, B. D.; Viovy, N.; Zaehle, S.; Zeng, N. *Earth Syst. Sci. Data* **2013**, *5*, 165–185.
13. Houghton, R. A. *Annu. Rev. Earth Planet. Sci.* **2007**, *35*, 313–347.
14. IPCC; IPCC: 2019.
15. IPCC; IPCC: Geneva, Switzerland, 2019.
16. IPCC; IPCC: Geneva, Switzerland, 2012.
17. Jane, S. F.; Hansen, G. J. A.; Kraemer, B. M.; Leavitt, P. R.; Mincer, J. L.; North, R. L.; Pilla, R. M.; Stetler, J. T.; Williamson, C. E.; Woolway, R. I.; Arvola, L.; Chandra, S.; DeGasperi, C. L.; Diemer, L.; Dunalska, J.; Erina, O.; Flaim, G.; Grossart, H.-P.; Hambright, K. D.; Hein, C.; Hejzlar, J.; Janus, L. L.; Jenny, J.-P.; Jones, J. R.; Knoll, L. B.; Leoni, B.; Mackay, E.; Matsuzaki, S.-I. S.; McBride, C.; Müller-Navarra, D. C.; Paterson, A. M.; Pierson, D.; Rogora, M.; Rusak, J. A.; Sadro, S.; Saulnier-Talbot, E.; Schmid, M.; Sommaruga, R.; Thiery, W.; Verburg, P.; Weathers, K. C.; Weyhenmeyer, G. A.; Yokota, K.; Rose, K. C. *Nature* **2021**, *594*, 66–70.
18. IPCC; IPCC: Geneva, Switzerland, 2011.
19. Lewis, J. D.; Clark, I. P.; Moore, J. N. *J. Phys. Chem. A* **2007**, *111*, 50–58.
20. Kannan, N.; Vakeesan, D. *Renew. Sustain. Energy Rev.* **2016**, *62*, 1092–1105.
21. Miller, E. L. Solar Hydrogen Production by Photoelectrochemical Water Splitting: The Promise and Challenge. *On Solar Hydro. & Nanotechnol.* **2010**, 1–35.
22. Huynh, M.; Bediako, D. K.; Nocera, D. G. *J. Am. Chem. Soc.* **2014**, *136*, 6002–6010.

23. Hunter, B. M.; Gray, H. B.; Müller, A. M. *Chem. Rev.* **2016**, *116*, 14120–14136.
24. Dau, H.; Limberg, C.; Reier, T.; Risch, M.; Roggan, S.; Strasser, P. *ChemCatChem* **2010**, *2*, 724–761.
25. Reier, T.; Oezaslan, M.; Strasser, P. *ACS Catal.* **2012**, *2*, 1765–1772.
26. Lee, Y.; Suntivich, J.; May, K. J.; Perry, E. E.; Shao-Horn, Y. *J. Phys. Chem. Lett.* **2012**, *3*, 399–404.
27. Kanan, M. W.; Nocera, D. G. *Science* **2008**, *321*, 1072–1075.
28. Kanan, M. W.; Yano, J.; Surendranath, Y.; Dincă, M.; Yachandra, V. K.; Nocera, D. G. *J. Am. Chem. Soc.* **2010**, *132*, 13692–13701.
29. Surendranath, Y.; Kanan, M. W.; Nocera, D. G. *J. Am. Chem. Soc.* **2010**, *132*, 16501–16509.
30. Burke, M. S.; Kast, M. G.; Trotochaud, L.; Smith, A. M.; Boettcher, S. W. *J. Am. Chem. Soc.* **2015**, *137*, 3638–3648.
31. Du, P. W.; Eisenberg, R. *Energ. Environ. Sci.* **2012**, *5*, 6012–6021.
32. Enman, L. J.; Stevens, M. B.; Dahan, M. H.; Nellist, M. R.; Toroker, M. C.; Boettcher, S. W. *Angew. Chem. Int. Ed.* **2018**, *57*, 12840–12844.
33. Fan, J.; Chen, Z.; Shi, H.; Zhao, G. *Chem. Commun.* **2016**, *52*, 4290–4293.
34. Gerken, J. B.; Landis, E. C.; Hamers, R. J.; Stahl, S. S. *ChemSusChem* **2010**, *3*, 1176–1179.
35. Ndambakuwa, W.; Ndambakuwa, Y.; Choi, J.; Fernando, G.; Neupane, D.; Mishra, S. R.; Perez, F.; Gupta, R. K. *Surf. Coat. Technol.* **2021**, *410*, 10.
36. Peng, Z.; Jia, D.; Al-Enizi, A. M.; Elzatahry, A. A.; Zheng, G. *Adv. Energy Mater.* **2015**, *5*, 1402031.
37. Smith, R. D. L.; Prévot, M. S.; Fagan, R. D.; Trudel, S.; Berlinguette, C. P. *J. Am. Chem. Soc.* **2013**, *135*, 11580–11586.
38. Wang, Y.; Yang, C.; Huang, Y.; Li, Z.; Liang, Z.; Cao, G. *J. Mater. Chem. A* **2020**, *8*, 6699–6708.
39. Xu, Y. F.; Gao, M. R.; Zheng, Y. R.; Jiang, J.; Yu, S. H. *Angew. Chem. Int. Ed.* **2013**, *52*, 8546–8550.
40. Zhu, C. Z.; Wen, D.; Leubner, S.; Oschatz, M.; Liu, W.; Holzschuh, M.; Simon, F.; Kaskel, S.; Eychmuller, A. *Chem. Commun.* **2015**, *51*, 7851–7854.
41. Reath, A. H.; Ziller, J. W.; Tsay, C.; Ryan, A. J.; Yang, J. Y. *Inorg. Chem.* **2017**, *56*, 3713–3718.
42. Fukuzumi, S.; Ohkubo, K. *Chem. A Eur. J.* **2000**, *6*, 4532–4535.
43. Blasco-Ahicart, M.; Soriano-López, J.; Carbó, J. J.; Poblet, J. M.; Galan-Mascaros, J. R. *Nat. Chem.* **2018**, *10*, 24–30.
44. Soriano-López, J.; Goberna-Ferrón, S.; Carbó, J. J.; Poblet, J. M.; Galán-Mascarós, J. R. In *Advanes in Inorganic Chemistry*; van Eldik, R., Cronin, L., Eds.; Vol. 69; Academic Press, 2017; pp. 155–179.
45. Folkman, S. J.; Soriano-Lopez, J.; Galán-Mascarós, J. R.; Finke, R. G. *J. Am. Chem. Soc.* **2018**, *140*, 12040–12055.
46. Gerken, J. B.; McAlpin, J. G.; Chen, J. Y. C.; Rigsby, M. L.; Casey, W. H.; Britt, R. D.; Stahl, S. S. *J. Am. Chem. Soc.* **2011**, *133*, 14431–14442.
47. Zidki, T.; Zhang, L.; Shafirovich, V.; Lymar, S. V. *J. Am. Chem. Soc.* **2012**, *134*, 14275–14278.
48. Zhang, R. R.; Zhang, Y. C.; Pan, L.; Shen, G. Q.; Mahmood, N.; Ma, Y. H.; Shi, Y.; Jia, W. Y.; Wang, L.; Zhang, X. W.; Xu, W.; Zou, J. J. *ACS Catal.* **2018**, *8*, 3803-+.
49. Pijpers, J. J. H.; Winkler, M. T.; Surendranath, Y.; Buonassisi, T.; Nocera, D. G. *Proc. Natl. Acad. Sci.U.S.A* **2011**, *108*, 10056–10061.
50. Jiao, F.; Frei, H. *Angew. Chem. Int. Ed.* **2009**, *48*, 1841–1844.
51. Chauhan, M.; Reddy, K. P.; Gopinath, C. S.; Deka, S. *ACS Catal.* **2017**, *7*, 5871–5879.
52. Brodsky, C. N.; Hadt, R. G.; Hayes, D.; Reinhart, B. J.; Li, N.; Chen, L. X.; Nocera, D. G. *Proc. Nat. Acad. Sci.U.S.A.* **2017**, *114*, 3855–3860.

53. Pegis, M. L.; Wise, C. F.; Martin, D. J.; Mayer, J. M. *Chem. Rev.* **2018**, *118*, 2340–2391.
54. Landon, J.; Demeter, E.; Inoglu, N.; Keturakis, C.; Wachs, I. E.; Vasic, R.; Frenkel, A. I.; Kitchin, J. R. *ACS Catal.* **2012**, *2*, 1793–1801.
55. Yin, Q.; Tan, J. M.; Besson, C.; Geletii, Y. V.; Musaev, D. G.; Kuznetsov, A. E.; Luo, Z.; Hardcastle, K. I.; Hill, C. L. *Science* **2010**, *328*, 342–345.
56. Surendranath, Y.; Dincă, M.; Nocera, D. G. *J. Am. Chem. Soc.* **2009**, *131*, 2615–2620.
57. Young, E. R.; Nocera, D. G.; Bulović, V. *Energ. Environ. Sci.* **2010**, *3*, 1726–1728.
58. Huang, Z.; Luo, Z.; Geletii, Y. V.; Vickers, J.; Yin, Q.; Wu, D.; Hou, Y.; Ding, Y.; Song, J.; Musaev, D. G.; Hill, C. L.; Lian, T. *J. Am. Chem. Soc.* **2011**, *133*, 2068–2071.
59. Wee, T.-L.; Sherman, B. D.; Gust, D.; Moore, A. L.; Moore, T. A.; Liu, Y.; Scaiano, J. C. *J. Am. Chem. Soc.* **2011**, *133*, 16742–16745.
60. Kim, H.; Park, J.; Park, I.; Jin, K.; Jerng, S. E.; Kim, S. H.; Nam, K. T.; Kang, K. *Nat. Commun.* **2015**, *6*, 8253.
61. Lv, H.; Song, J.; Geletii, Y. V.; Vickers, J. W.; Sumliner, J. M.; Musaev, D. G.; Kögerler, P.; Zhuk, P. F.; Bacsa, J.; Zhu, G.; Hill, C. L. *J. Am. Chem. Soc.* **2014**, *136*, 9268–9271.
62. Deng, X.; Tüysüz, H. *ACS Catal.* **2014**, *4*, 3701–3714.
63. Moysiadou, A.; Lee, S.; Hsu, C.-S.; Chen, H. M.; Hu, X. *J. Am. Chem. Soc.* **2020**, *142*, 11901–11914.
64. Chou, N. H.; Ross, P. N.; Bell, A. T.; Tilley, T. D. *ChemSusChem* **2011**, *4*, 1566–1569.
65. Nguyen, A. I.; Ziegler, M. S.; Oña-Burgos, P.; Sturzbecher-Hohne, M.; Kim, W.; Bellone, D. E.; Tilley, T. D. *J. Am. Chem. Soc.* **2015**, *137*, 12865–12872.
66. Carroll, G. M.; Zhong, D. K.; Gamelin, D. R. *Energ. Environ. Sci.* **2015**, *8*, 577–584.
67. Ertem, M. Z.; Cramer, C. J. *Dalton Trans.* **2012**, *41*, 12213–12219.
68. Ullman, A. M.; Brodsky, C. N.; Li, N.; Zheng, S.-L.; Nocera, D. G. *J. Am. Chem. Soc.* **2016**, *138*, 4229–4236.
69. Schilling, M.; Luber, S. *Front. Chem.* **2018**, 6.
70. Zhang, M.; de Respinis, M.; Frei, H. *Nat. Chem.* **2014**, *6*, 362–367.
71. Brunschwig, B. S.; Chou, M. H.; Creutz, C.; Ghosh, P.; Sutin, N. *J. Am. Chem. Soc.* **1983**, *105*, 4832–4833.
72. Feizi, H.; Bagheri, R.; Jagličić, Z.; Singh, J. P.; Chae, K. H.; Song, Z.; Najafpour, M. M. *Dalton Trans.* **2019**, *48*, 547–557.
73. Singh, A.; Chang, S. L. Y.; Hocking, R. K.; Bach, U.; Spiccia, L. *Energ. Environ. Sci.* **2013**, *6*, 579–586.
74. Menezes, P. W.; Indra, A.; Levy, O.; Kailasam, K.; Gutkin, V.; Pfrommer, J.; Driess, M. *Chem. Commun.* **2015**, *51*, 5005–5008.
75. Luo, G.-Y.; Huang, H.-H.; Wang, J.-W.; Lu, T.-B. *ChemSusChem* **2016**, *9*, 485–491.
76. Wang, D.; Ghirlanda, G.; Allen, J. P. *J. Am. Chem. Soc.* **2014**, *136*, 10198–10201.
77. Moreno-Hernandez, I. A.; MacFarland, C. A.; Read, C. G.; Papadantonakis, K. M.; Brunschwig, B. S.; Lewis, N. S. *Energ. Environ. Sci.* **2017**, *10*, 2103–2108.
78. Zhu, G.; Glass, E. N.; Zhao, C.; Lv, H.; Vickers, J. W.; Geletii, Y. V.; Musaev, D. G.; Song, J.; Hill, C. L. *Dalton Trans.* **2012**, *41*, 13043–13049.
79. Diaz-Morales, O.; Ferrus-Suspedra, D.; Koper, M. T. M. *Chem. Sci.* **2016**, 7, 2639–2645.
80. González-Flores, D.; Fernández, G.; Urcuyo, R. *J. Chem. Educ.* **2021**, *98*, 607–613.
81. Zhang, L.-H.; Yu, F.; Shi, Y.; Li, F.; Li, H. *Chem. Commun.* **2019**, *55*, 6122–6125.
82. Ng, J. W. D.; García-Melchor, M.; Bajdich, M.; Chakthranont, P.; Kirk, C.; Vojvodic, A.; Jaramillo, T. F. *Nat. Energy* **2016**, *1*, 16053.
83. Kalantarifard, S.; Allakhverdiev, S. I.; Najafpour, M. M. *Int. J. Hydrogen Energy* **2020**, *45*, 33563–33573.

84. Song, Y.-F.; Tsunashima, R. *Chem. Soc. Rev.* **2012**, *41*, 7384–7402.
85. Pope, M. T.; Muller, A. *Angew. Chem. Int. Ed. Engl.* **1991**, *30*, 34–48.
86. Long, D.-L.; Burkholder, E.; Cronin, L. *Chem. Soc. Rev.* **2007**, *36*, 105–121.
87. Long, D.-L.; Tsunashima, R.; Cronin, L. *Angew. Chem. Int. Ed.* **2010**, *49*, 1736–1758.
88. Hill, C. L.; Prosser-McCartha, C. M. *Coord. Chem. Rev.* **1995**, *143*, 407–455.
89. Sullivan, K. P.; Yin, Q.; Collins-Wildman, D. L.; Tao, M.; Geletii, Y. V.; Musaev, D. G.; Lian, T.; Hill, C. L. *Front. Chem.* **2018**, *6*, 1–10.
90. Rhule, J. T.; Hill, C. L.; Judd, D. A.; Schinazi, R. F. *Chem. Rev.* **1998**, *98*, 327–357.
91. Judd, D. A.; Nettles, J. H.; Nevins, N.; Snyder, J. P.; Liotta, D. C.; Tang, J.; Ermolieff, J.; Schinazi, R. F.; Hill, C. L. *J. Am. Chem. Soc.* **2001**, *123*, 886–897.
92. Hasenknopf, B. *Front. Biosci.* **2005**, *10*, 275–287.
93. Bareyt, S.; Piligkos, S.; Hasenknopf, B.; Gouzerh, P.; Lacôte, E.; Thorimbert, S.; Malacria, M. *J. Am. Chem. Soc.* **2005**, *127*, 6788–6794.
94. Yamase, T.; Tomita, K. *Inorg. Chim. Acta* **1990**, *169*, 147–150.
95. Misono, M. *Catal. Rev.: Sci. Eng.* **1987**, *29*, 269–321.
96. Misono, M.; Okuhara, T.; Mizuno, N. *Stud. Surf. Sci. Catal.* **1989**, *44*, 267–278.
97. Misono, M.; Ono, I.; Koyano, G.; Aoshima, A. *Pure Appl. Chem.* **2000**, *72*, 1305–1311.
98. Geletii, Y. V.; Botar, B.; Kögerler, P.; Hillesheim, D. A.; Musaev, D. G.; Hill, C. L. *Angew. Chem. Int. Ed.* **2008**, *47*, 3896–3899.
99. Sartorel, A.; Carraro, M.; Scorrano, G.; Zorzi, R. D.; Geremia, S.; McDaniel, N. D.; Bernhard, S.; Bonchio, M. *J. Am. Chem. Soc.* **2008**, *130*, 5006–5007.
100. Lv, H.; Geletii, Y. V.; Zhao, C.; Vickers, J. W.; Zhu, G.; Luo, Z.; Song, J.; Lian, T.; Musaev, D. G.; Hill, C. L. *Chem. Soc. Rev.* **2012**, *41*, 7572–7589.
101. Lauinger, S. M.; Yin, Q.; Geletii, Y. V.; Hill, C. L. In *Advances in Inorganic Chemistry*; Cronin, L., Eldik, R. V., Eds.; 1st ed.; Polyoxometallate Chemistry, Vol. 69; Elsevier: Oxford, UK, 2017; pp. 117–154.
102. Sartorel, A.; Bonchio, M.; Campagna, S.; Scandola, F. *Chem. Soc. Rev.* **2013**, *42*, 2262–2280.
103. Li, J.; Triana, C. A.; Wan, W.; Adiyeri Saseendran, D. P.; Zhao, Y.; Balaghi, S. E.; Heidari, S.; Patzke, G. R. *Chem. Soc. Rev.* **2021**, *50*, 2444–2485.
104. Shi, X.; Siahrostami, S.; Li, G.-L.; Zhang, Y.; Chakthranont, P.; Studt, F.; Jaramillo, T. F.; Zheng, X.; Nørskov, J. K. *Nat. Commun.* **2017**, *8*, 701.
105. Seh, Z. W.; Kibsgaard, J.; Dickens, C. F.; Chorkendorff, I. B.; Norskov, J. K.; Jaramillo, T. F. *Science* **2017**, *355*, 1.
106. Siahrostami, S.; Li, G.-L.; Viswanathan, V.; Nørskov, J. K. *J. Phys. Chem. Lett.* **2017**, *8*, 1157–1160.
107. Shaffer, D. W.; Xie, Y.; Concepcion, J. J. *Chem. Soc. Rev.* **2017**, *46*, 6170.
108. McKone, J. R.; Lewis, N. S.; Gray, H. B. *Chem. Mater.* **2014**, *26*, 407–414.
109. Xu, Z.; Hou, B.; Zhao, F.; Cai, Z.; Shi, H.; Liu, Y.; Hill, C. L.; Musaev, D. G.; Mecklenburg, M.; Cronin, S.; Lian, T. *Nano Lett.* **2021**, (Accepted 13 September 2021).
110. Xiang, X.; Fielden, J.; Rodriguez-Cordoba, W.; Huang, Z.; Zhang, N.; Luo, Z.; Musaev, D. G.; Lian, T.; Hill, C. L. *J. Phys. Chem. C* **2013**, *117*, 918–926.
111. Geletii, Y. V.; Huang, Z.; Hou, Y.; Musaev, D. G.; Lian, T.; Hill, C. L. *J. Am. Chem. Soc.* **2009**, *131*, 7522–7523.
112. Weinstock, I. A.; Barbuzzi, E. M. G.; Wemple, M. W.; Cowan, J. J.; Reiner, R. S.; Sonnen, D. M.; Heintz, R. A.; Bond, J. S.; Hill, C. L. *Nature* **2001**, *414*, 191–195.
113. Hill, C. L.; Brown, R. B., Jr. *J. Am. Chem. Soc.* **1986**, *108*, 536–538.
114. Mansuy, D.; Bartoli, J.-F.; Battioni, P.; Lyon, D. K.; Finke, R. G. *J. Am. Chem. Soc.* **1991**, *113*, 7222–7226.
115. Khenkin, A. M.; Hill, C. L. *J. Am. Chem. Soc.* **1993**, *115*, 8178–8186.

116. Toma, F. M.; Sartorel, A.; Carraro, M.; Bonchio, M.; Prato, M. *Pure Appl. Chem.* **2011**, *83*, 1529–1542.
117. Guo, S.-X.; Liu, Y.; Lee, C.-Y.; Bond, A. M.; Zhang, J.; Geletii, Y. V.; Hill, C. L. *Energ. Environ. Sci.* **2013**, *6*, 2654–2663.
118. Liu, Y.; Guo, S.-X.; Bond, A. M.; Zhang, J.; Geletii, Y. V.; Hill, C. L. *Inorg. Chem.* **2013**, *52*, 11986–11996.
119. Guo, S.-X.; Lee, C.-Y.; Zhang, J.; Bond, A. M.; Geletii, Y. V.; Hill, C. L. *Inorg. Chem.* **2014**, *53*, 7561–7570.
120. Sullivan, K. P.; Wieliczko, M.; Kim, M.; Yin, Q.; Collins-Wildman, D. L.; Mehta, A. K.; Bacsa, J.; Lu, X.; Geletii, Y. V.; Hill, C. L. *ACS Catal.* **2018**, *8*, 11952–11959.
121. Pope, M. T., Springer-Verlag: Berlin, 1983; Vol. 8, p 180.
122. Pettersson, L. Equilibria of polyoxometalates in aqueous solution. In *Topics in Molecular Organization and Engineering - Polyoxometalates: From Platonic Solids to Anti-Retroviral Activity*; Pope, M. T., Müller, A., Eds.; Kluwer Academic Publishers: Netherlands, 1993; pp. 27–40.
123. Sonnen, D. M.; Reiner, R. S.; Atalla, R. H.; Weinstock, I. A. *Ind. Eng. Chem. Res.* **1997**, *36*, 4134–4142.
124. Kanan, M. W.; Surendranath, Y.; Nocera, D. G. *Chem. Soc. Rev.* **2009**, *38*, 109–114.

Interface design, surface-related properties, and their role in interfacial electron transfer. Part I: Materials-related topics

Anna Kusior*, Anita Trenczek-Zajac*, Julia Mazurków, Kinga Michalec, Milena Synowiec, and Marta Radecka
AGH University of Science and Technology, Faculty of Materials Science and Ceramics, Krakow, Poland
*Corresponding authors: e-mail address: akusior@agh.edu.pl; anita.trenczek-zajac@agh.edu.pl

Contents

Abstract

Extensive research concerning the role of active centers in processes for the protection of life and the environment, is of particular importance. All photocatalytic processes are based on interface properties. Carefully designed architecture is thus of key importance in sorption processes. Wisely designed surfaces allow for control of both the mechanisms and kinetics of adsorption.

In this chapter, special attention is devoted to the development of contact centers for homo- and heterostructure through forming crystals with highly reactive facets, introducing hierarchical porosity, and designing interface in single- and multiple-core@shell materials. The discussion is based on our own studies concerning metal oxides and sulfides including Fe_2O_3, TiO_2, Cu_2O, SnO_2, Cu_2S, SnS_2, and CdS. The results are presented in the context of the background of the in-depth literature overview.

Advances in Inorganic Chemistry, Volume 79
ISSN 0898-8838
https://doi.org/10.1016/bs.adioch.2021.12.005

1. Sorption processes: The role of active centers in processes for the protection of life and the environment

Adsorption is the process of binding molecules, atoms, or ions to a surface or a phase boundary, resulting in local changes in concentration. Adsorption is distinguished from absorption, which is the process of penetration into a phase. Adsorption, absorption, and ion exchange are collectively referred to as sorption processes. Depending on the forces that act between the surface of the adsorbing solid-state (adsorbent) and molecules of adsorbed gas/liquid (adsorbate), we distinguish physical adsorption from chemical adsorption or chemisorption. In the case of physical adsorption, there are intermolecular forces, viz. van der Waals, that control physical absorption. In the case of chemisorption, the forces acting between the surface of a solid adsorbent and adsorbed molecules, are chemical forces. The opposite process to adsorption, absorption, or sorption, in general, is desorption. A classification of sorption processes that includes metal biosorption is illustrated in the diagram (Fig. 1).[1]

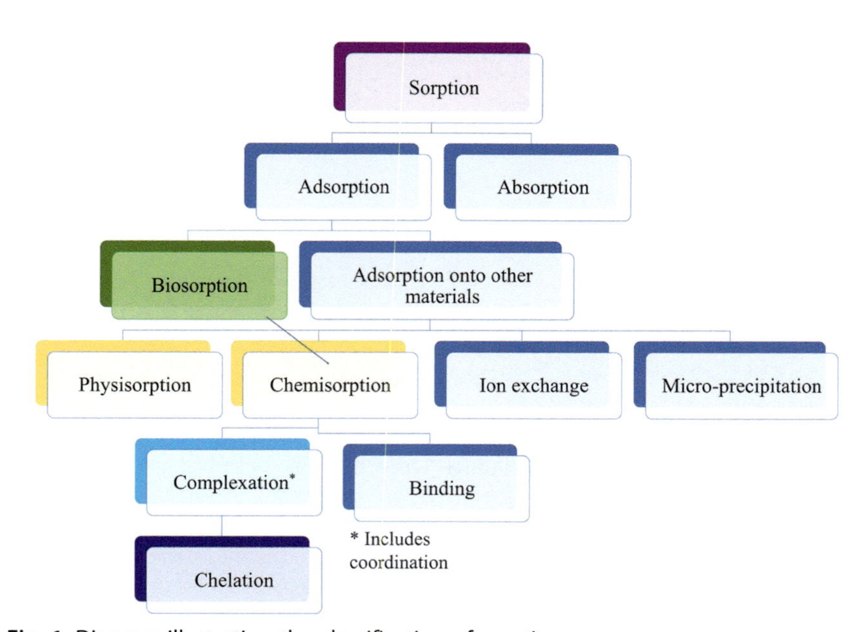

Fig. 1 Diagram illustrating the classification of sorption processes.

1.1 Mechanisms of adsorption

When designing materials for which adsorption processes play an essential role, it is important to know the adsorption capacity at constant temperature vs the adsorbate concentration. The equilibrium concentration of the adsorbent depends on the solute concentration and cannot exceed the maximum adsorption capacity of the adsorbent. The mechanism of the adsorption process that occurs in active solid centers proposed by Henning and Degel[2] is shown in Fig. 2.

In the case of porous materials, the adsorption mechanism is more complex, and both surface and bulk diffusion phenomena must be considered. Four mass transport steps that accompany adsorption for a porous adsorbent, were proposed by Weber[3]:

1) "bulk transport"—transport in solution-phase;
2) "film diffusion"—transport of adsorbed molecules from the bulk liquid phase to the external surface of the adsorbent;
3) "intraparticle diffusion"—diffusion of adsorbed molecules from the exterior to the pores of the adsorbent;
4) "adsorption"—physical and/or chemical adsorption of molecules at active sites of the adsorbent.

A typical adsorption process is presented in Fig. 3.

Taking into account, the kinetics of the individual processes and the fact that the first and last stages occur at a high velocity, it can be assumed that they are not significant for the entire adsorption process.[4–6]

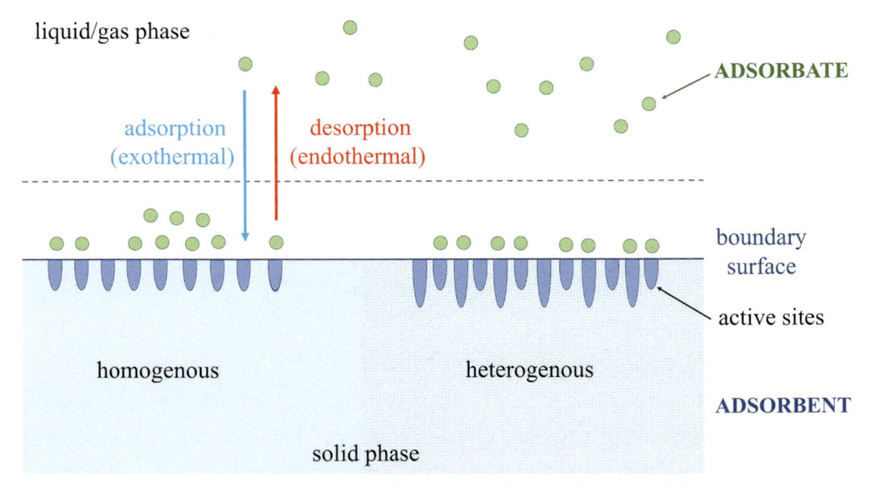

Fig. 2 Mechanism of adsorption and desorption for the liquid/solid system.

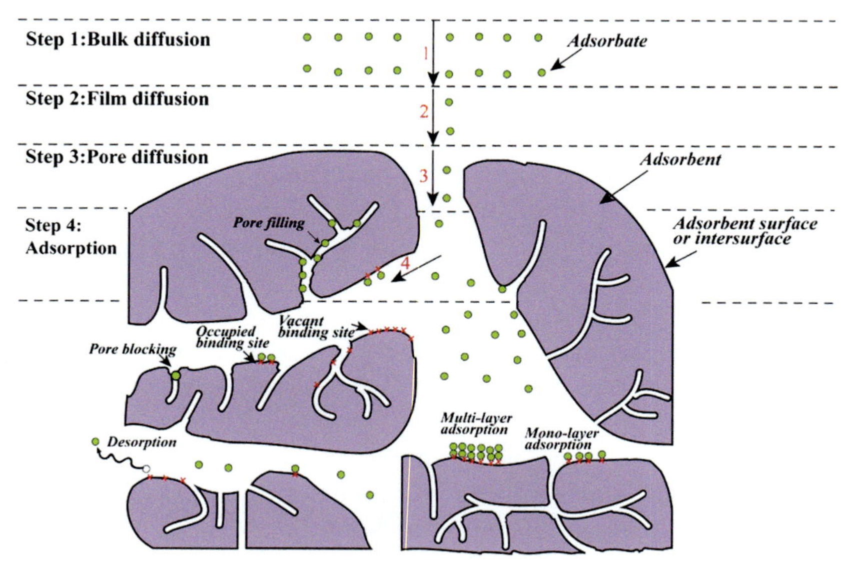

Fig. 3 Schematic diagram of the transport processes during adsorption by a porous adsorbent proposed by Wang et al.[4] *Republished with permission of RSC Pub, from Wang, L., Shi, C., Pan, L., Zhang, X., Zou, J. J. Nanoscale 2020, 12, 4790–4815. Copyright (2021), permission conveyed through Copyright Clearance Center, Inc.*

1.2 Kinetics of adsorption

The adsorption rate is one of the most important factors that characterize potential materials as adsorbents. Various kinetics models, such as pseudo-first-order and pseudo-second-order, are considered to investigate the adsorption kinetics of different dyes on the adsorbents (Table. 1). In advance of performing photocatalytic studies, it is worth knowing how the material will behave in the presence of various substances or dyes. Knowledge of the microstructure characteristics provides information on the phase, size, shape, oxidation state, or optical properties. On the basis of these data, the potential photocatalyst can be preclassified. However, it is important to determine the conditions of the experiment.

1.2.1 Application of the Lagergren model

The adsorption rate is one of the most important factors that describe potential materials suitable for the role of an adsorbent. To investigate the kinetics of adsorption of different dyes on the adsorbents, various kinetic models, such as pseudo-first-order and pseudo-second-order, are considered.

Table 1 Dye adsorption kinetic models.

Model		Description	Equation
Lagergen[7,8]	Pseudo-first-order $\mathbf{k_1}$ ($1\ h^{-1}$, $1\ min^{-1}$)	The adsorption rate	$\ln(q_e - q_t) = \ln q_e - k_1 t$ • q_e the amount of adsorbate adsorbed at time 0 ($mg\ g^{-1}$) • q_t is the amount of adsorbate adsorbed at time t ($mg\ g^{-1}$) • initial and end boundary conditions: (0 to t, 0 to q_t) linear equation is obtained • The plot of $\ln(q_e\text{-}q_t)$ vs t should give a straight line with slope—k_1 and intercept $\ln_{qe}$
	Pseudo-second order $\mathbf{k_2}$ ($g\ mg^{-1}\ h^{-1}$, $g\ mg^{-1}\ min^{-1}$)	Chemisorption is a rate-controlling step in the adsorption processes	$\frac{t}{q_t} = \frac{1}{k_2 q_e^2} + \frac{1}{q_e} t$ • initial and end boundary conditions (0 to t, 0 to q_t) linear equation is obtained • both factors k_2 and q_e can be determined from the slope and intercept of the plot of t/q_t vs t
Intra-particle diffusion[9]		Diffusion controls an adsorption process	$D_F = \frac{0.03 r_0}{t_{0.5}}$ • D_F is the diffusion coefficient ($cm^2\ s^{-1}$) • r_0 (cm) the average radius of the adsorbent (obtained from the SEM or TEM images) • $t_{0.5}$ is the time (min) necessary to complete half of the adsorption process • for D_F in the range of the 10^{-11} to $10^{-13}\ cm^2\ s^{-1}$ the rate-limiting step will be the intra-particle diffusion model • for D_F from 10^{-6} to 10^{-8}, the process will be controlled by the boundary layer diffusion,

Continued

Table 1 Dye adsorption kinetic models.—cont'd

Model	Description	Equation
Weber-Morris plot[8,10–13]	The nature of the rate-controlling step	$q_t = k_{id}t^{0.5} + C$ • C ($mol\ g^{-1}$) is a boundary layer thickness • k_{id} is the diffusion coefficient ($mol\ g^{-1}\ h^{-2}$) • the plot q vs $t^{0.5}$ given by multiple regions represents the external mass transfer followed by intraparticle diffusion (macro, meso, and micropores) • intra-particle rate coefficient k_{id} can be determined from the slope, and C from the intercept • if the data yield multilinear plots, the adsorption process is influenced by two or more steps • the first sharp portion is assigned to the mass transfer surrounding the particle, while the intra-particle diffusion dominates the second
Boyd[9,14]		$B_t = -\ln \frac{\pi^2(1-F)}{6} = -0.4977 - \ln(1-F)$ $F = \frac{q_t}{q_e}$ • B_t—time constant • F is fraction of the solute adsorbed at different times • for straight plot B_t vs time passing through the origin, the process is governed by the intra-particle diffusion model, otherwise by film diffusion (boundary layer)
Percentage of adsorption		$q_{ads} = \frac{100(C_0 - C_e)}{C_0}$ • q_{ads}/% • C_0 the initial sorbate concentration ($mg\ L^{-1}$) • C_e the equilibrium sorbate concentration ($mg\ L^{-1}$), • V the volume of the solution • w is the mass of the adsorbent
Equilibrium concentration		$q_e = \frac{(C_0 - C_e)V}{w}$

The pseudo-first-order model, described by Lagergren,[7] defines adsorption by diffusion through the boundary. However, the assumptions of the model are met only in a narrow measurement range. Under the initial and end boundary conditions (0–t, 0–q_t) the linear equation is obtained:

$$\ln(q_e - q_t) = \ln q_e - k_1 t \quad (1)$$

where q_e and q_t are the amounts of adsorbed adsorbate at time 0, and t (mg g^{-1}), respectively, and k_1 is defined as a rate constant of adsorption (1 h^{-1} or 1 min^{-1}). The plot of ln(q_e-q_t) vs t should give a straight line with slope k_1 and intercept ln q_e.[8] However, the pseudo-second-order model assumes that chemisorption is a rate-controlling step in the adsorption processes. Under initial and end boundary conditions (0–t, 0–q_t), a linear equation is obtained:

$$\frac{t}{q_t} = \frac{1}{k_2 q_e^2} + \frac{1}{q_e} t \quad (2)$$

where k_2 is the pseudo-second-order rate constant of adsorption (g mg^{-1} h^{-1} or g mg^{-1} min^{-1}). Both the factors k_2 and q_e can be determined from the slope and intercept of the plot of t/q_t vs t, respectively.[8] The best-fit model can be selected after the analysis of the R^2 regression correlation values.

1.2.2 Application of the intraparticle diffusion model

To confirm that diffusion controlled the adsorption process, the diffusion coefficient D_F (cm^2 s^{-1}), which depends greatly on the surface properties of the adsorbent,[8] should be studied. The value of the D_F can be calculated from the following equation:

$$D_F = \frac{0.03 r_0}{t_{0.5}} \quad (3)$$

where r_0 is the average radius (cm) of the adsorbent and $t_{0.5}$ is the time (min) required to complete half of the adsorption process. It is claimed that for D_F in the range of 10^{-11} to 10^{-13} cm^2 s^{-1}, the rate-limiting step will be the interparticle diffusion model. If the coefficient value varies from 10^{-6} to 10^{-8}, the process will be controlled by diffusion from the boundary layer.[9]

The nature of the rate-controlling step can also be evaluated by an interparticle diffusion model, also known as the Weber-Morris plot, which can be written in the form[8,10]:

$$q_t = k_{id} t^{0.5} + C \quad (4)$$

where C (mol g^{-1}) is the thickness of the boundary layer, and k_{id} is the diffusion coefficient (mol g^{-1} h^{-2}). Identification of the steps involved in the adsorption processes is necessary. The plot q vs $t^{0.5}$ given by multiple regions represents the external mass transfer followed by intraparticle diffusion (macro-, meso-, and micropores).[11,12] The intraparticle rate coefficient k_{id} can be determined from the slope, while C can be determined from the intercept. Moreover, if the data yield multilinear plots, the adsorption process is influenced by two or more steps. The first sharp increase is assumed to be assigned to the mass transfer surrounding the particle, while the second is dominated by the intraparticle diffusion.[13]

It is worth highlighting that the actual rate-controlling step involved in the dye adsorption process, can be further analyzed using the Boyd kinetic model for materials with double and more intraparticle diffusion plots. In a simplified form, it can be expressed as follows[9,14]:

$$B_t = -\ln\frac{\pi^2(1-F)}{6} = -0.4977 - \ln(1-F) \tag{5}$$

where B_t is a time constant and F is the fraction of the solute adsorbed at different times, defined as:

$$F = \frac{q_t}{q_e} \tag{6}$$

In general, for a straight line fitted to the linear range of the B_t vs time plot intersecting the x-axis, the process is governed by the intraparticle diffusion model, otherwise by film diffusion (boundary layer).

1.2.3 DLS measurements

Dynamic light scattering (DLS) is an indirect method that allows the probability of various dye adsorptions at the adsorbent surface to be analyzed and helps to match photocatalytic-active material to the pollutant. This technique analyzes the Brownian motion and relates the data to the size of the particle. Moreover, the particle size (hydrodynamic diameter, d_h) is calculated from the translational diffusion coefficient. With increasing particle size, slower Brownian movements are observed. It is worth highlighting that any change to the surface, affects the diffusion speed, which correspondingly changes the apparent size of the particle.[15] It can be assumed that in the presence of a dye in the solution, d_h will increase.

Fig. 4 presents the measured hydrodynamic diameter for hematite-based materials in a disc and cube shape after mixing the solution, the powder, and

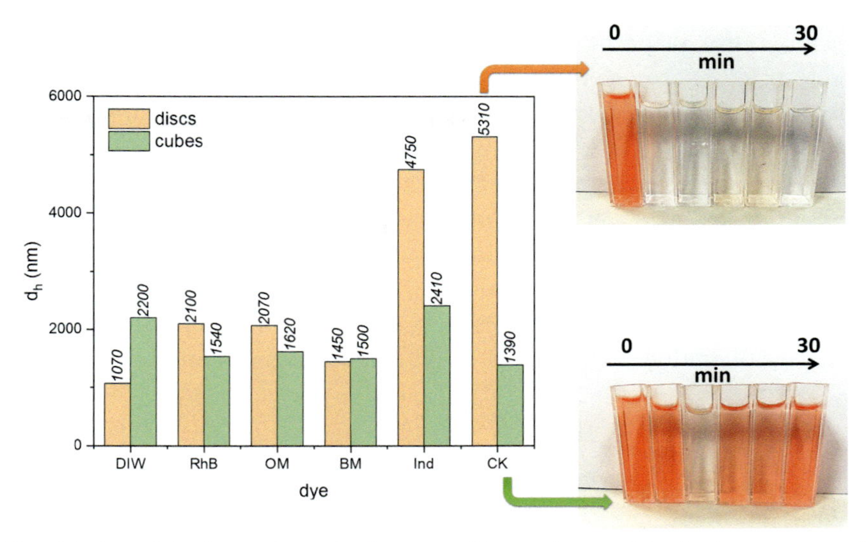

Fig. 4 Value of the hydrodynamic diameter of iron oxide particles in the presence of various dyes (DIW, distilled water; RhB, Rhodamine B; OM, methyl orange; BM, methylene blue; Ind, Indigo carmin; CK, carmin Kongo).

the dye in the dark for 30 min. For all disc samples, d_h is higher than that in distilled water, suggesting that adsorption takes place. The largest changes have been observed for Congo red and indigo carmine, which correlate well with kinetic studies. In contrast, almost all measured hydrodynamic diameters are smaller than those in a reference sample for cubes. In this case, it must be a given that a charge carries the dye. It is thought that dye molecules attract water dipoles to one another more strongly than the material, which may result in a reduction in d_h. However, nanomaterials can agglomerate in a water-based solution, and the presence of the dye may affect the breaking down of aggregates.

It should be kept in mind, that the purity of the sample, surface charge, and grain size, which in the case of nanomaterials may undergo strong agglomeration, needs to be taken into account in adsorption-desorption studies. Although this is an indirect method, it is fast and allows a distinct focus on the area of research.

1.3 Dye adsorption kinetic studies

Prior to commencing photocatalytic studies, it is worth ascertaining how the material will perform in the presence of various substances or dyes. Knowledge of the microstructure characteristics provides information on the phase, size, shape, oxidation state, or optical band gap. On the basis of the microstructure characteristics, such as phase, size, shape, or oxidation state, the potential photocatalyst can be selected. Nevertheless, it is important to determine initially the conditions to be applied for the experiment.

One of the possible ways is to analyze the adsorption-desorption kinetics. Fig. 5 presents the most important factors that affect the correctness of the process. First, the photodegradation of pollutants and, therefore, the decomposition rate depends on the ratio of the potential photocatalyst used and dye.[16] Fig. 5A shows the adsorption properties of titanium dioxide in the presence of methylene blue (MB) at different concentrations. In most cases,

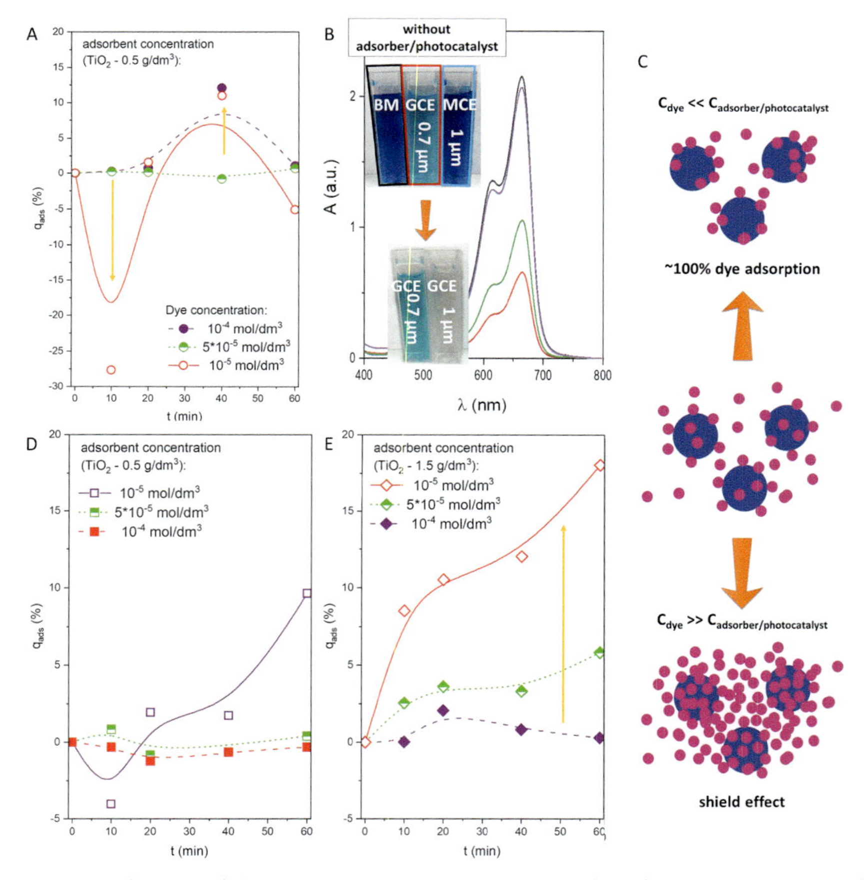

Fig. 5 (A) Influence of dye concentration (BM), (B) syringe filter fiber, (C) proportion of dye vs photocatalyst concentration, potential photocatalysts—TiO_2—with a concentration of (D) 0.5 g dm^{-3} and (E) 1.5 g dm^{-3} in the BF solution on the adsorption-desorption activity.

at the initial stage of the contact period, adsorption results reveal that uptake becomes slower near equilibrium.

Moreover, in between these two stages, the adsorption-desorption rate is found to be variable. Many vacant surface sites during the initial stage affect the adsorption processes. However, after a period, due to repulsive forces between the dye in the solution and the solid and bulk surface, some of the previously adsorbed molecules are desorbed. The time necessary to establish an equilibrium between these two opposite reactions depends on several features, such as the amount, size, and geometry of both the adsorbent and analyzed compounds, their surface properties, and ambient conditions (e.g., pH of the solution). In Fig. 5A, it can be observed that when the dye concentration increases, the probability of adsorption at the surface of the potential photocatalyst decreases[17,18]; it is closely related to the availability of binding sites on the adsorbent surface. An incorrectly selected amount of adsorbent relative to the impurity/dye ratio can disrupt proper sorption and, in future, catalytic processes. If the amount of adsorbent is higher than the quantity of dye molecules, even a complete decolorization of the solution can be observed (100% adsorption). However, here ($cBM = 10^{-5}$ M) strong fluctuations in the amount of adsorbed BM can be observed. It could be assumed that other reactions related to some impurities (or residues from the synthesis) take place on the surface of the potential photocatalyst. However, the reason for such a behavior can be more prosaic and could be assigned to experimental artifacts.

Fig. 5B presents the influence of various syringe filters on the absorbance data. Methylene blue was collected by filters based on fiberglass (GCE) and mixed cellulose esters (MCE). As shown, this type of dye strongly adsorbs on the GCE fibers, and even applying the filter with larger pores (1 μm) results in 100% adsorption. Appropriate selection of filters guarantees stable measurements and reliable data.

In addition to the selction of suitable equipment, it is significant that when a low dye concentration is used, such that almost all possible active centers are occupied by dye molecules; this is schematically presented in Fig. 5C. In the reverse situation, the low concentration of the adsorbent affects the presence of a considerable amount of dye molecules in the solution. This, in turn, affects the extent of the occurrence of shielding effects. As a result, the surface of the catalyst cannot be exposed, and the efficiency of the process decreases. Figs. 5D and E show the adsorption kinetics of basic fuchsine (BF) in the presence of different amounts of titanium dioxide. As can be observed, independently of the dye concentration, the shape of the curves changes with an increasing amount of TiO_2 in the solution. For a low number of BF molecules, the equilibrium of the adsorption-desorption processes is established in a short time. But when the concentration reached 10^{-4} M, the time used to analyze the adsorption-desorption process is insufficient. Therefore, if photocatalysis starts after 30 min, it would not be possible to distinguish whether changes in dye concentration are related to adsorption or photodegradation.

2. Developing the active contact centers

To enable a photocatalytic process, substrates/dyes must adsorb at the semiconductor surface. Moreover, the efficiency of the redox reaction depends on its average coverage ratio and rapid adsorption-desorption steps. The sorption mechanism of various molecules from a liquid phase to a solid-state phase can be considered as a reversible process with equilibrium being established between the solution and the solid phase through the multistep mechanism.[11] The process consists of the following stages:

- diffusion of molecules around the solid particles (external mass transfer),
- diffusion of molecules through the pores of the particle (intraparticle diffusion),
- adsorption (chemical or physical).

The main forces responsible for adsorption are van der Waals forces, hydrogen bonding, or dipole–dipole interactions.[19] Factors that influence the adsorption rate include the adsorbent surface area, interactions between adsorbent and adsorbate, the specific surface area, particle size and shape of the adsorbent, the temperature, and pH of the solution. One of the most important factors is the adsorbate-to-adsorbent ratio. The increase in the concentration of one of the elements affects the number of potential adsorption sites, driving forces, and mass transfer.[20,21]

It is well-known that various properties of nanosized materials may be different from their bulk counterparts. The rational design of novel materials with well-structured network topologies is one of the most significant and challenging fields of research in view of its potential in adsorption and separation processes.[22] Development in particle synthesis has begun to focus on heterogeneous nucleation, which facilitates size and morphology control, with various available forms such as cubes, discs, rods, or mesoporous structures. This increasing interest in the design, synthesis and properties of multifunctional particles, stemmed from their special structure–property relationship. Moreover, incorporating a pore hierarchy might make it possible to obtain material properties comparable to materials with uniform pore size. This is due to the increased mass transport through the pore channels of the material and the maintenance of a high specific surface area at the pore system level. By offering the advantage of a high specific surface area and well-ordered porous channel systems, mesoporous materials have been shown to be suitable candidates for adsorbents in applications such as removing heavy metal ions from waste water.[23]

2.1 Architectural structures: Homostructures formation

2.1.1 Exposed facets: Crystal-like geometry (Fe_2O_3, TiO_2, Cu_2O)

Shape-controlled metal oxide nanocrystals such as TiO_2, α-Fe_2O_3, and Cu_2O, have extensive applications in photocatalysis, hydrogen generation, photovoltaics, and gas sensing.[24–27] Performance in these areas of science depends mainly on the surface properties of nanomaterials, which can differ in shape, size, crystal structure, and exposed facets. Well-defined crystals with highly reactive facets can enhance, for example, the efficiency of redox reactions or gas adsorption on the specified surfaces. Therefore, it is important to understand their growth mechanism and influence of the shape on general properties.[28,29]

Nanocrystals during growth tend to minimize their surface energy and increase stability, which is manifested by low-Miller-index facets forming a final shape. According to the Wulff construction, which determines the equilibrium shape of a crystal, a vector drawn normal to a crystal face will be proportional to its surface energy. This means that crystallographic planes appearing at a low distance (d_i) from the center of the nanoparticles, possess the lowest energy and are the most thermodynamically favored. As a result, the high-energy planes are unstable and rarely observed.[30]

The TiO_2 anatase crystals minimize their surface energy forming a truncated bipyramid with the side facets made of low-energy {101} planes, which constitute up to 94% of the crystal area, and a truncated top made of high-energy {001} facets (Fig. 6A). {001} exposed facets are characterized by a high density of uncoordinated Ti atoms on the surface, which contributes to their high photocatalytic activity.[31,32] Unfortunately, as mentioned earlier, the highest surface energy leads to the lowest stability. Moreover, many other physicochemical properties of crystals, e.g., optical, magnetic, or (photo) chemical, depend on their shape. For example, excited charge carriers in the anatase nanocrystal accumulate in different crystallographic planes. The holes are transferred to the {001} planes, which gain oxidizing properties, and the electrons transfer to the {101} facets, allowing the reduction of chemical compounds. As a result, effective separation of the charge carriers

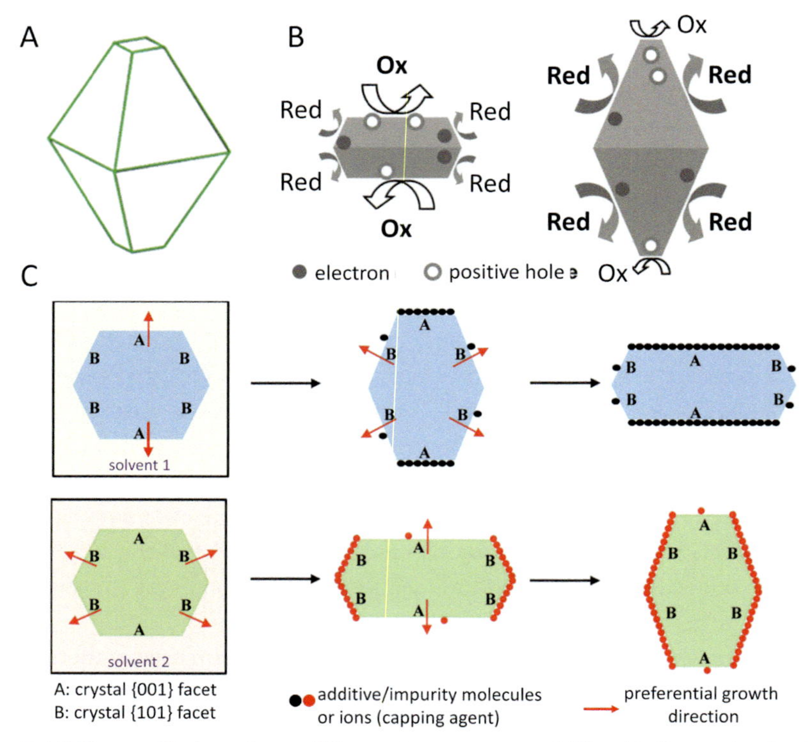

Fig. 6 (A) The equilibrium shape of the anatase nanocrystal, (B) oxidation and reduction processes on various crystallographic planes, and (C) the way to control the shape of nanocrystals using capping agents. *Panel (A) based on Lazzeri, M.; Vittadini, A.; Selloni, A. Phys. Rev. B - Condens. Matter Mater. Phys.* ***2001****,* 63, *1554091–1554099. Panel (B) adapted with permission from* J. Phys. Chem. C ***2009****,* 113, *8, 3062–3069. Copyright (2009) American Chemical Society). Panel (C) republished with permission of the Royal Society of Chemistry, from Nanoscale, Highly reactive {001} facets of* TiO_2*-basedcomposites: synthesis, formation mechanism and characterization, WJ Ong et al., vol. 6, 2014; permission conveyed through Copyright Clearance Center, Inc.*

occurs (Fig. 6B). This phenomenon is associated with a characteristic arrangement of atoms on particular surfaces, which can slightly change the level of the valence or conduction band level.[29,33]

One of the most widely used methods for the synthesis of shape-controlled nanocrystals is the hydrothermal method. The key factors affecting the final shape are the type of precursor, the time and temperature of the synthesis, as well as the reaction environment, i.e., the type of solvent, pH, and concentration of shape-controlling agents. Among these factors, the most important are capping agents, which lower the surface energy of high-energy planes by covering them with their molecules, which allows them to grow (Fig. 6C).[34] In our previous review,[30] the capping agents and their influence on the final shape of the anatase TiO_2, hematite α-Fe_2O_3, and cuprite Cu_2O nanocrystals were described.

The most common shapes of anatase nanocrystals are rods, sheets, cubes (or truncated cubes), cuboids (or truncated cuboids), bars, and bipyramids (or truncated bipyramids) obtained using the following surfactants that passivate the highly active planes: HF, DEA (diethanolamine), CTAB (hexadecyltrimethylammonium bromide), OH^-, OA (oleic acid), [bmim][BF_4] (1-butyl-3-methylimidazolium tetrafluoroborate), OLEA (oleylamine), NH_4F, urea, HCOOH, and NH_4Cl.[35–40]

The first shape-controlling agent used to reduce the surface tension of {001} facets of TiO_2 NCs (nanocrystals) was hydrofluoric acid.[41,42] Nevertheless, the use of HF as an F^- ion source, despite the success in obtaining high-energy facets, involved using a highly toxic and corrosive chemical compound. Therefore, researchers began to replace this compound with others that supply F^- ions into the reaction medium but are less problematic. The ionic liquid [bmim][BF_4] is a good example of a replacement for HF. Its double action is to stabilize {001} facets by F^- anions, and $[bmim]^+$ cations reduce the surface energy of {100} planes, which causes growth of NCs in the shape of cubes.[38] There is also a fluorine-free strategy for controlling the morphology that was successfully used in our research group. Changing the molar ratio of diethanolamine (DEA) to tetrabutylamine hydroxide (TBAH) allowed us to manipulate the growth of TiO_2 nanocrystals effectively. In the study,[43] a series of TBAH:DEA concentrations such as 1:1, 2:5, 1:10 were used. Diethanolamine selectively covers the high-energy {001} planes, resulting in decreasing the access of TiO_2 particles during crystallization. In this case, the crystal grows with the exposure of thermodynamically unstable facets. The method of increasing the amount of capping agent in the form of an amine, allowed to obtain the following shapes at a temperature of 215 °C: rod → cube → sheet (Fig. 7). In addition, inhibition of growth in the [001] direction caused an increase in the number of high-energy 001 facets.

In the case of hematite nanocrystals, a controlled environment of growth allows one to obtain the following shapes: cube, cuboid, thorhombic*, polyhedron, rhombohedron, octahedron, plate, concave, and icositetrahedron. The morphology of NCs can be easily modified because the low-Miller-index facets possess similar energy. Selective adsorption of metal

* Thorhomic—cubic geometry with concave faces. For an example illustration see Ref. 46.

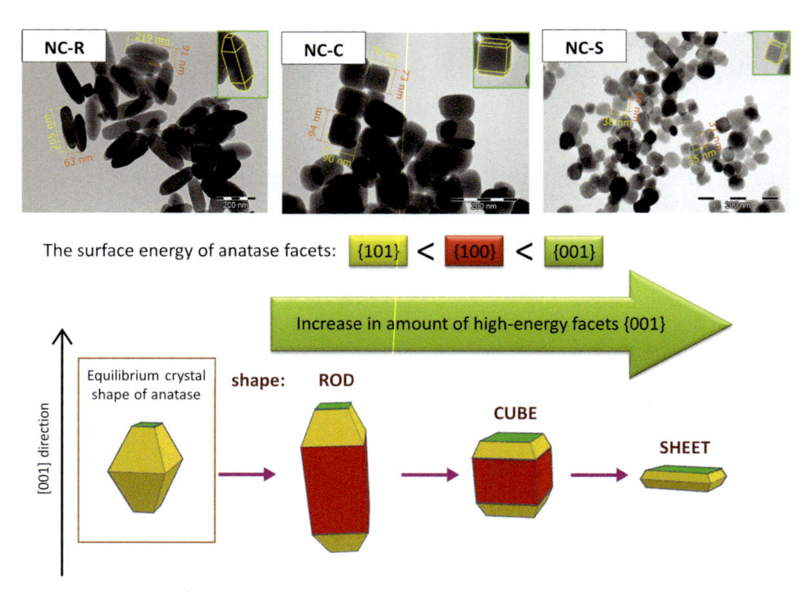

Fig. 7 TEM images of TiO_2 nanocrystals in the shape of rods (NC_R), cubes (NC_C) and sheets (NC_S). Evolution of the nanocrystal shape of anatase caused by an increase in the molar ratio of TBAH:DEA. *Reprinted from* Applied Surface Science, *vol. 473, Synowiec et al., Functionalized structures based on shape-controlled TiO_2, pp. 603–613. Copyright (2019) with permission from Elsevier.*

cations such as Ni^{2+}, Zn^{2+}, Cu^{2+}, and Al^{3+} on high-energy facets of α-Fe_2O_3 is known as a metal-ion-mediated route.[44–46] The capping agent in the form of Zn^{2+} ions is able to change surface energy in such a way that cube formation is thermodynamically favorable.[46] On the other hand, the use of Al^{3+} ions determines the plane and cube shape of hematite nanocrystals,[47] whereas applying Cu^{2+} cations is responsible for stabilizing concave and thorhombic NCs.[48]

In our team, we conducted detailed studies on the influence of Al^{3+} ions on the time-scale of the synthesis of the formation of iron oxide nanomaterials.[49] The introduction of aluminum cations into the reaction system gave rise to the formation of cubic nanocrystals. In the first stage, after 4 h of synthesis, a cubic shape is obtained with the existence of small particles on the crystal surface. After 8 h the pseudocubic form was formed with an adsorbed larger number of grains on the NC surface. Lengthening the reaction time to 16 h allowed the increase of the size of Fe_2O_3 cubes at the expense of the particles attached to the walls of the cube. Finally, after 32 h, only large, well-formed crystals were present in the SEM images (Fig. 8).

Fig. 8 The influence of synthesis time on α-Fe_2O_3 nanocrystals obtained by the hydrothermal method at 200 °C. *Reprinted from Kusior, A.; Michalec, K.; Jelen, P.; Radecka, M.* Appl. Surf. Sci. ***2019***, 476, *342–352. Copyright (2021), with permission from Elsevier.*

In contrast to TiO_2 and Fe_2O_3 nanocrystals, cuprite forms much more different shapes such as a cube, concave cube, octahedron, cuboctahedron, rhombic dodecahedron, truncated octahedra, concave octahedra, octopod, polyhedron, and star-shaped. As was seen with the earlier materials, the addition of capping agents is a crucial factor. The most common are AOT (sulfosuccinate ions), SDS (sodium dodecyl sulfate), PVP (poly (vinyl pyrrolidone)), glycine, and ammonia.[50–53] It should also be mentioned that copper oxide can easily change its oxidation state: $Cu \leftrightarrow Cu^{+} \leftrightarrow Cu^{2+}$; therefore, reducing agents need to be added to the reaction medium. The most popular are ascorbic acid, glucose, hydroxylamine, and hydrazine.[54,55]

In our group, we used the wet-chemical method to obtain Cu_xO nanocrystals from different precursors: $CuCl_2$ (sample Cu_Cl_2), $Cu(NO_3)_2$ (sample Cu_NO_3), and $CuSO_4$ (sample Cu_SO_4) (Fig. 9). The addition of ascorbic acid as a reducing agent was necessary to change the oxidation state of copper

Fig. 9 SEM images of copper oxides synthesized using various copper ion sources and with the addition of ascorbic acid.

from +2 to +1 to obtain Cu_2O. Amomg these copper species, Cu_2O was the best crystallized; it had smooth walls without undersized fragments, and it had the homogeneous shape of cubes, with large and small sizes. Nevertheless, the diffraction pattern for this sample also showed low-intensity peaks that originated from CuO.

2.1.2 Hierarchically porous materials

Recognition that the combination of meritorious properties brought by each porosity domain can significantly enhance the photocatalytic properties of the material, has triggered the development of hierarchically porous structures. According to the IUPAC, porous materials can be divided into three groups depending on the pore diameter (d): microporous ($d < 2$ nm), mesoporous ($2 < d < 50$ nm) and macroporous materials ($d > 50$ nm) (Fig. 10A). It has been established that micro- and mesoporosity introduce additional active sites and offer separation of guest molecules (reactants, intermediates, and products) based on shape and size selectivity. In contrast, macroporosity enhances mass transport, provides access to active sites, and induces light-harvesting mechanisms ensuring deeper light penetration (Fig. 10B).[56] Designing multimodal structures is greatly inspired by nature, especially plants. Synthesis strategies of artificial photocatalysts include templating-based methods and physical-chemical processes. Templates can serve as surfactants, polymers, emulsions, biomaterials, or colloidal crystals. Among the commonly applied physical-chemical processes, notable ones are the sol-gel method, phase separation, self-formation, freeze-drying, and selective leaching. Most often, the techniques mentioned above are combined to obtain materials with diversified porosity.

Moreover, theses methods are followed by post-treatment for the introduction of secondary porosity. In the literature, reports concerning mono-, dual- and multiporous materials can be found. This chapter will devote

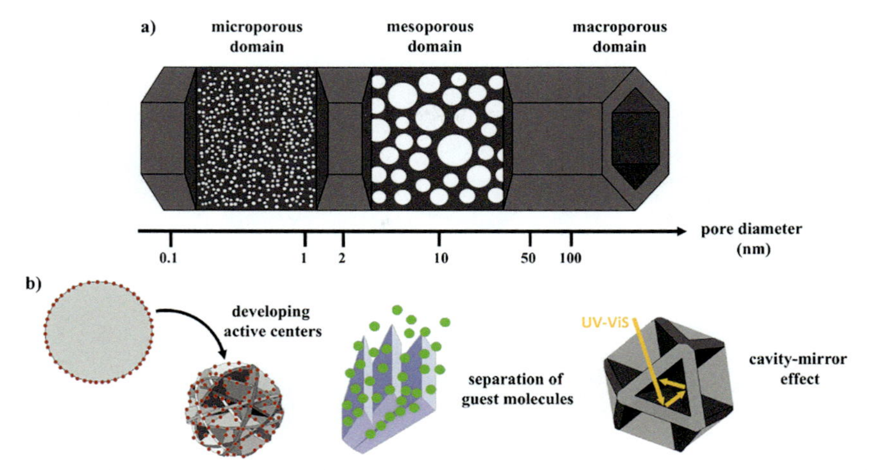

Fig. 10 Schematic illustration of (A) micro-, meso- and macroporosity in materials, and (B) mechanisms of light-harvesting characteristic for hierarchically porous structures.

special attention to the porous structures of TiO_2 and metal sulfides (CdS, ZnS, $Cu_{2-x}S$, SnS_2).

2.1.2.1 Titanium dioxide (TiO_2)

Titanium dioxide is one of the most widely studied materials for photocatalytic applications. Recently, special attention has been devoted to its mesoporous and hierarchically porous structures due to enhanced surface area, elevated pore volume, and expected nanoscale effects.[57] Mamghani et al. compared the photocatalytic performance of TiO_2 with different morphologies while keeping the crystallinity, composition, and surface area comparable.[58] Solid microspheres, mesoporous microspheres, hollow spheres, nanosheets, nanotubes, urchin-like, and hierarchically porous structures, were synthesized by varying the parameters of the hydrothermal method, such as precursors, solvent, reaction temperature, and time as well as calcination conditions. The best removal efficiencies applying methyl ethyl ketone (MEK) of 71.3% and 62.5% were obtained for nanosheets and hollow spheres, respectively, compared with only 35% for commercial P25 and 31.8% for solid microspheres (Fig. 11). In another article, Li et al. biomimetically synthesized hierarchical Sn-doped TiO_2 structures via the sol-gel method.[59] As a template, medical cellulosic cotton was used. The materials obtained possessed a microtubular morphology composed of nanofibers. Specific surface area, mean pore diameter, and average pore volume were changed depending on the level of tin present. The sample exhibiting the best properties contained an intermediate amount of

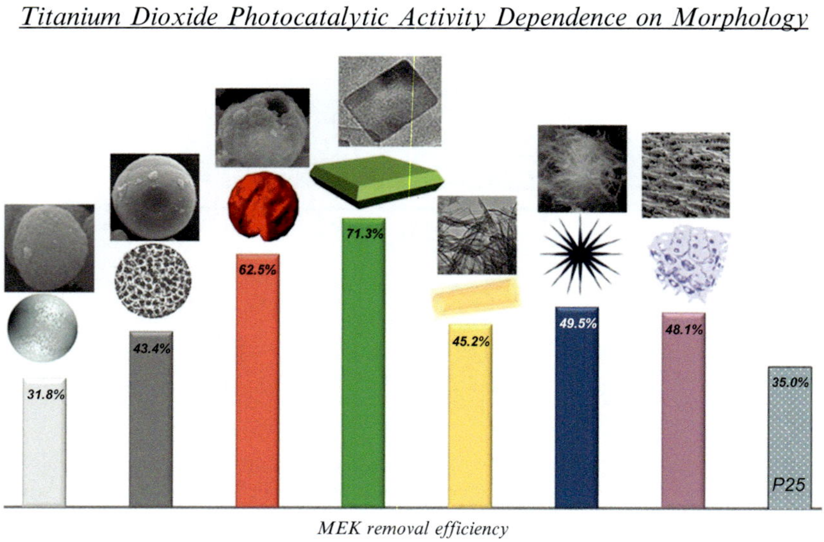

Fig. 11 Photocatalytic efficiency toward MEK removal for different TiO_2 morphologies. *Reprinted from Mamaghani, A.H.; Haghighat, F.; Lee, C.S.* Appl. Catal. B Environ., ***2021****, 269, 118735, Copyright (2021), with permission from Elsevier.*

Sn, i.e., 6.25 mol%. It also possessed the smallest band gap energy and the highest degradation efficiency of methylene blue. A template-based approach was also utilized by Khan et al. to prepare hierarchically porous TiO_2 films.[60] For this purpose, poly(methyl methacrylate) spheres were deposited on the FTO glass using the dip-coating method and served as a colloidal crystal template. Subsequently, as-prepared substrates were dipped into the precursor solutions of TiO_2 and heat-treated to remove the organic template. The hierarchical macroporous scaffolds obtained possessed embedded mesopores. As-prepared, TiO_2 films exhibited a photocurrent density of ca. 45 μA cm^{-2}.

2.1.2.2 Metal sulfides (ZnS, CdS, $Cu_{2-x}S$, SnS_2)

Metal sulfides have received significant attention in recent years as compounds for photocatalytic applications, especially hydrogen production and CO_2 reduction, due to narrow band gaps (compared to appropriate oxides) and more favorable positions of the conduction bands. Among these compounds, the most commonly reported photocatalysts are zinc, cadmium, and copper sulfides. Bhushan et al. synthesized different hierarchical mesoporous ZnS structures using hydro- and solvothermal routes.[61]

By adjusting the reaction parameters (e.g., zinc precursor, solvents, surfactant addition, reaction time, and temperature), raspberry-, ball-, and flower-like structures with a pore diameter in the range of 2–8 nm were obtained (Fig. 12). Materials removal efficiencies of Eosin Y were 95.8%, 94.0%, and 96.0%, respectively, after 90 min of illumination. It was attributed to the increased accessible surface area for raspberry-like (49.78 $m^2 g^{-1}$) and nanoflowers (39.15 $m^2 g^{-1}$) structures in comparison with nanoballs (46.61 $m^2 g^{-1}$). Our group has developed a synthetic procedure for hierarchically porous copper sulfides.[62] By varying the precursor ratio (copper chloride and thiourea) in a hydrothermal reaction, it was possible to obtain structures with differently assembled nanoplates (Fig. 12C). In the case of well-defined flower-like structures, the calculated band gaps were 2.2 and 2.8 eV, indicating their excellent prospects for photocatalytic applications. A similar synthetic approach was also utilized by Li et al. for SnS_2 preparation in the form of a tablet- and flower-like structures for CO_2 reduction.[63] The variable was the addition of an appropriate acid to the reaction solution—formic or acetic acid, respectively. Both powders had similar bandgap energies. However, flower-like structures possessed two-times larger specific surface areas. Moreover, such hierarchical morphology shortened the diffusion paths and subsequently enhanced carrier separation. It resulted in higher photocurrent density, 1.2 μA cm^{-2}, for flower-like materials compared to 0.6 μA cm^{-2} for tablet-like materials.

Another method, anion exchange, was applied by Xiang et al. for hierarchically porous CdS nanoflowers fabrication (Fig. 12B).[64] First, the cadmium hydroxide template was synthesized via the hydrothermal route and was further converted to cadmium sulfide using Na_2S. According to

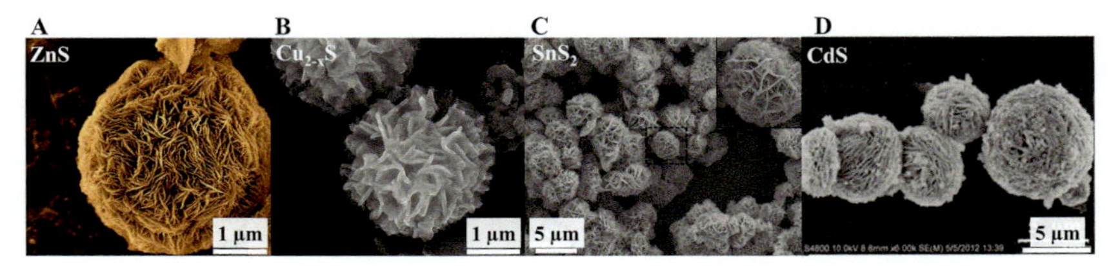

Fig. 12 SEM images of flower-like (A) ZnS, (B) $Cu_{2-x}S$, (C) SnS_2, and (D) CdS. *Panel (A) reprinted from Bhushan, M.; Jha, R. Appl. Surf. Sci.* ***2020****, 528, 146988. Copyright (2021), with permission from Elsevier. Panel (C) reprinted from Li, G.; Sun, Y.; Sun, S.; Chen, W.; Zheng, J.; Chen, F.; Sun, Z.; Sun, W. Adv. Powder Technol.* ***2020****, 31, 2505–2512. Copyright (2021), with permission from Elsevier. Panel (D) reprinted from Xiang, Q.; Cheng, B.; Yu, J. Appl. Catal. B Environ.* ***2013****, 138–139, 299–303. Copyright (2021), with permission from Elsevier.*

the IUPAC nomenclature, isotherms of type IV indicated that the flower-like structures obtained possessed mesopores and a specific surface area of 57.8 $m^2 g^{-1}$. As-prepared materials were utilized for hydrogen evolution, and the production rate reached 468.7 mol h^{-1}, whereas for CdS nanoparticles, it was three-times lower. Such an increase in photoactivity was explained by the enabled light-harvesting mechanisms and the increased number of active sites due to the hierarchical assemblies of nanosheets. Our group has developed a synthetic procedure for hierarchically porous copper sulfides.[62] By varying the precursor ratio (copper chloride and thiourea), it was possible to obtain structures with differently assembled nanoplates (Fig. 12C). In the case of well-defined flower-like structures, the calculated band gaps were 2.2 and 2.8 eV, indicating their excellent prospects for photocatalytic applications.

2.1.2.3 Other forms: Nanotubes

The synthesis of one-dimensional nanostructures continues to stimulate interest among researchers worldwide.[65–67] Due to their properties, these structures have a wide range of potential applications in modern medicine, aviation, electronics, or alternative energy sources. Many methods are used to obtain materials of this type, among which the following should be mentioned: electrochemical anodization, a hydrothermal technique, electrospinning, and porous matrix technologies (template).[66] However, particularly interesting methods are those based on the self-organization processes of matter (electrochemical anodization). This property makes it possible to obtain highly ordered structures, which in turn provides opportunities to create materials with unique properties. Among the transition-metal oxides, titanium dioxide is of particular interest. The prospects of the application of self-organized TiO_2 nanostructures in solar cells, Li-ion cells, photocatalytic, and photoelectrochemical processes are very promising. Furthermore, the study of materials derived from anodization is an important step in understanding and developing electrochemical processes such as oxidation, porosity enhancement, electrodissolution, and self-organization that take place at the liquid/solid interface.[65–68]

Table 2 shows a comparison of the properties of TiO_2 nanotubes obtained by different methods, with particular emphasis on the electrochemical anodization technique, the application of which enables the formation of ordered nanostructures parallel to each other and connected to the substrate. This is extremely important when these materials are used as electrodes.

Table 2 Comparison of the properties of TiO_2 nanotubes obtained by different methods, with particular emphasis on the electrochemical anodization technique.

	Hydrothermal	**Template**	**Electrospinning**	
Remarks	Single tubes or loose agglomerates	Tube arrays or loose agglomerates	Single tubes or loose agglomerates	
Length L	Up to several μm	nm to μm range	Significant length	
Diameter D	2–20 nm	10–500 nm	From few nm to few μm	
Wall thickness d	Few atomic layers		From few nm to few μm	
Anodization				
L: from 100 nm to 1 μm; D: 10–500 nm d: 2–80 nm		As growth: amorphous, after annealing transformation to anatase or mix anatase/rutile		
	1st generation	2nd generation	3rd generation	4th generation
Electrolyte	HF aqueous solutions	Aqueous solutions of fluoride salts	Organic solutions containing fluoride salts and small amount of water	Aqueous HCl/ H_2O_2 solutions containing no fluoride ions
Length	Up to 0.5 mm	Up to 0.5 mm	Up to 100–1000 mm	Up to several hundreds of nm
Remarks	Poorly self-organized and ribbed	Self-organized and ribbed	Self-organized and smooth	Poorly self-organized and ribbed

Additionally, the possibility of shape modification of TiO_2 nanotubes, whose growth occurs as a result of electrochemical anodization, is practically unlimited. The anodization method is characterized by a wide range of polarization parameters, electrolytes used and consequently allows one to obtain nanostructures differing in crystal structure or geometric parameters. The surface morphology of TiO_2 nanotubes depends on such process parameters as: type of electrolyte used (viscosity), ammonium fluoride and water content, anodization time, and applied potential difference.

The geometry of TiO_2 nanotubes can be described by three models proposed by Grimes[69] (Fig. 13A). The first case assumes a perfectly porous structure, where the individual pores share a common wall (Fig. 13A), case I—ideal pores). According to this model, the spatial distribution of

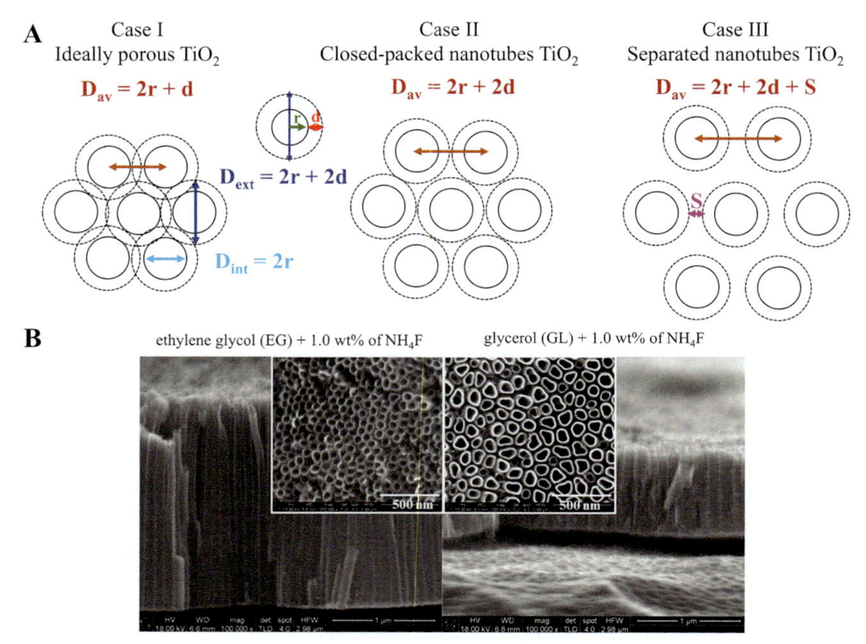

Fig. 13 (A) Geometric models of TiO_2 nanotubes and (B) SEM images of the surface and cross-section of TiO_2 nanotubes. Parameters of anodization or process: potential difference $\triangle V = 30$ V; time t: for EG $t = 120$ min; for GL $t = 60$ min.

nanotubes can occur in two variants: densely packed (case II—close-packed) or lose unconnected tubes (case III—separated nanotubes). On the basis of the analysis of the ratio r/D_{av}, it is possible to determine how far from a perfectly dense packing each arrangement of NTs is. D_{av} corresponds to the distance between the centers of two adjacent nanotubes and D to their diameter. Assuming that the thickness of the d-wall is much smaller than the radius (r), this ratio in a perfectly porous structure is $r/D_{av} = 0.5$. The geometry of TiO_2 nanotubes obtained by anodizing metallic titanium, and characterized in this work depending on the electrolyte used, follows this pattern: case II for NTs obtained in ethylene glycol (EG) electrolyte and case III for glycerine (GL), respectively. Fig. 13B shows an image of the surface and cross-section of TiO_2 nanotubes that were grown in solutions of ethylene glycol EG or glycerine (GL) with 1% wt. NH_4F. The anodization process was carried out at a potential difference of 30 V. The nanotubes obtained in the EG solution are characterized by the most packed structure $r/D_{av} = 0.48$ in contrast to the nanostructures in the GL solution, where this parameter is close to 0.30.

2.2 Architectural structures: Heterostructures formation

2.2.1 Core-shell structures formation: TiO_2@CdS and TiO_2@SnO_2 nanopowders

When the design is focused on the interface between at least two elements, the crucial role of composites should be highlighted. They can be considered anisotropic and inhomogeneous materials, as well as the matrix (one type of material) that surrounds particles, fibers, or reinforcement fragments (another material).[70] Nevertheless, their principal role is to obtain the unique properties that particle elements by themselves cannot achieve. This specious combination of two different materials into a support structure can positively affect its properties, among other things, by the occurrence of additional electron states in the excited band and separation of photocarriers. On the basis of the mutual position of the CB, conduction band, and VB, valence band, and the direction of electron charge transfer, three types of bonding can be distinguished (see Fig. 2, Part II).

An excellent example is TiO_2 sensitized by cadmium sulfide. TiO_2@CdS heterostructures (small amount of CdS deposited on the surface of TiO_2) can be classified into two groups. The first is cadmium sulfide nanoparticles adsorbed on the titanium dioxide surface, and the second is nanometric TiO_2 on the CdS surface. Despite differences in morphology, phase structure or fabrication method, efficient electron charge transfer, and their significant separation, are the determining factors for this system.[71,72]

The use of mesoporous substrate in the form of nanobranches,[73] nanoflowers,[71,72] nanofibers,[74] and hollow spheres type structures,[75] allows to increase the number of adsorption sites of the sensitizer, shortens the diffusion path of electron carriers, reduces recombination, and improves the light-absorbing properties. In the reverse case, when CB and VB bands of the other material, e.g., SnO_2, are located below the corresponding bands of titanium dioxide, it becomes the type II heterostructure.

TiO_2 and SnO_2 can crystallize in the rutile tetragonal structure and form solid solutions in the full range of compositions, above the so-called critical temperature.[76] Additionally, the size of the exciting gap for tin dioxide (E_g=3.8 eV) is larger than that for TiO_2 (E_g=3.0 eV for rutile, E_g=3.2 eV for anatase). The position of the energy levels affects the specific movement and separation of excited electron carriers, resulting in electrons accumulating in the conduction band of tin dioxide and electron holes in the valence band of TiO_2.[77]

2.2.2 Multiple-core@shell heterostructures

In recent years, multiple-core@shell heterostructures have attracted widespead attention due to the possibility of combining more different functionalities in one nanoparticle compared to single-core@shell ones. Overall, multiple-core@shell materials can be classified into four subcategories: (1) core@multishell, (2) multicore@single-shell, (3) multicore@multishell, and (4) multi-core@shell (Fig. 14). Depending on the core and shell morphologies, these materials may be categorized as spherical, partially nonspherical (i.e., either the core or shell exhibits nonspherical geometry), or complete nonspherical nanoparticles. Moreover, the core material may be partially or entirely removed, forming yolk-shell or hollow structures, respectively. The presence of void space in such materials results in more active sites available and contributes to the enhanced light-harvesting.[78–83] Therefore, the precise morphological control and conscious choice of core/shell configuration in multiple-core@shell heterostructures, are essential for obtaining a new class of materials with enhanced functionality for (photo)catalytic, energy storage, sensing, and biomedical applications.

Core@multishell nanomaterials (Fig. 14A) comprise a single core covered with multiple layers of shells, some of which may be removed (hollow shells). The materials with a hollow core encapsulated in hollow shells are defined as multishells.[78,82] Compared to single-core@shell materials, multilayered heterostructures exhibit improved optical/electrical properties,

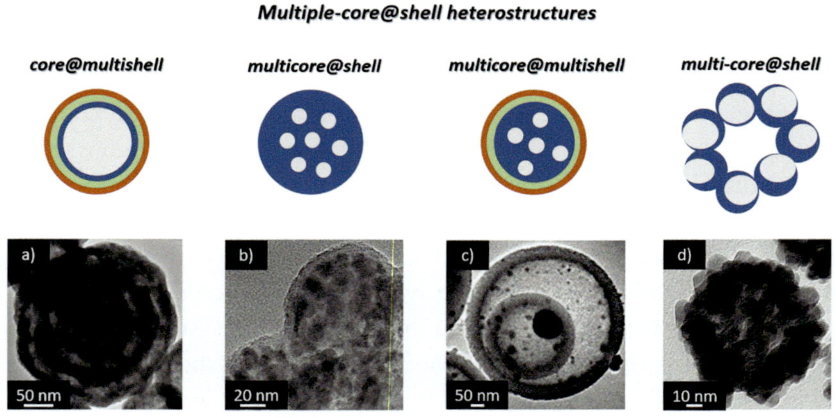

Fig. 14 Schematic representation of multiple-core@shell heterostructures with the exemplary TEM images: (A) core@multishell, (B) multicore@shell, (C) multicore@multishell, (D) multicore@shell. *Panel (A) republished with permission of John Wiley and Sons, from Xu, H.; Wang, W.; Angw. Chem. Int. Ed.* ***2007****, 46, 1489–1492. Copyright (2021), permission conveyed through Copyright Clearance Center, Inc. Panel (B) reprinted from Zhu, W.; Mi, J.; Fu, Y.; Cui, D.; Lü, C. Appl. Surf. Sci.* ***2021****, 538, 148087. Copyright (2021), with permission from Elsevier. Panel (C) republished with permission of RSC Pub, from Hu, Y.; Zheng, X. T.; Chen, J. S.; Zhou, M.; Li, C. M.; Lou, X. W. J. Mater. Chem.* ***2011****, 21, 8052–8056. Copyright (2021), permission conveyed through Copyright Clearance Center, Inc. Panel (D) Michalec, K.; Kusior, A.; Mikuła, A.; Radecka, M.* Mater. Res. Lett. ***2021****, 9, 445–451*

quantum yield, photoluminescence efficiency, (photo)stability, and better lattice matching between the layers.[79] Core@multishell structures are usually obtained via a layer-by-layer approach using sol-gel, solvothermal, chemical vapor deposition (CVD), vapor-liquid-solid (VLS), or successive ionic layer adsorption and reaction (SILAR) methods.[80,82,86] Multishells, in turn, may be synthesized using hard-, soft-, and self-templating techniques.[82] The first one involves the repeated process of covering a hard template with a target material and selective etching. In the soft-templating method, surfactant aggregates—such as micelles or polymer vesicles—are usually used. The last technique does not require template removal (Kirkendall growth, galvanic exchange, or Ostwald ripening processes).[82] Dong et al.[87] reported the synthesis of quintuple-shelled SnO_2 hollow microspheres for dye-sensitized solar cells (DSSCs) applications. The authors used alkali-treated carbonaceous microspheres and tin(IV) chloride as a template and SnO_2 as a precursor, respectively. After the Sn^{4+} ions adsorption, the microspheres were subjected to one-step thermal treatment to remove the template.

Moreover, it was found that depending on the $SnCl_4$ concentration, materials with a varying number of shells (one to five) can be obtained. The synthesis of multishelled structures with a tunable number of layers was also proposed by Zhang et al.[88] The authors used a Zn-based zeolitic imidazolate (ZIF-8) template that was subsequently etched and sulfurized to form ZnS multi-shells. Thereafter, the as-prepared structures were converted into ZnS-CdS multishelled cages via a cation-exchange reaction. The authors reported that the number of shells can be adjusted by changing the size of the ZIF-8 templates.

Multicore@shell nanostructures (Fig. 14B) consist of the same/different core nanoparticles encapsulated in a single shell/hollow shell.[78,80,89] Due to the collective interaction between cores, this class of materials offers a possibility of tuning their optical, electrical, catalytic, and magnetic properties.[78] Such structures may be obtained using sol-gel, hydro/solvothermal, microemulsion, or one-pot reverse micelle methods.[78,83] Zhang et al.[89] synthesized multicore@shell clustered carbon dots (CDs) via a hydrothermal route using citric acid (CA) and 5-amino-1,10-phenanthroline (Aphen) as reactants. The authors speculated that during the synthesis, CA molecules formed multiple small (5 nm) CDs that subsequently reacted with Aphen, which was accompanied by amidation and carbonization processes. The structures obtained were analyzed for white light-emitting diodes (WLEDs). Bian et al.[90] in turn, studied multicore@shell structures composed of Ni—Mg phyllosilicate nanotubes (PSNTS) covered with silica as catalysts for dry reforming of methane. The materials were prepared via a two-step process. First, Ni—Mg PSNTS were obtained via the hydrothermal route. Second, the as-prepared nanotubes were covered with silica using a modified Stöber method. The report demonstrated that the applied multicore@shell structure effectively inhibited the sintering of metal particles and prevented carbon deposition on the catalysts' surface.

In the literature, there are also various reports on complex multicore@multishell structures (Fig. 14C), which combine the advantages of applying multiple cores and shells. Li et al.[91] examined the catalytic properties of Ni@Ni-phyllosilicate@SiO_2 hollow spheres with multiple Ni nanoparticles for dry reforming of methane. First, multi-Ni@Ni phyllosilicate hollow spheres were prepared via the hydrothermal route using nickel nitrate hexahydrate as Ni precursor and SiO_2 spheres as a template. Second, the prepared structures were covered with silica using a microemulsion method. The applied SiO_2 outer shell inhibited the detaching of Ni particles from the Ni-phyllosilicate surface and prevented the growth of carbon nanotubes. These effects resulted in much higher carbon resistance and enhanced long-term stability compared to the Ni@Ni-phyllosilicate catalysts. The structures with multilevel architecture for catalytic applications were also proposed by Hu et al.[84] The authors successfully synthesized uncommon rattle-type Sn@SiO_2@SiO_2 hollow nanostructures with multiple Au nanoparticles incorporated into the void space between the silica shells. First, SnO_2@SnO_2 double-shell hollow spheres were prepared via a two-step hydrothermal method using SiO_2 nanospheres as a template. Second, the prepared SnO_2@SnO_2 species were dispersed and mixed in solutions containing ammonia, ethanol, tetraethyl orthosilicate (TEOS), and *N*-[3-(trimethoxysilyl)propyl]ethylenediamine (TSD). The intermediate organosilica layer was removed via etching in hydrofluoric acid. Subsequently, the Au nanoparticles were introduced to the prepared nanorattles by aging in $HAuCl_4$ solution. Finally, the SnO_2 layer was converted to a Sn core by annealing under reducing conditions. The report suggests that in the multilevel architectures obtained, gold particles may be responsible for catalytic reactions, whereas the Sn core, isolated by the inner shell, may provide another functionality, such as magnetism.

Our research group proposed both the synthesis, and the mechanism of formation of multicore@shell SnO_2@SnS_2 heterostructures (Fig. 14D), composed of multiple single-core@shell individual particles forming raspberry-like hollow spheres.[85] The structures were obtained via chemical conversion of SnO_2 mesoporous nanocrystals, applying different concentrations of a sulfide precursor (thioacetamide, TAA). We found that the multicore@shell structure is determined by the SnO_2 defective surface and can be obtained in a certain range of TAA concentrations. With an increasing amount of the precursor, the bonds that connect smaller individual SnO_2@SnS_2 particles are being weakened, which results in complete structural disintegration in the case of the highest TAA concentration. Moreover, an increasing molar ratio of TAA to SnO_2 leads to additional redox reactions (S^{2-}/S^0, S^0/S^{6+}, Sn^{4+}/Sn^{2+}) occurring in the system. Due to the combination of applying the type-II heterojunction and mesoporous hollow-sphere morphology, the proposed SnO_2@SnS_2 multicore@shells may be promising candidates for visible-light-driven photocatalysts, adsorbents, and sensing materials.

2.3 Intimate platform (ITO/Me template)

Immobilized nanosized semiconductors offer greater application prospects than powdered ones due to the required complicated separation procedures for the latter.[92] However, the preparation of the appropriate coating remains a challenging step. The resulting interface between the photocatalytic coating and the substrate can very significantly affect the performance of photoactive electrodes. The most commonly applied deposition techniques can be divided into two major groups, namely, solution-based and gas-phase methods. The techniques of the first group are electrodeposition (ED), electrophoretic deposition (EPD), electrochemical anodization (EA), dip coating (DC), chemical bath deposition (CBD), spin coating (SC), doctor blade (DB), and spray pyrolysis (SP) (Fig. 15). They offer the possibility to deposit powdered materials or simultaneously synthesize and deposit materials from precursors at a relatively low cost. The other approach involves the transition of precursors or compounds into the gas phase and subsequent resublimation on the substrate. To this group belong different variations of physical vapor deposition (PVD), i.e., magnetron sputtering (MS) and chemical vapor deposition (CVD), i.e., atomic layer deposition (ALD) or aerosol-assisted chemical vapor deposition (AACVD). These techniques ensure high homogeneity and controlled morphology of the coatings; however, they require sophisticated equipment.

Solution-based techniques offer the possibility to design the interface between photoactive material and the substrate by appropriate adjustment of process parameters such as solvent, precursor type and concentration,

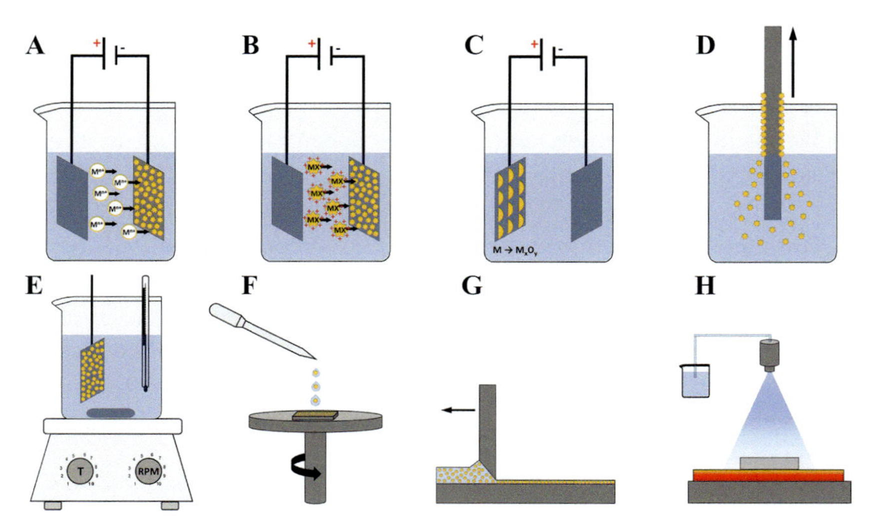

Fig. 15 Schematic illustrations of solution-based deposition methods: (A) electrodeposition, (B) electrophoretic deposition, (C) electrochemical anodization, (D) dip coating, (E) chemical bath deposition, (F) spin coating, (G) doctor blade, and (H) spray pyrolysis.

additives, solid load, the distance between electrodes, deposition time, and voltage, depending on the selected method. Our group investigated the influence of the solvent (water or ethanol) and concentration of precursor on coatings morphology, deposited using successive ionic layer adsorption and reaction (SILAR) method.[71] Coatings of CdS and PbS quantum dots (QDs) obtained on TiO_2 substrates from ethanol-based solutions were generally more uniform, whereas higher precursor concentrations led to particle agglomeration. The observed increase in photocurrent for CdS QDs was 200% and for PbS QDs—100% compared with unmodified TiO_2. In another report, we described the preparation of TiO_2@Cu_2O heterostructures by varying the deposition time of copper(I) oxide.[93] First, the titanium foil was electrochemically anodized in ethylene glycol, and ammonium fluoride (NH_4F) solution for 120 min at 30 V. A subsequent step involved electrodeposition of Cu_2O from copper sulfate, lactic acid, with sodium hydroxide to yield a solution (pH 12) in a three-electrode system with an applied voltage of −0.34 V at different times (30–180 s). It was demonstrated that a shorter deposition time is beneficial in terms of the resulting interface. The discontinuous layer enabled both absorptions in the UV (exposed TiO_2) and visible ranges (deposited Cu_2O). Enesca studied the effect of hydrophilic (HL) and hydrophobic (HB) additives on

SnO_2-TiO_2 and ZnO-TiO_2 coatings morphology obtained by the doctor blade method.[94] Deposited pastes were composed of ethanol, sodium maleat-methylmethacrylate (HL additive) or sodium maleat-vinyl acetate (HB additive), triton X100, and commercial powders of TiO_2 Degussa P25, and SnO_2 or ZnO. Higher photocatalytic activity towards acetaldehyde was observed for homogeneous layers obtained from the HL-based pastes. In another investigation, Falsetti et al. prepared different Bi_2O_3 layers on a glass substrate using a homemade spin coating device by varying the number of drops.[95] In the case of 5 and 10 deposited drops, the coatings were composed of Bi_2SiO_5 due to Si diffusion from the substrate. An increased number of drops to 20 promoted the formation of β-Bi_2O_3, and the crystallite sizes were larger. Moreover, this coating was thicker, characterized by higher wettability, and exhibited better photocatalytic performance toward Rhodamine B. It was explained by the presence of β-Bi_2O_3, which aided in separating photogenerated carriers, larger crystallites reducing the number of recombination sites, and better contact with water molecules due to slightly higher wettability. The influence of substrate temperature on the physical properties of Cu_2FeSnS_4 films deposited by spray pyrolysis, was investigated by Nefzi et al.[96] The polycrystalline and stoichiometric film was obtained at 240 °C, whereas lower (160 °C, 200 °C) and higher (280 °C) temperatures led to the deviation in atomic ratios and the presence of secondary phases as well as a decrease of thickness of coatings. It resulted in the lowest resistivity and the highest volume carrier concentration of the former, subsequently leading to the best photocatalytic performance towards methylene blue.

Gas-phase techniques offer the most precise control over the shape of the interface between the photocatalyst and the substrate on the atomic scale. Among PVD methods, special attention has been devoted to magnetron sputtering (MS) for photoactive electrode preparation. Our group has prepared TiO_{2-x} layers on indium tin oxide (ITO) glass by varying current titanium emission line intensity in the MS process.[97] For this purpose, optical emission spectroscopy (OES) was applied. Increased deviation from stoichiometry resulted in higher charge carrier concentration; however, this led to crystallographic disorder and reduced the photocurrent slightly. Another PVD method, vacuum thermal evaporation, and condensation has been utilized for Cu_2O/TiO_2 composite coatings fabrication by Morozov et al.[98] As a precursor, copper acetylacetonate was used, and the variable was the process temperature (80–200 °C). With increased temperature, the copper content in the sample also increased. However, the highest photocatalytic activity was observed for the relatively low Cu_2O concentration

(1.70 wt% at 160 °C). Above this optimal amount, the photocatalytic performance was lower. Similar observations were also reported in Ref. [93], as discussed in the previous paragraph (Fig. 16). It was explained by the shielding effect of copper(I) oxide, leading to the decreased light penetration depth compared to pure TiO_2. In the case of CVD techniques, the most commonly encountered methods for photocatalyst coatings preparation, are atomic layer deposition (ALD) and aerosol-assisted vapor deposition (AACVD). The effect of the number of deposition cycles on titanium dioxide nucleation in the ALD process was studied by Wang et al.[99] By increasing the ALD cycles, it was possible to ensure the crystallinity of TiO_2 films, which was beneficial in terms of improving the photocatalytic activity toward methyl orange degradation. In an article by Taylor et al., the issue of titanium dioxide precursor stabilization in solutions for the AACVD method has been raised.[100] For this purpose, various acetylacetone to titanium isopropoxide ratios have been used, and the physical and catalytic properties of the resulting coatings have been investigated. In the presence of the stabilizing agent, in addition to anatase, a rutile phase and carbon were also noted. Moreover, the coatings were composed of nanoparticles. It resulted in better photocatalytic performance. However, the highest activity was observed in the case of the optimal titanium to acetyl acetate ratio being equal to one. All the results mentioned above indicate that the adjustment of deposition parameters is a tool for designing the substrate/photocatalyst interface and thus tuning the photocatalytic performance.

3. Conclusions and future perspectives

Designing the surface structure of materials is an important factor for the formation of the synergetic effect, consisting of tuning light-harvesting, control of electron carriers transfer, and engineering of active sites. Especially in the case of the active sites, materials engineering has shown an important contribution to improving the photocatalytic activity. Due to the alteration of the surface adsorption and activation of specific reactants, it may be possible to control the location of chemical reactions.

So far, design strategies are based on the control of dimension, composition, phases, electronic coupling, and geometrical changes. Although significant achievements have been made in this field, the challenge remains; the full exertion of the adsorption process, and therefore, photocatalytic activity.

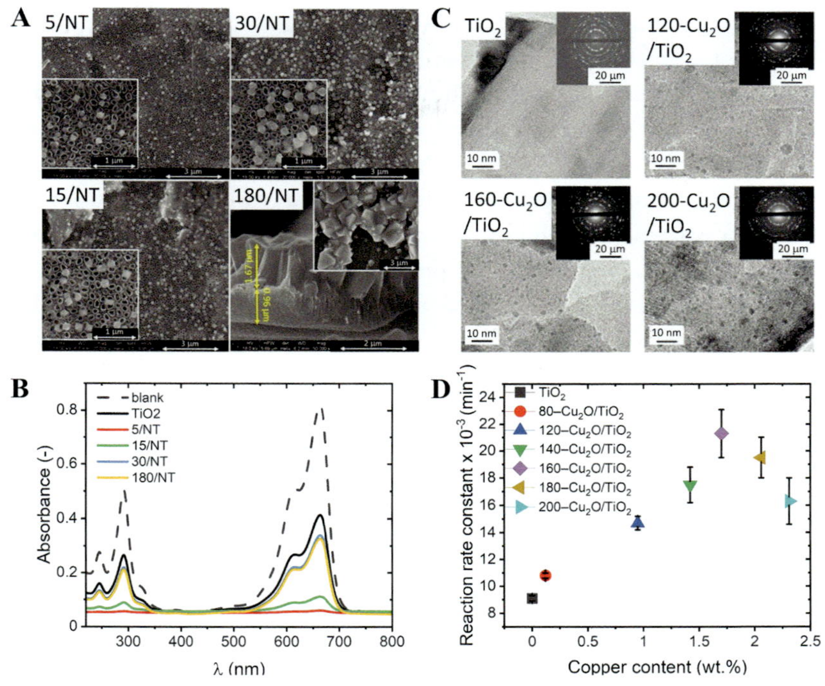

Fig. 16 (A) SEM images of TiO_2-NT@Cu_2O heterostructures obtained by the electrodeposition method at different times, (B) comparison of removal efficiencies for methylene blue (MB) for different samples,[93] (C) TEM images of Cu_2O/TiO_2 heterostructures obtained by the PVD method at different temperatures, and (D) comparison of reaction rate constants towards the degradation of phenol for samples with a different copper content. *Panel (C) reprinted by permission from Springer Nature, Morozov, A. N., Phyo, T. Z., Denisenko, A. V., Kryukov, A. Y. Pet. Chem.* ***2021****,* 61*, 951–958. Copyright (2021).*

Generally, it is assumed that the surface of the semiconductor is not a highly active site responsible for the reduction and oxidation reactions simultaneously. Nevertheless, due to the presence of the coordination-unsaturated atoms at the surface of the material, the possibility of the thermodynamic tendency to bind various molecules increases. Therefore, the surface structure determines its photocatalytic performance. Polycrystalline materials may expose different facets, edges, and defects, affecting various active sites at the particle surface. As a result, the adsorption mechanism is a sum of the individual performance of all the active sites. This kind of complexity and inadequate uniformity of the surface makes it very challenging to identify the structure and density of different types of active centers.

Although it may seem that almost everything has been accomplished in the field of surface design, there are still a few questions that need to be answered. Particular attention should be paid to selecting appropriate parameters of the synthetic process and the possibility of *in situ* control of structure growth.

Recently, it has been possible to observe a veritable revolution of a spectacular variety of building blocks of different shapes, compositions, and functionalities. However, the precise control of the surface parameters is still limited. Moreover, the real mechanism of the adsorption and the photocatalytic activity of the materials behind the structural properties remain elusive. For this reason, it is important that the process of surface design, and thus the creation of active site densities, should be based on three pillars: controlled synthesis, *in situ* characterization, and supported by theoretical calculations. Only then will it be possible to describe the relationship between surface parameters and adsorption capacity, and the photocatalytic activity will be elucidated more accurately.

This review has addressed the selected surface modifications of semiconductor materials to enhance their photocatalytic and photoelectrochemical efficiencies. The following issues were taken into consideration: the process of adsorption, with particular emphasis on the role of active sites, and the development of active contact centers by the architecture of the formation of both homo- and heterostructures. Several synthetic techniques were mentioned. Representative results of the authors' studies and those from the literature were cited, and examples devoted to titanium dioxide, iron oxides, copper, and tin compounds in the form of crystal-like, nanotubes, hierarchically porous materials, as well as their core-shell heterostructures, were presented. The role of the substrate in photoelectron-catalytic processes was highlighted.

Acknowledgments

Work for the purpose of this review was performed within the framework of the project no. 2016/21/B/ST8/00457 National Science Center, Poland. K.M. acknowledges The Ministry of Education and Science in Poland (project number DI2018 008148).

References

1. Robalds, A.; Naja, G. M.; Klavins, M. *J. Hazard. Mater.* **2016**, *304*, 553–556.
2. Henning, K. D.; Degel, J. *Activated carbon for solvent recovery*. http://www.activatedcarbon.com/solrec2.html.
3. Weber, W., Jr. *J. Environ. Eng.* **1985**, *110*, 899–917.
4. Wang, L.; Shi, C.; Pan, L.; Zhang, X.; Zou, J. J. *Nanoscale* **2020**, *12*, 4790–4815.
5. Tran, H. N.; You, S. J.; Hosseini-Bandegharaei, A.; Chao, H. P. *Water Res.* **2017**, *120*, 88–116.

6. Abebe, B.; Murthy, H. C. A.; Amare, E. *J. Encapsulation Adsorpt. Sci.* **2018**, *08*, 225–255.
7. Lagergren, S. *K. Sven. Vetenskapsakademiens Handl.* **1898**, *24*, 1–39.
8. Doğan, M.; Özdemir, Y.; Alkan, M. *Dyes Pigm.* **2007**, *75*, 701–713.
9. Debnath, S.; Ghosh, U. C. *Chem. Eng. J.* **2009**, *152*, 480–491.
10. Huang, J.; Cao, Y.; Liu, Z.; Deng, Z.; Wang, W. *Chem. Eng. J.* **2012**, *191*, 38–44.
11. Wang, S.; Li, H. *Dyes Pigm.* **2007**, *72*, 308–314.
12. Mittal, A.; Krishnan, L.; Gupta, V. K. *Sep. Purif. Technol.* **2005**, *43*, 125–133.
13. Srivastava, V. C.; Swamy, M. M.; Mall, I. D.; Prasad, B.; Mishra, I. M. *Colloids Surf. A Physicochem. Eng. Asp.* **2006**, *272*, 89–104.
14. Liu, S.-S.; Lee, C.-K.; Chen, H.-C.; Wang, C.-C.; Juang, L.-C. *Chem. Eng. J.* **2009**, *147*, 188–193.
15. Patravale, V.; Dandekar, P.; Jain, R. Nanoparticulate. *Drug Deliv.* **2012**, 87–121.
16. Nam, W.; Kim, J.; Han, G. *Chemosphere* **2002**, *47*, 1019–1024.
17. Salehi, M.; Hashemipour, H.; Mirzaee, M. *Am. J. Environ. Eng.* **2012**, *2*, 1–7.
18. Rauf, M. A.; Meetani, M. A.; Hisaindee, S. *Desalination* **2011**, *276*, 13–27.
19. Yagub, M. T.; Sen, T. K.; Afroze, S.; Ang, H. M. *Adv. Colloid Interface Sci.* **2014**, *209*, 172–184.
20. Azeez, F.; Al-Hetlani, E.; Arafa, M.; Abdelmonem, Y.; Nazeer, A. A.; Amin, M. O.; Madkour, M. *Sci. Rep.* **2018**, *8*, 1–10.
21. Xiong, L. B.; Li, J. L.; Yang, B.; Yu, Y. *J. Nanomater.* **2012**, *2012*, 831524.
22. Hartmann, M.; Schwieger, W. *Chem. Soc. Rev.* **2016**, *45*, 3311–3312.
23. Zhu, Y. P.; Liu, Y. L.; Ren, T. Z.; Yuan, Z. Y. *Nanoscale* **2014**, *6*, 6627–6636.
24. Yu, J.; Dai, G.; Xiang, Q.; Jaroniec, M. *J. Mater. Chem.* **2011**, *21*, 1049–1057.
25. Bai, S.; Wang, L.; Li, Z.; Xiong, Y. *Adv. Sci. News* **2017**, *4*, 160021.
26. Kang, D.; Kim, T. W.; Kubota, S. R.; Cardiel, A. C.; Cha, H. G.; Choi, K. S. *Chem. Rev.* **2015**, *115*, 12839–12887.
27. Zeng, W.; Liu, T.; Gou, Z.; Lin, L. *Phys. E Low-Dimensional Syst. Nanostructures* **2012**, *44*, 1567–1571.
28. Liu, C.; Lu, H.; Zhang, J.; Gao, J.; Zhu, G.; Yang, Z.; Yin, F.; Wang, C. Sensors Actuators. *B Chem.* **2018**, *263*, 557–567.
29. Murakami, N.; Kurihara, Y.; Tsubota, T.; Ohno, T. *J. Phys. Chem. C* **2009**, *113*, 3062–3069.
30. Kusior, A.; Synowiec, M.; Zakrzewska, K.; Radecka, M. *Crystals* **2019**, *9*, 163.
31. Lazzeri, M.; Vittadini, A.; Selloni, A. *Phys. Rev. B - Condens. Matter Mater. Phys.* **2001**, *63*, 1554091–1554099.
32. Maisano, M.; Dozzi, M. V.; Selli, E. *J. Photochem. Photobiol. C Photochem. Rev.* **2016**, *28*, 29–43.
33. Liu, C.; Tong, R.; Xu, Z.; Kuang, Q.; Xie, Z.; Zheng, L. *RSC Adv.* **2016**, *6*, 29794–29801.
34. Ong, W. J.; Tan, L. L.; Chai, S. P.; Yong, S. T.; Mohamed, A. R. *Nanoscale* **2014**, *6*, 1946–2008.
35. Roy, N.; Park, Y.; Sohn, Y.; Leung, K. T.; Pradhan, D. *ACS Appl. Mater. Interfaces* **2014**, *6*, 16498–16507.
36. Li, J.; Yu, Y.; Chen, Q.; Li, J.; Xu, D. *Cryst. Growth Des.* **2010**, *10*, 2111–2115.
37. Li, J.; Cao, K.; Li, Q.; Xu, D. *CrstEngComm* **2012**, *14*, 83–85.
38. Zhou, Y.; Ding, E. Y.; Li, W. D. *Mater. Lett.* **2007**, *61*, 5050–5052.
39. Zhao, X.; Jin, W.; Cai, J.; Ye, J.; Li, Z.; Ma, Y.; Xie, J.; Qi, L. *Adv. Funct. Mater.* **2011**, *21*, 3554–3563.
40. Ye, L.; Liu, J.; Tian, L.; Peng, T.; Zan, L. *Appl. Catal. Environ.* **2013**, *134–135*, 60–65.
41. Liu, L.; Jiang, Y.; Zhao, H.; Chen, J.; Cheng, J.; Yang, K.; Li, Y. *ACS Catal.* **2016**, *6*, 1097–1108.

42. Yu, J.; Qi, L.; Jaroniec, M. *J. Phys. Chem. C* **2010**, *114*, 13118–13125.
43. Synowiec, M.; Micek-Ilnicka, A.; Szczepanowicz, K.; Różycka, A.; Trenczek-Zajac, A.; Zakrzewska, K.; Radecka, M. *Appl. Surf. Sci.* **2019**, *473*, 603–613.
44. Wu, W.; Yang, S.; Pan, J.; Sun, L.; Zhou, J.; Dai, Z.; Xiao, X.; Zhang, H.; Jiang, C. *CrstEngComm* **2014**, *16*, 5566–5572.
45. Gao, F.; Liu, R.; Yin, J.; Lu, Q. *Sci. China Chem.* **2014**, *57*, 114–121.
46. Yang, S.; Zhou, B.; Ding, Z.; Zheng, H.; Huang, L.; Pan, J.; Wu, W.; Zhang, H. *J. Power Sources* **2015**, *286*, 124–129.
47. Liu, J.; Yang, S.; Wu, W.; Tian, Q.; Cui, S.; Dai, Z.; Ren, F.; Xiao, X.; Jiang, C. *ACS Sustain. Chem. Eng.* **2015**, *3*, 2975–2984.
48. Liang, H.; Jiang, X.; Qi, Z.; Chen, W.; Wu, Z.; Xu, B.; Wang, Z.; Mi, J.; Li, Q. *Nanoscale* **2014**, *6*, 7199–7203.
49. Kusior, A.; Michalec, K.; Jelen, P.; Radecka, M. *Appl. Surf. Sci.* **2019**, *476*, 342–352.
50. Zhong, Y.; Li, Y.; Li, S.; Feng, S.; Zhang, Y. *RSC Adv.* **2014**, *4*, 40638–40642.
51. Gao, H.; Zhang, J.; Li, M.; Liu, K.; Guo, D.; Zhang, Y. *Curr. Appl. Phys.* **2013**, *13*, 935–939.
52. Pang, H.; Gao, F.; Lu, Q. *Chem. Commun.* **2009**, *9*, 1076–1078.
53. Sun, S.; Kong, C.; You, H.; Song, X.; Ding, B.; Yang, Z. *CrstEngComm* **2012**, *14*, 40–43.
54. Kim, M. C.; Kim, S. J.; Han, S. B.; Kwak, D. H.; Hwang, E. T.; Kim, D. M.; Lee, G. H.; Choe, H. S.; Park, K. W. *J. Mater. Chem. A* **2015**, *3*, 23003–23010.
55. Shang, Y.; Sun, D.; Shao, Y.; Zhang, D.; Guo, L.; Yang, S. *Chem. - A Eur. J.* **2012**, *18*, 14261–14266.
56. Yang, X. Y.; Chen, L. H.; Li, Y.; Rooke, J. C.; Sanchez, C.; Su, B. L. *Chem. Soc. Rev.* **2017**, *46*, 481–558.
57. Zhang, W.; Tian, Y.; He, H.; Xu, L.; Li, W.; Zhao, D. *Natl. Sci. Rev.* **2020**, *7*, 1702–1725.
58. Mamaghani, A. H.; Haghighat, F.; Lee, C. S. *Appl. Catal. Environ.* **2020**, *269*, 118735.
59. Li, J.; Shi, J.; Li, Y.; Ding, Z.; Huang, J. *Ceram. Int.* **2021**, *47*, 8218–8227.
60. Khan, H.; Samanta, S.; Seth, M.; Jana, S. *J. Mater. Sci.* **2020**, *55*, 11907–11918.
61. Bhushan, M.; Jha, R. *Appl. Surf. Sci.* **2020**, *528*, 146988.
62. Kusior, A.; Jelen, P.; Mazurkow, J.; Nieroda, P.; Radecka, M. *J. Therm. Anal. Calorim.* **2019**, *138*, 4321–4329.
63. Li, G.; Sun, Y.; Sun, S.; Chen, W.; Zheng, J.; Chen, F.; Sun, Z.; Sun, W. *Adv. Powder Technol.* **2020**, *31*, 2505–2512.
64. Xiang, Q.; Cheng, B.; Yu, J. *Appl. Catal. Environ.* **2013**, *138–139*, 299–303.
65. Roy, P.; Berger, S.; Schmuki, P. *Angew. Chem. Int. Ed.* **2011**, *50*, 2904–2939.
66. Lee, K.; Mazare, A.; Schmuki, P. *Chem. Rev.* **2014**, *114*, 9385–9454.
67. Regonini, D.; Bowen, C. R.; Jaroenworaluck, A.; Stevens, R. *Mater. Sci. Eng. R Reports* **2013**, *74*, 377–406.
68. Abbas, W. A.; Abdullah, I. H.; Ali, B. A.; Ahmed, N.; Mohamed, A. M.; Rezk, M. Y.; Ismail, N.; Mohamed, M. A.; Allam, N. K. *Nanoscale Adv.* **2019**, *1*, 2801–2816.
69. Yoriya, S.; Bao, N.; Grimes, C. A. *J. Mater. Chem.* **2011**, *21*, 13909.
70. Amir, S. M. M.; Sultan, M. T. H.; Jawaid, M.; Ariffin, A. H.; Mohd, S.; Salleh, K. A. M.; Ishak, M. R.; Md Shah, A. U.; Durab. Life Predict. *Biocomposites, Fibre-Reinforced Compos. Hybrid Compos.* **2018**, 367–388.
71. Trenczek-Zajac, A.; Kusior, A.; Lacz, A.; Radecka, M.; Zakrzewska, K. *Mater. Res. Bull.* **2014**, *60*, 28–37.
72. Trenczek-Zajac, A.; Kusior, A.; Radecka, M. *Int. J. Hydrogen Energy* **2016**, *41*, 7548–7562.
73. Liu, C.; Li, Y.; Wei, L.; Wu, C.; Chen, Y.; Mei, L.; Jiao, J. *Nanoscale Res. Lett.* **2014**, *9*, 1–8.

74. Chaguetmi, S.; Mammeri, F.; Pasut, M.; Nowak, S.; Lecoq, H.; Decorse, P.; Costentin, C.; Achour, S.; Ammar, S. J. *Nanoparticle Res.* **2013**, *15*, 1–10.
75. Li, J.; Lin, C. J.; Lai, Y. K.; Du, R. G. *Surf. Coat. Technol.* **2010**, *205*, 557–564.
76. Zakrzewska, K. *Thin Solid Films* **2001**, *391*, 229–238.
77. Kusior, A.; Zych, L.; Zakrzewska, K.; Radecka, M. *Appl. Surf. Sci.* **2019**, *471*, 973–985.
78. Purbia, R.; Paria, S. *Nanoscale* **2015**, *7*, 19789–19873.
79. Ghosh Chaudhuri, R.; Paria, S. *Chem. Rev.* **2012**, *112*, 2373–2433.
80. Singh, R.; Bhateria, R. *Environ. Geochem. Health* **2021**, *43*, 2459–2482.
81. Srdic, V.; Mojic, B.; Nikolic, M.; Ognjanovic, S. *Process. Appl. Ceram.* **2013**, *7*, 45–62.
82. Xu, H.; Wang, W. *Angw. Chem. Int. Ed.* **2007**, *46*, 1489–1492.
83. Das, S.; Pérez-Ramírez, J.; Gong, J.; Dewangan, N.; Hidajat, K.; Gates, B. C.; Kawi, S. *Chem. Soc. Rev.* **2020**, *49*, 2937–3004.
84. Hu, Y.; Zheng, X. T.; Chen, J. S.; Zhou, M.; Li, C. M.; Lou, X. W. *J. Mater. Chem.* **2011**, *21*, 8052–8056.
85. Michalec, K.; Kusior, A.; Mikuła, A.; Radecka, M. *Mater. Res. Lett.* **2021**, *9*, 445–451.
86. Gawande, M. B.; Goswami, A.; Asefa, T.; Guo, H.; Biradar, A. V.; Peng, D. L.; Zboril, R.; Varma, R. S. *Chem. Soc. Rev.* **2015**, *44*, 7540–7590.
87. Dong, Z.; Ren, H.; Hessel, C. M.; Wang, J.; Yu, R.; Jin, Q.; Yang, M.; Hu, Z.; Chen, Y.; Tang, Z.; Zhao, H.; Wang, D. *Adv. Mater.* **2014**, *26*, 905–909.
88. Zhang, P.; Guan, B. Y.; Yu, L.; Lou, X. W. *Chem* **2018**, *4*, 162–173.
89. Zhang, T.; Zhao, F.; Li, L.; Qi, B.; Zhu, D.; Lü, J.; Lü, C. *ACS Appl. Mater. Interfaces* **2018**, *10*, 19796–19805.
90. Bian, Z.; Suryawinata, I. Y.; Kawi, S. *Appl. Catal. Environ.* **2016**, *195*, 1–8.
91. Li, Z.; Jiang, B.; Wang, Z.; Kawi, S. *J. CO_2 Util.* **2018**, *27*, 238–246.
92. Obregón, S.; Amor, G.; Vázquez, A. *Adv. Colloid Interface Sci.* **2019**, *269*, 236–255.
93. Trenczek-Zajac, A.; Banas-gac, J.; Radecka, M. *Materials (Basel).* **2021**, *14*, 1–16.
94. Enesca, A. *Front. Chem.* **2020**, *8*, 1–8.
95. Falsetti, P. H. E.; Soares, F. C.; Rodrigues, G. N.; Del Duque, D. M. S.; de Oliveira, W. R.; Gianelli, B. F.; de Mendonça, V. R. *Mater. Today Commun.* **2021**, *27*, 102214.
96. Nefzi, C.; Souli, M.; Cuminal, Y.; Kamoun-Turki, N. *Mater. Sci. Eng. B Solid-State Mater. Adv. Technol.* **2020**, *254*, 114509.
97. Radecka, M.; Brudnik, A.; Kulinowski, K.; Kot, A.; Leszczyński, J.; Kanak, J.; Zakrzewska, K. *J. Electron. Mater.* **2019**, *48*, 5481–5490.
98. Morozov, A. N.; Phyo, T. Z.; Denisenko, A. V.; Kryukov, A. Y. *Pet. Chem.* **2021**, *61*, 951–958.
99. Wang, J.; Yin, Z.; Hermerschmidt, F.; List-Kratochvil, E. J. W.; Pinna, N. *Adv. Mater. Interfaces* **2021**, *8*, 2100759.
100. Taylor, M.; Pullar, R. C.; Parkin, I. P.; Piccirillo, C. *J. Photochem. Photobiol., A* **2020**, *400*, 112727.

CHAPTER ELEVEN

Interface design, surface-related properties, and their role in interfacial electron transfer. Part II: Photochemistry-related topics

Anita Trenczek-Zajac*, Anna Kusior*, Julia Mazurków, Kinga Michalec, Milena Synowiec, and Marta Radecka
AGH University of Science and Technology, Faculty of Materials Science and Ceramics, Krakow, Poland
*Corresponding authors: e-mail address: anita.trenczek-zajac@agh.edu.pl; akusior@agh.edu.pl

Contents

Abstract

The photoactivity of heterostructures is determined by the morphology and physicochemical properties of the constituents of the surface. Therefore, it is essential to consider their architecture in the interfacial design process. The properties of the heterostructure component determine both the photoactivity and charge carrier transfer. This review presents the results of our work on heterostructured systems such as TiO_2@SnO_2 nano-powders, TiO_2@Fe_2O_3 nano-crystals, TiO_2@Cu_2O nano-materials, as well as multiple-core@shell heterostructures and substrates for photoelectrodes in comparison with the results of other authors.

Advances in Inorganic Chemistry, Volume 79
ISSN 0898-8838
https://doi.org/10.1016/bs.adioch.2021.12.010

1. Surface design of sensitizer

1.1 Projecting photoactive materials

The fundamentals of heterogeneous photocatalysis in the initial stages, absorption of the energy quantum, and generation of the electron-hole pair, are the same regardless of other chemical reaction pathways.[1] If the charge carrier pair does not undergo an unfavorable recombination reaction, the further fate of the carrier pairs depends on the type of photochemical process in which they participate (Fig. 1):

- Photocatalytic decomposition of water or impurities in the liquid phase—redox reactions with particles adsorbed on the semiconductor surface: the electron hole participates in the oxidation reaction, and the electron participates in the reduction reaction.
- Photoelectrochemical decomposition of water or photoelectrocatalytic decomposition of contaminants in the liquid phase; an external potential difference applied to the cell (external support) causes forced separation of the pair of charge carriers, thus limiting their recombination. The

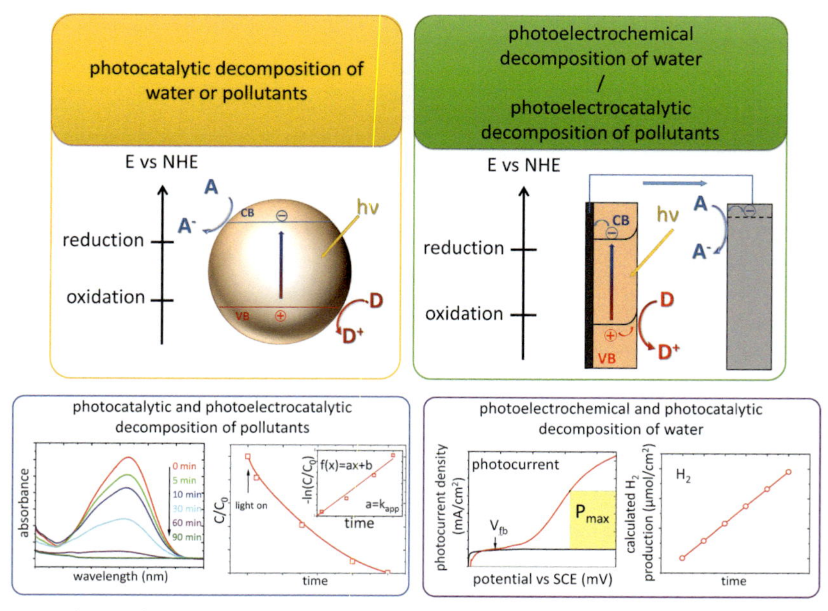

Fig. 1 Scheme for the principle of operation: photocatalytic decomposition of water or pollutants (left-hand side), photoelectrochemical decomposition of water or photoelectrocatalytic decomposition of pollutants (right-hand side), and data received to be analyzed.

electron is transported by an external circuit from the photoelectrode to the counter-electrode. At the photoelectrode, there is an oxidation reaction, and at the counter-electrode, there is a reduction reaction.

In both cases, the condition of suitable positions of redox levels of water or contaminants (E_{redox}) in relation to the edges of the valence band (E_{VB}) and the conduction band (E_{CB}) must be fulfilled: $E_{CB} > E_{redox} > E_{VB}$.[2]

The type of data obtained for analysis depends on the type of photochemical process and the products of the photochemical reactions (Fig. 1). When an inorganic pollutant is decomposed, either photocatalytically or photoelectrocatalytically, control of the progress of the degradation reaction is performed by measuring the absorbance, the change of which results directly from a change in the concentration of the pollutant. Depending on the kinetics of the degradation reaction, it is possible to determine a pseudo-first or pseudo-second order kinetic model.[3–5] During the photoelectrochemical hydrogen production process, current-voltage (I-V) characteristics can be analyzed to calculate the H_2 production rate.[6] On the other hand, the photocatalytic hydrogen production process allows for direct measurement of the H_2 production rate.

Semiconductors for photochemical processes must meet some specific requirements. In addition to the aforementioned, appropriate alignment of E_{CB} and E_{VB} with respect to the position of E_{redox} of the water or contaminant, the most crucial aspects include: absorption of light from a wide spectral range, high mobility and long diffusion path of charge carriers, and stability under photochemical process conditions.[7] There are also criteria determined by the presence or absence of an external potential difference during the process. In the photocatalytic process, which proceeds without external support, nanomaterials in the form of powders (substrate-free) with different morphologies and shapes are applicable, such as 0D—nanoparticles,[8,9] 1D—nanowires, nanorods, nanobelts, nanofibers, and nanotubes,[10–17] 2D—nanosheets, nanoflakes,[9,17–19] 3D—spheres, shells, nanoflowers, and hierarchical superstructures,[17,20,21] and nanocrystals.[9,11,12,22–25] The photocatalyst in a photoelectrochemical and photoelectrocatalytic process must provide charge transport via an external circuit. Forced electron flow by an external voltage from the anode to the cathode, therefore requires the presence of a substrate in the photoanode. However, the group of nanomaterials capable of acting as photoanodes is characterized by an equally wide variety of morphologies: 0D—nanoparticles,[26] 1D—nanowires, nanorods, nanobelts, nanofibers and nanotubes,[17,25,27] 2D—nanosheets, nanoflakes,[17] 3D—porous spheres, shells, nanoflowers, and hierarchical superstructures.[17,20,28–30]

1.2 Interfacial location

Developing new solutions based on known materials with different production techniques and working principles, can improve chemical stability and catalytic activity, thus improving all aspects related to the "golden triangle" issues related to efficiency, cost, and stability. A widely used approach to promote photocatalyst performance, is to combine the semiconductor with a sensitizer or co-catalyst to form a hybrid structure. While the sensitizer harvests light, the co-catalyst is not the light-harvesting component for generating photoinduced charge carriers. It is responsible for steering the charging kinetics by trapping them (the charge carriers) to promote electron-hole separation and to serve as a highly reactive active site.

A promising approach to resolving this contradiction is to integrate the semiconductor with the light sensitizer or co-catalyst,[31,32] which can work as a supporting structure. In the first case, light is harvested by the sensitizer instead of the semiconductor. The photogenerated charge carriers are then injected into the conduction or valence bands for reduction or oxidation reactions. Another approach to promote the performance of the photocatalyst, involves the design of the co-catalyst hybrid structure. In such a form, the co-catalyst is responsible for (i) trapping charge carriers to promote the separation of the charge carriers and (ii) serving as a highly active reaction site to supply the trapped charges for the redox reaction on their surface.[33] In contrast to the sensitizer, the co-catalyst as an alternative reaction site may suppress the photocorrosion of semiconductors resulting from electron or hole accumulation and thus increase the stability of the photocatalysts. Inefficient interfacial charge transfer or a slow surface reaction may bring about a limited number of charge carriers or lead to their accumulation on the co-catalyst side, which affects the reduction of potential differences. Therefore, designing the sensitizer/co-catalyst surface and interface, represents a more straightforward strategy. In the case of a bare semiconductor, which acts as a light-harvesting center and surface reaction site simultaneously, the tailored surface would affect the light absorption and tune surface reactions, making it challenging to assess the contribution of the surface, designed for photocatalytic performance.

Therefore, the following issues must be outlined.

- Architectural structures—The simplest structure is based on reduction/oxidation co-catalysts loaded alone on a light-harvesting semiconductor.[34] In other cases, both types of co-catalyst can be placed in the same photocatalysts,[35] or alternatively, connection with plasmonic metals/sensitizers can inject charge carriers into the semiconductor.[36,37]

It is worth noting that direct contact is not necessary between the co-catalyst and the semiconductor. A conductive component can also serve as the charging bridge (tandem electron transfer);

- The surface design of sensitizer/co-catalysts, including surface composition, facets, phase, and defects, to realize the high absorption, adsorption, or activation ability for specific reactant molecules tailoring the surface of the sensitizer/co-catalyst, can be recognized as of great importance for enhancing the charge transfer. Moreover, by controlled chemical conversion of the bare semiconductor, surface specificity (trapping sites) may affect the accumulation of electrons or holes on the reactive surface for redox reactions. As the adsorption and activation behavior is key to enhanced performance, the selection of the sensitizer/co-catalysts depends on the type of reaction.[38] In the case of photocatalytic water splitting, a new approach is the formation of the core-shell co-catalyst, as well as the incorporation of other elements.
- Interface design, with an emphasis on interfacial composition, location and facets.

  So far, the interface design has been claimed as the adjustment parameter and optimization of the contact between various components. However, due to the controlled synthesis and deposition of the co-catalyst, we can influence:
- The efficient charge transfer, for example, a similar composition, MoS_2-CdS of the two sides of the interface favors the covalent junction with low defect density[39];

  Separation of charge carriers—e.g., by deposition of the reduction and oxidation co-catalyst at the surface, this anisotropic platform facilitated the migration of charge carriers—electrons and holes are differently transferred; moreover, the back reaction cannot occur.[23,40];
- Stability of the photocatalyst.[6,30,41]

  It is believed that the interfacial electron-hole transfer and surface activation reaction will improve charge trapping and thus enhance photocatalytic performance.

1.3 Formation of heterostructures

The mechanism of charge carrier transfer in a heterostructure that is a combination of two semiconductors (1 & 2), depends on their electronic structure and the position of E_{VB}s and E_{CB}s relative to each other. According to Marschall[42] and Yu et al.,[43] there are four types of heterostructures (Fig. 2)

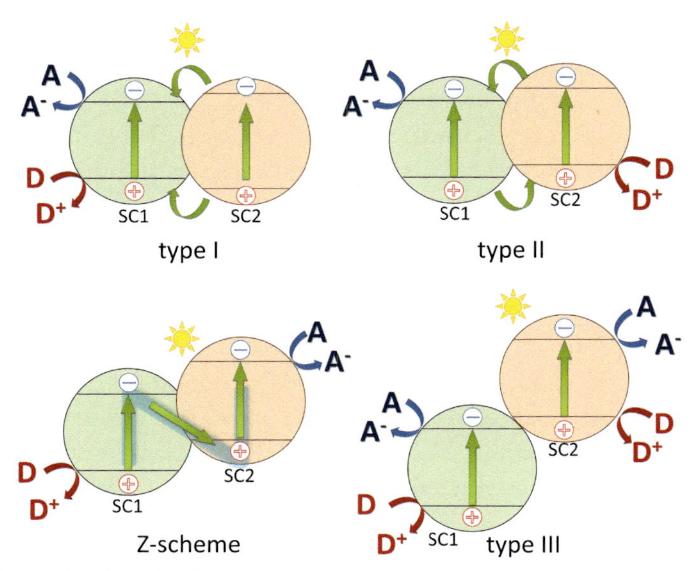

Fig. 2 Types of heterostructures.

that differ in the relative position of the edges of the conduction band (E_{CB}) and the valence band (E_{VB}), charge carrier transfer, and location of redox reactions:

- Type I—$E_{VB(2)}$ below $E_{VB(1)}$ and $E_{CB(2)}$ above $E_{CB(1)}$. Photoexcited holes and electrons from semiconductor 2 (SC2) are transported to semiconductor 1 (SC1) and redox reactions with adsorbed particles take place on the surface of SC1;
- Type II—$E_{VB(2)}$ above $E_{VB(1)}$ and $E_{CB(2)}$ above $E_{CB(1)}$. Photoexcited holes are transported from SC1 and transported to SC2, and electrons are transported in the opposite direction. Redox reactions with adsorbed particles are separated; oxidation occurs on the surface of SC2 and reduction—SC1;
- *Z*-scheme—$E_{VB(2)}$ above $E_{VB(1)}$ and $E_{CB(2)}$ above $E_{CB(1)}$. Electrons are transported from CB1 to VB2 and undergo recombination. Electrons from CB2 with a higher negative potential and holes from VB1 with a higher positive potential participate in the oxidation reaction in SC2 and SC1, respectively;
- Type III—$E_{VB(2)}$ and $E_{CB(2)}$ above $E_{CB(1)}$. There is no charge-carrier transfer between the semiconductors. Redox reactions with adsorbed particles take place on the surface of both semiconductors.

2. Photoactive structures

2.1 TiO_2@SnO_2 nanopowders

Studies on the combination of these two oxides have shown that even a small addition of TiO_2 to tin dioxide (TiO_2@SnO_2 heterostructure where a small amount of SnO_2 is deposited on the surface of TiO_2) increases the number of adsorption sites for chemisorption and that both electrical and photocatalytic properties are not degraded. However, depending on the assumed grain shape (Fig. 3A), the influence of structural complexity should be taken into account.[23]

In the most straightforward system, where both materials that make up the system are approximately spherical (0D-0D), there are a certain number of active sites, and more importantly a certain number of contacts between the materials that ensure charge transfer. The more complicated the system, the higher the 0D-3D or 3D-3D systems, then the number of active sites is higher and the number of possible sorption processes is almost unlimited. However, such arrangements are difficult to define, and the mechanisms involved are difficult to analyze in detail.

One of the possible ways to describe the influence of the applied shape of the semiconductor and sensitizer in the form of a composite, is the calculation of the intra-particle diffusion coefficient. From the multi-linearity (two or more steps), the number of processes that affect photocatalytic activity can be assumed.

Core-shell circuits have been developed to provide a specific charge flow that can be controlled. Core-shell structures are nanoparticles (core) enclosed by an outer envelope (shell),[44] consisting of at least two different materials. The first information on this type of form, concerns the noble metal@metal oxide system. The choice of Au or Ag, for example, was due to the strong optical signal resulting from the surface plasmon resonance.[45] The complexity of the system was analyzed by the pioneers in this field, Liz-Marzana et al.[46] The aim of that work was to change the properties of colloidal gold. The vast majority of colloidal gold is of spherical form, but depending on the application possibilities, the structure may be different.[23,47] In the core/shell system, the interior can be uniform or porous (Fig. 3B). When the interior is removed by the calcination or dissolution process, the properties and structures change. A form completely devoid of the core structure, is called a hollow sphere, while an intermediate structure, is called a yolk-shell. The situation is similar when considering the outer layer. It can be continuous, delaminated, or in the form of embedded nanoparticles and form blocks or mixed systems.

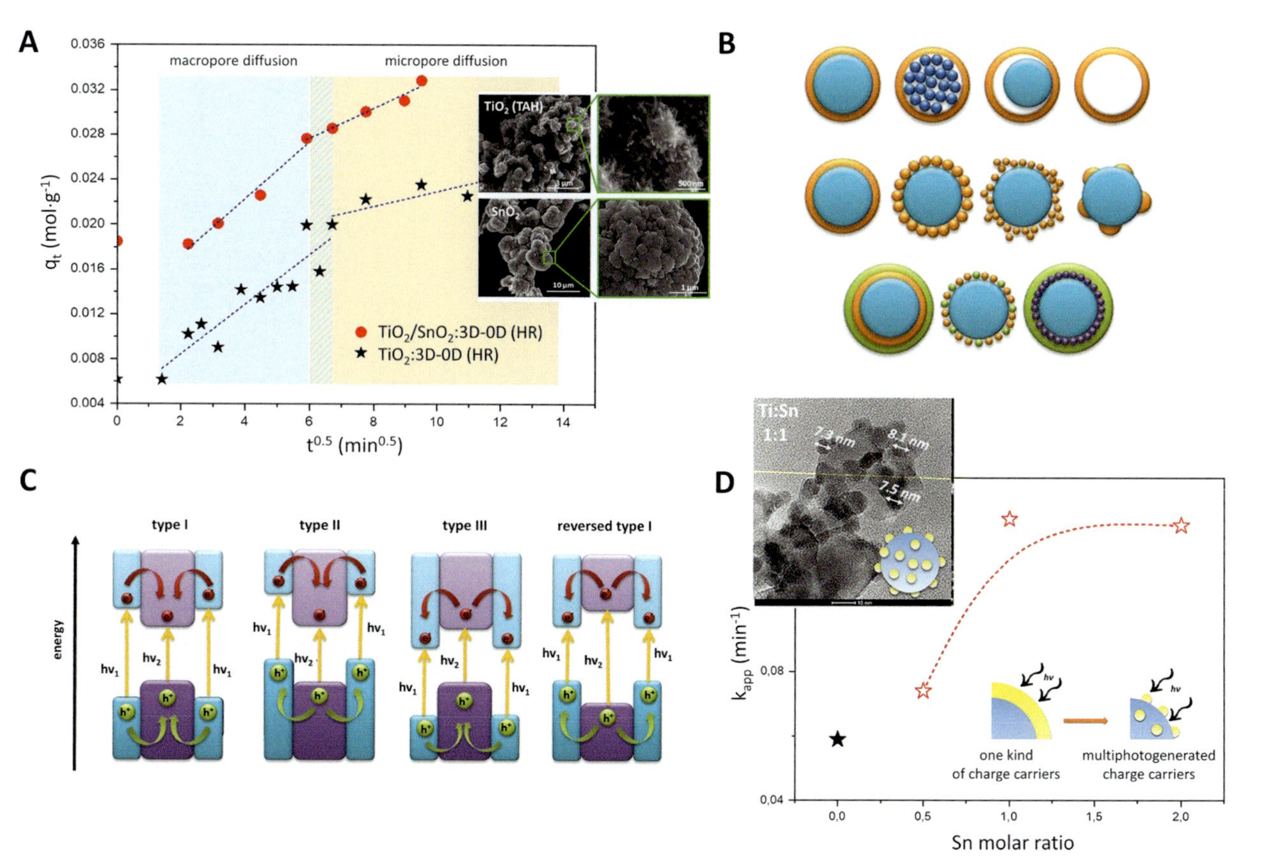

Fig. 3 (A) The inter-particle diffusion rate of 3D-3D nanostructured materials of TiO_2 and TiO_2/SnO_2 synthesized by hydrothermal reactions, HR. (B) Schematic representation of core-shell forms with (C) possible charge transfer pathways and (D) calculated pseudo-first-rate constant of the reaction k_{app} as a function of the Sn molar ratio in the case of core-shell materials. *Reprinted from Kusior, A.; Zych, L.; Zakrzewska, K.; Radecka, M.* Appl. Surf. Sci. ***2019***, 471, *973–985, Copyright (2021), with permission from Elsevier.*

In a two-semiconductor system, electrons and holes can be generated in two different materials and used for selective reduction and oxidation processes.[23,47] Chetri et al.[48] analyzed combinations of titanium and tin dioxide systems. In the first case, titanium dioxide was the core of the structure, whereas titanium dioxide was the outer layer in the second case. In the TiO_2@SnO_2 arrangement, the excited electron from the valence band to the conduction band in TiO_2, migrates to the CB band of tin dioxide. There it participates in the reaction with O_2 molecules adsorbed on the surface. Holes become "trapped" inside the core and participate in the conversion of OH^- groups to their hydroxyl radicals. In the reverse situation of SnO_2@TiO_2, the photocatalytic activity of the system decreases. The results of this work highlight the significant influence of the direction of carrier tunneling. However, when the shell layer is not continuous, the number of possible photogenerated charge carriers increases (Fig. 3C and D).[23] Moreover, reduction and oxidation processes may take place on the surfaces of titanium dioxide and tin dioxide.

2.2 TiO_2@Fe_2O_3 nanocrystals

Titanium dioxide-based materials, such as TiO_2@Fe_2O_3 and TiO_2@SnO_2, are important photocatalysts due to their unique physicochemical properties. Pure TiO_2 absorbs only the UV range of light; therefore, it is important to extend the absorption to visible light, which predominates in the spectrum of sunlight. Iron oxide (Fe_2O_3) with a bang gap equal to 2.2 eV is a great candidate for this purpose. The TiO_2@Fe_2O_3 heterojunction is a promising material; therefore, knowledge of the relative positions of the valence and conductive bands of these two materials is extremely important. In the literature, there are several theories about band configuration and hence electron and hole separation.

Mendiola-Alvarez et al.[49] proposed that covering titanium dioxide with iron oxide results in the formation of a Type I heterojunction where both electrons and holes are transferred to the conduction and valence bands of Fe_2O_3, respectively. However, this assumption was not verified experimentally. Therefore, in this case, recombination processes cannot be avoided. Kodan et al.[50] using UV–vis spectroscopy, XPS valence band spectra, and work function measurements for both materials separately, confirm Type I heterojunction creation when Fermi level alignment occurs (Fig. 4A).

On the other hand, Peng et al.[51] measured the work function for both materials independently. Knowing the relative position of the bands, they

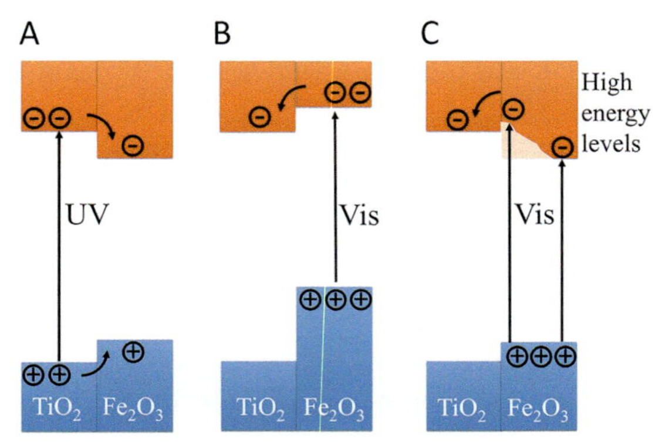

Fig. 4 Schematics of three types of $TiO_2@Fe_2O_3$ heterojunctions of: (A) Type I, (B) Type II, (C) Type I with high energy levels in the Fe_2O_3 conduction band.

concluded that after contact of these two materials, a Type II heterojunction is created, where electrons formed under visible light in Fe_2O_3 can be transferred to the conduction band of TiO_2. However, the issue of how the Fermi level alignment occurs has not been determined. This type of heterojunction has been suggested in many cases (Fig. 4B).[52–54]

However, there are also reports in which it is accepted that although the $TiO_2@Fe_2O_3$ compound forms a Type I heterojunction, it behaves favorably for electron transfer. It is claimed that in the Fe_2O_3 conduction band (CB_{Fe2O3}) there are higher levels to which electrons can be transported. Higher levels of iron oxide are located above the TiO_2 conduction band (CB_{TiO2}), so excited e^- can be injected into titanium dioxide unhindered (Fig. 4C).[55–57]

The use of the heterostructure discussed for the photocatalytic decomposition of dyes, is based on the absorption of visible light by a semiconductor photocatalyst (PC) according to the equation:

$$PC + h\nu \rightarrow e^-_{CB} + h^+_{VB} \tag{1}$$

The charge carriers are then transported to the surface of the photocatalyst, where the formation of reactive oxygen species (ROS) takes place through the reaction of electrons and holes with oxygen and water molecules.

$$O_2 + e^- \rightarrow {}^\bullet O_2^- \tag{2}$$

$$H_2O + h^+ \rightarrow {}^\bullet OH + H^+ \tag{3}$$

$$H_2O_2 + e^- \rightarrow {}^\bullet OH + {}^-OH \tag{4}$$

Finally, reactive oxygen species ($\dot{O}_2^-$, $^{\bullet}OH$) can oxidize dye molecules. In addition, electrons and holes, which are also active, can attack the dye to complete the photodecomposition process, as shown in the following reaction:

$$Dye + {}^{\bullet}O_2^- / {}^{\bullet}OH / e^- / h^+ \rightarrow intermediate\ products \rightarrow CO_2 + H_2O \quad (5)$$

For titanium dioxide, the location of the conduction band above the $O_2/\dot{O}_2^-$ level (Fig. 5), allows the reduction of oxygen to superoxide radicals by excited electrons, while the holes in the valence band can be used for the oxidation of H_2O to hydroxyl radicals. On the other hand, the valence band of iron oxide lies above the $H_2O/^{\bullet}OH$ level, which prevents water oxidation, and the location of the conduction band prevents the formation of $\dot{O}_2^-$ radicals. Only the addition of hydrogen peroxide to the photocatalytic system allows one to generate the $^{\bullet}OH$ radicals by electrons from the conduction band of Fe_2O_3.

Therefore, a small amount of H_2O_2 was added to the photocatalytic system consisting of $TiO_2@Fe_2O_3$ and Rhodamine B in an investigation by the authors.[12] Photocatalysis with commercial titanium dioxide powder P25, bare TiO_2 nanocrystals (NC-R) and covered with Fe_2O_3 (P25-F (H), NC-R-F (H)), was performed with and without hydrogen peroxide (Fig. 6A).

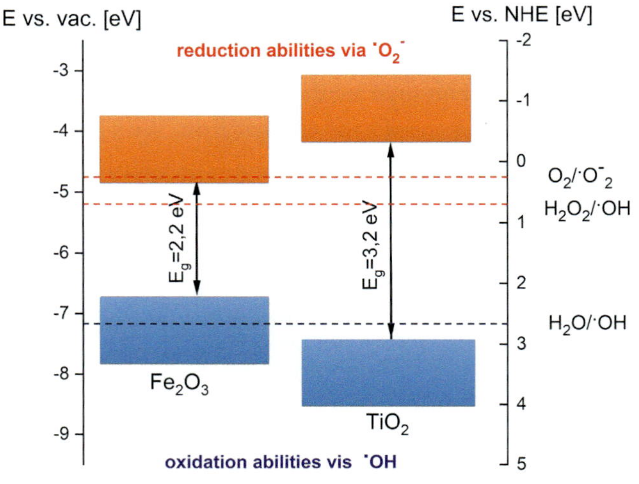

Fig. 5 Energy band diagram of separate metal oxides of TiO_2 and Fe_2O_3. *Based on Chiu, Y. H.; Chang, T. F. M.; Chen, C. Y.; Sone, M.; Hsu, Y. J.* Catalysts ***2019***, *9, 430.*

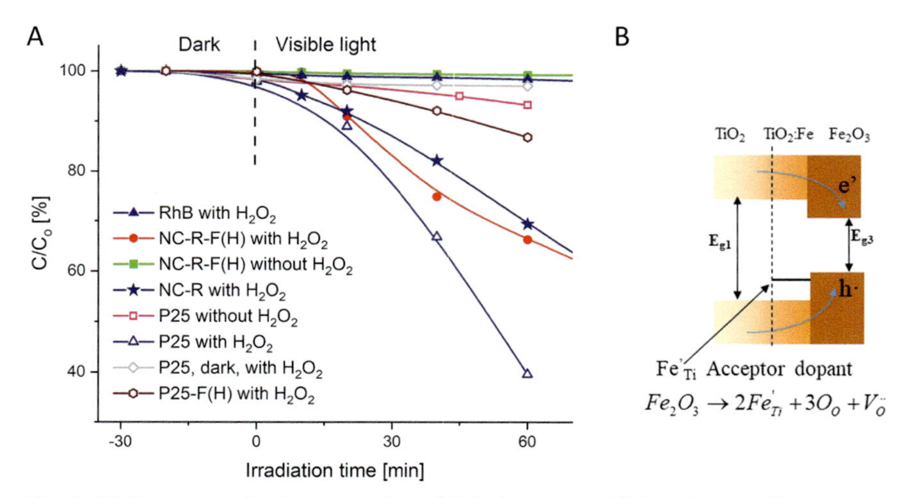

Fig. 6 (A) Photocatalytic decomposition of RhB dye under visible radiation, (B) electronic band diagram of TiO_2@Fe_2O_3 heterostructure.[12] *Reprinted from Synowiec, M.; Micek-Ilnicka, A.; Szczepanowicz, K.; Różycka, A.; Trenczek-Zajac, A.; Zakrzewska, K.; Radecka, M.* Appl. Surf. Sci. ***2019****, 473, 603–613, Copyright (2019) with permission from Elsevier.*

The heterostructure NC-R-F(H) does not decompose Rhodamine B (RhB) without H_2O_2 present; only a small amount of this compound allows for complete degradation of the dye after 270 min. The positive influence of H_2O_2 on photodegradation shows a large decrease in RhB concentration after 60 min using P25 powder. The iron oxide coating causes deterioration of the photo properties of P25, contrary to a slight improvement for TiO_2 nanocrystals. Furthermore, it was shown that the covering of TiO_2 nanocrystals by iron oxides, may result in an intermediate TiO_2:Fe layer. The formation of a heterojunction allows the transfer of charges from TiO_2 to Fe_2O_3 and the incorporation of Fe ions into the structure of TiO_2 according to the following equation:

$$Fe_2O_3 \rightarrow 2Fe'_{Ti} + 3O_O + V_O^{\cdot} \quad (6)$$

The additional acceptor level created within the TiO_2 band gap, lowered its value (Fig. 6B).

2.3 TiO_2@Cu_2O nanostructures

Copper(I) oxide, Cu_2O is another semiconductor that belongs to the group of transition metal oxides. It has a band-gap energy of 2.14 eV and therefore absorbs solar light from the visible range. Therefore, it could be used successfully in photochemistry, but its main drawback is the lack of electrochemical stability under illumination conditions. Fig. 7 shows the current-voltage characteristics without illumination (dark current) and under illumination with white light (photocurrent) of the Cu_2O layer deposited by the electrochemical method on a titanium substrate. In the range of 0–0.3 V, there are distinct peaks due to the redox reaction of copper.[58] This indicates the instability of the Cu_2O photoanode in the photoelectrochemical cell under the measurement conditions. This is due to the relative positions of redox potentials of the Cu_2O degradation (cathodic, $E_{c,d}$ and anodic, $E_{a,d}$ decomposition potentials) in the band gap.[59]

The problem of photoelectrochemical instability affects many narrow-band semiconductors, in addition to Cu_2O, MoS_2, GaAs, $BiVO_4$, ZnO and CdS[60] are also affected. However, existing research indicates that such a semiconductor gains stability when combined with a stable broadband semiconductor such as TiO_2.[6,29,30,61] There are many possibilities to engineer the morphology of the TiO_2@Cu_2O heterostructure, including nanopowder@nanopowder,[62] nanoparticles@nanoctystals,[63] nanowires@ nanoparticles,[64] thin film@thin film,[65] and nanocrystals@nanoparticles.[66] However, it appears attractive to use TiO_2 nanotubes, TiO_2-NT as a base material, with a dispersed deposit on Cu_2O nanoparticles on it. There are many studies, one also carried out in our group,[7] that highlight the advantages

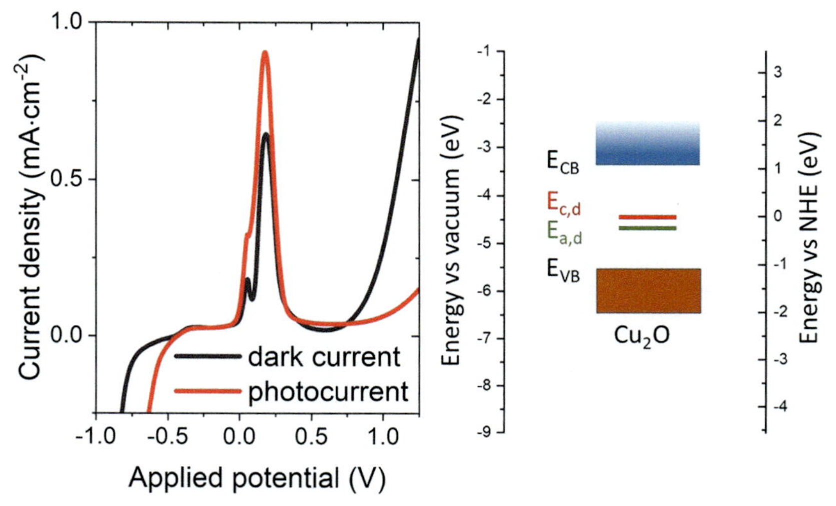

Fig. 7 Current-voltage characteristics of the Cu_2O photoanode and the edges of the relative position of the valence and conduction band vs redox potentials of oxide decomposition (J. Banas-Gac, in preparation).

of such a solution in photocatalysis, photoelectrochemistry, and photoelectrocatalysis.[62,67–72] The employment of TiO_2-NT, in addition to the aforementioned advantages of nanomaterials, also brings the rapid and directed transport of carriers in opposite directions, significantly reducing their recombination, high specific surface area, and high surface energy.[68,73] To ensure even higher photochemical activity of the heterostructure, it is required to meet the morphological criterion. It is necessary to ensure many quadruple points, namely places where access to light and electrolytes is provided to all components of the heterostructure that remain in direct contact with each other.[7] If nanotubes are considered, this requirement would be fulfilled if Cu_2O was dispersed on the surface of TiO_2. Fig. 8 presents two types of TiO_2@Cu_2O heterostructures—one with TiO_2 nanotubes on the Ti substrate, TiO_2-NT/S[61] and the second without substrate, TiO_2-NT/W[70]—that meet the requirement; Cu_2O is evenly spread on the surface of TiO_2 nanotubes and access to light and electrolytes is ensured for both components.

The first step in the formation of TiO_2@Cu_2O heterostructures is the preparation of TiO_2-NT, which can be achieved by, among others,[74–81] the solvo- and hydrothermal method, the anodization technique (electrochemical oxidation), and template-based methods. The next step is the deposition of Cu_2O on TiO_2-NT by means of, e.g., photodeposition,

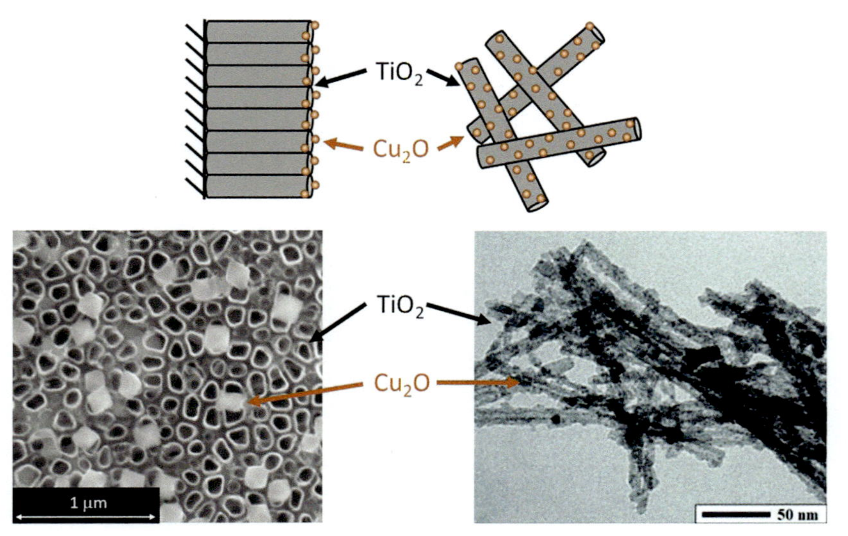

Fig. 8 TiO_2 nanotubes grown on titanium substrate[61] (left side) and TiO_2 nanotubes (substrate-free) partially covered with Cu_2O[70] (right side).

impregnation, the hydrothermal method, and chemical bath or electrochemical deposition.[61,64,67,70–72,82–87]

Pham et al. have successfully used TiO_2-NT/W@Cu_2O in the degradation of methylene blue.[70] As shown by their investigations (Fig. 9A), the presence of uniformly distributed nanometric Cu_2O particles on the TiO_2-NT/W surface, allows one to increase the degradation process efficiency twofold, accompanied by an almost threefold increase in the pseudo-first-order rate constant. Interestingly, the photoactivity of TiO_2@Cu_2O in phenol degradation (Fig. 9B) was found to be even 30% higher than that of TiO_2-NT/W. Trang et al. investigated the photoactivity of TiO_2-NT/W@Cu_2O in glycerol H_2 production through a photocatalytic process.[71] The presence of Cu_2O nanoparticles on the surface of the nanotubes resulted in a threefold increase in the hydrogen generation rate from 15.2 up to 48.7 mmol h^{-1} g_{cat}^{-1} (Fig. 9C). Wang et al. observed a similar correlation in the case of Rhodamine B.[72] TiO_2-NT/S@Cu_2O allowed the photocurrent density to be 2.7 times higher than that of TiO_2 nanotubes, 1.83 and 0.67 mA cm^{-2}, respectively (Fig. 9D). These authors attributed this significant photoactivity to the formation of an n-p junction at the interface between n-type TiO_2-NT and p-type Cu_2O, and the establishment of an electric field at the interface that separates charge carriers and thus limits recombination. However, in the reports, no evidence was offered for the p-type conductivity of Cu_2O.

Wang et al. also conducted a comparative study of the efficiency of different photochemical processes with and without external voltage.[72] Photocatalysis boosted by an additional potential difference (0.5 V) resulted in an amazing 80% increase in efficiency from 5.00% for photocatalysis to 84.29% for photoelectrocatalysis. A similar effect was observed when comparing photochemical and photoelectrochemical hydrogen generation. Applying a potential difference of 1 V resulted in an almost 2.5-fold increase in the registered photocurrent from 1.76 to 4.00 mA cm^{-2}.

Studies on the photochemical activity of heterostructures TiO_2-NT/S@Cu_2O were also carried out in our group.[61] Heterostructures that met the morphological criterion, were characterized by an increase in the photocurrent density by 50% in the photoelectrochemical cell, from 0.325 up to 0.472 mA cm^{-2} (Fig. 10). In the case of the photoelectrocatalytic decomposition of methylene blue, the increase in activity was even greater, i.e., two times. To explain the reason for such occurrences of heterostructures, it is necessary to know their energy diagrams. Because the electronic structure of a semiconductor depends on features such as the type of conductivity (p-type or n-type), band-gap energy, and band edge position, the key is

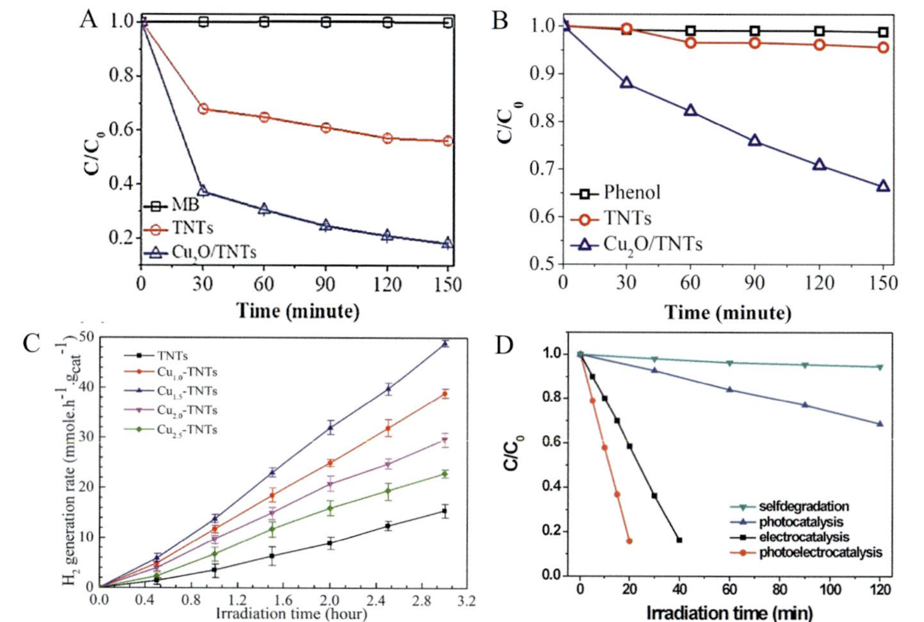

Fig. 9 Photocatalytic activity of the TiO_2-NT and TiO_2-NT@Cu_2O heterostructures toward: (A) photocatalytic decomposition of MB,[70] (B) photocatalytic decomposition of phenol (supplementary materials[70]), (C) photocatalytic production of H_2 from glycerol,[71] (D) photocatalytic and photoelectrocatalytic decomposition of Rhodamine B[72] (https://pubs.acs.org/doi/10.1021/acsomega.8b03404, further permissions related to the material excerpted should be directed to the ACS). *Panel (C) reprinted from Trang, T. N. Q.; Tu, L. T. N.; Man, T. V.; Mathesh, M.; Nam, N. D.; Thu, V. T. H.* Compos. Part B Eng. ***2019***, 174, *106969, Copyright (2021), with permission from Elsevier.*

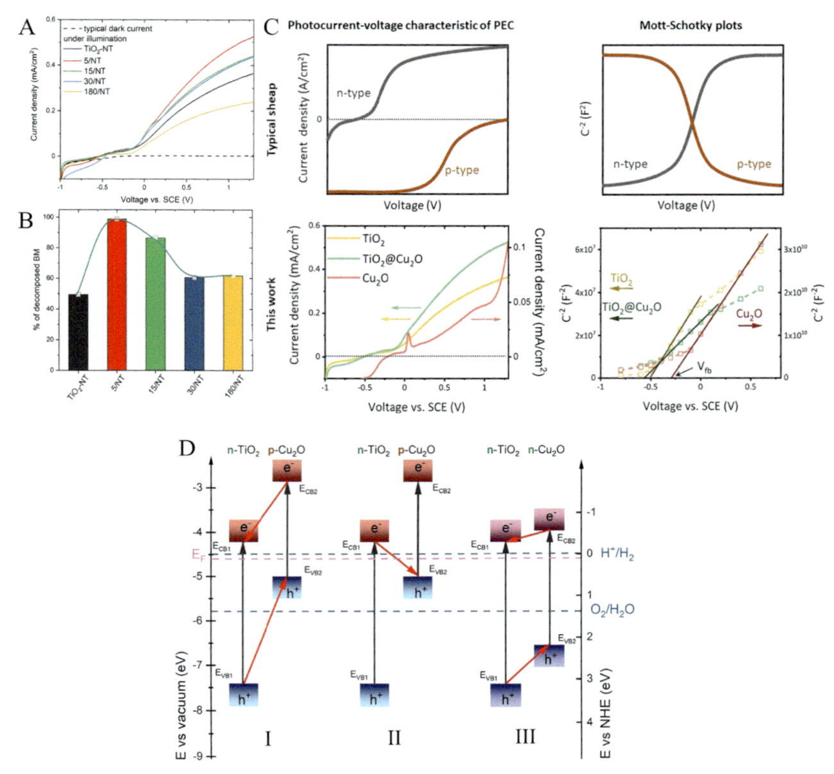

Fig. 10 (A) Photoelectrochemical and (B) photoelectrocatalytic activity of the TiO_2-NT and TiO_2-NT@Cu_2O heterostructures for the generation of H_2 and the decomposition of methylene blue, respectively.[61] (C) Examples of theoretical and experimental results of photoelectrochemical and electrochemical measurements for the determination of the type of conductivity.[61] (D) Possible TiO_2@Cu_2O heterostructures: p-n heterojunction (I), *Z*-Scheme (II), and n-n heterojunction (III). ECB, semiconductor conduction band, EVB, semiconductor valence band, 1, 2—the first and second semiconductor, EF—Fermi level.[61]

to determine the type of conductivity of the material. For this purpose, our investigations included determination of the energy of the band gap (based on reflectance spectral dependence measurements) and the conductivity type of both heterostructure components. The methods that make determination of conductivity type possible include photoelectrochemical measurements (current-voltage (I-V) characteristics) and impedance spectroscopy (Mott-Schotky (M-S) plots). Fig. 10C displays theoretical results of these two methods for a hypothetical photoanode (n-type) and photocathode (p-type). Our experimental results on TiO_2@Cu_2O[61] show that the M-S plots possess a negative slope and the photocurrent density measured in photoelectrochemical cells increases along with increase in the potential. These results demonstrate the n-type conductivity of copper(I) oxide. Since Cu_2O showed n-type conductivity and a flat band potential of Cu_2O (identified as the edge of the conduction band), we proposed an energy diagram of the heterostructure of n-n-type TiO_2@Cu_2O.

It is also beneficial for photocatalysis to use nanocrystals with selected exposed facets to form TiO_2-NC@Cu_2O and TiO_2@Cu_2O-NC heterostructures. Xiaong et al. conducted research on the photocatalytic activity of TiO_2-NC@Cu_2O.[88] Titanium dioxide nanocrystals with co-exposed {101} and {001} facets, were coated with Cu_2O nanoparticles. Then, investigations on the photocatalytic activity toward reduction of CO_2 and production of H_2 and CH_4 were carried out. Both the yield and the selectivity of the TiO_2-NC@Cu_2O heterostructures in CH_4 were found to increase fourfold compared with those for TiO_2-NC (Fig. 11A). The marked improvement in direct reduction of CO_2 was attributed to two effects: (i) increased electron density due to additional light absorption and (ii) increased probability of multi-electron reactions of CH_4 formation.

Li et al. used TiO_2-NC@Cu_2O heterostructures composed of octahedron-like TiO_2 nanocrystals and Cu_2O nanoparticles to generate hydrogen from a methanol solution.[89] The heterojunction with a 0.19 Cu-Ti ratio demonstrated a hydrogen generation rate of 24.83 mmol g^{-1} h^{-1}, which was almost three orders of magnitude higher than that of TiO_2-NC (0.03 mmol g^{-1} h^{-1}) (Fig. 11A). A 17-fold improvement was also observed in visible light.

The n-p heterojunction TiO_2@Cu_2O-NC with copper(I) oxide nanocrystals of different shapes and titanium dioxide nanoparticles, was synthesized and studied by Liu et al.[90] Heterostructural photocatalysts based on cubes, cubic-octahedral, and octahedral-based heterostructural photocatalysts (Fig. 11C), demonstrated better photocatalytic activity in visible light toward degradation of methylene blue and 4-nitrophenol than those of Cu_2O and TiO_2. A significant difference was found between the photoactivity of various TiO_2@Cu_2O-NC shapes: cubes (85%) > octahedral (70%) > cubic-octahedral (60%). Further investigations showed the correlation between the catalytic results and both photoinduced carrier separation and charge-transfer behavior. The results of surface photovoltage spectroscopy (SPS) correlated well with the type of facet configuration. However, the increase in the photocatalytic activity of the heterostructures compared to TiO_2, was attributed to the formation of an n-p junction between the semiconductors, and enhanced separation of charge carriers.

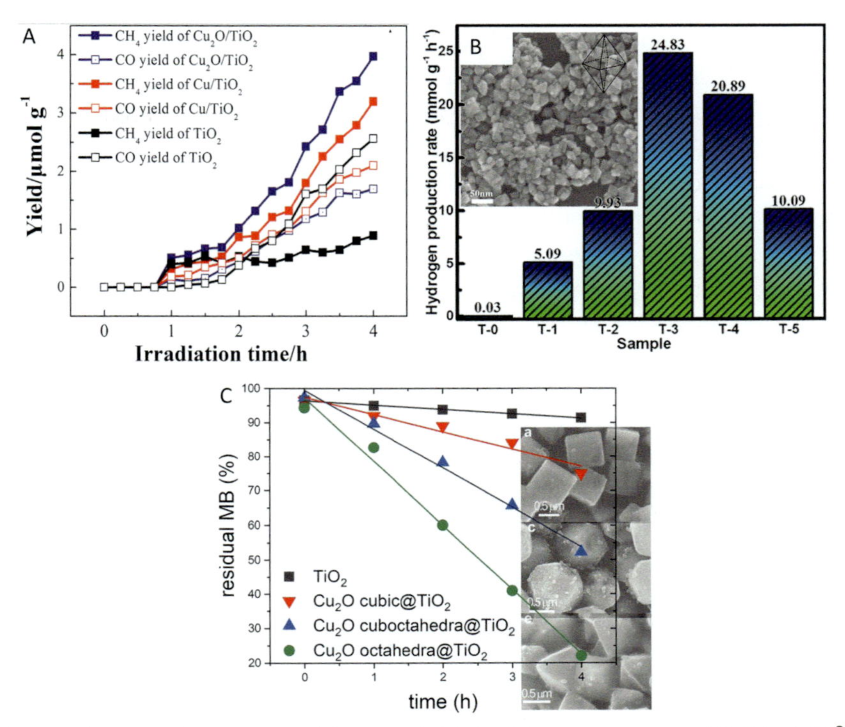

Fig. 11 Photocatalytic activity of the TiO_2-NC@Cu_2O heterostructures. (A) CH_4 yield,[88] (B) H_2 generation rate from a methanol solution on pristine TiO_2 (T-0) and TiO_2-NC@Cu_2O catalysts, amid SEM image of catalyst T-3[89] (https://pubs.acs.org/doi/10.1021/acsomega.8b03404, further permissions related to the material excerpted should be directed to the ACS.), (C) the percentage of residue MB vs time for TiO_2 nanoparticles and Cu_2O-NC@TiO_2 heterostructures. *Panel (A) reprinted from Xiong, Z.; Lei, Z.; Kuang, C. C.; Chen, X.; Gong, B.; Zhao, Y.; Zhang, J.; Zheng, C.; Wu, J. C. S.* Appl. Catal. B Environ. ***2017****, 202, 695–703, Copyright (2021), with permission from Elsevier. Panel (C) based on Liu, L.; Yang, W.; Sun, W.; Li, Q.; Shang, J. K.* ACS Appl. Mater. Interfaces ***2015****, 7, 1465–1476.*

2.4 Multiple-core@shell heterostructures

Single-core@shell heterostructures, which consist of two different materials (e.g., TiO_2@SnO_2, TiO_2@Fe_2O_3, TiO_2@Cu_2O), exhibit significantly improved photoactivity and photostability compared with the separate individual components. Therefore, to enhance further photochemical properties, more complex multiple-core@shell systems have been proposed.

To date, core@multishell nanostructures (i.e., a single core enclosed by multiple layers of shells) have been extensively studied for applications related to the optical and electronic properties of materials.[91] Wen et al.[92] investigated lanthanide-doped $NaYbF_4$@Na(*Yb,Gd*)F_4@$NaGdF_4$ core@shell@shell nanoparticles as NIR up-converting materials (i.e., materials capable of converting NIR excitation to UV and vis emissions). The core material was designed to harvest NIR light, while the shell layers were used to tune the optical emission. In the inner shell layer, the upconverter ions (Er^{3+}, Ho^{3+}, Tm^{3+}) were applied, whereas the outer shell served as a protective layer. Core@double-shell luminescent materials were also analyzed by Talapin et al.[93] The introduction of the inner-shell layer (CdS, ZnSe), which separated the CdSe core and the ZnS outer shell, significantly improved the photoluminescence quantum efficiency and photostability of the CdSe@CdS@ZnS and CdSe@ZnSe@ZnS systems. These effects were attributed to better lattice matching between the layers than in CdSe@ZnS materials. Huang et al.[94] developed lanthanide-doped core@multishell nanostructures for anti-counterfeiting and information encryption applications. The combination of four different shell layers resulted in the system with dual-modal up-converting and down-shifting emissions. Downshifting is a process in which high-energy photons are converted into low-energy ones, while upconversion is the opposite process. To date, both strategies have been applied in commercial anticounterfeiting. The proposed core@multishell materials exhibited tunable upconverting and downshifting emissions simultaneously, which was believed to improve security. Due to the presence of Nd, Yb, and Ce ions, the nanoarchitectures could be excited by three different wavelengths (254, 980, or 808 nm), and emit red, green, and blue light (due to Er, Tm, Eu, Tb activators). Furthermore, spatial separation of dopants in the shells prevented adverse interactions between lanthanide ions and contributed to the higher efficiency of the system. Dong et al.[95] analyzed hollow SnO_2 microspheres (Fig. 12A) with a varying number of shells (from one to five) for applications in dye-sensitized solar cells (DSSC). Short-circuit current densities (J_{sc}) and light conversion efficiencies (η) increased from single-shelled ($J_{sc}=12.54$ mA cm^{-2}, $\eta=5.21\%$) to quintuple-shelled materials ($J_{sc}=17.62$ mA cm^{-2}, $\eta=7.18\%$). The structures comprising more shells were characterized by enhanced active surface area and light-trapping properties, resulting in better performance. Zhang et al.,[98] in turn, investigated the effect of the number of layers in ZnS-CdS multi-shells on their photoelectrochemical performance. It prevailed that the double-, triple-, and quadruple-shells showed increased photocurrent density values compared to those of the single-shelled structures. This effect was attributed to the multiple light reflections between the shells,

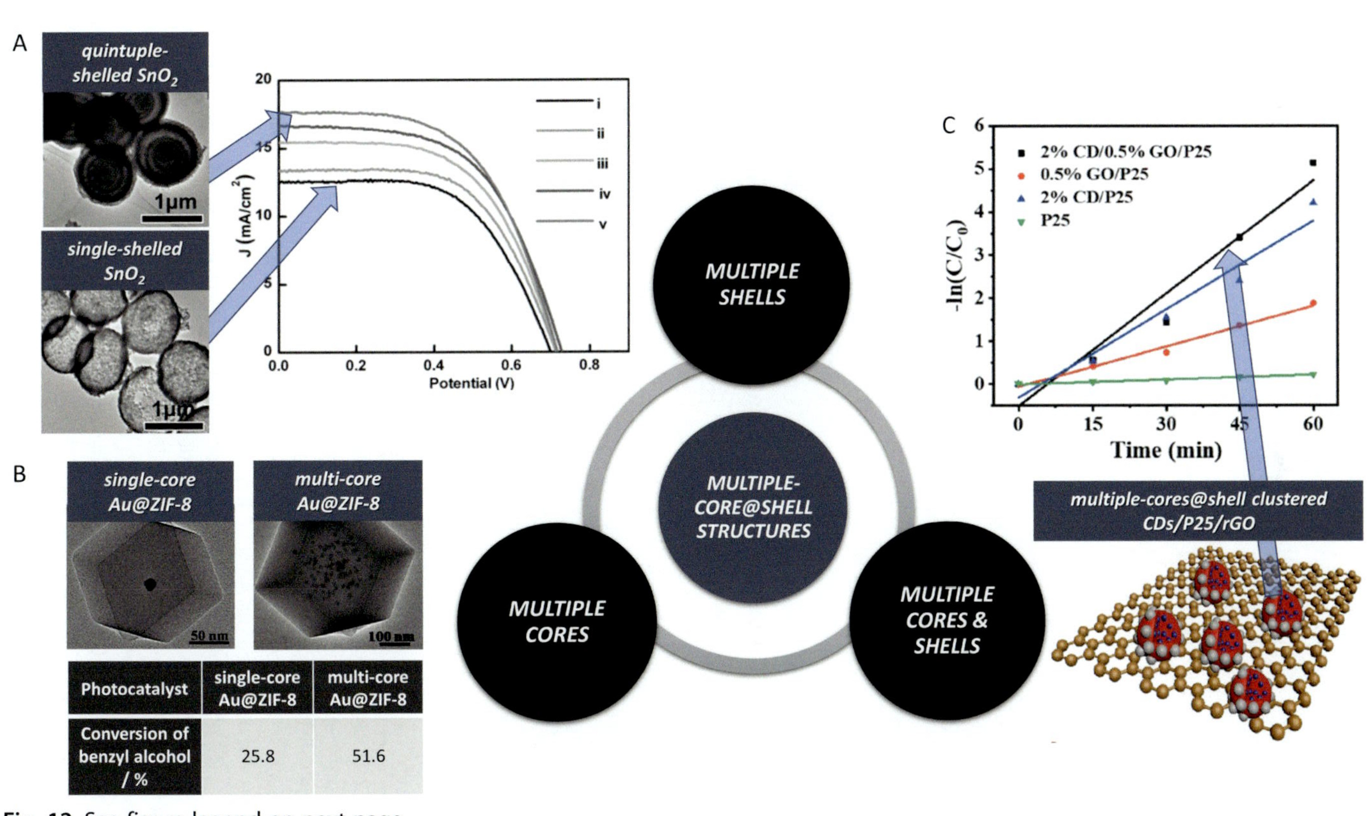

Photocatalyst	single-core Au@ZIF-8	multi-core Au@ZIF-8
Conversion of benzyl alcohol / %	25.8	51.6

Fig. 12 See figure legend on next page.

which resulted in enhanced light-harvesting. The best photoelectrochemical performance was obtained for the triple-shelled materials. However, the further increase in the number of layers led to lower photocurrent densities, which might be the result of limited mass transfer.

The multicore@shell structure (i.e., single or multiphase cores encapsulated in a single shell) may be beneficial for enhancing the separation of photogenerated charge carriers, increasing the number of active sites, and tuning the optical/electrical properties, which may contribute to improved photocatalytic efficiency.[99,100] Zhang et al.[101] examined the photoluminescence properties of nitrogen-doped multicore@shell carbon dots for white-light-emitting diodes (WLEDs). The materials obtained displayed simultaneous tricolor (blue, green, red) emissions. Moreover, these multicore@shell structures exhibited concentration-induced enhancement of fluorescence, which led to hindrance of the self-quenching effect caused usually by the aggregation of carbon dots. Studies demonstrating the influence of core number on photocatalytic activity were carried out by Chen et al.[96] After 24 h of illumination, only 25.8% of benzyl alcohol was photo-oxidized using Au@ZIF-8 single core@shell structures, while for multicore@shell materials—51.6% (Fig. 12B). The better photocatalytic activity was attributed to the plasmonic coupling between multiple cores.

Structures that simultaneously contain multiple cores and shells (multicore@multishell materials) are also promising candidates for photochemistry-related applications. For example, Zhu et al.[102] analyzed the photocatalytic properties (Fig. 12C) of multicore@shell clustered carbon dots (CDs) covered with P25 particles and deposited on reduced graphene oxide (rGO). After 45 min of visible light irradiation, the prepared CDs/P25/rGO compound decomposed 98.6% of rhodamine B (RhB) and showed the highest

Fig. 12 Photocatalytic/photoelectrochemical properties of various multiple-core@shell structures: (A) J-V curves of DSSCs based on hollow SnO_2 spheres with an increasing number of shells (from 1 to 5) along with TEM images of single- and quintuple-shelled materials[95]; (B) photocatalytic conversion efficiency of benzyl alcohol obtained using single- and multiple-core Au@ZIF-8 photocatalysts along with TEM images of the structures[96]; (C) photocatalytic decomposition kinetics of RhB using P25, CDs/P25, rGO/P25, and CDs/rGO/P25 along with a schematic representation of the latter structure.[97] *Panel (A) © 2013 WILEY-VCH Verlag GmbH & Co. KGaA, Weinheim. Panel (B) Republished with permission of Royal Society of Chemistry, from Chen, L.; Peng, Y.; Wang, H.; Gu, Z.; Duan, C.* Chem. Commun. ***2014****, 50, 8651–8654; permission conveyed through Copyright Clearance Center, Inc. Panel (C) Reprinted (adapted) with permission from Magnan, H.; Stanescu, D.; Rioult, M.; Fonda, E.; Barbier, A.* J. Phys. Chem. C ***2019****, 123, 5240–5248. Copyright (2021) American Chemical Society.*

pseudo-first order rate constant value ($k_1 = 0.087\ min^{-1}$) compared with P25 ($0.003\ min^{-1}$) rGO/P25 ($0.031\ min^{-1}$), and CDs/P25 ($0.068\ min^{-1}$). In addition, the authors analyzed the antibacterial properties of the samples. Again, the best antibacterial efficiency against *E. coli* showed CDs/P25/rGO (100% after 60 min of visible light illumination), while P25, rGO/P25 and CDs/P25 inactivated 15.03%, 64.05% and 75.63% of the bacteria, respectively. The best photocatalytic performance of the prepared composite was attributed to the synergistic effect of multicore@shell CDs combined with P25 and rGO, effective separation of photogenerated charge carriers, and enhancement of electrical conductivity.

The reports presented, demonstrate that the application of multiple cores and/or shells in single nanoparticles may result in enhanced light-harvesting, photocatalytic efficiency, photoelectrochemical performance, tunable optical and electrical properties, more active sites, and increased stability of the systems. Therefore, the combination of the properties of different materials in multiple core@shell heterostructures is a promising strategy for obtaining materials with multiple functionalities for various applications.

2.5 Substrates for photoelectrodes

Strategies for improving the photo-electrode performance in PECs mainly concern optimizing the properties of the deposited film photocatalysts. However, the selection of an appropriate substrate material is also of key importance. Since various substrates have different optical and electrical properties, it is crucial to fit a photocatalytically active material to the applied platform to achieve the highest PEC efficiency.[103,104] In general, two main approaches for improving the performance of the photoelectrode can be distinguished through the design of the substrate properties. These are morphology control and heterostructure engineering (Fig. 13).

In our previous papers,[28,30,105] we analyzed the influence of the TiO_2 substrate form (1D, 2D, and 3D nanostructures) on the photoelectrochemical properties of CdS/TiO_2, MoS_2/TiO_2, GO/TiO_2 and bare TiO_2 photoanodes. The results (Fig. 14A) revealed that electrodes composed of 3D flower-shaped TiO_2 (NF) nanostructures showed higher photocurrent density values (I_{ph} up to $120\ \mu A \bullet cm^{-2}$), longer relaxation times (τ up to 64 s), and more negative flat-band potential values (V_{FB} ranging from -0.61 to -0.91 V) than those comprising 1D nanotubes (NT) (I_{ph} up to $50\ \mu A\ cm^{-2}$, τ up to 18 s, and V_{FB} from -0.54 to -0.72 V).[28] The better photocatalytic performance of flower-like electrodes was attributed to a

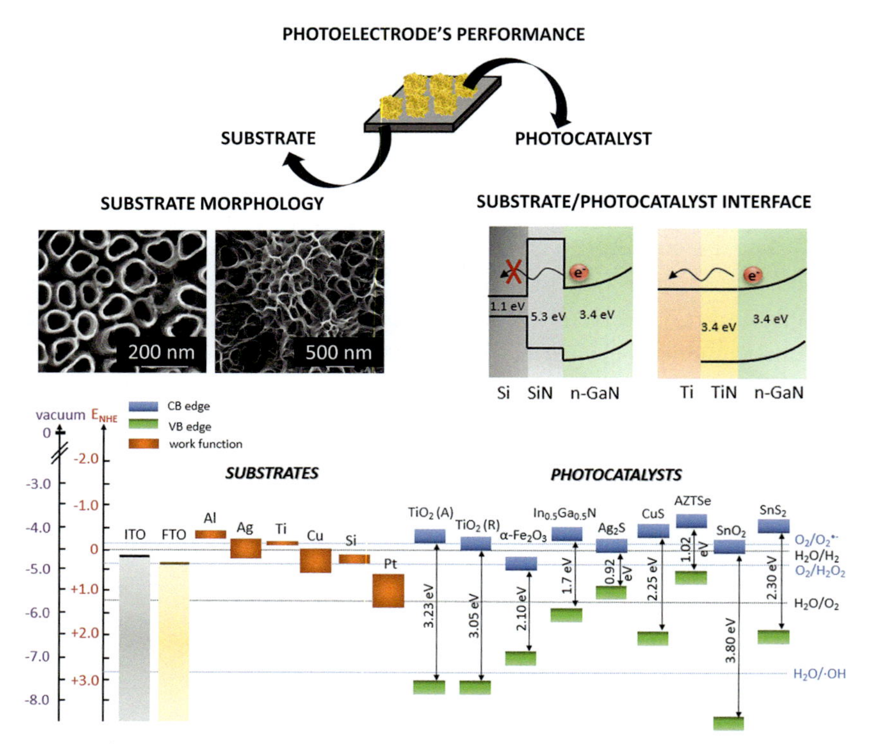

Fig. 13 Schematic representation of the enhancement of the photoelectrode performance of the properties of the photoelectrode by designing the substrate properties.[28,104]

thicker depletion layer that affects the better separation of photogenerated charge carriers. Furthermore, we carried out studies with the purpose of determining the effect of the applied substrate (2D titanium foil with a native layer of titanium oxides, 2D TiO_2 thin films, and 3D TiO_2 nanoflowers) on the morphology of the deposited cadmium sulfide.[30] The well-developed surface of the flower-like TiO_2 substrate yielded the production of evenly distributed CdS 0D nanoparticles, which was beneficial for the photoelectrochemical activity of the electrodes. The MoS_2/TiO_2 and GO/TiO_2 photoanodes comprising TiO_2 substrates in the form of layers (TiO_2-L) and nanotubes (TiO_2-NT), respectively, were also analyzed.[105] The results revealed that both TiO_2-L and TiO_2-NT electrodes met the requirements set for photo-anodes in PECs, while TiO_2-NT was characterized by slightly better flat-band potential values (−462 and −442 mV for TiO_2-NT and TiO_2-L, respectively).

The influence of substrate morphology on the Ti/TiO_2 and Ti/Fe_2O_3 photo-anodes was also studied by Bialuschewski et al.[107] In this work, the authors proved that laser texturing of the Ti platforms, regardless of the

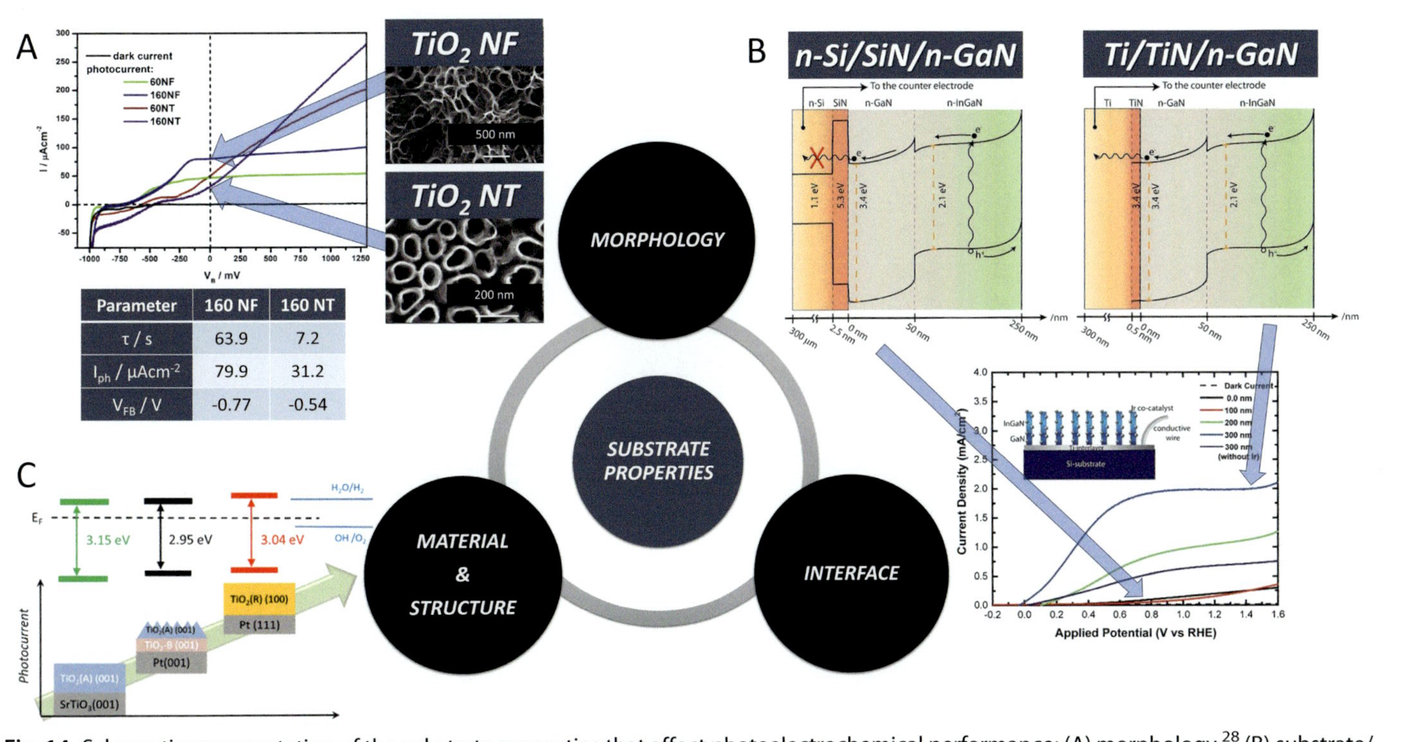

Parameter	160 NF	160 NT
τ / s	63.9	7.2
I_{ph} / μAcm^{-2}	79.9	31.2
V_{FB} / V	-0.77	-0.54

Fig. 14 Schematic representation of the substrate properties that affect photoelectrochemical performance: (A) morphology,[28] (B) substrate/photocatalyst interface,[106] (C) material and structure.[97] *Panel (A) Reprinted from Kusior, A.; Wnuk, A.; Trenczek-Zajac, A.; Zakrzewska, K.; Radecka, M.* Int. J. Hydrogen Energy ***2015***, 40, *4936–4944, Copyright (2021), with permission from Elsevier. Panel (B) republished with permission of the Royal Society of Chemistry, from Ebaid, M.; Min, J. W.; Zhao, C.; Ng, T. K.; Idriss, H.; Ooi, B. S.* J. Mater. Chem. A ***2018***, 6, *6922–6930, Copyright (2021); permission conveyed through Copyright Clearance Center, Inc. Panel (C) Reprinted (adapted) with permission from Magnan, H.; Stanescu, D.; Rioult, M.; Fonda, E.; Barbier, A.* J. Phys. Chem. C ***2019***, 123, *5240–5248, Copyright (2021) American Chemical Society.*

film material, results in better photoelectrochemical water-splitting performance compared to the untreated ones. The improved photoactivity was attributed to the enhanced light harvesting of the periodically patterned substrates.

In addition to controlling the morphology of the applied platform, judicious choice of substrate material is another important factor that affects the performance of photoelectrodes. Desai et al.[108] studied the effect of different transparent substrates (glass, indium tin oxide—ITO and fluorine tin oxide—FTO) on the formation of thin Bi_2Se_3 films obtained by the arrested precipitation technique (APT). Depending on the applied platform (glass, ITO and FTO), the authors obtained Bi_2Se_3 layers of different shapes (spongy balls, nanospheres, and nanofibers, respectively). Photoelectrochemical measurements revealed that the films deposited on FTO showed the best photocatalytic performance. Henry et al.,[103] in turn, investigated the properties of $Ag_2ZnSnSe_4$ (AZTSe) thin films prepared on metallic substrates (Al, Cu, Ag). The highest photoelectrochemical efficiency of the Ag/AZTSe electrode was attributed to the larger particle size of the AZTSe electrode, and thus the reduced amount of grain boundaries, which resulted in better conductivity. It is worth noting that in the case of Cu and Ag substrates, the authors observed the presence of secondary phases, such as CuSe and Ag_2Se, which affected the optical and electrical properties of the electrodes. The Al substrate did not react with the chalcogenide.

The presence of additional phases formed at the substrate/photocatalyst interface can have a substantial impact on the overall performance of the photoelectrode, which was discussed by Ebaid et al.[106] (Fig. 14B). The application of different substrates (Ti and Si) for the InGaN layers resulted in the formation of intermediate phases (TiN and SiN_x, respectively). In the case of the titanium platform, the Ti/TiN/InGaN heterojunction formed was beneficial for electron transport due to the comparable work functions of Ti (4.3 eV), TiN (4.25 eV), and InGaN (4.1 eV). For the silicon substrate, in turn, the presence of the SiN_x insulating layer led to the conduction band offset in the InGaN/SiN_x heterojunction, which blocked the electron transport. Lin et al.[104] proposed applying the Si/MXene platform to improve the photoelectrochemical performance of the Si/MXene/InGaN photoanode due to the formation of the Type II heterojunction and this results in enhanced migration of the photogenerated electrons.

Magnan et al.[97] also investigated the influence of the substrate material (Pt, Nb:$SrTiO_3$) on the photoelectrochemical properties of TiO_2 films (Fig. 14C). In addition, the authors examined the effect of the crystallographic structure of the applied platform (Pt(111), Pt(001)). The TiO_2 film deposited on Pt(111) crystallized in the rutile structure (TiO_2-R), while those obtained for Nb:$SrTiO_3$ and Pt(001) were in the form of anatase (TiO_2-A). Moreover, in the case of the latter, a 3 nm brookite layer (TiO_2-B) was formed between TiO_2-A and the substrate. The measured photocurrent values for the Pt(111)/TiO_2 photoanode were approximately 1.4 and 5.5 times higher than for Pt(001)/TiO_2 and Nb:$SrTiO_3$/TiO_2, respectively. Surprisingly, the rutile-based electrode showed the best photocatalytic performance. The authors concluded that this finding is connected with the Pt(111)/TiO_2-R band being beneficial for hydrogen generation, presence of defects at the Pt(001)/TiO_2-B/TiO_2-A interface and Nb:$SrTiO_3$/TiO_2 morphology.

The results demonstrate that, in order to obtain the highest photoelectrochemical performance of the electrodes, it is crucial to select judiciously the substrate material and control its surface properties. These factors not only affect the morphology and phase composition of the deposited photocatalysts, but also contribute to better conductivity, enhanced light harvesting, and obtaining heterojunctions that improve PEC efficiency.

3. Summary and future perspectives

Application of heterogeneous photocatalysis/photoelectrochemistry provides a very promising method for the degradation of environmental contamination in liquids/gases, decomposition of water for generation of hydrogen or photoelectrocatalytic decomposition of contaminants in the liquid phase.

In this work, we focused on the application of heterostructured systems. This review presents the results of our work on: TiO_2@SnO_2 nanopowders, TiO_2@Fe_2O_3 nanocrystals, TiO_2@Cu_2O nanomaterials, multiple-core@shell heterostructures, and substrates for photoelectrodes.

Among various green chemistry technologies with low environmental impact for renewable technology, catalysis has been regarded as the center of such greener technologies. For the design of novel and efficient photocatalysts, it is crucial to maximize the interfacial area. The photocatalyst is involved in a complex series of photophysical and electrocatalytic processes, which can be divided into the following components: (i) photon absorption, (ii) excitation separation, (iii) carrier diffusion, (iv) carrier transport, (v) catalytic efficiency, and (vi) the mass transfer of reactants and products. However, the critical process on which everything depends is light absorption. Semiconductor materials are considered to be unique substrates for an assembly of multiple components, offering the creation of multitasking interfaces. The charge carrier transfer mechanism, rate, and efficiency of photocatalytic processes depend on the selected sensitization challenge/co-catalyst application. Incorporating various molecules at the surface offers many potential advantages, such as light-harvesting, maneuvering electrons, or maximizing the density of active centers.

The general trend in maximizing the efficiency of photocatalysis and photoelectrocatalysis processes is based on multi-sensitization/co-deposition. On the one hand, such solutions positively influence the quality and quantity of reactions occurring on the surface. On the other hand, explaining the mechanisms responsible for the electron injection, their transfer, and the stability of the system becomes problematic. A fundamental understanding of the interactions and their physicochemical properties is crucial for the design of multicomponent materials. Therefore, new materials or novel structures with high active site density and light-harvesting opportunities need to be designed to allow accurate determination of the degree of photodecomposition of contaminants.

Further exploration in designing an effective interface remains the same challenge as in recent years, in the field of photocatalysis and focuses on:

- obtaining new materials, e.g., patchwork, where the most prominent group studied possesses Janus-like structures;
- mimicking the nature and development of the strategies based on bio-inspired artificial photo processes.

Due to the extraordinary effects of anisotropic interfaces of Janus-like particles, significant efforts have been devoted to selective surface functionalization. Combining at least two types of materials with such structure can be specially tailored to a synergetic effect, resulting in a substantial increase in their desired efficiency. Moreover, such a form affects the asymmetric flow of electrons, or chemically modified surfaces can perform a localized chemical reaction while products are released asymmetrically.

Despite the traditional research focused on active materials and synthesis, considerable attention has been paid to rediscovering the best photocatalytic system that was ever discovered, namely nature.

Green plants or bacteria developed almost perfect light-harvesting strategies for natural photosynthesis. Moreover, in natural photosynthesis, solar energy is converted to chemical energy due to light absorption by antenna protein pigment complexes containing chlorophyll, followed by multistep photoinduced energy transfer from the absorption site to the reaction center. In a chain of topo-tactically attached crystallites, excitation may occur at one crystallite at any position in the chain. The redox reactions are finally realized at another crystallite upon migrating the photogenerated charge carriers within the chain. The idea assumed that chains of the nano-crystallites could receive the electromagnetic waves from the light irradiation source, and transform them into the electronic currency in which the photogenerated holes and electrons are transported to other crystallites where they induce the specific redox reactions.

Despite the success of many studies devoted to improving photocatalysis and photoelectrocatalysis processes through interface engineering, there are still many challenges to overcome.

Acknowledgments

Work for the purpose of this review was performed within the framework of the project no. 2016/21/B/ST8/00457 National Science Center, Poland. K.M. acknowledges The Ministry of Education and Science in Poland (Project Number DI2018 008148).

References

1. Fujishima, A.; Honda, K. *Nature* **1972**, *238*, 37–38.
2. Hashimoto, K.; Irie, H.; Fujishima, A. *Jpn. J. Appl. Phys.* **2005**, *44*, 8269.
3. Han, J.; Zhu, G.; Hojamberdiev, M.; Peng, J.; Zhang, X.; Liu, Y.; Ge, B.; Liu, P. *New J. Chem.* **2015**, *39*, 1874–1882.
4. Ho, Y. S.; McKay, G. *Process Biochem.* **1999**, *34*, 451–465.
5. Sopyan, I.; Hafizah, N.; Jamal, P. *Indian J. Chem. Technol.* **2011**, *18*, 263–270.
6. Trenczek-Zajac, A.; Banas, J.; Radecka, M. *Int. J. Hydrogen Energy* **2018**, *43*, 6824–6837.
7. Bak, T.; Nowotny, J.; Rekas, M.; Sorrell, C. C. *Int. J. Hydrogen Energy* **2002**, *27*, 991–1022.
8. Trenczek-Zajac, A. *New J. Chem.* **2019**, *43*, 8892–8902.
9. Liu, Y.; Li, Z.; Green, M.; Just, M.; Li, Y. Y.; Chen, X. *J. Phys. D. Appl. Phys.* **2017**, *50*, 193003.
10. Kusior, A.; Banas, J.; Trenczek-Zajac, A.; Zubrzycka, P.; Micek-Ilnicka, A.; Radecka, M. *J. Mol. Struct.* **2018**, *1157*, 327–336.
11. Kusior, A.; Michalec, K.; Jelen, P.; Radecka, M. *Appl. Surf. Sci.* **2019**, *476*, 342–352.
12. Synowiec, M.; Micek-Ilnicka, A.; Szczepanowicz, K.; Różycka, A.; Trenczek-Zajac, A.; Zakrzewska, K.; Radecka, M. *Appl. Surf. Sci.* **2019**, *473*, 603–613.
13. Rahimi, K.; Yazdani, A. *Mater. Sci. Semicond. Process.* **2018**, *80*, 38–43.
14. Baruah, S.; Mahmood, M. A.; Myint, M. T. Z.; Bora, T.; Dutta, J. *Beilstein J. Nanotechnol.* **2010**, *1*, 14–20.
15. Chacko, D. K.; Madhavan, A. A.; Arun, T. A.; Thomas, S.; Anjusree, G. S.; Deepak, T. G.; Balakrishnan, A.; Subramanian, K. R. V.; Sivakumar, N.; Nair, S. V.; Nair, A. S. *RSC Adv.* **2013**, *3*, 24858–24862.
16. Liang, H. C.; Li, X. Z.; Nowotny, J. *Solid State Phenom.* **2010**, *162*, 295–328.
17. Reghunath, S.; Pinheiro, D.; KR, S. D. *Appl. Surf. Sci. Adv.* **2021**, *3*, 100063.
18. Jo, Y. K.; Lee, J. M.; Son, S.; Hwang, S. J. *J. Photochem. Photobiol. C Photochem. Rev.* **2019**, *40*, 150–190.
19. Kanakkillam, S. S.; Shaji, S.; Krishnan, B.; Vazquez-Rodriguez, S.; Martinez, J. A. A.; Palma, M. I. M.; Avellaneda, D. A. *Appl. Surf. Sci.* **2020**, *501*, 144223.

20. Cho, S.; Ahn, C.; Park, J.; Jeon, S. *Nanoscale* **2018**, *10*, 9747–9751.
21. Zhang, S.; Sun, Y.; Li, C.; Ci, L. *Solid State Sci.* **2013**, *25*, 15–21.
22. Kusior, A.; Synowiec, M.; Zakrzewska, K.; Radecka, M. *Crystals* **2019**, *9*, 163.
23. Kusior, A.; Zych, L.; Zakrzewska, K.; Radecka, M. *Appl. Surf. Sci.* **2019**, *471*, 973–985.
24. Li, Y. F.; Liu, Z. P. *J. Am. Chem. Soc.* **2011**, *133*, 15743–15752.
25. Liu, B.; Ning, L.; Zhao, H.; Zhang, C.; Yang, H.; Liu, S. *Phys. Chem. Chem. Phys.* **2015**, *17*, 13280–13289.
26. Deng, S.; Zhang, B.; Choo, P.; Smeets, P. J. M.; Odom, T. W. *Nano Lett.* **2021**, *21*, 1523–1529.
27. Seo, D. B.; Bae, S. S.; Kim, E. T. *Nanomaterials* **2021**, *11*, 1377.
28. Kusior, A.; Wnuk, A.; Trenczek-Zajac, A.; Zakrzewska, K.; Radecka, M. *Int. J. Hydrogen Energy* **2015**, *40*, 4936–4944.
29. Radecka, M.; Kusior, A.; Trenczek-Zajac, A.; Zakrzewska, K. *Adv. Inorg. Chem.* **2018**, *72*, 145–183. Academic Press.
30. Trenczek-Zajac, A.; Kusior, A.; Radecka, M. *Int. J. Hydrogen Energy* **2016**, *41*, 7548–7562.
31. Sridharan, M.; Maiyalagan, T. *Chem. Eng. J.* **2021**, *424*, 130393.
32. Kusior, A.; Kollbek, K.; Kowalski, K.; Borysiewicz, M.; Wojciechowski, T.; Adamczyk, A.; Trenczek-Zajac, A.; Radecka, M.; Zakrzewska, K. *Appl. Surf. Sci.* **2016**, *380*, 193–202.
33. Bai, S.; Jiang, J.; Zhang, Q.; Xiong, Y. *Chem. Soc. Rev.* **2015**, *44*, 2893–2939.
34. Bai, S.; Wang, X.; Hu, C.; Xie, M.; Jiang, J.; Xiong, Y. *Chem. Commun.* **2014**, *50*, 6094–6097.
35. Maeda, K.; Xiong, A.; Yoshinaga, T.; Ikeda, T.; Sakamoto, N.; Hisatomi, T.; Takashima, M.; Lu, D.; Kanehara, M.; Setoyama, T.; Teranishi, T.; Domen, K. *Angew Chem. Int. Ed.* **2010**, *49*, 4096–4099.
36. Tanaka, A.; Sakaguchi, S.; Hashimoto, K.; Kominami, H. *ACS Catal.* **2013**, *3*, 79–85.
37. Zheng, Z.; Tachikawa, T.; Majima, T. *J. Am. Chem. Soc.* **2014**, *136*, 6870–6873.
38. Yang, J.; Wang, D.; Han, H.; Li, C. *Acc. Chem. Res.* **2013**, *46*, 1900–1909.
39. Zong, X.; Yan, H.; Wu, G.; Ma, G.; Wen, F.; Wang, L.; Li, C. *J. Am. Chem. Soc.* **2008**, *130*, 7176–7177.
40. Mubeen, S.; Lee, J.; Singh, N.; Krämer, S.; Stucky, G. D.; Moskovits, M. *Nat. Nanotechnol.* **2013**, *8*, 247–251.
41. Trenczek-Zajac, A.; Kusior, A.; Lacz, A.; Radecka, M.; Zakrzewska, K. *Mater. Res. Bull.* **2014**, *60*, 28–37.
42. Marschall, R. *Adv. Funct. Mater.* **2014**, *24*, 2421–2440.
43. Yu, J.; Wang, S.; Low, J.; Xiao, W. *Phys. Chem. Chem. Phys.* **2013**, *15*, 16883–16890.
44. Li, J.; Lin, C. J.; Lai, Y. K.; Du, R. G. *Surf. Coatings Technol.* **2010**, *205*, 557–564.
45. Hao, P.; Zhao, Z.; Tian, J.; Sang, Y.; Yu, G.; Liu, H.; Chen, S.; Zhou, W. *Acta Mater.* **2014**, *62*, 258–266.
46. Liz-Marzán, L. M.; Giersig, M.; Mulvaney, P. *Langmuir* **1996**, *12*, 4329–4335.
47. Pan, J.; Hühne, S. M.; Shen, H.; Xiao, L.; Born, P.; Mader, W.; Mathur, S. *J. Phys. Chem. C* **2011**, *115*, 17265–17269.
48. Chetri, P.; Basyach, P.; Choudhury, A. *Chem. Phys.* **2014**, *434*, 1–10.
49. Mendiola-Alvarez, S. Y.; Hernández-Ramírez, A.; Guzmán-Mar, J. L.; Maya-Treviño, M. L.; Caballero-Quintero, A.; Hinojosa-Reyes, L. *Catal. Today* **2019**, *328*, 91–98.
50. Kodan, N.; Agarwal, K.; Mehta, B. R. *J. Phys. Chem. C* **2019**, *123*, 3326–3335.
51. Peng, L.; Xie, T.; Lu, Y.; Fan, H.; Wang, D. *Phys. Chem. Chem. Phys.* **2010**, *12*, 8033–8041.

52. Liu, J.; Yang, S.; Wu, W.; Tian, Q.; Cui, S.; Dai, Z.; Ren, F.; Xiao, X.; Jiang, C. *ACS Sustainable Chem. Eng.* **2015**, *3*, 2975–2984.
53. Mei, Q.; Zhang, F.; Wang, N.; Yang, Y.; Wu, R.; Wang, W. *RSC Adv.* **2019**, *9*, 22764–22771.
54. Zhang, J.; Kuang, M.; Wang, J.; Liu, R.; Xie, S.; Ji, Z. *Chem. Phys. Lett.* **2019**, *730*, 391–398.
55. Li, X.; Lin, H.; Chen, X.; Niu, H.; Liu, J.; Zhang, T.; Qu, F. *Phys. Chem. Chem. Phys.* **2016**, *18*, 9176–9185.
56. Huang, R.; Liang, R.; Fan, H.; Ying, S.; Wu, L.; Wang, X.; Yan, G. *Sci. Rep.* **2017**, *7*, 1–8.
57. Tilgner, D.; Friedrich, M.; Verch, A.; de Jonge, N.; Kempe, R. *ChemPhotoChem* **2018**, *2*, 349–352.
58. Chen, K.; Xue, D. *Appl. Sci. Converg. Technol.* **2014**, *23*, 14–26.
59. Bard, A. J.; Stratmann, M.; Licht, D. *Encyclopedia of Electrochemistry*; Wiley-VCH, 2007.
60. Jiang, C.; Moniz, S. J. A.; Wang, A.; Zhang, T.; Tang, J. *Chem. Soc. Rev.* **2017**, *46*, 4645–4660.
61. Trenczek-Zajac, A.; Banas-Gac, J.; Radecka, M. *Materials (Basel)* **2021**, *14*, 3725.
62. Janczarek, M.; Endo, M.; Zhang, D.; Wang, K.; Kowalska, E. *Materials (Basel)* **2018**, *11*, 2069.
63. Aguirre, M. E.; Zhou, R.; Eugene, A. J.; Guzman, M. I.; Grela, M. A. *Appl. Catal. B Environ.* **2017**, *217*, 485–493.
64. Yuan, W.; Yuan, J.; Xie, J.; Li, C. M. *ACS Appl. Mater. Interfaces* **2016**, *8*, 6082–6092.
65. Sawicka-Chudy, P.; Sibiński, M.; Pawełek, R.; Wisz, G.; Cieniek, B.; Potera, P.; Szczepan, P.; Adamiak, S.; Cholewa, M.; Głowa, L. *AIP Adv.* **2019**, *9*, 1–12.
66. Cheng, Y.; Gao, X.; Zhang, X.; Su, J.; Wang, G.; Wang, L. *New J. Chem.* **2018**, *42*, 9252–9259.
67. Koiki, B. A.; Orimolade, B. O.; Zwane, B. N.; Nkosi, D.; Mabuba, N.; Arotiba, O. A. *Electrochim. Acta* **2020**, *340*, 135944.
68. Liao, Y.; Deng, P.; Wang, X.; Zhang, D.; Li, F.; Yang, Q.; Zhang, H.; Zhong, Z. *Nanoscale Res. Lett.* **2018**, *13*, 221.
69. Xing, C.; Zhang, Y.; Liu, Y.; Wang, X.; Li, J.; Martínez-Alanis, P. R.; Spadaro, M. C.; Guardia, P.; Arbiol, J.; Llorca, J.; Cabot, A. *Nanomaterials* **2021**, *11*, 1399.
70. Pham, V. V.; Bui, D. P.; Tran, H. H.; Cao, M. T.; Nguyen, T. K.; Kim, Y. S.; Le, V. H. *RSC Adv.* **2018**, *8*, 12420–12427.
71. Trang, T. N. Q.; Tu, L. T. N.; Man, T. V.; Mathesh, M.; Nam, N. D.; Thu, V. T. H. *Compos. Part B Eng.* **2019**, *174*, 106969.
72. Wang, M.; Sun, L.; Lin, Z.; Cai, J.; Xie, K.; Lin, C. *Energy Environ. Sci.* **2013**, *6*, 1211–1220.
73. Wehrenfennig, C.; Palumbiny, C. M.; Snaith, H. J.; Johnston, M. B.; Schmidt-Mende, L.; Herz, L. M. *J. Phys. Chem. C* **2015**, *119*, 9159–9168.
74. Fu, Y.; Mo, A. *Nanoscale Res. Lett.* **2018**, *2018* (13), 187.
75. Zwilling, V.; Aucouturier, M.; Darque-Ceretti, E. *Electrochim. Acta* **1999**, *45*, 921–929.
76. Hui, K. C.; Suhaimi, H.; Sambudi, N. S. *Rev. Chem. Eng.* **2021**, 1–28.
77. Abbas, W. A.; Abdullah, I. H.; Ali, B. A.; Ahmed, N.; Mohamed, A. M.; Rezk, M. Y.; Ismail, N.; Mohamed, M. A.; Allam, N. K. *Nanoscale Adv.* **2019**, *1*, 2801–2816.
78. Liu, N.; Chen, X.; Zhang, J.; Schwank, J. W. *Catal. Today* **2014**, *225*, 34–51.
79. Ge, M.; Li, Q.; Cao, C.; Huang, J.; Li, S.; Zhang, S.; Chen, Z.; Zhang, K.; Al-Deyab, S. S.; Lai, Y. *Adv. Sci. News* **2017**, *4*, 1600152.
80. Wan, T.; Ramakrishna, S.; Liu, Y. *J. Appl. Polym. Sci.* **2018**, *135*, 45649.
81. Hou, X.; Aitola, K.; Lund, P. D. *Energy Sci. Eng.* **2020**, *9*, 921–937.
82. Abidi, M.; Hajjaji, A.; Bouzaza, A.; Trablesi, K.; Makhlouf, H.; Rtimi, S.; Assadi, A. A.; Bessais, B. *J. Photochem. Photobiol. A* **2020**, *400*, 112722.

83. Molenda, Z.; Grochowska, K.; Karczewski, J.; Ryl, J.; Darowicki, K.; Rysz, J.; Cenian, A.; Siuzdak, K. *Mater. Res. Express* **2019**, *6*, 1250b6.
84. Hou, Y.; Li, X. Y.; Zhao, Q. D.; Quan, X.; Chen, G. H. *Appl. Phys. Lett.* **2009**, *95*, 093108.
85. Huang, X.; Liu, Z. *J. Nanomater.* **2013**, *2013*, 517648.
86. Tsui, L. K.; Zangari, G. *Electrochim. Acta* **2014**, *128*, 341–348.
87. Kang, Y. S.; Kim, C. W. *Cryst. Growth Des.* **2018**, *18*, 6929–6935.
88. Xiong, Z.; Lei, Z.; Kuang, C. C.; Chen, X.; Gong, B.; Zhao, Y.; Zhang, J.; Zheng, C.; Wu, J. C. S. *Appl. Catal. B Environ.* **2017**, *202*, 695–703.
89. Li, G.; Huang, J.; Chen, J.; Deng, Z.; Huang, Q.; Liu, Z.; Guo, W.; Cao, R. *ACS Omega* **2019**, *4*, 3392–3397.
90. Liu, L.; Yang, W.; Sun, W.; Li, Q.; Shang, J. K. *ACS Appl. Mater. Interfaces* **2015**, *7*, 1465–1476.
91. Gawande, M. B.; Goswami, A.; Asefa, T.; Guo, H.; Biradar, A. V.; Peng, D. L.; Zboril, R.; Varma, R. S. *Chem. Soc. Rev.* **2015**, *44*, 7540–7590.
92. Wen, H.; Zhu, H.; Chen, X.; Hung, T. F.; Wang, B.; Zhu, G.; Yu, S. F.; Wang, F. *Angew Chem. Int. Ed.* **2013**, *52*, 13419–13423.
93. Talapin, D. V.; Mekis, I.; Go, S.; Kornowski, A.; Benson, O.; Weller, H. *J. Phys. Chem. B* **2004**, *108*, 18826–18831.
94. Huang, H.; Chen, J.; Liu, Y.; Lin, J.; Wang, S.; Huang, F.; Chen, D. *Small* **2020**, *16*, 2000708.
95. Dong, Z.; Ren, H.; Hessel, C. M.; Wang, J.; Yu, R.; Jin, Q.; Yang, M.; Hu, Z.; Chen, Y.; Tang, Z.; Zhao, H.; Wang, D. *Adv. Mater.* **2014**, *26*, 905–909.
96. Chen, L.; Peng, Y.; Wang, H.; Gu, Z.; Duan, C. *Chem. Commun.* **2014**, *50*, 8651–8654.
97. Magnan, H.; Stanescu, D.; Rioult, M.; Fonda, E.; Barbier, A. *J. Phys. Chem. C* **2019**, *123*, 5240–5248.
98. Zhang, P.; Guan, B. Y.; Yu, L.; Lou, X. W.; David. *Chem* **2018**, *4*, 162–173.
99. Das, S.; Pérez-Ramírez, J.; Gong, J.; Dewangan, N.; Hidajat, K.; Gates, B. C.; Kawi, S. *Chem. Soc. Rev.* **2020**, *49*, 2937–3004.
100. Purbia, R.; Paria, S. *Nanoscale* **2015**, *7*, 19789–19873.
101. Zhang, T.; Zhao, F.; Li, L.; Qi, B.; Zhu, D.; Lü, J.; Lü, C. *ACS Appl. Mater. Interfaces* **2018**, *10*, 19796–19805.
102. Zhu, W.; Mi, J.; Fu, Y.; Cui, D.; Lü, C. *Appl. Surf. Sci.* **2021**, *538*, 148087.
103. Henry, J.; Mohanraj, K.; Sivakumar, G. *J. Phys. Chem. C* **2019**, *123*, 2094–2104.
104. Lin, J.; Wang, W.; Li, G. *Adv. Funct. Mater.* **2020**, *30*, 2005677.
105. Trenczek-Zajac, A.; Banas, J.; Radecka, M. *RSC Adv.* **2016**, *6*, 102886–102898.
106. Ebaid, M.; Min, J. W.; Zhao, C.; Ng, T. K.; Idriss, H.; Ooi, B. S. *J. Mater. Chem. A* **2018**, *6*, 6922–6930.
107. Bialuschewski, D.; Hoppius, J. S.; Frohnhoven, R.; Deo, M.; Gönüllü, Y.; Fischer, T.; Gurevich, E. L.; Mathur, S. *Adv. Eng. Mater.* **2018**, *20*, 1–8.
108. Desai, N. D.; Ghanwat, V. B.; Khot, K. V.; Mali, S. S.; Hong, C. K.; Bhosale, P. N. *J. Mater. Sci. Mater. Electron.J. Mater. Sci. Mater. Electron.* **2016**, *27*, 2385–2393.

Index

Note: Page numbers followed by "*f*" indicate figures and "*t*" indicate tables.

H

I

Q

R

S

Printed in the United States
by Baker & Taylor Publisher Services